T0298706

New Concepts in Polymer Science

Elementorganic Monomers: Technology, Properties, Applications

New Concepts in Polymer Science

Previous titles in this book series:

New Concepts in Polymer Science

Elementorganic Monomers: Technology, Properties, Applications

L.M. Khananashvili, O.V. Mukbaniani and G.E. Zaikov

CRC Press
Taylor & Francis Group
Boca Raton London New York

CRC Press is an imprint of the
Taylor & Francis Group, an **informa** business

CRC Press
Taylor & Francis Group
6000 Broken Sound Parkway NW, Suite 300
Boca Raton, FL 33487-2742

© 2006 by Taylor & Francis Group, LLC
CRC Press is an imprint of Taylor & Francis Group, an Informa business

No claim to original U.S. Government works

Version Date: 20120727

International Standard Book Number: 978-9-00-415260-1 (Hardback)

Visit the Taylor & Francis Web site at
http://www.taylorandfrancis.com

and the CRC Press Web site at
http://www.crcpress.com

Contents

Preface

The chemical industry in our country and abroad is rapidly developing. It is only natural that the young industry of elementorganic monomers, oligomers and polymers should develop at the same rate. The numerous valuable and sometimes unique properties of these substances account for their wide application in various industries, households, medicine and cutting-edge technologies. That is why contemporary industry produces more than 500 types of silicone monomers, oligomers and polymers, to say nothing of other elementorganic compounds. The synthesis of these elementorganic compounds is based on many different reactions.

New fields of science and technology, the use of high and ultra low temperatures, high pressures and high vacuum, the developments in electrification, mechanical engineering, radio engineering and radio electronics, the design of supersonic aeroplanes and artificial Earth satellites – all this calls for new materials with valuable performance characteristics. It is well known that the surface of load-carrying machine parts at very high speeds can heat to 300°C upwards. Long-term resistance to such temperatures can be found only in polymers with chains made up of thermostable fragments. Particularly interesting in this respect are elementorganic polymers with inorganic and organo-inorganic molecular chains.

Elementorganic polymers are not only highly thermostable, but also perform well under low temperatures, sunlight, humidity, weather, etc. Besides, their physics and chemistry change little in a wide temperature range. Thus, these polymers (especially silicones) are widely and effectively used in the electrical, radio, coal, mechanical rubber, aircraft, metallurgical, textile and other industries. They are of great utility not only in industry, but also in households and in medicine, where their merits can hardly be overestimated.

Silicone polymers can render various materials unwettable (hydrophobic), which can be used in the manufacture of waterproof clothes, shoes, and construction materials. Silicone antifoam agents destroy foam, which is difficult to deal with in many spheres (in pharmaceutics, as well as in sugar-refining, wine-making and other food industries). They are indispensable even in contemporary medicine: these substances help to eliminate blood foaming during major surgeries, when great amounts of blood have to be drawn from the body. In this case, if surgical instruments are treated with silicone oligomers, there is no danger of tiny air bubbles (thrombi) entering blood and causing immediate death. Silicone oligomers are also widely used in the production of hydraulic fluids and lubricants to assure the performance of devices in a wide temperature range (from -120-140 to 250-350°C).

Each year sees more and more silicone elastomers used in the production of thermostable rubbers, and silicone polymers of various composition used for nonmetal composites. At present there are few industries where silicone polymers and materials based on them are not used to a greater or lesser extent. With that in mind, it is small wonder that the silicone industry is gathering such momentum all over the world.

Other elementorganic compounds, e.g. organoaluminum compounds, are extremely valuable components of the Ziegler-Natta catalysts, widely used in the production of stereoregular polymers. They are also used in the synthesis of higher fatty alcohols, carboxylic acids, α-olefines, cycloolefines and other important compounds. Organotin compounds are increasingly used as catalysts, as stabilizers for polymers and polymer-based materials, etc. Organolead compounds, tetraalkyl derivatives in particular, are used as antiknock substances in engine fuels. Organophosphorus compounds have found a wide application as pesticides, plasticizers and fire-resistant agents in polymers

This list, by no means complete, testifies to a constantly growing role of elementorganic compounds in industry, economy and households, which is to promote further development of the technology of elementorganic compounds.

The need to publish this book arose with the scientific and technical developments of the last decade, the reconstruction and technical renovation of existing factories, as well as fundamental changes in some syntheses of elementorganic monomers and polymers. Moreover, nowadays it is essential to train highly-skilled chemical engineers with a comprehensive knowledge of current chemistry, of the production technology of elementorganic monomers and polymers, and of their characteristics and applications.

1. A review of elementorganic compounds

1.1. History of elementoorganic compounds

The history of organic chemistry is full of examples when certain subdisciplines, which had not enjoyed much attention before, began to thrive owing to a discovery of an unexpected practical application for a class of compounds, or of their new properties. Sulfamides are connected with one of these examples. The use of sulfamide preparations as valuable therapeutic agents caused an intensive development of this branch of chemistry, and thousands of new sulfamide preparations were synthesized in a short space of time. At present elementorganic chemistry is at a similar stage of rapid development, which is demonstrated by many examples. Organophosphorus chemistry, which had long been only of theoretical interest, also soars due to the wide application of organic derivatives in various spheres of economy. Organic titanium and aluminum chemistry was boosted in 1954, when Ziegler discovered the ability of aluminorganic compounds mixed with titanium tetrachloride to polymerize ethylene, as well as in 1955, when Natta found the stereospecific polymerisation of unsaturated compounds in the presence of various complex catalysts.

Silicone chemistry also develops by leaps and bounds. The first silicon- and carbon-containing compound, ethyl ether of orthosilicon acid, was obtained by the French scientist Ebelmen in 1844. Subsequently, in 1963, Friedel and Crafts synthesized the first silicone compound with a Si-C bond, tetraethylsilane. At the initial stages of silicone chemistry, the researchers' attention was attracted to silicon, the closest counterpart of carbon. It seemed that silicon could give rise to as large a subdiscipline as chemistry itself. Yet, it was found that silicon, unlike carbon, does not form stable molecular chains from successively bonded Si atoms, and the interest for organic silicone derivatives dropped.

However, high-molecular chemistry could not limit itself only to the use of carbon and organogenic elements (oxygen, halogens, nitrogen, sulfur) for constructing polymer molecules. It logically aimed to involve other elements of the periodic table, based on assumptions that replacing carbon in the backbone with other elements would radically change the polymer properties.

Silicon was the first element used by K.A.Andrianov (1937), M.M.Koton (1939) and some time later by Y.Rokhov (1941) to build inorganic main chains of large molecules, which consist of alternating silicon and oxygen atoms surrounded by organic radicals. Thus appeared a new class of silicone polymers, currently known as polyorganosiloxanes or silicones. This was the first demonstration of how silicone compounds could be used to synthesise polymers with inorganic molecular chains and lateral organic groups. This stage became pivotal in silicone polymer chemistry and initiated active research not only into silicone polymers, but also into other elementorganic high-molecular compounds.

Recently, elementorganic polymers have drawn the interest of various industries, especially apparatus building, as well as mechanical, aircraft and rocket engineering, where thermostability is an essential requirement. Consider power engineering: new applications for power-producing units require more electric equipment to be produced and, consequently, extremely large amounts of copper, magnetic materials, etc. Moreover, due to the development of aircraft, fleet and rocket engineering, as well as the electrification of underground mining, it becomes necessary to reduce the weight and dimensions of electric equipment. Hence, engineers seek to design powerful, but lightweight and modest-sized electrotechnical devices. To solve these problems, one should obviously increase the current density, which sharply raises the operating temperature of a machine or apparatus. Since polymers are essential for manufacturing any power-producing units, we should take into account that as dielectrics they are the first to receive the heat radiated by conductive elements. In such cases, thermostability of polymer materials is particularly important.

The introduction of nuclear power into power engineering makes the requirements to dielectrics even more rigid. Specifically, there is a demand for dielectrics which can operate at 180-200 °C long-term, and sustain short-term temperatures of 250-350 °C upwards. Another example can be found in contemporary aircraft engineering. Nowadays the speed of aircraft accelerates at incredible rates. While landing, the aviation tyres of such high-speed aeroplanes develop a temperature up to 320 °C and higher. At the same time, high-speed aeroplanes should be also protected from heat released when they move through the atmosphere at high speed, which is also becoming more difficult. Thermostable polymers can also help to solve the problems of space exploration.

As stated above, polyorganosiloxanes were the first representatives of high-molecular compounds with inorganic main molecular chains surrounded by organic groups. These polymers opened a new field for chemistry to develop without copying natural substances or materials, since polymers of this composition are unknown in nature and are completely lab-developed. The studies of elementorganic high-molecular compounds intensified in the post-war years, and nowadays are conducted in all industrial and developing countries. The number of publications and patents in the field is greater every year, with new theoretical and practical papers emerging all the while. At the same time, the manufacture of elementorganic polymers and monomers is on the rise, and the world production of silicone monomers and polymers increases day by day.

Researchers of polymer synthesis focus their attention on 45 elements of the periodic table. The most important elements used to construct polymer chains are listed below:

Group II	Mg, Zn	Group V	N, P, V, As, Sb, Bi
Group III	B, Al	Group VI	O, S, Cr, Se, Mo
Group IV	C, Si, Ti, Ge, Zr, Sn, Pb	Group VIII	Fe, Co, Ni

Indeed, it has been established that many of them (B, Al, Si, Ti, Sn, Pb, P, As, Sb, Fe) can combine with oxygen and nitrogen to form inorganic chains of polymer molecules with lateral organic and organosiloxane groups. Some of these polymers have already found an industrial application. It should be expected that in the nearest future the design of new synthesis techniques gives us new elementorganic polymers with important properties.

1.2. Chemistry and technology of elementorganic compounds

Elementorganic compounds considerably differ in their properties and structure from both organic and inorganic compounds, occupying an intermediate position. Elementorganic compounds are rarely found in nature and are obtained synthetically.

In the chemistry of living things the role of elementorganic compounds is not quite clear; however, it is safe to say that compounds of silicon, phosphorus and other elements play a significant part in the life and metabolism of highly evolved living organisms, human beings in particular. The body of a man or animal contains silicon compounds in various forms, including silicone and complex compounds, which can be dissolved in organic solvents. Nevertheless, as far as silicone compounds are concerned, there is only one known example of their presence in nature, a specific ether of orthosilicon acid $Si(OC_{34}H_{69})_4$ in bird feathers. Organophosphorus compounds, especially ethers of phosphoric and polyphosphoric acids, are essential in the chemistry of living things. I.e., adenosine triphosphate (ATP) can be found in living tissues and plays a vital role as an energy source.

Table 1. Electronegativity scale of elements (X)

Li	Be	B				H							C	N	O	F
1.0	1.5	2.0				2.1							2.5	3.0	3.5	4.0
Na	Mg	Al											Si	P	S	Cl
0.9	1.2	1.5											1.8	2.1	2.5	3.0
K	Ca	Se	Ti	V	Cr	Mn	Fe	Co	Ni	Cu	Zn	Ga	Ge	As	Se	Br
0.8	1.0	1.3	1.5	1.6	1.6	1.5	1.8	1.9	1.9	1.9	1.6	1.6	1.8	2.0	2.4	2.8
Rb	Sr	Y	Zr	Nb	Mo	Te	Ru	Rh	Pd	Ag	Cd	In	Sn	Sb	Te	I
0.8	1.0	1.2	1.4	1.6	1.8	1.9	2.2	2.2	2.2	1.9	1.7	1.7	1.8	1.9	2.1	2.5
Cs	Ba	La-Lu	Hf	Ta	W	Re	Os	If	Pt	Au	Hg	Tl	Pb	Bi	Po	At
0.7	0.9	1.0-1.2	1.3	1.5	1.7	1.9	2.2	2.2	2.2	2.4	1.9	1.8	1.9	1.9	2.0	2.2
Fr	Ra	Ac	Th	Pa	U	Np-No										
0.9	0.9	1.1	1.3	1.4	1.4	1.4-1.3										

We can single out several fundamental characteristics which distinguish elementorganic compounds from carbon compounds.

1. *Different elective affinity of elements as compared to carbon.* Electropositive elements (Si, B, AI, P) have a considerably larger affinity to electronegative elements than carbon. In other words, silicon, boron, aluminum, phosphorus and other elements form weaker bonds with electropositive elements (H, Si, B, Al, As, Sb, Bi, etc.), and stronger bonds with electronegative elements (O, N, CI, Br, F, etc.) than carbon.

The analysis of the electronegativity of various elements (Table 1) shows that carbon (X_C=2.5) is roughly in the middle between the most electronegative element, fluorine (X_F=4.0) and the most electropositive elements, cesium and francium (X_{Cs}=0.7, X_{Fr}=1.7). The half-sum of the electronegativities of these elements is X_{HS} =2.35; consequently, the C atom has the least tendency to give or receive electrons, i.e. form positive or negative ions. This means that carbon in compounds is less ionised in comparison with electropositive or electronegative elements. For example, if the Si-CI bond is 30-50% ionised, the C-CI bond is approximately 6% ionised. Hence, carbon has the least susceptibility to an electrophilic or nucleophilic attack, and the C-C bond is much stronger than the E-E bond (e.g., B-B, Si-Si, AI-AI, P-P, As-As), and vice versa, for example, the C-O bond, with its electronegativity half-sum of X_{HS}=3.0, is not so strong as As-O ($\overline{X}_{HS}$=2.5), Si-O ($\overline{X}_{HS}$=2.65), Si-N ($\overline{X}_{HS}$= 2.4) and other bonds.

Table 2. Energy of certain single bonds in B, Si, P, As and C atoms

Bond	Bond energy, KJ/mol	Bond	Bond energy, KJ/mol	Bond	Bond energy, KJ/mol
C-C	344	B-C	312	B-O	460
B-B	225	Si-C	290	Si-O	432
Si-Si	187	P-C	272	P-O	360
P-P	217	C-O	350	As-O	311
As-As	134				

The comparison of the bond energies in boron, silicon, phosphorus and arsenic atoms with that of carbon atoms supports this idea (Table 2).

As shown in Table 2, As-As, Si-Si, P-P and B-B bonds have the least energy, which makes them more prone to thermolysis than P-C, Si-C, B-C and C-C bonds. As-O, P-O, Si-O and B-O possess an even greater energy; that is why most important elementorganic oligomers and polymers with a practical application are characterised by the presence of siloxane, alumoxane, boronsiloxane and other groups of the kind in their chains.

Thus, the difference of elementorganic compounds from inorganic compounds is largely caused by the relative weakness of E-E bonds in comparison with C-C bonds, and, vice versa, by the considerable strength of E-O-E bonds in comparison with C-O-C bonds. Therefore, elementorganic compounds have a greater tendency to condensation reactions, forming elementoxane polymer chains, whereas organic monomers form carbonoxane polymer chains (C-O-C) by have a very inferior thermostability. For example, polyoxomethylenes $(CH_2-O)_n$ are easily broken up at 150-170°C, whereas the thermostability of polyorganosiloxanes with branched and especially ladder molecular chains exceeds 500°C. This is why molecules of elementorganic polymers contain thermostable elementoxane and other bonds, where elements with positive and negative polarisation alternate.

2. *Hightened reactivity of functional groups* (e.g. CI, Br, OH, OR, OCOR, NH$_2$, SH) at the atoms of silicon, aluminum, titanium, phoshorus and other elements in comparison with their reactivity binded with oxygen. This is due to the fact that the silicon atom is one and a half times bigger than the carbon atom: it has a covalent radius of 0.117 nm, whereas the radius of the carbon atom is only 0.077 nm. It follows that functional groups of the Si atom are much more distanced from each other than

those of the C atom (e.g. the distances between CI atoms in $SiCI_4$ and CCI_4 are 0.329 and 0.298 nm correspondingly); hence, the central silicon atom is less shielded. Besides, silicon, like other elements, is more electropositive than carbon and is therefore more prone to the nucleophilic attack. Thus, functional groups at these elements are more active in various reactions.

3. *Different formation of bond types.* Multiple σ,π-bonds are essential in organic chemistry, because unsaturated compounds are the main monomers for multiple bond polymerisation, whereas in elemenorganic chemistry the role of σ,π-bonds is insignificant due to the fact that the formation of stable compounds with multiple E-E bonds has not been established with reliability. This demonstrates that electropositive elements, unlike carbon, seem to have little capability to form stable compounds with multiple bonds.

As we know, organic compounds with two hydroxyl groups attached to one carbon atom are rare; on the contrary, more electropositive elements can retain two and even three hydroxyl groups. When an attempt is made to obtain organic compounds with two hydroxyl groups, they usually detach a water molecule and form aldehydes or ketones. Di- and trihydroxy-derivatives of electropositive elements are not capable of such a transformation (because the atoms in these elements[1] cannot form a strong double bond), which is why elementorganic chemistry does not have substances similar to organic aldehydes or ketones.

Thus, while in organic chemistry the multiple p_π—p_π bond is very valuable in the synthesis of many classes of high-molecular compounds, these bonds in elementorganic chemistry can participate in the formation of macromolecules only if they are part of the groups surrounding the elementorganic chain. In all other cases (especially for elementorganic macromolecules with inorganic chains) multiple chains do not participate in reactions. At the same time, unlike carbon compounds, many compounds of electropositive elements have an important d_π—p_π covalent bond, when atoms of electropositive elements (Si, B, AI, P) accept electrons owing to the presence of free orbitals. For example, in silicone compounds such a bond may occur between silicon and oxygen atoms (or halogen, nitrogen and carbon atoms in the aromatic nucleus). The above mentioned features of elementorganic compounds are what determines their specific chemical characteristics.

---

[1] In 1952 K.A.Andrianov and N.N.Sokolov used the mass spectrographic technique to prove the possibility of the existence of extremely unstable dialkyloxysilanes $R_2Si=O$ (Ac.Sci.USSR, 1952. Vol.82, p.909).

As far as the manufacture of elementorganic monomers and polymers is concerned, we can highlight the following features.

1. great variety of technological processes for their synthesis;
2. corrosive synthesis environment and hence rigid requirements to the choice of equipment;
3. often relatively small tonnage of the product and hence the periodic synthesis of some elementorganic compounds, which raises the costs;
4. wide range of products (e.g. Dow Corning alone manufactures about 2500 different silicone monomers, oligomers and polymers);
5. relatively high prices for elementorganic products.

Nevertheless, we should stress that the demand for elementorganic monomers and polymers grows with each year, which is due to their valuable characteristics. Sometimes the introduction of negligible amounts of certain elementorganic compounds into various compositions or composites considerably improves their performance characteristics. That is why, in spite of their relatively high costs, the use of elementorganic compounds is beneficial both technically and economically.

1.3. Prospects of elementorganic polymer chemistry and technology

One of the most important contemporary features of scientific and technological advance is a wide use of polymers and polymer-based materials virtually in all spheres of economy and in households; moreover, the application range of synthetic materials is wider every year. Thus, further development of economy requires an increased production of various polymer materials with valuable properties. Many polymer materials are based on synthetic elementorganic oligomers and polymers used in the manufacture of plastics, sealants and rubbers; paint, anticorrosive and other coatings; insulating, lubricating and construction materials, etc. Nowadays it is hard to find an industry which does not use elementorganic compounds, because their valuable technical characteristics are combined with convenient and highly productive techniques to process them into materials and products of various shapes and sizes. All this promises a big future to elementorganic oligomers and polymers.

Carbon-chain superpolymers (with chains consisting only of carbon atoms) are as a rule not heat- and weather-resistant enough; that is why synthetic chemists have always aimed to synthesise new, more heat- and weather-resistant polymers. This aim was one of the reasons for creating high-molecular compounds with chains made up of various atoms (Si, Al, B, Ti, etc.) and oxygen or nitrogen.

The production scale of elementorganic compounds, especially elementorganic oligomers and polymers, as well as extremely diverse requirements to the materials based on these compounds in different economic spheres (medicine, transportation, agriculture, etc.) and advances in aeroplane and rocket building, microelectronics, radio and electrical engineering pose new problems to elementorganic chemistry. Among these problems are:

1. Expanding the polymer operating temperature range to produce nonmetallic materials (plastics, rubbers, fibres, paint coatings, etc.);
2. Improving mechanical characteristics of synthesised polymers and materials based on them;
3. Improving their physical and chemical inertness (i.e. resistance to weather effects, light, radiation, liquids);
4. Creating the polymers which can be the basis of nonflammable materials with a required set of technical characteristrics.

These problems can be successfully solved by finding techniques to synthesise new polymers with a varied molecular structure, by modifying known polymers, as well as by polymer alloying, i.e. adding small amounts of substances with a composition different from that of the polymer.

Because high-molecular compounds in the form of various nonmetallic materials are so widely used, especially in crowded places, a serious problem nowadays is to create nonmetallic materials which do not sustain combustion or are totally nonflammable.

Polymer heat resistance largely depends on their macromolecular structure, and their thermal-oxidative stability and nonflammability depend on the type and number of the organic groups surrounding the chains. Hence the importance of combining the optimal lattice density in macromolecules with a convenient type and number of lateral organic groups. In this case the greatest effect can be expected from polyorganosiloxanes with a branched (I), ladder (II) and spirocyclic (III) molecular structure, or from polyorganosiloxanes where these structures are combined with methyl and phenyl lateral groups. Methyl groups oxidise easier than phenyl groups, but, if they are replaced by oxygen, the weight loss of the polymer is insignificant. In I and II structure polymers with methyl groups the carbon content does not exceed 8-12%. These polymers have already been used in the technology for obtaining nonflammable fibreglass and asbestos plastics.

$$
\mathrm{I} \quad\quad\quad \mathrm{II} \quad\quad\quad \mathrm{III} \quad m=1\text{-}3
$$

Polymers with structures II and III are of great interest in the design of heat resistant nonmetallic materials. These blocks can polymerise without emitting volatile matter and, consequently, ensure contact molding of glass and asbestos plastics, whereas copolymers with a combination of structures II and III seem to be able to transform into latticed polymers with cyclic or polycyclic silicone groups in cross-link sites between linear sections. In this case the cross-link sites, depending on the size and structure of the cycle, may be subject to conformational transitions when stressed; consequently, the rigidity of the cross-link sites will be very different from the one typical of latticed polymers. These polymers commonly have carbon atoms in the cross-link sites and are currently studied. At present we distinguish polymers with various structures of the macromolecular chain (Diagram 1).

Of particular interest is the study of synthesis reactions of linear polymers, the chains of which consist of flexible linear sections and rigid mono- and polycyclic fragments:

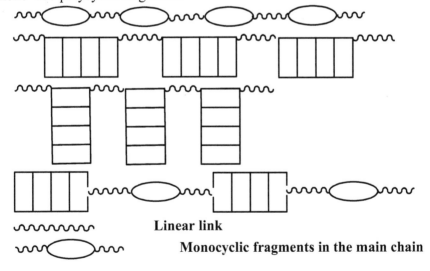

Linear link

Monocyclic fragments in the main chain

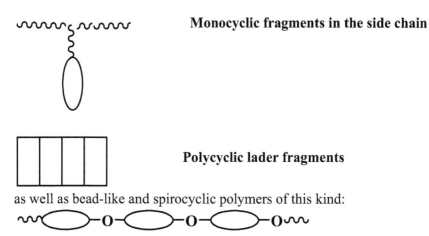

Monocyclic fragments in the side chain

Polycyclic lader fragments

as well as bead-like and spirocyclic polymers of this kind:

In this case the combination of fragments with flexible and rigid molecular chains in linear chains can help to synthesise elastomers and plastomers with a higher thermostability.

To improve mechanical and some other specific properties, one might find interesting the synthesis of comb-shaped polymers with a uni- and bidirectional molecular structure:

Wide opportunities for the controlled variation of properties are offered by the synthesis of block copolymers with organo-inorganic main molecular chains, as well as by the synthesis of block copolymers containing besides silicon other elements in the form of various groups (spirotitaniumsiloxane, spiroironsiloxane, phosphonitrilsiloxane, carboransiloxane), which will undoubledly lead to the creation of new technically valuable materials.

The results that have been achieved at present still do not meet the demands of our economy in the production scale of elementorganic oligomers and polymers. We need to increase the production pace to use more elementorganic compounds in various fields of technology and in households. Most polymers with these molecular chains have already been synthesised in labs, but only few of them have been implemented in industry. Therefore, the nearest and most urgent task of chemical engineers is to unite their efforts with scientists in order to design and implement convenient and accessible technological processes of the synthesis of elementorganic polymers. In the nearest future chemical engineers should pay close attention to the design and implementation of new ways of obtaining elementorganic, silicone and elementosilicone polymers in particular, by finding efficient techniques to synthesise block oligomers (prepolymers) of a given composition and their subsequent transformation into polymers with optimal characteristics.

Diagram 1. Major macromolecular architectures

Major Macromolecular Architectures

I	II	III	IV
Linear	Cross-Linked	Branched	Dendritic

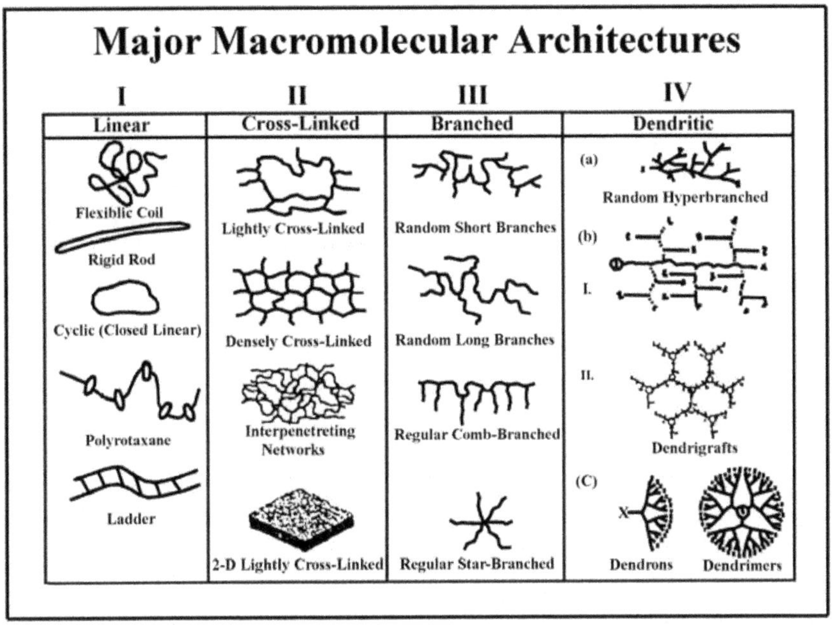

The synthesis of silicone polymers from prepolymers will allow one to obtain not only polyorganosiloxanes, but also polyelementorganosiloxanes of a more regular structure (unlike the existing processes of hydrolytic co-condensation of various organochlorosilanes which form polymers of a static composition). Therefore, the polymers obtained in this way will have improved chemical and physicochemical properties, and materials based on them will have valuable performance characteristics. Moreover, a new technique for obtaining polymers based on block oligomers will help to build harmless and wasteless industries, which is especially important from the ecological point of view.

A promising area is the design of modern ceramic and glass ceramic materials based on high purity elementorganic compounds $Si(OR)_4$, $Al(OR)_3$, $B(OR)_3$, etc.), because traditional materials (metals, metal-based alloys, plastics, etc.) do not meet contemporary technical requirements to products designed to operate under extreme conditions.

We understand modern ceramics as all strong inorganic materials which are processed at high temperatures and have superior physicochemical and heat resistance characteristics. The range of their application is very wide. For instance, modern ceramics is used as a substrate for catalysts, as well as in the production of ball bearings and various elements for electronic equipment, atomic power plant and thermonuclear fusion equipment. Refractory ceramics is one of the best heat insulators for aircraft, rocket, space technology, etc. Let us give just one example of the advisability and economic feasibility of the use of modern ceramics in turbine and diesel engines instead of doped heat resistant alloys. The use of ceramics in this case helps to increase the operating temperature in the combustion chamber up to 1400-1500°C without any additional cooling of the given products, which reduces the fuel consumption almost twofold.

Glass-ceramic materials can be used to produce elements of fibre optics, translucent ceramics, photochromic and laser glasses, oxide conductive glasses, etc.

The advantages of modern ceramic and glass-ceramic materials are realised only if they are produced not by the traditional technique, but using the recent so-called *sol-gel technology.* One of the most important *sol-gel* methods is based on the hydrolitic condensation of tetraethoxysilane in the presence of water-soluble metal salts, or on hydrolitic cocondenstation of tetraethoxysilane and alcoxides of aluminum and other metals, followed by hydrolysate gelating and processing the gels at 500-600°C. We should note here that in order to obtain modern ceramics and glass-ceramics, one needs compounds with an exceptionally small amount of foreign impurities, since impurities interfere with obtaining ceramic materials of required quality.

Aerogel possesses excellent heat insulating properties: felt-tip pens lying on Aerogel are protected from the flames below and do not melt. Aerogel, which is 99.9% air and 0.1% silicone-dioxide gel, is subjected to maximum desiccation. This preserves its original size and shape, because normal evaporation can destroy the gel. Of all materials known, Aerogel is the least dense (it is only 3 times denser than air), but is a unique insulator. Its insulating properties are 39 times higher than those of fibreglass plastic; at the same time its density is 1000 times less than that of glass, which also has a silicone structure. Aerogel can sustain temperatures up to 1400°C. A man-sized aerogel block, which weighs not more than 400 g, supports up to half a ton of weight.

Aerogel is a special materials with extreme micron porosity. It consists of separate particles of several nanometers, interconnected in a high-porosity branched structure. It was made on the basis of gel consisting of colloid silicone, the structural parts of which are filled with solvents. Aerogel is subjected to high temperature under pressure which rises to the critical point; it is very strong and easily endures stress both at lift-off and in the space environment. This material has already been tried in space by Spacelab II and Eureca shuttles, as well as by the American Mars Path-finder Rover.

The growing amount of research of the application of modern ceramics in various industries, as well as their virtually unlimited possibilities suggest that in the nearest decade these materials will be widely used in various spheres of economy. Recently a lot of interest has been drawn to the synthesis of silicone and other elementorganic highly branched oligomers with a so-called *dendrimer structure*. Oligomers of this kind are obtained by multistage synthesis. For example, highly branched oligomethylsiloxanes of the given structure are obtained in the following way. The first stage is the reaction between methyltrichlorosilane and sodiumoximethyldiethoxisilane.The second stage is the selective replacement of ethoxyl groups with chlorine atoms using sulfuryl chloride.These reactions yield a second, and then a third "generation" of dendrimers, which will eventually contain 22 silicon atoms.

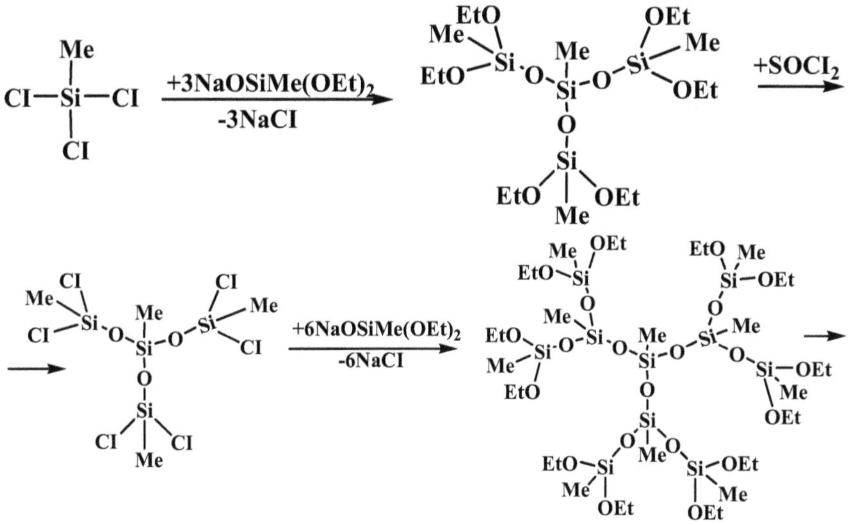

We should also develop the processes of producing filled polymers during their synthesis, which is economically feasible and justifiable. Researchers have already developed a process to obtain filled conductive silicone rubbers by the technique stated above.

In the conclusion we should say that the chemistry of synthetic elementorganic polymers is a young science and still has a lot to discover. The possibilities of elementorganic polymer chemistry, and consequently of their production development, are truly unlimited. If originally synthetic polymers appeared as a result of emulating natural compounds and as their substitutes, nowadays we have many polymers which resulted from scientific and engineering creativity and have no counterparts in nature.

2. Chemistry and technology of silicone monomers

Monomer silicone compounds, especially organochlorosilanes, hydroxyorganosilanes, alkylalkoxysilanes with amide groups in the alkyl radical, and acyloxyorganosilanes, have recently found an independent practical application, but their role as parent substances for the production of silicone oligomers and polymers is particularly important.

The greatest importance for the synthesis of silicone oligomers and polymers is attached to the following monomers:

1. organochlorosilanes and their chlorination products;
2. ethers of orthosilicon acid, i.e. tetraalkoxy(aroxy)silanes;
3. substituted ethers of orthosilicon acid, i.e. alkyl- and arylalkoxysilanes;
4. acyloxyderivatives of organosilanes, i.e. alkyl- and arylacetoxysilanes.

2.1. Preparation of organochlorosilanes

There are a number of techniques to prepare organochlorosilanes; the most widespread are the following:

1. techniques based on the application of metalorganic compounds;
2. techniques based on the interaction of chlorine hydrocarbon derivatives with free carbon (direct synthesis);
3. techniques based on the replacement of hydrogen atoms in hydridechlorosilanes with alkyl, alkenyl and aryl radicals.

2.1.1. Techniques based on the use of metalorganic compounds

The synthesis of alkyl(aryl)chlorosilanes, based on the use of metalorganic compounds, can be carried out with the help of organomercury, organozinc, organosodium, organolithium, organoaluminum and organomagnesium compounds.

In 1863 Friedel and Crafts were the first to use metalorganic compounds to synthesise silicone monomers. Industry focused on organomagnesium synthesis (the Grignard technique), in which alkyl- or arylhalogenide interacts with metallic magnesium to form organomagnesium compounds:

$$\textbf{RX + Mg} \longrightarrow \textbf{RMgX} \tag{2.1}$$

where R = C_2H_5, C_6H_5 etc., and X = CI, Br or F.

After that, the organomagnesium compound reacts with silicon tetrachloride, forming corresponding alkyl(aryl)chlorosilanes:

$$\textbf{4SiCI}_4 + \textbf{10RMgX} \longrightarrow \textbf{RSiCI}_3 + \textbf{R}_2\textbf{SiCI}_2 + \textbf{R}_3\textbf{SiCI} + \tag{2.2}$$

$$\textbf{+R}_4\textbf{Si + 10MgXCI}$$

The process of organomagnesium synthesis yields a mixture of alkyl(aryl)chlorosilanes with various numbers of radicals attached to the silicon atom. However, by regulating the ratio of the parent components and changing the conditions of the process, one can shift the reaction towards the preferential formation of a certain monomer.

This process is also possible in a hydrocarbon medium with tetraethoxysilane or diethyl ether (1%) as a catalyst, both in two and one stages, by the interaction of the mixture of silicon tetrachloride and alkyl- or arylhalogenide with metallic magnesium in a solvent medium (toluene or xylene).

Speaking of the economical characteristic of preparing alkyl- or arylchlorosilanes by the Grignard technique, we should note that this method is less efficient than direct synthesis for the preparation of pure ethyl- and phenylchlorosilanes; as for methylchlorosilanes, it is virtually unusable, because original methylchloride is a gas under normal conditions; besides, methylmagnesiumchloride reacts with silicon tetrachloride too energetically, which makes it hard to regulate the process. Nevertheless, the organomagnesium technique has practical value for obtaining a range of organochlorosilanes, especially those with different radicals at the silicon atom, such as methylphenylchlorosilane, ethylphenyldichlorosilane, etc.

Preparation of methylphenyldichlorosilane

Methylphenyldichlorosilane is synthesised in two stages. First, chlorobenzene reacts with metallic magnesium in a xylene medium to form phenylmagnesiumchoride:

$$\textbf{C}_6\textbf{H}_5\textbf{CI + Mg} \longrightarrow \textbf{C}_6\textbf{H}_5\textbf{MgCI} \tag{2.3}$$

then, methyltrichlorosilane reacts with phenylmagnesiumchloride suspension to form methylphenyldichlorosilane:

$$CH_3SiCI_3 + C_6H_5MgCI \longrightarrow CH_3(C_6H_5)SiCI_2 + MgCI_2 \qquad (2.4)$$

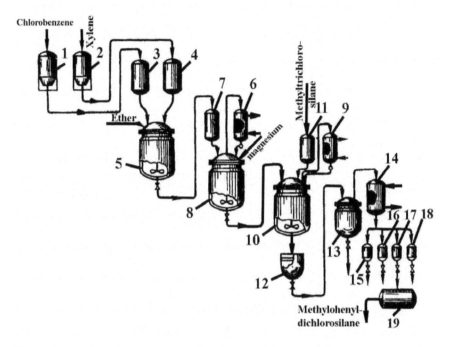

Fig. 1. Production diagram of methylphenyldichlorosilane by organomagnesium-synthesis: *1, 2* - dehydration boxes; *3, 4, 7, 11* - batch boxes; *5* - agitator; *6, 9, 14* - coolers; *8* - reactor of phenylmagnesiumchloride synthesis; *10* - reactor of methylphenyldichlorosilane synthesis; *12* -- nutsch filter; *13* -- vacuum distillingtank; *15-18-* collectors; *19-* container.

It should be noted that the latter reaction is complicated: a deeper phenylation also forms methyldiphenyldichlorosilane, whereas pyrolytic processes form diphenyl, diphenylbenzene and other substances.

The production process of methylphenyldichlorosilane by organomagnesium synthesis (Fig. 1) comprises three main stages: the production of phenylmagnesiumchloride; the synthesis of methylphenyldichlorosilane; the extraction of marketable methylphenyldichlorosilane.

Stored chlorobenzene and xylene are fed into dehydration boxes *1* and *2* to dehydrate with burnt calcium chloride. After drying, chlorobenzene and xylene are sent correspondingly into batch boxes *3* and *4* and from there into agitator *5*. Metal magnesium is degreased (by soaking in toluene or xylene), dried, cut on a planing machine into ribbon chipping 0.07 -0.1

mm thick and loaded through a hatch into a clean and dried reactor *8*. The reactor is heated with vapour; at 110 °C intensively agitated magnesium chipping out of agitator *5* through batch box *7* receives a small part of reactive mixture consisting of chlorobenzene, xylene and diethyl ether (1-3% of the total amount of chlorobenzene and xylene). Due to exothermicity, the temperature in the apparatus rises up to 125-135 °C.

At 125 °C vapour is stopped, and reactive mixture is gradually introduced into the apparatus at such speed that the temperature does not exceed 123±2 °C. Then the reactive mixture is mixed for 3 hours at 120-126 °C, after which the heating is switched off and at agitation an additional amount of xylene is added so as to obtain a 30-35% suspension of phenylmagnesiumchloride in xylene; then the mixture is agitated for 1 more hour at 20-25 °C.

The reactive mixture is sampled to determine phenylmagnesiumchloride content; after that the finished suspension is sent into reactor *10*. The agitator is switched on and at 18-25 °C the reactor receives methyltrichlorosilane at such speed that the temperature in the apparatus does not exceed 75-80 °C. After introducing the whole of methyltrichlorosilane at agitation within 1 hour the reactive mixture is cooled down to 18-20 °C and agitated at this temperature for 3 more hours. After sampling and filtering the product with nutsch filter *12* the mixture is distilled.

Apart from methylphenyldichlorosilane, the reaction also yields some methyldiphenyldichlorosilane, diphenyl, diphenylbenzene and other substances. The product is distilled in vacuum distilling tank *13*, separating several distillates: distillate I (up to 135 °C) contains a small quantity of the main product, unreacted methyltrichlorosilane and chlorobenzene, as well as xylene; distillate II (up to 180 °C) always contains small quantities of the main product and xylene; distillate III (180-300°C) generally contains the main product -- methylphenyldichlorosilane; distillate IV (more than 300°C) is usually separated under a residual pressure of 20 GPa and vapour temperature of 140 °C and is a mixture of methyldiphenylchlorosilane, diphenyl, diphenylbenzene and other impurities. To obtain pure methylphenyldichlorosilane, the products of distillation are subsequently rectified.

To increase the yield of the target product, it is advisable to subject methyldiphenylchlorosilane in distillate IV to disproportionation according to the reaction

$$CH_3(C_6H_5)_2SiCI \xrightarrow[-C_6H_6]{+HCI} CH_3(C_6H_5)SiCI_2 \tag{2.5}$$

The reaction should be conducted in the presence of catalytic amounts of $AICI_3$ at 220°C, passing through methyldiphenylchlorosilane anhydrous hydrogen chloride at the speed of 0.2 m³/min. The unreacted methyltrichlorosilane and chlorobenzene, as well as xylene, are regenerated and returned to production, and the intermediate higher distillates extracted at rectification are used to obtain oligomethylphenylsiloxanes.

Methylphenyldichlorosilane is a colourless liquid (the boiling point is 204°C) with a specific pungent odour characteristic of organochlorosilanes; it fumes in air. It is also easily hydrolysed by water.

Technical methylphenyldichlorosilane should meet the following requirements:

Colour	From colourless to purple
Density at 20 °C, g/cm³	1.1750-1.1815
Content, %:	
of distillate at 196-204 °C, not less than	93
of chlorine	36.9-37.6

Methylphenyldichlorosilane is a source monomer in the synthesis of oligomethylphenylsiloxanes, various silicone elastomers and polymers used in the production of varnishes.

Similarly to methylphenyldichlorosilane, one can obtain phenylethyldichlorosilane and other organochlorosilanes with different radicals at the silicon atom.

Production of tris(γ-trifluoropropyl)chlorosilane

Tris(γ-trifluoropropyl)chlorosilane is synthesised in three stages. First of all, tris(γ-trifluoropropyl)silane is obtained by the Grignard technique in the dibutyl ether medium in the presence of a catalyst, ethyl bromide:

$$3CF_3\text{-}CH_2\text{-}CH_2CI + HSiCI_3 + 3Mg \longrightarrow \qquad (2.6)$$
$$\longrightarrow (CF_3\text{-}CH_2\text{-}CH_2)_3SiH + 3MgCI_2$$

The process occurs through the formation of a salt, trifluoropropylmagnesiumchloride, which does not react fully; therefore, to destroy the unreacted salt, the reactive mixture is neutralised by 2-3% hydrochloric acid:

$$3CF_3\text{-}CH_2\text{-}CH_2MgCI + H_2O \xrightarrow{+HCI} \qquad (2.7)$$
$$\longrightarrow CF_3\text{-}CH_2\text{-}CH_3 + Mg(OH)CI \xrightarrow[-H_2O]{+HCI} MgCI_2$$

Finally, by the chlorination of tris(γ-trifluoropropyl)silane with gaseous chlorine in the carbon tetrachloride medium, we obtain tris(γ-trifluoropropyl)chlorosilane:

$$(CF_3\text{-}CH_2\text{-}CH_2)_3SiH + Cl_2 \xrightarrow[\text{-HCl}]{} (CF_3\text{-}CH_2\text{-}CH_2)_3SiCl \quad (2.8)$$

Raw stock: trifluorochloropropane (the boiling point is 45.0 °C, $d_4^{20}=1.2900$); metallic magnesium; trichlorosilane [not less than 95% (vol.) of the distillate 31-34°C, 78-80% Cl⁻, 0.75-0.78% H_2]; dibutyl ether (the boiling point is 140-143 °C, $d_4^{20} = 0.768 \div 0.770$, $n_D^{20} = 1.3960 \div 1.5000$); ethyl bromide(the boiling point is 38.4 °C, $d_4^{20} = 1.4555$); carbon tetrachloride (the boiling point is 76 °C, $d_4^{20} = 1.590\text{-}5\text{-}1.597$).

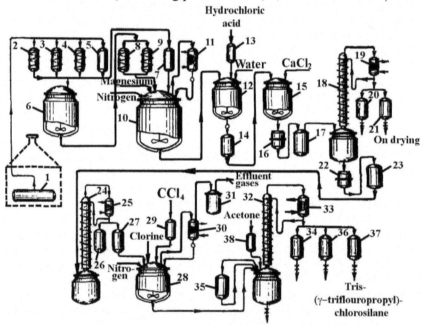

Fig. 2. Production diagram of tris(γ-trifluoropropyl)chlorosilane: 1 - tank, 2-5, 7, -9, 13, 29, 38 - batch boxes; 6 - agitator; 10 - reactor; 11, 30 - coolers; 12 - hydrolyser; 14, 17, 23, 35 - collectors; 15 - dehydrator; 16, 22 - nutsch filters; 18, 24, 32 - rectification towers; 19, 25, 33 - condensers; 20, 21,26, 27, 34, 36, 37 - receptacles ; 28 - chlorinator; 31 - hydraulic gate.

The industrial production of tris(γ-trifluoropropyl)chlorosilane (Fig.2) also comprises three main stages: the synthesis of tris(γ-trifluoropropyl)silane ; the hydrolysis of its synthesis and extraction products; the chlorination of tris(γ-trifluoropropyl)silane and the rectification of chlorination products.

Raw stock is sent to the apparatus in tanks *1* installed in the drafting device (one is shown in the figure). Then it is sent by nitrogen flow to the batch boxes: trifluoromonochloropropane to batch box *2*, trichlorosilane to batch box *3*, ethyl bromide to batch box *4*, and dibutyl ether (dehydrated with burnt calcium chloride and filtered) to batch box *5*. Before the synthesis begins, working mixtures I and II are prepared in apparatus *6*. Mixture I consists of trichlorosilane and ethyl bromide, and mixture II consists of trichlorosilane, dibutyl ether, trifluoromonochloropropane and ethyl bromide. Mixture I is sent to batch box *8*, and mixture II is sent to batch box *9*. All batch boxes and apparatus *6* have jackets or coils (on their external walls) to be cooled with Freon at -15 - -20°C.

Tris(γ-trifluoropropyl)silane is synthesised in reactor *10*, which is fashioned with an agitator and a jacket. Above the reactor there is backflow cooler *17* cooled by Freon. The synthesis takes place under atmospheric pressure. First, the reactor is loaded through a hatch with freshly prepared magnesium chipping; after that, the hatch is closed and the chipping is dried in nitrogen flow at ~80 °C. After nitrogen blowing, the process is continued by adding dibutyl ether, switching on the agitator and heating the reactive mixture up to 35 °C with hot water sent into the jacket of the apparatus. After that, all mixture I necessary to start the reaction self-flows into the reactor from batch box *8*. When the mixture is added, the temperature of the reactive mixture rises to 80-105 °C. After the temperature stops rising, the jacket is filled with water, and the temperature in the reactor is reduced down to 25-26 °C. If mixture I does not raise the temperature, it is necessary to fill the jacket with hot water and heat the reactive mixture up to 75-80 °C.

At 25-26 °C reactor *10* receives mixture II from batch box *9* at such speed that the temperature in the apparatus does not exceed 26°C. After all mixture II is added, the reactive mixture is heated, first to 30-35 °C, then (after 3.5-4 hours of standing) to 45-50 °C. At this temperature the mixture also stands for 3.5-4 hours; then it is cooled to 20 °C with water sent into the jacket.

The pastelike mixture is sent by nitrogen flow into hydrolyser 12, which has been filled with hydrochloric acid (2-3%), prepared in advance and cooled to 4 °C. From batch box *13*, the concentrated hydrochloric acid self-flows into the hydrolyser, which has been filled with a necessary amount of water. The hydrolyser has an agitator and a jacket for cooling with Freon. After the reactive mixture is treated with hydrochloric acid, the unreacted magnesium turns into magnesium chloride, which dissolves in the acid, and the formed salts (magnesiumfluoroalkylchlorides) are destroyed. After the agitation in the hydrolyser, the reaction products are settled.

The water layer, which contains magnesium chloride, is neutralised with alkaline solution, and the organic layer is poured into collector *14*, and then is sent by nitrogen flow (0.3 MPa) to dehydrator *15* with calcium chloride, and to the nutsch filter *16*. The filtered organic layer is poured from the filter into collector *17* and from there by nitrogen flow (0.07 MPa) is sent to the rectification tower tank *18*, where dibutyl ether is distilled from tris(γ-trifluoropropyl)silane. The jacket of the tank is filled with a heat carrier like ditolylmethane or a silicone heat carrier like 1,2-bis(triphenoxysiloxy)benzene. Tower *18* has an external coil, also filled with a heat carrier, which is connected to the tank jacket. Dibutyl ether is distilled in the tank at 125 °C (76°C on top of the tower) and the residual pressure of 66-120 GPa.

The vapours which escape the tower are condensed in condenser *19*. First, we distil the distillate which boils out at a temperature below 70 °C and contains low-boiling by-products. It is collected and used later to flash the tower tank. Then we distil dibutil ether, which is collected and sent to dry. The product remaining in the tank, which mostly consists of tris(γ-trifluoropropyl)silane, is cooled, filtered of tarry matter and poured into collector *23*. The product collected there after several operations is sent by nitrogen flow (0.07 MPa) into rectification tower tank 24, where tris(γ-trifluoropropyl)silane is distilled from tank residues. The jacket of tank is filled with ditolylmethane or 1,3-bis(triphenoxysiloxy)benzene, and the tank coil is filled with vapour.

Tris(γ-trifluoropropyl)silane is distilled in the tank at 130-160°C (75-80°C on top) and the residual pressure of 6.7-13.3 GPa. First, we distil the distillate which boils out at a temperature below 75 °C and contains dibutyl ether (it is collected into receptacle *26*); then receptacle *27* is filled with separated tris(γ-trifluoropropyl)silane, which is sent by nitrogen flow (0.3 MPa) depending on the refraction index either for chlorination in apparatus *28*, or back into collector 23.

The chlorination of tris(γ-trifluoropropyl)silane is achieved in the carbon tetrachloride medium. First, carbon tetrachloride from batch box *29* and tris(γ-trifluoropropyl)silane from receptacle *27* self-flow into chlorinator *28*. Gaseous chlorine passes through a receiver and a glass cotton filter (not shown on the diagram) and is also fed into the chlorinator. The process is conducted at atmospheric pressure and at the temperature of 23-26 °C, which is supported by water sent into the apparatus jacket. Above the chlorinator there is backflow cooler *30* cooled by Freon. The discharged hydrogen chloride with escaping gases is sent into hydraulic gate *31*.

After the chlorination, hydrogen chloride is removed from the reactive mixture by nitrogen flow; the reactive mixture is sent to the rectification section (into tower tank *32*). Carbon tetrachloride and intermediate distillate are the first to be distilled in the tower, at 120 °C in the tank and at 30 °C and 0.1 MPa on top. The vapours of carbon tetrachloride which escape the tower are condensed and collected in receptacle *34*; the condensate is sent back to chlorinate.

After the separation of tetrachloride the tank is cooled. The concentrate of tris(γ-trifluoropropyl)chlorosilane is poured from the tank into collector *35*. As soon as it is accumulated, the concentrate is sent back to the tower tank; then the tank is connected to the vacuum line to distil the intermediate distillate, a mixture of carbon tetrachloride, unreacted tris(γ-trifluoropropyl)silane and chlorination by-products. The distillation is conducted at 160-170°C in the tank and at 80 °C and a residual pressure of 6.7--13.3 GPa on top. The intermediate distillate is collected in receptacle *36*. After the intermediate distillate is distilled, the temperature in the tank rises to 220 °C. This also distills tris(γ-trifluoropropyl)chlorosilane, which is collected in receptacle *37*. The distillation of the target product is conducted at 170-220°C in the tank and at 80 °C and a residual pressure of 6.7--13.3 GPa on top. Since tris(γ-trifluoropropyl)chlorosilane crystals melt at 20 °C, all communications are heated with hot water. Tank residue is cooled to 40-50 °C and then dissolved with acetone poured into the tank. After the dissolution is complete, the contents of the tank are used in the production of varnishes.

Tris(γ-trifluoropropyl)chlorosilane is a colourless liquid with a specific pungent odour. It fumes in air and is easily hydrolysed with water. The main constants of the product are:

Relative density d_4^{20}	1.390-1.392
Viscosity at 20 °C, MPa • s	6.98
n_D^{20}	1.3670-1.3680
Temperature, °C	
of boiling	206
of solidification	~20
Chlorine content, %	9.8-10.2

Tris(γ-trifluoropropyl)chlorosilane is a parent monomer in the synthesis of high-temperature α,ω-tris(γ-trifluoropropylsiloxy)organosiloxane liquids.

The reviewed technique of organomagnesium synthesis has significant drawbacks. The process develops in the necessary direction only in a narrow temperature range and requires solvents, which results in the low efficiency of equipment. The process makes use of metallic magnesium sub-sequently turned into magnesium chloride, an unusable waste product, which requires an additional filtering stage. As for the synthesis of or-ganohalogensilanes with the help of other metalorganic compounds, of particular interest may be organolithium compounds; in certain cases their use is appropriate. However, the exceptionally high sensitivity of organo-lithium compounds to oxygen requires the synthesis to be conducted in in-ert gas, which considerably complicates the technological process.

2.1.2. Techniques based on the interaction of chlorine derivatives of hydrocarbon with free carbon

Expanding applications for polyorganosiloxanes obtained from alkyl- and arylchlorosilanes called for simpler and more efficient techniques to syn-thesise these valuable monomers.

In the early 1940's appeared a technique for the direct synthesis of al-kyl(aryl)chlorosilanes exposing free silicon to halogenhydrocarbons in the presence of copper or silicon-copper alloys.

The reaction is complicated, but the process can be generally presented in the following way:

$$Si + 2RHal \longrightarrow R_2SiHal_2 \qquad (2.9)$$

$$2Si + 4RHal \longrightarrow RSiHal_3 + R_3SiHal \qquad (2.10)$$

where R = CH3, C_2H_5, C_6H_5 , etc. The choice of conditions for the proc-ess and the economical evaluation of the direct synthesis of organohalo-gensilanes are usually based on the yield of difunctional monomer (the first reaction) as the most valuable product. However, in practice, when silicon (or its metal alloys) interact with alkyl- or arylhalogenides, the re-action gives a great variety of products: liquid (R_2SiHal_2, $RSiHal_3$, R_3SiHal, $SiHal_4$, $SiHHal_3$, $RSiHHal_2$, $R_2SiHHal$), gaseous (RH, H_2) and solid (carbon).

The process can also occur without a catalyst, but then is much slower; the main products are halogensilanes with high halogen content ($SiHal_4$, $SiHHal_3$, $RSiHal_3$). On the other hand, in the presence of a catalyst the process develops at a good speed and with high yields of the main prod-ucts. Various metals have been tried as catalysts for direct synthesis, such as nickel, chrome, platinum, antimony, lead, aluminum, zinc, iron, silver, and copper. Some of them (Ni, Cr, Pt) gave no positive results, others (Al, Zn) displaced the reaction towards forming high-alkyl compounds. Good results have been shown by copper (in the synthesis of alkylchlorosilanes) and copper or silver (in the synthesis of phenylchlorosilanes).

Production of methyl-, ethyl and phenylchlorosilanes

The main raw stock for the direct synthesis of *methyl-, ethyl and phenyl-chlorosilanes* is correspondingly methylchloride, ethylchloride and chloro-benzene, as well as copper-silicon alloy or mechanical mixture of silicon and copper powders (so-called *contact mass*).

Methylchloride. Industrially it is obtained by chlorinating methane or reacting methyl alcohol with hydrogen chloride in the presence of dehydrating substances.

When methane is chlorined directly, alongside with methylchloride we find products of a deeper chlorination, such as methylenechloride, chloroform, and carbon tetrachloride. However, creating certain conditions, one can direct the process toward the prevailing formation of a desired product. Thus, methylchloride is obtained at considerable mole excess of methane and relatively high temperature (>400 °C); on the other hand, any temperature rise to 500-550 °C should be avoided because it may cause an explosion with the release of hydrogen chloride and free carbon:

$$CH_4 + 2Cl_2 \longrightarrow 4HCl + C \qquad (2.11)$$

Thus, methylchloride is obtained by chlorinating methane at 400-450 °C in the presence of a catalyst (metal chlorides precipitated on pumice) with a 10-fold surplus of methane. In these conditions, 80-85% of the incoming chloride is used up to form methylchloride.

The second technique for producing methylchloride is based on the interaction of methyl alcohol and hydrogen chloride in the presence of dehydrating substances (for example, powder anhydrous aluminum chloride or zinc chloride on coal):

$$CH_3OH + HCl \xrightarrow[-H_2O]{} CH_3Cl \qquad (2.12)$$

Methylchlorosilanes are more efficiently produced from methylchloride obtained in this way, because this product contains considerably lesser amounts of impurities, which inhibit direct synthesis.

Methylchloride is a colourless gas with an ether odour (the boiling point is -24.2 °C, the solidification point is -97.6 °C, the density at boiling temperature is 0.992 g/cm^3, the latent evaporation heat is 427 KJ/kg); 100 g of water dissolve 0.74 g of methylchloride.

Ethylchloride. At present the main industrial technique to produce ethylchloride is the hydrochlorination of ethylene.

$$CH_2=CH_2 + HCI \longrightarrow C_2H_5CI \qquad (2.13)$$

The reaction is usually conducted at -15 °C in the presence of a catalyst, powder anhydrous aluminum chloride. The parent gases (hydrogen chloride and ethylene) should be dried thoroughly, because humidity hydrolyses aluminum chloride, which increases catalyst consumption and causes the corrosion of equipment.

The process is conducted in a vertical steel apparatus filled with the catalyst suspended in liquid ethylchloride. This mixture is treated by hydrogen chloride and ethylene, while the contents of the reactor are intensively agitated. With the formation of ethylchloride, the volume of the liquid in the apparatus grows; therefore, the surplus of ethylchloride is constantly withdrawn from the reaction zone. Liquid ethylchloride, leaking from the reactor with catalyst particles, as well as dissolved hydrogen chloride, is vapourised, washed in a scrubber with a 10% alkaline solution, dried with sulfuric acid and condensed. The reaction gases, laden with ethylchloride vapours, are washed with water from hydrogen chloride, dried with concentrated sulfuric acid and sent into an absorber, where ethylchloride is extracted with kerosene. By distillation and subsequent condensation, ethylchloride is extracted from the obtained solution.

Ethylchloride can also be produced by ethane chlorination. In this case the reaction is conducted at 450°C and the volume ratio of ethane and chlorine is 8:1. Ethane reacts with chlorine much easier than methane, which makes it possible to use even natural gas, which contains only 10% of ethane and 90% of methane. Under such conditions ethane is almost completely chlorinated, whereas the formation of chlorine derivatives of methane is virtually ruled out.

Finally, ethylchloride can be obtained by a combined technique from a mixture of ethane and ethylene. The process is based on combined subsequent reactions of substitutuve chlorination of ethane and hydrochlorination of ethylene with hydrogen chloride obtained from the first reaction:

$$C_2H_6 + CI_2 \xrightarrow[-HCI]{} C_2H_5CI \qquad (2.14)$$

$$CH_2=CH_2 + HCI \longrightarrow C_2H_5CI \qquad (2.15)$$

This technique is more economical in comparison with other ways of producing ethylchloride due to a fuller use of chlorine and the reduction of costs to build an installation for the synthesis of concentrated hydrogen chloride.

Ethylchloride is a colourless liquid with a pleasant odour (the boiling point is 12.5°C, the solidification point is -138.7 °C); it is combustible. The pressure of saturated ethylchloride vapours is 0.113 MPa at 15°C. Its evaporation heat is 387 KJ/kg. It dissolves well in organic solvents and only to some extent in water (100 g water at 0°C dissolve only 0.45 g of ethylchloride).

Chlorobenzene. At present it is obtained by continuous catalytic chlorination of benzene:

$$C_6H_6 + Cl_2 \xrightarrow[-HCI]{} C_6H_5CI \qquad (2.16)$$

In the process reactive mixture is heated approximately to the boiling point of benzene (76-83 °C); there is a surplus of benzene for chlorination. At this temperature some of the chlorobenzene formed evaporates. The evaporation uses a lot of heat released during the reaction; the rest is intensively withdrawn, and chlorinators, which work when the reactive mixture is boiling, are hightly efficient. The process is catalysed by iron chloride in the amount of 0.01-0.015% (mass) of benzene. To avoid the formation of polychlorides, chlorination is stopped when 50-68% of benzene remain unchanged. In this case polychlorides account for not more than 3.5-4.5% of the chlorobenzene amount.

The chlorination of benzene can be conducted in a steel tower with a head of steel and ceramic rings, lined with acidproof tiles. The upper (expanded) part of the tower is a mist extractor; the lower part of the apparatus receives benzene and gaseous chlorine.

Chlorobenzene is a colourless motile liquid with a weak scent of almonds (the boiling point is 132.1°C, the solidification point is -45.2 °C; the density is 1.113 g/cm^3); it is combustible. Its combustion liberates soot and hydrogen chloride.

The main requirement to methylchloride, ethylchloride and chlorobenzene is the absence of impurities, by-products and especially moisture. With even the slightest amount of liquid entering the reaction zone, the products start to hydrolyse and condense, the activity of the contact mass or copper-silicon alloy decreases, and the process subsides. That is why the technology of direct synthesis usually provides for a device to dehydrate alkyl- and arylchlorides. For this purpose one can pass methyl- or ethylchloride through the tower sprayed with sulfuric acid, or use other dehydrating substances (burnt CaCI$_2$, AI$_2$O$_3$ and zeolites, e.g. burnt klinoptilolite).

A good dehydrating substance for chlorobenzene is silicon tetrachloride, which easily reacts with water in chlorobenzene forming a sediment, sili-cagel. Moisture content in chlorobenzene (as well as in methylchloride and ethylchloride) after drying should not exceed 0.02%.

The direct synthesis of organochlorosilanes makes use of copper-silicon alloy or a mixture of silicon and copper powders (contact mass).

Silicon is a brittle dark grey crystalline material (the melting point is 1480-1500°C, the density is 2.49 g/cm^3). Industrial contact masses are made from silicon with various impurity content, obtained by the reducing fusion of quartzites. The silicon is not subjected to additional purification, and it contains 1-3% of impurities. Contact mass is prepared from Si-0 (99% Si), Si-1 (98% Si) and Si-2 (97% Si) silicon with various impurity content (Ca, Mg, Al, Fe, etc.).

The chemical activity of contact mass largely depends on the purity of silicon. Pure silicon brands (>99% Si) are less reactive, which makes them uneconomical for this purpose; if the purity of silicon is less than 97%, it reduces the yield of the main product in the direct synthesis of organochlorosilanes.

The particle size of metallic silicon also has a noticeable effect on the chemical activity of contact mass: in case of coarse grinding, the reactivity of powder is lower than required; in case of fine grinding the reactivity rises sharply, and the heat has to be intensively diverted from the reaction zone.

Copper in contact mass plays the role of a catalyst. Pure copper is obtained by the electrolysis of copper sulfate. For direct synthesis, we use the copper of two brands, Mo and M$_1$ with 99.95-99.9% of Cu. The total impurity content (Bi, Sb, As, Fe, Ni, Pb, etc.) should not exceed 0.05-0.1%. To ensure high activity of contact mass, it is necessary to use copper powders with complex surfaces. Good results in direct synthesis are also obtained when using fine copper prepared by the mechanical spraying of copper powder or deposition of copper from copper salts.

Copper-silicon alloy or contact mass used for the direct synthesis of organohalogensilanes should be highly active and selective. The yield of reaction products largely depends on copper content: when there is not enough copper (less than 1-2%), the synthesis of alkyl- and especially arylhalogensilanes is slow. When copper content is increased to 3-5%, the yield of the reaction products considerably increases. At subsequent increase of copper content (more than 10%) the yield does not grow, whereas the product composition deteriorates due to the formation of compounds with high chlorine content (SiCl$_4$, HSiCl$_3$, RSiCl$_3$). Thus, increasing copper content up to more than 10% does not increase the yield of methylchlorosilanes and has a negative effect on the composition of the reaction products. This seems to be due to copper shielding the surface of silicon. A similar situation is observed in the direct synthesis of ethylchlorosilanes.

For the direct synthesis of phenylchlorosilanes, copper content in contact mass should be significantly bigger than for the synthesis of methyl- and ethylchlorosilanes. Good yields of phenylchlorosilanes are attained if catalysed by silver (10% of silicon), but owing to its high cost and scarcity, industrial preference is given to copper.

However, the catalytic activity of contact mass is determined not only by copper concentration. Crucially important is also the structure of the crystal lattices of silicon, copper and the intermetallic compound Cu_3Si, which is formed during the alloying of silicon with copper or in the synthesis, since the structural failure seems to facilitate silicon or copper entering the reaction. The structure of the crystal lattice is largely determined by the way contact mass is prepared.

Preparation of contact mass. There are three techniques for this:

1. mechanical mixture of silicon and copper powders;
2. chemical deposition of copper on silicon particles;
3. alloying of silicon and copper powders;

1. Thin silicon and copper powders are mixed, pelleted at ≈500 MPa and for 2-4 hours held in hydrogen atmosphere at the temperature up to 1050 °C (hydrogen reduction is used to increase the catalytic activity of contact mass).

2. Silicon and univalent copper chloride powders are mixed and pelleted. The pellets are dried and treated with hydrogen at 300 °C. Copper chloride is reduced, and silicon particles are covered with free copper:

$$2Cu_2CI_2 + Si \longrightarrow 4Cu + SiCI_4 \qquad (2.17)$$

The contact mass prepared by this technique is as a rule highly active.

3. The process is conducted in a high-frequency electric furnace for 2 hours at 1200-1400 °C in a reducing medium (e.g. in hydrogen atmosphere). After that, the alloy is quickly poured off and cooled to avoid copper liquation (deposition). The alloy can be poured in thin layers into water-cooled casting molds or onto a roller machine with rollers cooled with water from the inside. The alloy from the furnace flows through a graphite pan onto a swinging feeder, and then into the space between the rotating rollers. There the alloy quickly cools, forming thin crystal films. After that the films are sealed by the rollers, forming a band. The thickness of the band is regulated by changing the speed of roller rotation and temperature. The formed band is stripped off the roller with a band stripper and is sent

through a chute into a special kubel. There the alloy is cooled for 3 hours and ground.

The analysis of alloy microstructure shows that they are a multiphase system consisting of intermetallides Cu_3Si, $FeSi_2$, calcium silicide, multicomponent phase Fe + Si + Ti + (Mn) + (V) + (P), complex silica oxides, metallurgical cinder and a matrix, silicon. However, initially only the intermetallic compound Cu_3Si actively interacts with alkylchlorides. That is why it would be interesting to realise the direct synthesis of organochlorosilanes using a specially prepared silicon intermetallide; this can be expected to lower the costs, decrease the temperature and decrease the inductive period of the reaction.

Activation of contact mass. In order to improve the contact properties of the alloy or powder mixture, increase the yield of the synthesis products and shift the reaction towards the preferential formation of more valuable substances (R_2SiCl_2, $RSiHCl_2$), the contact mass is subjected to additional activation.

There are several activation techniques. One of the most widespread techniques is the thermal treatment of the contact mass in hydrogen flow or hydrogen-nitrogen mixture at 1050 °C for several hours. According to the second technique, the mass is submerged into a 30% solution of bivalent copper chloride for 1 minute, which changes into univalent copper chloride:

$$CuCl_2 + Cu \longrightarrow Cu_2Cl_2 \qquad (2.18)$$

Then the mass is heated up to 260-270 °C in alkylchloride flow; Cu_2Cl_2 is reduced with free copper depositing on the surface of particles.

The third technique is the activation of the mass in the solid phase with halogenides of group II metals . For this purpose, one can introduce zinc chloride, 0.1% of the mass content (if a great amount of $ZnCl_2$ is introduced, the mass aggregates, which makes it impossible to support the necessary temperature mode; this, in its turn, reduces the productivity of the process and the dichloride content in the reactive mixture). After that the mass is dried for 4-5 hours in nitrogen flow at 200 °C, and for 1.5-2 more hours the reactor is put into the operating mode, i.e. the temperature is increased up to 300-320 °C (also in nitrogen flow). In the operating mode zinc chloride melts (melting point 260 °C) and intensively mixes with copper-silicon alloy, which contributes to the cementation of the alloy particles and to the uniform distribution of $ZnCl_2$ throughout the contact mass. This, in its turn, helps to preserve the catalytic activity of contact mass thoughout the whole process.

The latter activation technique uses more silicon in the process of direct synthesis (70-75% against 30-40% for nonactivated mass) and considerably increases the yield of dialkyldichlorosilanes. For example, if for the normal mass during the synthesis of methylchlorosilanes the yield of dimethyldichlorosilane varies from 30 to 45% in time, for the mass activated by zinc chloride the yield grows up to 60-75%. However, this activation technique also has disadvantages:

1. there are considerable temperature changes (up to 60-80 °C) during the synthesis; this, however, has little effect on the yield of dimethyldichlorosilane and the overall activity of the process.
2. alkylchloride feeding, i.e. the process of synthesis as such, takes 25-30 hours (as against the common 35-40 hours), after which the process subsides;
3. if for some reason the temperature in the reactor rises over 370-380 °C, the zone temperature grows sharply (up to 500 °C and more) and cannot be regulated; in this case the synthesis has to be stopped.

Raw stock for the direct synthesis of methylchlorosilanes, methylchloride, has such impurities as moisture, methyl alcohol, oxygen, sulfur dioxide, methylenechloride, dimethyl ether, carbon oxide and dioxide, etc. Most of them negatively affect the synthesis of methylchlorosilanes: harmful impurities are chemisorbed on the active centres of contact mass and foul the copper catalyst, which naturally inhibits the reaction of methylchloride with contact mass. A similar situation is observed in the direct synthesis of ethylchlorosilanes.

A way to counteract the effect of harmful impurities in alkylchlorides is to introduce substances which would be adsorbed on the active centres of contact mass at the initial stage of synthesis, thus opposing undesirable chemisorption and harmful substances fouling the catalyst. Methyl- or ethylchlorosilanes were found to be active substances of the kind; moreover, their ability to activate contact mass grows in the following sequence:

$$HSiCl_3 > SiCl_4 > RSiCl_3 > R_2SiCl_2 \qquad (2.19)$$

This gave rise to another technique for activating contact mass by treating it with organochlorosilane vapours in the gaseous phase at 290-300 °C. This technique in comparison with the former is characterised by the following positive features:

1. the feeding of alkylchloride (i.e. direct synthesis as such) takes more time, about 30-35 hours;

2. dialkyldichlorosilanes are yielded almost constantly (60-80% within 30-35 hours);

3. a negligible amount of silicon tetrachloride is formed, which allows to extract pure trimethylchlorosilane from the intermediate distillate by direct rectification in industrial conditions; it is known that in the direct synthesis of methylchlorosilanes the formed $SiCl_4$ and $(CH_3)_3SiCl$ are hard to separate, because their boiling points are very close (57.7 and 57.3°C);

4. the temperature of the process can be regulated easily and precisely.

5. All the advantages mentioned make this technique for the activation of contact mass the most promising choice for industry.

Selectivity of contact mass. Since the most important products in the synthesis of organochlorosilanes are diorganodichlorosilanes, it is natural that the increase of their yield receives much attention. The selective formation of diorganodichlorosilanes when organic chlorine derivatives interact with contact mass is connected with the purity of the reactants, silicon above all. High yield of dimethyldichlorosilane in the direct synthesis of methylchlorosilanes largely depends on the presence of noticeable quantities of aluminum in contact mass. The yield of dimethyldichlorosilane in the presence of aluminum as a rule decreases due to the formation of trimethylchlorosilane; in the reaction with pure (semiconductor) silicon in the presence of copper trimethylchlorosilane is virtually not formed.

The formation of trimethylchlorosilane due to the reduction of the quantity of dimethyldichlorosilane and the absence of trimethylchlorosilane in reactions based on pure silicon seem to testify that trimethylchlorosilane is a product of dimethyldichlorosilane disproportioning, which occurs in the conditions of synthesis under the influence of impurities (first of all, aluminum and its compounds):

$$2(CH_3)_2SiCl_2 \rightleftarrows (CH_3)_3SiCl + CH_3SiCl_3. \qquad (2.20)$$

The reaction of dimethyldichlorosilane disproportioning under the influence of aluminum chlorides sharply accelerates on the surface of copper in the presence of methylchloride. In this connection, the process of direct synthesis can encounter the conditions in which the formation of trimethylchlorosilane due to dimethyldichlorosilane disproportioning occurs at noticeable speed, and the quantity of trimethylchlorosilane in the mixture formed can be increased to 20%. At the same time, another reaction accelerates on the surface of copper reacting with methylchloride, the methylation of dimethyldichlorosilane with methylchloride in the presence of aluminum or zinc:

$$3(CH_3)_2SiCl_2 + 3CH_3Cl + 2Al \longrightarrow 3(CH_3)_3SiCl + 2AlCl_3 \qquad (2.21)$$

Thus, the methylation of dimethyldichlorosilane becomes very significant in the processes of direct synthesis, whereas the total amount of trimethylchlorosilane, which is formed according to the reactions of disproportioning and methylation, may reach 60-65%. Thus, in the direct synthesis of methylchlorosilanes the introduction of significant amounts of AI or its compounds into contact mass reduces the yield of dimethyldichlorosilane and respectively increases the yield of trimethylchlorosilane.

This example shows the importance of so-called activators, or *promoters*, for increasing the activity and selectivity of contact mass. These additives can sharply activate the reaction and shift it in a certain direction. Various substances have different effects on the activity of contact mass. For example, antimony has a positive effect on the direct synthesis of organochlorosilanes and increases the total yield of methylchlorosilanes, whereas lead and bismuth reduce the formation of these substances. However, the positive effect of a promoter manifests itself only in a certain concentration, exceeding which transforms a positively acting additive into poison or an inhibitor of the reaction. For example, in a 0.002--0.005% concentration antimony is a promoter of the direct synthesis of methylchlorosilanes; on the other hand, in a concentration higher than 0.005% it becomes poison.

Apart from antimony, there are other good promoters of the direct synthesis of methylchlorosilanes, which increase the yield of dimethyldichlorosilane, such as arsenic and zinc chloride. If it is necessary to increase the yield of alkylhydridechlorosilanes, one should use univalent copper chloride, cobalt, and titanium. The addition of tin or lead into contact mass increases the yield of dimethyldichlorosilane up to 70%; the yield of ethyldichlorosilane is increased to 50-80% when contact mass receives 0.5-2% of calcium silicide (Ca_2Si). In the synthesis of phenylchlorosilanes effective promoters are zinc, cadmium, mercury or their compounds. In particular, the introduction of zinc oxide (up to 4%) into contact mass may increase the diphenyldichlorosilane content up to 50%, and the introduction of a mixture of zinc oxide and cadmium chloride, even up to 80%.

Thus, we can make the following important conclusion: the composition of a complex reactive mixture in the direct synthesis of methyl-, ethyl- and phenylchlorosilanes noticeably changes depending on the introduction of promoting additives into contact mass. Since the demand for the products of direct synthesis wavers quite considerably (depending on the monomer market condition), it is very important for industry to know how this or that promoter affects the yield of each polymer.

At present, our country manufactures alloys of A, B, C and D brands, with the composition given below, %:

Component	A	B	C	D
Cu	5-7	5-7	9-12	9-12
Fe, not more than	0.7	0.7	0.7	0.7
Al	0.6-1.2	≤0.7	≤0.7	≤0.7
Sb	-	0.002-0.005	0.002-0.005	-
Pb, not more than	0.002	0.002	0.002	0.002
Si	The rest			

Alloy A is used for the synthesis of methylchlorosilanes and other alkylchlorosilanes; alloy B is used for the synthesis of methylchlorosilanes with an increased yield of dimethyldichlorosilane; alloy C is used for the synthesis of methyl- and ethylchlorosilanes; alloy D is used for the synthesis of phenyltrichlorosilane.

Grinding of copper-silicon alloy. Copper-silicon alloy before direct synthesis is subjected to grinding. The alloy enters the plant in the shape of fragments or plates weighing 60-80 kg. First, they are broken into pieces up to 75 mm in size, then crushed to 15-25 mm on a jaw crusher and dumped into the bottom of an elevator. The elevator feeds the alloy into a bin, and from the bin the alloy is sent through a branch pipe into ball grinders. The ground alloy is sent to the bin and from there to the synthesis reactor.

The alloy ground in this way should have the following sieve composition: Fraction I is not less than 5% of 0.5-0.25 mm grains; fraction II is not less than 80% of 0.25-0.075 mm grains; fraction III is not less than 15% of grains smaller than 0.075 mm. In case there is a great amount of fraction III grains (i.e. fine grains), the yield of the products is decreased, because the gas flow of alkylchloride (or arylchloride) in the process of synthesis carries away a lot of alloy in the form of dust, which reduces the efficiency of the reactor and reduces the output of dialkyl(diaryl)dichlorosilane. Predominance of fraction III grains can be caused by an overload of balls in the ball grinder or by a nonoptimal mesh gap in the drum of the grinder.

The conditions of the reaction are also important for the yield and composition of direct synthesis products. Thus, the growth of temperature increases the transformation degree of alkyl- or arylchloride, but reduces the yield of dialkyl(diaryl)dichlorosilanes and increases the amount of products with big chlorine content ($SiCl_4$, $HSiCl_3$, $RSiCl_3$). This is explained by the fact that the increase of temperature accelerates the processes of destruction. That is why the direct synthesis of organochlorosilanes should be conducted at as low temperatures as possible, i.e. at the minimal temperature when the process still gives a satisfactory yield of the target product. The minimal temperature for the synthesis of methyl-, ethyl and phenylchlorosilanes is different. Thus, the reaction of silicon with ethylchloride in the presence of copper occurs at a noticeable speed already at 240-260 °C, with methylchloride, at 260-280 °C, and with chlorobenzene, only at 450-500 °C.

It follows from all the above-mentioned facts that the direct synthesis of methyl-, ethyl and phenylchlorosilanes is a complex heterophase process which depends on many factors and forms a compex reactive mixture. For example, in the direct synthesis of methylchlorosilanes there are about 130 compounds found and characterised. This does not mean, however, that in this or other definite synthesis all the 130 products are formed. The composition of the mixtures formed and the transformation degree of alkyl-chlorides and chlorobenzene in the synthesis of methyl-, ethyl and phenyl-chlorosilanes depend on the synthesis conditions, the type of the reactor used and many other factors. In spire of the complexity of the process and the variety of its products, the reaction of direct synthesis can nevertheless be directed (towards a preferential formation of a main product), changing the conditions for the preparation of contact mass, introducing various promoters into contact mass and changing the reaction conditions.

Preparation of methylchlorosilanes

Because direct synthesis in this case is very complex, the mechanism of this process has not been fully established. However, the following way of methylchlorosilane formation with the catalytic effect of copper on the reaction of methylchloride with silicon is the most probable.

It is supposed that the contact mass used in direct synthesis consists of two phases, free silicon and the intermetallic compound Cu_3Si (η-phase). At the initial stage at synthesis temperature methylchloride interacts with the silicon atom from the intermetallic compound:

$$2CH_3Cl + Cu_3Si \longrightarrow (CH_3)_2SiCl_2 + 3Cu \qquad (2.22)$$

Then, under the catalytic influence of the releasing active copper, methylchloride reacts with free silicon from the contact mass:

$$2CH_3Cl + Si \xrightarrow{\ Cu\ } (CH_3)_2SiCl_2 \qquad (2.23)$$

$$2CH_3Cl + Si \xrightarrow{\ Cu\ } CH_3SiHCl_2 + \overset{\bullet}{C}H_2 \qquad (2.24)$$

$$(2.25)$$

$$3CH_3CI + Si \xrightarrow{Cu} CH_3SiCI_3 + 2\overset{\bullet}{C}H_3$$

$$4CH_3CI + 2Si \xrightarrow{Cu} CH_3SiCI_3 + (CH_3)_3SiCI \qquad (2.26)$$

The free radicals formed in these reactions undergo subsequent transformations:

$$(2.27)$$

$$2 \cdot \overset{\bullet}{C}H_2 \longrightarrow C_2H_4$$

$$(2.28)$$

$$2 \overset{\bullet}{C}H_3 \longrightarrow CH_4 + C + H_2$$

When hydrogen chloride is fed into the reaction zone, the yield of methylphenyldichlorosilane increases.

$$CH_3CI + Si + HCI \xrightarrow{Cu} CH_3SiHCI_2 \qquad (2.29)$$

Apart from the main products (dimethyldichlorosilane, methyldichlorosilane, methyltrichlorosilane and trimethylchlorosilane) there are various amounts of such by-products as trichlorosilane, silicon tetrachloride, dimethylchlorosilane and products with a boiling point of more than 70.2°C, i.e. the products of condensation and pyrolysis, which contain fragments like

$$\overset{\diagdown}{\diagup}Si-O-Si\overset{\diagup}{\diagdown} \qquad \overset{\diagdown}{\diagup}Si-Si\overset{\diagup}{\diagdown} \qquad \overset{\diagdown}{\diagup}Si-CH_2-Si\overset{\diagup}{\diagdown}$$

gaseous products (hydrogen, methane, ethylene) and a solid product (carbon). The formation of gaseous and solid substances results from the pyrolysis of methylchloride, methyl and methylene radicals. All these substances are produced in various ratios depending on the type of contact mass activator and the quantity of the supplied hydrogen chloride. It should also be noted that the formation of carbon and its accumulation in contact mass are one of the reasons for reducing its activity, increasing the synthesis temperature, intensification of pyrolytic processes and deterioration of the composition of the reactive mixture.

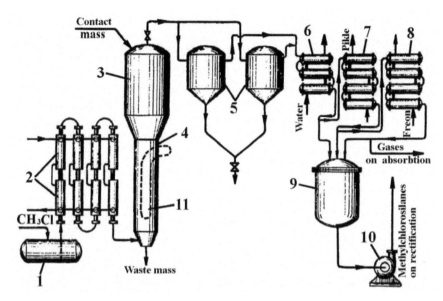

Fig. 3. Production diagram of methylchlorosilanes by direct synthesis technique: *1* - container; *2* - evaporator; *3* - separator; *4* - reactor; *5* - filters; *6-8* – condensers; *9* - collector; *10* - centrifugal pump; *11* - Field tube.

In the production of methylchlorosilanes by the direct synthesis technique (Fig. 3) reactor *4* is loaded with contact mass and subjected to electric heating; at 200 °C nitrogen is fed through the heater at the speed of 8-12 m³/h (not shown in the figure) for drying the mass. The temperature in the reactor gradually rises to 340 °C. After this temperature is achieved, nitrogen flow is stopped and followed by gaseous methylchloride.

The interaction of methylchloride with contact mass occurs in the fluidized layer in reactor *4*, built like a tower. Its upper expanded section *3* acts like a separator for separating small particles dragged by the flow. Inside the tower there is Field tube *11* with four ribs for better heat exchange; the tube is fed with water to absorb the escaping heat. Stored methylchloride is sent to container *7*, which is cooled with salt solution from the outside; from there it is sent by nitrogen flow into evaporator *2*, which is heated with vapour (1 MPa). From the evaporator, methylchloride vapours through a backflow gate are sent into the lower part of the reactor. In the course of the process the following parameters are established: the temperature in the evaporator is 120-150 °C; the temperature in the lower part and in the middle of the reactor is 320-340 °C; the temperature on top of the reactor is 250-300 °C; the pressure in the reactor is 0.4-0.5 MPa.

The reaction products and unreacted methylchloride are sent to separator *3* and from there, after passing filters *5* for disposal of the small particles of contact mass, enter in series condensers *6*, *7* and *8*. The condensers receive water, salt solution (-15 °C) and Freon (-50 °C) correspondingly. The condensate is collected in collector 9 and then pumped into the rectification section. If there is a necessity to increase the yield of methyldichlorosilane, in addition to methylchloride the reactor is supplied with hydrogen chloride. The noncondensed gases (N_2, H_2, CH_4 , etc.) and other so-called effluent gases pass through the water absorber (to extract methylchloride), the calcium chloride tower (not shown in the diagram) and through the fire-resistance device enter the atmosphere.

The cooled discharge mass is loaded out of the reactor into the bin and transported into the baking department.

Distillation of unreacted methylchloride and rectification of methylchlorosilane mixture

The direct synthesis of methylchlorosilanes forms a condensate of the following composition: 50-80% of methylchlorosilanes and 20-50% of methylchloride (unreacted).

In some cases we also find liquid products of thermal decomposition of methylchloride such as 2-methyl- and 3-methylpropane, 2-methyl- and 3-methylpentane, 2-methyl- и 3-methylhexane, ethylidenechloride, etc. Their appearance signals that the temperature of the process is not regulated satisfactorily.

First, unreacted methylchloride is distilled from the mixture; then, the condensate obtained in the process of direct synthesis is rectified. The rectification can be carried out in packed or tray towers made of ordinary steel (there is no corrosion if moisture does not enter the system).

The diagram of the distillation of unreacted methylchloride and rectification of methylchlorosilane mixture is given in Fig. 4.

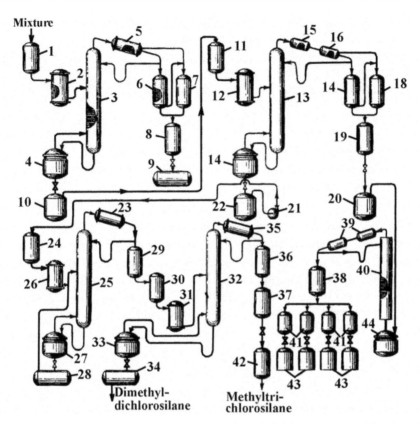

Mixture

Dimethyl-dichlorosilane

Methyltri-chlorosilane

Fig. 4. Diagram of the distillation of unreacted methylchloride and rectification of methylchlorosilane mixture: *1, 11, 24, 30* - pressure containers; *2, 12, 26, 31* – heaters; *3, 13, 25, 32, 40*- rectification towers; *4, 14, 27, 33, 44* – rectification tower tanks; *5, 15, 16, 23, 35, 39*- refluxers; *6, 7, 17, 18, 29, 36, 38*- cooling condensers; *8, 19, 37, 41*-receptacles; *9, 10, 20, 22, 28, 34, 42, 43* – tanks; *21* - centrifugal pump.

Distillation is conducted at the excess pressure of 0.5-0.55 MPa. From pressure container *1* the mixture is constantly fed into heater *2*, from where at 50-60 °C it is sent to the feeding plate of rectification tower *3*. In the tower, methylchlorosilanes and methylchloride are separated. Methylchlorosilanes from tank *4* are collected into collector *10*. The temperature in the tank is maintained within 145-155 °C with vapour (1 MPa) fed into the tank jacket. Methylchloride is condensed in refluxer *5*, which is cooled with Freon (-50 °C); from there part of methylchloride is returned into tower *3*, and the rest through cooler *6* is collected into receptacle *8*. The uncondensed methylchloride from refluxer *5* and cooler *6* is sent into condenser *7*, and from there is poured into receptacle *8* and collector *9*. The distilled methylchloride is sent back to the synthesis. The distillation of methylchloride is finished when two successive analyses show that the CH$_3$CI content in the tank does not exceed 3%. The tank residue, methyl-chlorosilane mixture, is collected in collector *10*.

The average composition of the mixture obtained in methylchlorosilane synthesis is (in %):

Methylchloride (un-reacted)	1 -3	Silicon tetrachloride	1-3
Trichlorosilane	0.5-1	Methyltrichlorosilane	13-30
Dimethylchlorosilane	0.5-1	Dimethyldichlorosilane	50-65
Methyldichlorosilane	1-5	Tank residue	5-10
Trimethylchlorosilane	0.3-5		

Note. The average composition may change in a wide range depending on the conditions of the reaction and promoters used to activate contact mass.

The first rectification stage. From collector *10* the mixture of methyl-chlorosilanes is periodically fed into pressure container *11*, from where at 50-65 °C it is sent through heater *12* (by self-flow) onto the feeding plate of rectification tower *13*. From the tower the tank liquid (methyltrichloro-silane, dimethyldichlorosilane and tank residue) flows into tank *14*, where the temperature of 80-90 °C is maintained, and from there is continously poured into collector *22*. After the tower, vapours of the head fraction at a temperature below 58 °C, consisting of the rest of methylchloride, di- and trichlorosilane, dimethylchlorosilane, methyldichlorosilane and the azeotropic mixture of silicon tetrachloride and trimethylchlorosilane are sent into refluxer *15*, cooled with water, and into refluxer *16*, cooled with salt solution (-15 °C). After that, through cooler 17 the condensate is gath-ered in receptacle 19. Volatile products, which did not condense in reflux-ers *15* and *16*, are sent into condenser *18* cooled with Freon (-50 °C). There they condense and also flow into receptacle *19*. As soon as it is ac-cumulated, the condensate is sent from receptacle *19* into collector *20*.

The second rectification stage. From collector *22* the mixture, which consists of 20-35% of methyltrichlorosilane, 45-60% of dimethyldichloro-silane and 20-25% of high-boiling tank residue, is pumped with pump *21* into pressure container *24* to the second rectification stage. From this con-tainer the mixture is sent into heater *26*, from where at 60-65 °C it is sent onto the feeding plate of rectification tower *25*. There a high-boiling resi-due containing up to 10% of dimethyldichlorosilane is extracted in the form of tank liquid. This residue is poured from tank *27* into collector *28* and is sent to rectification again; the mixture of methyltrichlorosilane and dimethyldichlorosilane vapours after passing through tower *25* enters at 70 °C refluxer *23*. From the refluxer part of the condensate is returned to reflux the tower, and the rest is sent through cooler *29* into pressure container *30* of the third rectification stage.

The third rectification stage. From pressure container *30* the mixture of 25-40% of methyltrichlorosilane and 60-75% of dimethyldichlorosilane is sent through heater *31* at 70-65 °C to the feeding plate of rectification tower *32*. Dimethyldichlorosilane in the form of tank liquid flows into tank *33*, where the temperature of 85-90 °C is maintained, and from there is continously poured into collector *34*. Methyltrichlorosilane vapours at 65-67 °C are sent into refluxer *35*. Part of the condensate is returned to reflux the tower, and the rest is sent through cooler *36* into receptacle 37 and poured into collector *42* as soon as methyltrichlorosilane is accumulated.

Separation of head fractions. The head fraction, which is obtained at the first stage of continuous rectification of methylchlorosilanes, from collector *20* self-flows into tank *44*. The temperature in the tank in the beginning of the process is maintained at 60-70 °C, and at the end it should be from 90 to 95 °C. Vapours from the tank rise up tower *40* and enter refluxers *39*, cooled with water and salt solution (-15 °C); from there, part of condensate is returned to reflux tower *40*, and the rest is sent through cooler *38* into receptacles *41* and fed into collectors *43*.

The temperature of the vapours in the upper part of tower *40* is 31-35 °C during the distillation of trichlorosilane, 38-44 °C for methyldichlorosilane, 53-57 °C for the azeotropic mixture of silicon tetrachloride and trimethylchlorosilane, 58-60 °C for trimethylchlorosilane.

It should be kept in mind during the rectification that methylchlorosilanes are easily hydrolysed even under the influence of humidity in air, and the hydrogen chloride formed in the hydrolysis corrodes the equipment. Thus, all equipment and communications should be absolutely dry, and the obtained products should be collected, separated and transported in the absence of moisture. If these conditions are observed, it is possible to produce all technological equipment from common brand steel.

Methylchlorosilanes are difficult to separate due to the closeness of some of their boiling points. It is especially difficult to separate pure dimethyldichlorosilane (the boiling point is 70.2 °C) devoid of methyltrichlorosilane (the boiling point is 66.1 °C), because the difference of their boiling points is only 4.1 °C. It is known that the efficiency of separating reactive mixtures depends on the number of theoretical plates in the rectification towers; moreover, in distillation there is a certain dependence between the number of theoretical plates and the difference in the boiling points of the components. For precise distillation and compete separation of methyltrichlorosilane from dimethyldichlorosilane, one needs a rectification tower with the efficiency of 60-80 theoretical plates.

As for the separation of trimethylchlorosilane (the boiling point is 57.3 °C) and silicon tetrachloride (the boiling point is 57.7 °C), this is a gruelling task, since they form an azeotropic mixture which cannot be separated by simple rectification. The separation can be achieved with the help of physical (azeotropic rectification) or chemical techniques (hydrolysis, etherification).

Azeotropic rectification makes use of organic additives, acetonitrile CH_3CN (the boiling point is 81.5 °C) or acrylonitrile $CH_2=CHCN$ (the boiling point is 79 °C), which combine with silicon tetrachloride and trimethylchlorosilane to form azeotropic mixtures with different boiling points. For acetonitrile: 90.6% $SiCl_4$ + 9.4% CH_3CN (the boiling point is 49.1 °C) and 92.6% $(CH_3)_3SiCl$ + 7.4% CH_3CN (the boiling point is 56.5 °C). For acrylonitrile: 11% $CH_2=CHCN$ + 89% $SiCl_4$ (the boling point is 51.2 °C) and 7% $CH_2=CHCN$ + 93% $(CH_3)_3SiCl$ (the boiling point is 57 °C).

Acetonitrile is more suitable for separation than acrylonitrile, because the azeotropic mixtures formed by acetonitrile differ more in their boiling point from chlorosilanes; besides, after settling the azeotropic mixture of silicon tetrachloride with acetonitrile splits. The upper layer at 25 °C consists of about 70% of acetonitrile, whereas the lower layer has only 2.7% of it. This gives an opportunity to separate almost pure silicon tetrachloride from the lower layer (with only 0.5% CH_3CN remaining in $SiCl_4$). To separate the azeotropic additive $SiCl_4$ + CH_3CN one can use rectification towers with the efficiency of 30-60 theoretical plates. The upper layer of the additive with a higher acetonitrile content is used to reflux the tower. It gives trimethylchlorosilane in pure form or (in case an excess of acetonitrile is used) or in the form of an azeotropic mixture with acetonitrile.

The distillation of the azeotropic mixture $SiCl_4$ + CH_3CN from the lower layer, which consists of 70% of $SiCl_4$, follows the extraction of silicon tetrachloride.

Azeotropic mixtures can also be separated in constantly operating towers (Fig.5). The azeotropic mixture $SiCl_4$ + $(CH_3)_3SiCl$, which contains acetonitrile, is sent into the middle section of the first tower 1. Trimethylchlorosilane, which is taken out of the lower section of the tower, through cooler 6 is sent into collector 7; the distillate from the tower is partly returned through cooler 2 into the tower in the form of reflux, and partly sent into separating flask 3 for splitting. The upper layer, which consists of acetonitrile approximately to the extent of 2/3, is returned into tower 1 as reflux; the lower layer feeds the second tower 4. Part of the distillate from tower 4 is returned into the tower (reflux), and the main quantity is also sent for splitting in flask 3.

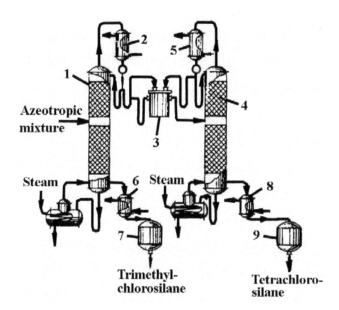

Fig. 5. Diagram of continuous separation of azeotropic mixture of trimethylchlorosilane and silicon tetrachloride, which contains acetonitrile. *1* is the tower for separating trimethylchlorosilane; *2, 5, 6, 8* are the coolers; *3* is the separating (Florentine) flask; *4* is the tower for separating silicon tetrachloride; *7, 9* are the collectors of the tower section *4*. Pure silicon tetrachloride is is collected through cooler *8* in collector *9*.

From the lower part of tower *4* pure silicon tetrachloride is collected through cooler *8* in collector *9*.

Of the chemical methods to separate the azeotropic mixture of silicon tetrachloride and trimethylchlorosilane, etherification is the most convenient one. This method is based on the different activity of components in the reactions of partial etherification by alcohols or phenols. Silicon tetrachloride is the first to interact with alcohols or phenols, eventually forming tetraloxy- or tetraaroxysilanes:

$$SiCl_4 \xrightarrow[-HCl]{+ROH} ROSiCl_3 \xrightarrow[-HCl]{+ROH} (RO)_2SiCl_2 \qquad \text{etc.} \qquad (2.30)$$

Trimethylchlorosilane reacts with alcohols or phenols much more slowly, and at certain quanitities of phenol or alcohol does not participate in the reaction altogether. Methyl alcohol cannot be used to separate azeotropic mixture, because in this case there is an active secondary reaction between methyl alcohol and liberated hydrogen chloride with the formation of methylchloride and water. Water hydrolyses the reaction products, and their yields are very low. Furthermore, methoxysilanes obtained with methyl alcohol are very toxic. That is why it is more convenient to use n-butil, isobutil or ethyl alcohol to separate the azeotropic mixture of trimethylchlorosilane and silicon tetrachloride. After that, pure trimethyl-chlorosilane is extracted from etherification products in a packed tower with the efficiency of 8-10 theoretical plates when using butil alcohols, or in a tower with the efficiency of 20-25 theoretical plates when using ethyl alcohol.

Methylchlorosilanes are transparent, colourless, motile liquids, which fume in air. They can be easily hydrolysed by water and humidity in air.

Table 3. Physicochemical properties of methylchlorosilanes

Substance	Boiling point, °C	Melting point, °C	d_4^{20}	n_D^{20}
CH_3SiHCl_2	40.6	-93.0	1.1050	1.4000
CH_3SiHCl_2	66.1	-77.8	1.2769	1.4124
$(CH_3)_2SiHCl$	38.0	-11.0	0.8660	-
$(CH_3)_2SiCl_2$	70.2	-86.0	1.0715	1.4002
$(CH_3)_2SiCl$	57.3	-57.7	0.8581	1.3888

They have a specific pungent odour characteristic of chloroanhydrides. They can be easily dissolved in common organic solvents.

Methylchlorosilanes are widely used in the manufacture of various sili-cone oligomers and polymers. Thus, dimethyldichlorosilane is used in the manufacture of silicone liquids, elastomers and polymers for varnishes and plastics. Methyltrichlorosilane is used for producing silicone plastics and varnishes; methyldichlorosilane and trimethylchlorosilane are used to manufacture various silicone liquids. Besides, methyldichlorosilane is a raw material for producing silicone monomers with various organic radi-cals attached to the silicon atom, such as methylphenyldichlorosilane, vi-nylmethyldichlorosilane, allylmethyldichlorosilane, methylnonyldichloro-silane, etc.). Tank residues mixed with methyldichlorosilane are used in the manufacture of silicone waterproofing liquids.

Some physicochemical properties of methylchlorosilanes are given in Table 3.

Preparation of ethylchlorosilanes

The process of ethylchlorosilane preparation by copper-catalysed ethyl-chloride reaction with silicon is very complex, and the mechanism of this process has not been fully established. However, similarly to the synthesis of methylchlorosilanes, the direct synthesis of ethylchlorosilanes most probably happens due to intermetallic compound Cu_3Si. At the first stage at synthesis temperature Cu_3Si reacts with ethylchloride

$$Cu_3Si + 2C_2H_5CI \longrightarrow (C_2H_5)_2SiCI_2 + 3Cu \qquad (2.31)$$

and under the influence of liberated active copper ethylchloride goes on to interact with free silicon:

$$2C_2H_5CI + Si \xrightarrow{Cu} \begin{cases} (C_2H_5)_2SiCI_2 & (2.32) \\ C_2H_5SiHCI_2 + C_2H_4 \end{cases}$$

The free radicals formed in this reaction undergo subsequent transformations:

$$3C_2H_5CI + Si \xrightarrow{Cu} C_2H_5SiCI_3 + 2C_2H_4 + 2H\cdot \qquad (2.33)$$

$$4C_2H_5CI + 2Si \xrightarrow{Cu} C_2H_5SiCI_3 + (C_2H_5)_3SiCI \qquad (2.34)$$

$$3C_2H_5CI + Si \xrightarrow{Cu} C_2H_5SiCI_3 + C_2H_5\cdot + C_2H_4 + H\cdot \qquad (2.35)$$

If hydrogen chloride is introduced into the reaction zone, the yield of ethyldichlorosilane increases:

$$C_2H_5\cdot \begin{cases} \xrightarrow{H\cdot} C_2H_6 & (2.36) \\ \longrightarrow 2C + 2.5H_2 \end{cases}$$

Apart from the products mentioned, secondary reactions also form various amounts of trichlorosilane, silicon tetrachloride, as well as gases (hydrogen, ethylene, ethane, etc.) and carbon. They also form compounds with fragments containing

$$\rangle Si-O-Si\langle \qquad \rangle Si-CH_2-CH_2-Si\langle \qquad \rangle Si-(CH_2CH_2)_2\cdot Si\langle$$

end groups or C_2H_5, H, Cl atoms. These products constitute the tank residues that boil above 143 °C.

The basic diagram of ethylchlorosilane production by direct synthesis is similar to the production diagram of methylchlorosilanes given in Fig.3. After loading the reactor with contact mass, electrical heating is switched on and at 200 °C nitrogen is fed through the heater. Thus, the mass is dried for 5-6 hours, and then the temperature is raised to 300 °C. The synthesis is conducted at 300-360 °C and 0.4-0.5 MPa.

The synthesis forms a mixture of 60-80% of ethylchlorosilanes and 20-40% of unreacted ethylchloride. This mixture (condensate) is then sent to rectification. The diagram of the distillation of unreacted ethylchloride and the rectification of ethylchlorosilanes is similar to the diagram given in Fig.4. After the distillation of unreacted ethylchloride at 0.2--0.25 MPa the ethylchlorosilane mixture is rectified. The mixture has the following average composition (%):

Ethylchloride (unreacted)	1-3	Diethylchlorosilane	2-3
Trichlorosilane	8-10	Diethyldichlorosilane	20-30
Silicon tetrachloride	5-8	Triethylchlorosilane	3-5
Ethyldichlorosilane	20-30	Tank residue	5-12
Ethyltrichlorosilane	30-35		

Note. The average composition may vary in a wide range depending on the conditions of the reaction and on the promoters used to activate contact mass.

The mixture is sent onto the feeding plate of the rectification tower at 60-70 °C. The rectification of ethylchlorosilanes, like that of methylchlorosilanes, is conducted stagewise. The first stage is the distillation of the head fraction, which consists of trichlorosilane, silicon tetrachloride and unreacted ethylchloride. At the second stage, ethyldichlorosilane is distilled when the tank temperature is 80-90 °C (in the beginning of the distillation) and 130-135 °C (at the end of the distillation). At the third stage, when the tank temperature is 130 °C (in the beginning of the distillation) and 180 °C (at the end of the distillation), ethyltrichlorosilane and diethyldichlorosilane are distilled. The fourth rectification stage is meant to separate triethylchlorosilane from tank residue; the tank temperature for this should be 180-200 °C..

Ethylchlorosilanes are motile, colourless liquids with a specific chloroanhydride odour. They can be easily hydrolysed with water and humidity in air, and fume in air. They can be dissolved in organic solvents.

Ethylchlorosilanes are used to obtain silicone oligomers and polymers. For example, ethyltrichlorosilane and diethyldichlorosilane are used to obtain polyphenylethylsiloxanes for varnishes. Diethyldichlorosilane is used to obtain ethylsiloxane liquids and elastomers. Ethyldichlorosilane and tank residue are used to prepare silicone waterproofing liquids. Besides, ethyldichlorosilane can be used for producing silicone monomers with various organic radicals attached to the silicon atom, such as phenylethyldichlorosilane, vinylethyldichlorosilane, etc.).

Some physicochemical properties of ethylchlorosilanes are given in Table 4.

Table 4. Physicochemical properties of ethylchlorosilanes

Substance	Boiling point, °C	Melting point, °C	d_4^{20}	n_D^{20}
$C_2H_5SiHCl_2$	75.0	-107.0	1.0887	1.4160
$C_2H_5SiCl_3$	98.8	-105.6	1.2338	1.4254
$(C_2H_5)_2SiHCl$	99.2	-	0.8880	1.4183
$(C_2H_5)_2SiCl_2$	129.0	-96.5	1.0517	1.4315
$(C_2H_5hSiCl$	143.0	-	0.8970	1.4315

Preparation of phenylchlorosilanes

Similarly to methyl- and ethylchlorosilanes, the process of phenylchlorosilane production by the copper-catalysed reaction of chlorobenzene and silicon is very complex. Unlike the processes mentioned, the direct synthesis of phenylchlorosilanes is carried out at higher temperatures (500-650 °C, depending on the activity of contact mass), which is largely due to the high temperature of chlorobenzene dissociation.

The mechanism of phenylchlorosilane formation has not been fully established either; however, similarly to methyl- and ethylchlorosilanes, it is most probable that initially intermetallic compound Cu_3Si in contact mass interacts with chlorobenzene. To intensify the reaction and increase the output of the target product, the reaction is fed with hydrogen chloride in addition to chlorobenzene.

$$Cu_3Si + C_6H_5CI + 2HCI \longrightarrow C_6H_5SiCI_3 + H_2 + 3Cu \quad (2.37)$$

The liberated active copper catalyses the direct synthesis of phenylchlorosilanes:

$$Si + C_6H_5CI + HCI \xrightarrow{\text{Cu}} C_6H_5SiHCI_2 \quad (2.38)$$

$$Si + 2C_6H_5CI \xrightarrow{\text{Cu}} (C_6H_5)_2SiCI_2 \quad (2.39)$$

$$Si + 3C_6H_5CI \xrightarrow{\text{Cu}} C_6H_5SiCI_3 + C_6H_5\cdot + H\cdot + C_6H_6 \quad (2.40)$$

$$Si + 4C_6H_5CI \xrightarrow{Cu} C_6H_5SiCI_3 + (C_6H_5)_3SiCI \qquad (2.41)$$

Silicon can also react with hydrogen chloride, forming silicon tetrachloride and trichlorosilane.

$$Si + 4HCI \xrightarrow{Cu} SiCI_4 + 2H_2 \qquad (2.42)$$

$$Si + 3HCI \xrightarrow{Cu} HSiCI_3 + 2H\cdot \qquad (2.43)$$

Besides, the free phenyl radicals formed in the reaction of chlorobenzene with silicon can undergo a sequence of transformations:

$$C_6H_5\cdot \xrightarrow{+H\cdot} C_6H_6 \qquad 2C_6H_5\cdot \longrightarrow \begin{array}{l} \longrightarrow C_6H_5\text{-}C_6H_5 \quad (2.44) \\ \longrightarrow 12C + 5H_2 \end{array}$$

Thus, the direct synthesis of phenylchlorosilanes produces a complex mixture, which, apart from phenyltrichlorosilane, diphenyldichlorosilane, phenyldichlorosilane and triphenylchlorosilane, also contains silicon tetrachloride, trichlorosilane, benzene, solid products (diphenyl and carbon) and a gaseous product (hydrogen). It also forms high-boiling polyolefines, which are part of tank residue and can deposit on contact mass, reducing its activity. It should be kept in mind that the production of phenylchlorosilanes requires silicon with a minimal impurity of aluminum, because the aluminum chloride formed contributes to the detachment of the phenyl group from phenylchlorosilanes at higher temperature. The harmful effect of aluminum chloride is counteracted by the addition of metal salts to contact mass, which form a nonvolatile and nonreactive complex with aluminum chloride.

The technological diagram of the direct synthesis of phenylchlorosilanes is similar to the production diagram of methylchlorosilanes given in Fig. 3. Phenylchlorosilanes can be synthesised from the mixture of freshly prepared and discharged (after the synthesis of methyl- or ethylchlorosilanes) contact mass in the ratio of 1:1. However, the discharge mass is full of carbon (contains 4-20% of C), which sharply decreases its activity. That is why the discharge mass should be preactivated by baking out carbon and residues of synthesis products in special furnaces. During the baking, carbon interacts with oxygen in air, changing into volatile carbon dioxide.

The discharge copper-silicon mass can be conducted in a furnace lined with fireproof brick with three horizontal tubular sections placed one above another and fashioned with screws. The baking is conducted at 700

°C in the presence of oxygen. The discharge mass passes from top to bottom through all the sections and regenerates; after that, it is sieved into two fractions. The fraction which passed through the sieve is mixed with freshly prepared mass and is sent to the synthesis of phenylchlorosilanes. The fraction which did not pass through the sieve is sent to metal recycling plants.

The same method can be used for baking contact mass after the synthesis of phenylchlorosilanes; after that, the mass can be sent to metal recycling plants to extract copper for copper extraction.

Contact mass (the mixture of freshly prepared and regenerated mass in the ratio of 1:1) enters reactor *4* (see Fig.3). The reactor is electically heated; at 200 °C the alloy is dried by passing nitrogen through it at the speed of 12-15 m³/h during 5-8 hours, with the temperature gradually increasing. At 550 °C the nitrogen flow is stopped and followed by chlorobenzene and hydrogen chloride from evaporator *2*. Phenylchlorosilanes are synthesised when the temperature in the evaporator does not exceed 150 °C, the temperature in the reactor is 550-620 °C, the excess pressure in the evaporator and reactor is below 0.07 MPa.

The direct synthesis of phenylchlorosilanes forms a condensate of the following average composition, (%):

Trichlorosilane	3-8	Phenyldichlorosilane	5-12
Silicon tetrachloride	4-10	Phenyltrichlorosilane	35-40
Benzene	3-7	Tank residue	7-12
Chlorobenzene (unreacted)...	12-18		

Note. The average composition may vary in a wide range depending on the conditions of the reaction and on the promoters used to activate contact mass.

The condensate is then rectified. In order to extract pure phenyltrichlorosilane and other products of direct synthesis, there is multistage rectification, the diagram of which is similar to the one given in Fig. 4.

The first rectification stage gives the following fractions:

1. below 67 °C is the head fraction (trichlorosilane, silicon tetrachloride, a mixture of SiCl₄ and benzene);
2. 67-79 °C (a mixture of silicon tetrachloride and benzene);
3. 79-84 °C (mainly benzene), which can be used to flush the equipment at subsequent rectification stages and to prepare reactive mixtures for the synthesis of various polyorganosiloxanes.
4. 84-136 °C (a mixture of benzene and chlorobenzene). This fraction can be further used to prepare the reactive mixture for phenylchlorosilane synthesis by high-temperature condensation.

The tank residue from the first rectification stage is sent to the second stage.

The head fraction, separated below 67 °C, is rectified to obtain the following fractions:

1. below 40 °C (in the higher part of the tower) is mainly trichlorosilane;
2. 40-55 °C (intermediate), containing a mixture of trichlorosilane and silicon tetrachloride. It can be used for phenylchlorosilane synthesis by high-temperature condensation.
3. 55-60 °C (mainly silicon tetrachloride);
4. 60-80 °C (intermediate), containing a mixture of silicon tetrachloride and benzene. It can be used in phenylchlorosilane synthesis by high-temperature condensation to suppress the secondary process of reduction.

Tank residues after the distillation of the head fraction can be sent back into the tank of the rectification tower for repeated distillation.

The second rectification stage is carried out under the residual pressure of 345 to 50 GPa; the following fractions are separated:

1. below 85 °C (mainly chlorobenzene);
2. 85-110 °C (mainly phenyldichlorosilane);
3. 110-113 °C, enriched with phenyltrichlorosilane with the following average composition: 1-2% of benzene, 7-9% of chlorobenzene, 1-2% of phenyldichlorosilane and 87-91% of phenyltrichlorosilane.

The tank residue after the second rectification stage can be directed to subsequent stages to extract diphenyldichlorosilane and triphenylchlorosilane.

The fraction enriched with phenyltrichlorosilane is then subjected to rectification in vacuum (under the residual pressure of 50-345 GPa) and separate the following fractions:

1. 80-120 °C (chlorobenzene with an impurity of benzene);
2. 120 - 140 °C - phenyldichlorosilane;
3. below 150 °C (intermediate), containing a mixture of phenyldichlorosilane and phenyltrichlorosilane. This fraction can be sent to repeated rectification;
4. 150-151 °C - phenyltrichlorosilane (49.5-50.5% of chlorine, $d_4^{20} = 1.317 \div 1.324$).

After this, the tank residue can be directed to the third rectification stage to extract diphenyldichlorosilane. It is, however, uneconomical to distil diphenyldichlorosilane from the condensate which actually contains only 7-12% of it.

To increase the diphenyldichlorosilane content in the condensate, it is advisable to conduct the direct synthesis of phenylchlorosilanes not with copper-silicon alloy but with a mechanical mixture of silicon and copper powders, promoted by zinc oxide. The introduction of zinc oxide seems to inhibit the undesirable reactions of diphenyl and benzene formation, creating favourable conditions to attach phenyl radicals to the silicon atom, i.t. to form diphenyldichlorosilane.

Besides, it is advisable to carry out the direct synthesis of phenylchlorosilanes not in hollow reactors in the fluidised layer, but in mechanically agitated reactors where the contact time of chlorobenzene and contact mass increases approximately 10-fold; this seems to have a favourable effect on the yield of diphenyldichlorosilane. Thus, the direct synthesis of phenylchlorosilanes with the mechanical mixture of silicon and copper promoted by zinc oxide and cadmium chloride produces a condensate, which after the separation of unreacted chlorobenzene contains 25-30% of phenyltrichlorosilane and 50-55% of diphenyldichlorosilane. This condensate is rectified to extract phenyltrichlorosilane by the technique described above; at the third rectification stage it yields diphenyldichlorosilane.

The third rectification stage is carried out in vacuum and entails the separation of three fractions.

The first fraction consisting of benzene, chlorobenzene and phenyltrichlorosilane is separated in two stages. First, when the temperature of the tower top is 60-80 °C and residual pressure is 210 GPa, the distillation gives low-boiling components, benzene and chlorobenzene; then, with the pressure increased to 80 GPa and the temperature on top increased to 195 °C the distillation produces high-boiling components, phenyltrichlorosilane and to some extent diphenyl. The end of the separation of low-boiling substances is determined by the reduction in chlorine content [to 26-25% (mass)] and the fraction density [to 1.200 g/cm^3].

The second (intermediate) fraction containing phenyltrichlorosilane, diphenyl and diphenyldichlorosilane is separated when the temperature of the tower top is 198-207 °C (the temperature depends on the composition of the raw stock and the pressure in the system) and residual pressure is 50-80 GPa. At the end of the separation of the intermediate fraction its chlorine content should not be lower than 27.3%, the density should not be lower than 1.214 g/cm^3, and the 301-308 °C fraction content (diphenyldichlorosilane as such) should not be lower than 90%. The intermediate fraction can be sent for repeated rectification.

The third (target) fraction mostly consists of technical diphenyldichlorosilane. It is separated when the temperature of the tower top is below 215 °C and residual pressure is 50 GPa. The separation is stopped when the distillate contains 27% of chlorine and has the density of 1.214 g/cm^3.

After the separation of diphenyldichlorosilane, the tank heating is switched off and the pressure in the system becomes atmospheric. Then the tank is unloaded. If it is necessary, the tank residue is sent to the fourth rectification stage to extract triphenylchlorosilane. The rectification of triphenylchlorosilane is similar to the rectification of phenyltrichlorosilane and diphenyldichlorosilane.

Organochlorosilanes can be also rectified in horizontal rectification towers (Fig. 6). These towers are peculiar in that they use rotating impellers 2, which spray liquid through vapour (in vertical towers, on the contrary, pressurised vapour passes through liquid) and thus mix liquid and vapour. Besides, horizontal towers use new constructions of water drain devices and breaker blades. Due to the impellers, horizontal rectification towers are more convenient to operate than vertical ones. Horizontal towers do not have to be very long, because there is no liquid loss in them, like in vertical apparatuses. Moreover, horizontal towers allow for vacuum rectification with very slight pressure differences, because there is no necessity of an increased pressure for mass exchange.

As we know, the efficiency of a rectification tower depends on the intensity of the mass exchange of vapour and liquid. In the horizontal tower liquid participates in the mass exchange with vapour three times. The impellers of a horizontal tower are similar to the impellers of a centrifugal pump. The liquid which falls on the impeller blades is thrown off in the shape of tiny droplets. Grates 7, which separate the tower, do not let the droplets into adjoining sections, whereas vapours can enter freely. Three impellers are usually equivalent to one theoretical plate, regardless of the viscosity of the rectifying liquid. Each impeller is 152 mm wide. Consequently, the separation that requires 3.5 m in a bubble-cap rectification tower will take up only 46 cm in a horizontal tower. The diameter of a horizontal tower can be smaller than that of vertical towers. Besides, if vertical towers need a certain minimum of vapour feeding maintained (to avoid liquid leakage through the plates), horizontal towers have no such requirements. It should be noted however that if the circumferential speed of the impeller is 9 m/s, a horizontal tower needs somewhat more energy than a vertical tower needs for reflux to be sent to the top of the tower. On the whole, horizontal rectification towers can compete with vertical towers economically, if liquid is distilled at a temperature below its boiling point under atmospheric pressure and if the diameter of the tower is less than 3 m.

Phenylchlorosilanes (except for triphenylchlorosilane) are colourless, transparent liquids, which fume in air and are easily hydrolysed. They have a specific chloroanhydride odour. Triphenylchlorosilane has the form of colourless crystals which melt at 97 °C. Phenylchlorosilanes are easily dis-solved in organic solvents.

Phenylchlorosilanes are widely used in the manufacture of various sili-cone oligomers and polymers. For example, phenyltrichlorosilane, phenyldichlorosilane and diphenyldichlorosilane are used in the synthesis of polyalkylphenylsiloxanes to produce plastics and varnishes.

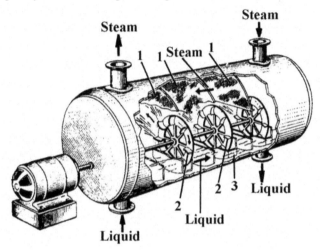

Fig. 6. Horizontal rectification tower: *1* - grates; *2* - impellers; *3* - pan.

Table 5. *Physicochemical properties of phenylchlorosilanes*

Substance	Boiling point, °C	d_4^{20}	n_D^{20}
$C_6H_5SiHCl_2$	181.5 (at 1000 GPa)	1.2115	1.5246
$C_6H_5SiCl_3$	201.5	1.3144	1.5222
$(C_6H_5)_2SiCl_2$	305.2	1.2216	1.5819
$(C_6H_5)_2SiHCl$	143.0 (at 13.3 GPa)	1.1180	1.5810
$(C_6H_5)_3SiCl$	378.0 (melting point. 97 °C)	-	-

Diphenyldichlorosilane is used to obtain elastomers and liquids, and triphenylchlorosilane is used for liquids. Some properties of phenylchloro-silanes are given in Table 5.

It is possible to produce all technological equipment for the industrial direct synthesis of alkyl- and arylchlorosilanes from common brand steel, since raw stock and reaction products in the absence of moisture do not corrode materials. The equipment is generally unsophisticated; there is no need for special constructive solutions, apart from the reactor, which should be manufactured in view of the peculiarities of the direct synthesis of organochlorosilanes.

The process of the direct synthesis of organochlorosilanes is hetero-phase. However, it differs from many other heterophase processes in that solid silicon mixed with the catalyst is one of the parent components and is thus constantly spent during synthesis. It means that, as the reaction prod-ucts form, the amount of the solid phase continually decreases. At the same time, the ratio of silicon and catalyst in contact mass also changes, until silicon is spent completely, which naturally impairs the conditions of the reaction. That is why there are several requirements to the construction of contact apparatuses for the effective direct synthesis of alkyl- and aryl-chlorosilanes. So, the contact apparatus should provide for:

1. a possibility for conducting a continuous process while keeping constant the ratio of silicon and catalyst in contact mass (i.e. it is necessary to en-sure continuous and unintermittent supply of silicon);
2. the regeneration of contact mass directly in the reactor or its gradual withdrawal from the reaction zone to regenerate in a separate apparatus. If the process is continuous, regeneration is necessary, since contact mass is contaminated with the products formed, which impairs the proc-ess;
3. the steadiness of temperature in the reaction zone throughout the whole process, because changes of temperature can abruptly change the com-position of the condensate;
4. steady and constant heat removal , because the process of direct synthe-sis is exothermal.

The diagrams of the direct synthesis of alkyl- and arylchlorosilanes dif-fer mostly in the construction of the contact apparatus. At present there are two basic types of contact apparatuses known: *stationary reactors*, in which contact mass is in the immobile layer, and *nonstationary reactors*, in which contact mass is mobile.

Stationary reactors can be either horizontal with gridlike shelves inside, or vertical tower or tubular types. However, these apparatuses have some serious drawbacks. Specifically, due to the absence of agitation in them, contact mass in the process of sythesis gets baked, which disrupts the op-erating cycle and reduces the degree of the usable transformation of reac-tants. The degree of silicon use in these apparatuses is very low. Besides, the operations of loading and unloading contact mass in stationary reactors are very cumbersome and labour-consuming. All the features mentioned naturally exclude the possibility of industrial application of these appara-tuses in the continuous processes of the direct synthesis of alkyl- and aryl-chlorosilanes.

At present the direct synthesis of organochlorosilanes is conducted in nonstationary reactors, in which the reactive mass is agitated, either with an agitator (mechanical agitation) or with the help of a fluidised layer created by the high speed of loading alkyl- or arylhalogenide into the reactor.

Nonstationary reactors with mechanical agitation

There are two known types of reactors with mechanical agitation, vertical (or tubular) and horizontal.

A vertical reactor (Fig.7) is an apparatus, the main part of which is taken up by reaction camera 7, built like a cylinder with a lid and bottom. From top to bottom the reactor is divided into three zones, or rings. Each ring is surrounded by jacket 2 for heating or cooling; there is a heat carrier circulating in the jacket. In reaction camera 7 there is agitator 3 which makes 30-60 rotations per minute and is built like a bandlike screw; such an agitator ensures the removal of contact mass from the walls of the reactor and its agitation. Contact mass gradually moves down the reactor, continuously agitated. Discharge contact mass is loaded off through a choke in the reactor bottom.

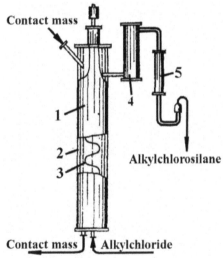

Fig. 7. Vertical reactor with an agitator: *1* - reaction camera; *2* - jacket; *3* - agitator; *4* - filter; *5* - cooler

Gaseous alkylchloride (or arylchloride) is fed through a special distribution device into the lower zone of the reactor. The products formed are withdrawn from the reactor through a side tube and are sent to purification in filter *4*, and then into cooler system *5* (the diagram shows one) for condensation.

Such reactors are doubtlessly superior to stationary apparatuses. First and foremost, the reactors with agitators offer an improved mixing of gaseous chlorine derivative with contact mass, which increases its conversion degree. They also help to considerably reduce the baking of contact mass, which increases the degree of silicon use. Such reactors can be also used for a continuous process of organochlorosilane production.

However, vertical reactors with mechanical agitation do have drawbacks:

1. the escaping heat is difficult to withdraw;
2. fresh portions of contact mass will always mix with discharge or deactivated mass, which decreases the reactor efficiency and impairs the product composition;
3. the mixing of gaseous alkylchloride (or arylchloride) with contact mass is still unsatisfactory; consequently, the output of condensate by a volume unit of the apparatus is relatively low.

However, in small-scale productions vertical apparatuses with mechanical agitation can be successfully used for the direct synthesis of alkyl- or arylchlorosilanes.

The horizontal reactor of the "rotating drum" type (Fig.8) does not have certain drawbacks characteristic of the vertical reactor. Particularly, rotating drums provide for a more thorough mixing of gaseous chlorine derivative with contact mass, because they increase the time of contact between the phases (10 times in comparison with the fluidised layer); consequently, the degree of chlorine derivative also grows. In the production of phenylchlorosilanes they create favourable conditions to increase the yield of diphenyldichlorosilane.

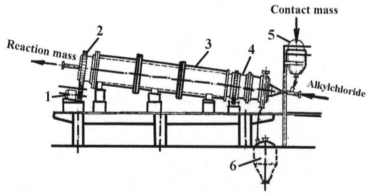

Fig. 8. Horizontal reactor with an agitator («rotating drum" type): *1* - electric motor; *2* - tooth gear; *3* - jacket; *4* - casing; *5* - replenishment bin; *6* - discharge contact mass bin

Due to the good mixing of contact mass with organochloride, the large surface of phase contact and the considerable time the chlorine derivative stays in the reaction zone, the reactors of this type give a 90% conversion degree and 80% product yield.

Nonstationary fluidised layer reactors

At present the direct synthesis of alkyl- and arylchlorosilanes is often carried out in apparatuses which operate using the phenomenon of fluidising. Turbulent movement of components in such a reactor guarantees good contact of reactants with contact mass, as well as steady temperature. Reactors with the fluidised layer are cylindrical apparatuses of various diameter with heat exchange elements. Fig.9 features a reactor with a heat exchange element in the form of a Field tube, and Fig.10 shows a reactor with a heat exchange element in the form of a small-diameter tube bundle.

Reactors with Field tubes should as a rule have a small diameter (400 mm), because a Field tube ensures that the heat is efficiently withdrawn from the whole surface of the apparatus only if the size is small. The Field tube is situated axially in the fluidised layer of contact mass and longitudinally selects the layer from top to bottom. There is cold tap water circulating in the tube (river water can not be tolerated, because it forms scale deposits in the tube and reduces the heat transfer coefficient).

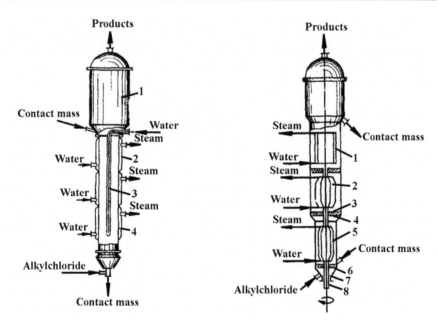

Fig. 9. Vertical reactor with a Field tube: *1* - expander; *2* - jacket; *3* - Field tube; *4* - casing.

Fig. 10. Vertical reactor with a heat exchange element in the form of a tube bundle: *1, 2, 5* - heat exchange surfaces; *3* - immobile board with shelves; *4* - redistribution grate with shelves; *6* – redistribution grate with shelves; *7* - conical bottom; *8* – shaft.

It should be kept in mind, however, that such reactors do not ensure highly efficient heat removal. The feeding of cold water into the Field tube brings about a condensation of methyl- and ethylchlorosilanes on the tube surface and the sticking of small particles of contact mass and carbon (which is formed as a result of pyrolysis of alkylchlorides); this impairs heat removal and buckles the tube. Temperature changes in such reactors amount to ±150 °C. In case external heating is switched on to reduce the temperature during freezing, the presence of stagnation zones in contact mass results in the overheating of reactor walls. This, in combination with external electric heating, destroys the walls by siliconising the material.

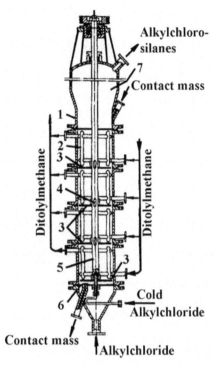

Fig. 11. Vertical four-section reactor with rotating grates and immobile perforated grate: 1 - casing; 2 – heat exchanger; 3 - redistribution grates ; 4 - key; 5 -shaft; 6 – perforated grate for preliminary gas distribution; 7 – separation chamber.

For reactors of a larger diameter (e.g. more than 600 mm) it is more advisable to use a bundle of small tubes as heat exchangers. Such a distribution of heat exchange surfaces virtually does not inhibit fluidising and ensures that the heat is efficiently withdrawn from the whole surface of the apparatus. It should be also kept in mind that tube bundles allow one to select the reaction space and place rotating gas distribution devices between sections; this considerably increases the coefficient of heat transfer and ensures a more uniform gas distribution in the reaction zone.

There are also several constructions of vertical reactors for the direct synthesis of organochlorosilanes, i.e. apparatuses with rotating distribution and redistribution grates.

Fig.11 shows a four-section apparatus with a stationary (fixed) nonremovable perforated grate 6 for preliminary gas distribution. Shaft 5 rests on this grate; removable perforated redistribution grates 3 are mounted on it with keys 4 The supporting bearing of the shaft is washed with the flow of cold alkylchloride. The low temperature of gas inhibits synthesis and eliminates the coking of contact mass on grate *6*. The rotation speed of grates *3* is 20-60 rotations per minute. The main alkylchloride flow at 150-200 °C enters the apparatus from below, under grate 6. The reaction products exit the apparatus through separation zone 7. The radiated heat is withdrawn in heat exchangers *2*, with liquid ditolylmethane passed through them.

Direct synthesis can also be conducted in a five-section 300 mm reactor (Fig.12) without an immobile supporting grate.

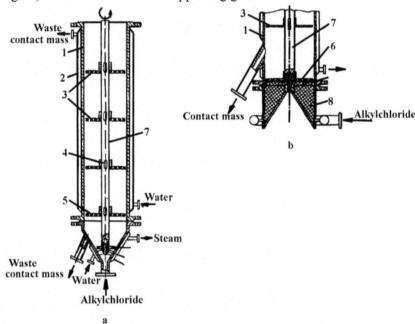

Fig. 12. Five-section reactor with rotating grates: *a* - sectional view; b - lower part; *1* - casing; *2* – jacket; *3* – redistribution grates: *4* - key; *5* - shaft support; *6* - rotating distribution grate; 7- shaft; *8* - head.

In the last section, the conical bottom with a cooling jacket, there are shaft support *5* and rotating perforated distribution grate 6, which at the same tume plays the role of support in case the gas supply abruptly ceases. The radiated heat is withdrawn by a heat exchanger, which circulates through jacket *2*. The conical lower part of this reactor is filled with head 8, which prevents contact mass from entering the gas supply line.

For such heat intensive processes as the direct synthesis of organochlorosilanes, it seems advisable to make the first gas flow sections of a smaller size than the next ones, in order to improve hydrodynamic conditions in the sections with the maximal concentration of reactants.

Two coupled synthesis reactors turned out to be an efficient means to increase the yield of target products and the conversion degree of alkyl- and arylchlorides.

Table 6 lists some thermophysical characteristics of various reactors for the direct synthesis of organochlorosilanes. It is seen that at approximately equal specific surface in reactors with the fluidised layer and mechanical agitation the average heat intensities corresponding to factual heat generation differ by degree, and the maximal heat resistances differ even more.

Table 6. Thermophysical characteristics of reactors for the direct synthesis of organochlorosilanes

Characteristics	Reactor with fluidised layer		Rotating drum with agitator (diameter 1600 mm)
	diameter 400 mm	diameter 600 mm	
Reaction volume, m^3	0.38	0.87	22
Heat exchange surface, m^2	1.4	3.2	48
Specific surface, m^2/m^3	3.70	3.64	2.18
Heat intensity, $KJ/(m^2 \cdot h)$			
average	71500	126000	13400
maximal	84000	218000	14250
Overall heat intensity coefficient, $KJ/(m^2 \cdot h\ K)$	218-335	755-920	42-75

The comparison of average and maximal heat intensities shows that a 400 mm reactor and a rotating drum operate in extreme heat modes, whereas a 600 mm reactor, taking into consideration the conditions of heat removal, gives an opportunity to increase the efficiency by 60%.

Assessing on the whole the method of the production of alkyl- and arylchlorosilanes based on the interaction of alkyl- and arylchlorides with free silicon (i.e. direct synthesis) , we should say that this method in comparison with metalorganic synthesis is more efficient, especially for the production of methyl- and phenylchlorosilanes. As for unsaturated chlorosilanes (vinyl- and allylchlorosilanes) and organochlorosilanes with higher radicals (hexyl-, heptyl-, octyl- and nonylchlorosilanes), no direct synthesis technique has yet been developed.

2.1.3. Techniques based on the replacement of hydrogen atoms in hydridechlorosilanes with alkyl, alkenyl and aryl radicals

For all techniques based on the replacement of hydrogen atoms in hydridechlorosilanes with various organic radicals, the main raw stock is hydrogen-containing chlorosilanes, and, consequently, these techniques are based on the success of the classical direct synthesis technique, because the production of alkyl- and arylchlorosilanes by direct synthesis, as stated above, produces rather large quantities of alkyl- and arylchlorosilanes, which contain a hydrogen atom at the silicon atom (*hydrideorganochlorosilanes*), for example:

$$\underset{H}{\overset{CH_3}{>}}SiCl_2 \qquad \underset{H}{\overset{C_2H_5}{>}}SiCl_2 \qquad \underset{H}{\overset{C_6H_5}{>}}SiCl_2$$

These substances are very promising for the synthesis of organochlorosilanes with various radicals at the silicon atom, i.e. by direct alkylation (or arylation) of hydrideorganochlorosilanes with chlorine derivatives of unsaturated and aromatic hydrocarbons.

There are three techniques based on the replacement of hydrogen atoms in hydridechlorosilanes with alkyl, alkenyl and aryl radicals.

1. a technique based on high-temperature condensation of hydridechlorosilanes with chlorine derivatives of unsaturated and aromatic hydrocarbons;
2. a technique based on dehydration of hydridechlorosilanes when they interact with aromatic hydrocarbons;
3. a technique based on the bonding of hydridechlorosilanes with unsaturated hydrocarbons (hydrosilylation).

Trichlorosilane is of certain interest in the production of many organochlorosilanes with one unsaturated, aromatic or higher alkyl group at the silicon atom; thus, below we consider its production technique.

The production of trichlorosilane in the fluidised layer by continuous technique.

Trichlorosilane can be obtained by direct synthesis from free silicon and anhydrous hydrogen chloride in the fluidised layer:

$$Si + 3HCI \xrightarrow[-H_2]{} HSiCI_3 \qquad (2.45)$$

The reaction is accompanied by secondary processes of the formation of silicon tetrachloride and dichlorosilane:

$$Si + 4HCI \xrightarrow[-2H_2]{} SiCI_4 \qquad (2.46)$$

$$Si + 2HCI \longrightarrow H_2SiCI_2 \qquad (2.47)$$

besides, it can form insignificant amounts of polychlorosilanes:

$$nSi + 2nHCI \longrightarrow \left[\begin{array}{cc} H & CI \\ | & | \\ -Si-Si- \\ | & | \\ CI & H \end{array} \right]_n \qquad (2.48)$$

Temperature has a profound effect on the direct synthesis of trichlorosilane. 280-320 °C are optimal; when temperature is raised above 320 °C, the content of silicon tetrachloride in reaction products is increased; if the temperature falls below 280 °C, the amount of dichlorosilane and polychlorosilanes grows. The synthesis of trichlorosilane is negatively affected by moisture; that is why it is necessary to dry raw stock and equipment thoroughly.

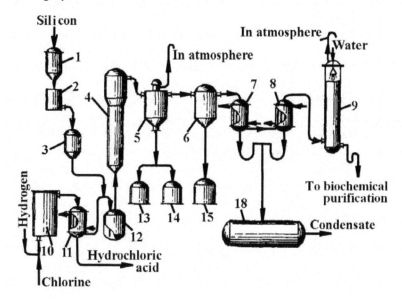

Fig. 13. Production diagram of trichlorosilane: *1*- bin; *2* – vacuum drying chamber; *3* - feeder; *4* - reactor; *5* - cyclone; *6* -filter; *7, 8, 11* – coolers; *9* - scrubber; *10*- furnace; *12*- receiver; *13-16*- collectors.

Raw stock is Si-1 and Si-2 and hydrogen chloride. Trichlorosilane production (Fig.13) comprises three main stages: the preparation of raw stock and equipment; the synthesis of trichlorosilane; the rectification of trichlorosilane.

Ground silicon is dried before being sent into reactor *4* . The drying can be conducted in a vacuum drying chamber *2* at 100-120 °C and the residual pressure of 400-470 GPa. Hydrogen chloride is obtained from hydrogen and gaseous chlorine in furnace *10*. After hydrochloric acid is separated in cooler *11*, cooled by salt solution, the hydrogen chloride is sent through receiver *12* into the lower part of the reactor. Before the beginning of the synthesis the system should be checked for hermiticity. For this purpose, the excess pressure of 0.1 MPa is created in the system by nitrogen and held for 30 minutes. The pressure fall during this time must not exceed 0.005-0.01 MPa. After that, the reactor and the filters are blown with nitrogen (the supply speed is 1-2 m^3/h) at 200-300 °C for 2 hours. The nitrogen should be freed from oxygen and dried.

Trichlorosilane is synthesised in fluidised layer reactors, similar to the apparatuses for the direct synthesis of alkyl- and arylchlorosilanes. For example, it can be a vertical steel cylindrical apparatus with a gas distribution device in the form of a conical bottom. The upper (expanded) part of the tower (expander) separates small particles of silicon carried from the fluidised layer by gas flow. The expander has filters from porous metal inside. (Steel 3). The reactor and expander are electrically heated. Trichlorosilane can also be synthesised in vertical section reactors.

The silicon after drying in apparatus *2* is sent by hydrogen chloride flow into reactor through feeder *3*. The speed of hydrogen chloride supply is regulated so that silicon remains in a suspended state. The temperature in the beginning of the synthesis increases up to 350-360 °C due to the exothermicity of the reaction, and then is automatically regulated within 280-320 °C. If the necessary speed of hydrogen chloride feeding and maintaining the given level of silicon in the reactor is strictly observed, the synthesis as a rule happens without electric heating.

The reaction products are purified from the entrained silicon and dust particles in cyclone *5* and filter *6* and are sent to condense in coolers *7* (salt solution) and *8* (ammonium). The condensate is collected in collector *16*, and the effluent gases (unreacted hydrogen chloride and hydrogen) after washing with water in scrubber *9* are released into the atmosphere. The condensate has the following average composition: 0.5-1% of H_2SiCl_2 (the boiling point is 8.3 °C), not less than 90% of $HSiCl_3$ (the boiling point is 31.8 °C) and not less than 10% of $SiCl_4$ (the boiling point is 57.7 °C). From collector *16* it is sent to rectify.

Rectification can be conducted in packed towers filled with Raschig rings. The rectification separates three fractions. Fraction I, which consists of trichlorosilane with a small amount of dichlorosilane, is separated at 35 °C at the top of the tower. Fraction II (a mixture of trichlorosilane and silicon tetrachloride) is separated at 35-36 °C at the top of the tower (this fraction can later be sent into the tower tank for repeated rectification. Fraction III (tank residue) mostly consists of silicon tetrachloride. Subsequent rectification of fraction II yields $HSiCl_3$ (95-100%) and $SiCl_4$ (to 5%). For obtaining phenyltrichlorosilane by high-temperature condensation, this condensate can be used without rectification.

Trichlorosilane can also be synthesised by using contact mass (3-6% of copper) instead of silicon; hydrogen chloride can be substituted by a mixture of hydrogen chloride with hydrogen. In this case there is a condensate which consists of 94.5% of trichlorosilane and 5.5% of silicon tetrachloride.

Trichlorosilane is a colourless transparent liquid (the boiling point is 31.8°C, the density is 1.38 g/cm^3). It dissolves well in organic solvents, fumes in air and is very sensitive to moisture in air. It is combustible and forms an explosive mixture with air.

Technical trichlorosilane should meet the following requirements: not less than 98% (vol.) of trichlorosilane (the 31-35 °C fraction), 78-79% (vol.) of chlorine and 0.75-0.78% (vol.) of hydrogen.

Trichlorosilane is used as raw stock in the synthesis of phenyltrichlorosilane, vinyl-, allyl-, hexyl- and nonylchlorosilanes, etc., as well as in the synthesis of triethoxysilane used in the production of highly pure semiconductor silicon.

The technique of high-temperature condensation of hydridechlorosilanes with chlorine derivatives of olefines and aromatic hydrocarbons.

Foreign references contain no information about the use of high-temperature condensation for industrial production of silicone monomers; thus, our country has a priority (Corresponding Member of the USSR Academy of Sciences A.D.Petrov, Corresponding Member of the RF Academy of Sciences E.A.Chernyshev) in the development of scientific foundations and industrial implementation of high-temperature condensation to obtain silicone monomers with aromatic radicals, unsaturated groups or mixed radicals attached to the silicon atom.

The synthesis of silicone monomers based on the interaction of hydridechlorosilanes with alkyl-, alkenyl- or arylchlorides, is a convenient technique, which has recently found a wide application.

In this technique a gaseous mixture of reactants is sent at atmospheric pressure through a hollow tube heated to 500-650 °C; the contact time is 10-100s. Within this time the original components condense and form corresponding organochlorosilanes:

$$HSiCI_3 + RCI \xrightarrow[-HCI]{} RSICI_3 \qquad (2.49)$$

$$RHSiCI_2 + R'CI \longrightarrow RR''SiCI_2 \qquad (2.50)$$

where R and R' are alkyl, alkenyl or aryl radicals.

The process seems to be based on the radical mechanism; apart from the main products, it forms correspondingly silicon tetrachloride and alkyl(aryl)trichlorosilanes:

$$HSiCI_3 + RCI \longrightarrow SiCI_4 + RH \qquad (2.51)$$

$$RHSiCI_2 + R'CI \longrightarrow RSiCI_3 + R'H \qquad (2.52)$$

However, using *tret*-butylperoxide or diazomethane as an initiator, one can shift the process towards the preference of the main reactions.

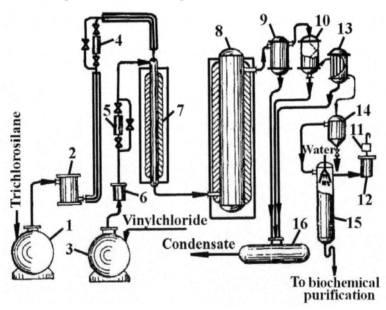

Fig. 14. Production diagram of vinyltrichlorosilane: *1, 3* - tanks; *2* –evaporator; *4, 5* - rotameters; *6, 12* - towers with CaCl$_2$; *7* - heater; *8* – reactor; *9* - expander; *10, 13* - coolers; *11*- fire-resistant apparatus; *14* – buffer capacity; *15* - absorber; *16* – collector.

The high-temperature condensation technique is important for the synthesis of unsaturated organochlorosilanes, phenyltrichlorosilane and organochlorosilanes with various radicals at the Si atom.

Production of vinyltrichlorosilane by continuous technique

The synthesis of vinyltrichlorosilane by the direct technique (the interaction of vinylchloride with contact mass) does not give high yields; low output also characterises the synthesis of vinyltrichlorosilane by dehydrochlorinating of trichloro-β-chloroethylsilane in the presence of a catalyst:

$$ClCH_2\text{-}CH_2\text{-}SiCl_3 \xrightarrow[\text{-HCl}]{} CH_2\text{=}CHSiCl_3 \qquad (2.53)$$

The technique for producing vinyltrichlorosilane by attaching trichlorosilane to acetylene in the presence of a catalyst can be promising, but the conditions of this reaction have not been studied well enough. The most acceptable technique for obtaining vinyltrichlorosilane by high-temperature condensation (HTC) of trichlorosilane with vinylchloride is:

$$HSiCl_3 + CH_2\text{=}CHCl \xrightarrow[\text{-HCl}]{} CH_2\text{=}CH\text{-}SiCl_3 \qquad (2.54)$$

This process also forms by-products. Thus, the reductions results in silicon tetrachloride and ethylene:

$$HSiCl_3 + CH_2\text{=}CHCl \xrightarrow{\hspace{1cm}} SiCl_4 + CH_2\text{=}CH_2 \qquad (2.55)$$

and the pyrolysis of vinylchloride in the conditions of synthesis (560-580 °C) forms hydrogen, methane, ethane and carbon.

Raw stock: trichlorosilane (fraction 31-35 °C containing 78-79% of chlorine and 0.75-0.78% of hydrogen) and vinylchloride (the boiling point is 13.9 °C). The production process of vinyltrichlorosilane (Fig. 14) comprises three main stages: the preparation of raw stock and equipment; the synthesis of vinyltrichlorosilane; the rectification of vinyltrichlorosilane.

Before the synthesis all the equipment should be checked for hermiticity; for this purpose, the excess pressure of 0.1 MPa is created in the system and held for 30 minutes. The pressure fall during this time must not exceed 0.005-0.01 MPa. Then the system is dried by passing nitrogen through it at 200 °C for 2 hours. Vinylchloride enters the reaction from tank *3* through drying tower *6* with $CaCl_2$, and trichlorosilane enters the reaction from tank *1* through evaporator *2*, where the temperature is maintained at 70 °C.

Vinyltrichlorosilane is synthesised in hollow tubular reactor *8*, which is built as a vertical cylindrical steel apparatus with thermocouples. Before the reactor, there is heater 7. First, the heater is heated to 300 °C, and the reactor is heated to 560 °C. Only after that the parent components are fed into the system. From tank *1* trichlorosilane passes evaporator *2* and through rotameter *4* in the form of vapours enters the heater, where it is mixed with vinylchloride vapours, fed from tank *3* through tower *6* and ro-tameter *5*. The mixture of trichlorosilane and vinylchloride vapours is heated to 300 °C and sent into reactor *8*.

The optimal synthesis conditions are the following: the temperature of 560-580 °C, the $HSiCl_3$ and $CH_2=CHCl$ mole ratio of 1:1, the contact time of about 30 s. After that, the gases from the reactor pass through the ex-pander, where they are purified from solid particles, and are sent to con-dense in subsequently joined coolers *10* (salt solution) and *13* (ammo-nium). The condensate from these coolers flows into collector *16*, cooled with salt solution through a jacket (-40 °C), and the effluent gases through buffer container *14* are sent into tower *15*, which is washed with water. Then. through calcium chloride tower *12* and fire-resistant apparatus *11*.they are sent into the atmosphere. Apart from unreacted vinylchloride and hydrogen chloride, the effluent gases contain approximately 5% of hy-drogen, up to 12% of ethylene and up to 3% of ethane.

In the process of synthesis part of vinylchloride is pyrolised; as a result, the reactor walls are covered with deposited soot , which sharply decreases the yield of vinyltrichlorosilane. That is why it is necessary to clean the walls of the apparatus from soot, which contains 30-75% of carbon and 10-15% of silicon, periodically (approximately every 500 hours of operation). After separating vinylchloride, the condensate from collector *16* is sent to rectify in order to extract the target product, vinyltrichlorosilane. The aver-age composition of the condensate is:

Volatile sub-stances	4-5	Silicon tetrachloride	2-4
Trichlorosilane (unreacted)	18-20	Vinyltrichlorosilane	60-65
		Tank residue	10-12

Volatile substances including the unreacted vinylchloride.

Rectification can be conducted in conventional towers filled with Ra-schig rings. This allows to separate the following fractions: fraction I, con-sisting mostly of trichlorosilane, is separated at below 35 °C at the top of the tower; fraction II (mostly silicon tetrachloride) is separated in the 55-59°C range; fraction III (a mixture of silicon tetrachloride and vinyltrichlo-rosilane) is separated below 88.5 °C; fraction IV, the target fraction con-taining vinyltrichlorosilane, is separated in the 88.5-91 °C range.

Vinyltrichlorosilane is a colourless transparent liquid (the boiling point is 90.6 °C) with a specific chloroanhydride odour. It is also easily hydrolysed by water, fumes in air and dissolves well in organic solvents.

Technical vinyltrichlorosilane should meet the following requirements: not less than 97% (vol.) of the 88.5-91 °C fraction, 65.8-67.1% (vol.) of chlorine, $d_4^{20} = 1.264$-1.273.

Vinyltrichlorosilane is used as raw stock for various silicone products. For example, hydrolytic cocondensation of alkyl(aryl)chlorosilanes with vinyltrichlorosilane and subsequent polymerisation in the presence of benzoyl or kumyl peroxides gives thermostable polymers. Besides, vinyltrichlorosilane is used for the surface treatment (dressing) of glass cloth and other glass products.

Preparation of vinylmethyldichlorosilane

The synthesis of vinylmethyldichlorosilane, like the synthesis of vinyltrichlorosilane, is achieved by a HTC reaction:

$$CH_3SiHCl_2 + CH_2{=}CHCI \xrightarrow{-HCl} CH_3(CH_2{=}CH)SiCl_2 \quad (2.56)$$

Here we also observe the secondary reaction of reduction, which forms methyltrichlorosilane and ethylene:

$$CH_3SiHCl_2 + CH_2{=}CHCI \longrightarrow CH_3SiCl_3 + CH_2{=}CH_2 \quad (2.57)$$

whereas vinylchloride at synthesis temperature is pyrolysed to form hydrogen, methane, ethane and carbon.

Raw stock: methyldichlorosilane (not less than 98% of 40-44 °C fraction, 61.3-62.5% of chlorine, 0.86-0.89% of active hydrogen) and vinylchloride (gas, the boiling point is -13.9 °C). The technological diagram of the production of vinylmethyldichlorosilane is similar to the production diagram of vinyltrichlorosilane given in Fig. 14. The optimal synthesis conditions are the following: The reactor temperature of 560-580 °C, the CH_3SiHCl_2 and $CH_2{=}$ mole ratio of 1:0.8, the contact time of 30 s. The reaction forms a condensate of the following average composition, (%):

Volatile substances	3-5	Silicon tetrachloride	4-6
Trichlorosilane	10-15	Methyltrichlorosilane	6-10
		Vinylmethyldichlorosilane	30-35
Methyldichlorosilane (unreacted)	20-23	Tank residue	8-12

Volatile substances including the unreacted vinylchloride.

After separating volatile products, the condensate is rectified in towers filled with Raschig rings. This allows to separate the following fractions:

fraction I (trichlorosilane) in the 31-35 °C range ; fraction II (methyldi-chlorosilane) at 40-44 °C; fraction III, consisting of silicon tetrachloride, in the 55-59 °C range; fraction IV (methyltrichlorosilane) at 65-67°C, frac-tion V (vinylmethyldichlorosilane) in the 90-93.5 °C range.

Vinylmethyldichlorosilane is a colourless transparent liquid (the boiling point is 93 °C) with a specific chloroanhydride odour. It is also easily hy-drolysed with water, fumes in air and dissolves well in organic solvents.

Technical vinylmethyldichlorosilane should meet the following re-quirements: not less than 98% (vol.) of the 91-93.5 °C fraction, 50.70-51.55% (vol.) of chlorine, $d_4^{20} = 1.078 \div 1.090$.

Vinylmethyldichlorosilane is used to synthesise polyvinylmethyldi-methylsiloxane elastomers, as well as various silicone polymers for var-nishes.

Preparation of methylthyenildichlorosilane

The synthesis of methylthyenildichlorosilane (MTDCS) is conducted by HTC of trichlorosilane and chlorothiophene. The reaction is conducted at 540-590 °C, atmospheric pressure, the reactant contact time of 25-30 s and the mole ratio of original components of $1:1 \div 1.5:1$ according to the scheme:

$$\text{(2.58)}$$

There are simultaneous secondary reactions, which form thiophene, me-thyltrichlorosilane and bis(methyldichlorosilyl)thiophene:

$$\text{(2.59)}$$

$$\text{(2.60)}$$

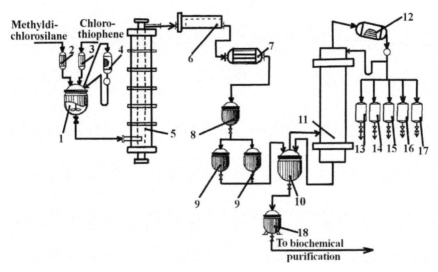

Fig. 15. Production diagram of methylthyenildichlorosilane: *1* - agitator; *2, 3* - batch boxes; *4, 7* - coolers; *5* - synthesis reactor; *6* - filter; *8, 13-17* – receptacles; *9, 18* - containers; *10* - tank; *11* - rectification tower; *12* - refluxer.

Reactor 5 is a vertical copper tube inserted into a casing tube with electric winding around it. The higher surface of the casing tube, which consists of four or five sections, is wrapped around with nichrome wire. Thus, the reactor from top to bottom is divided into four or five temperature zones. The casing of the reactor is filled with chamotte brick for insulation. The contact time of the components in the reactor depends on the volume rate of the supplied mixture and the reaction volume of the reactor. The reaction volume is the part of the reactor where the temperature is 580-590 °C.

Due to the thermal destruction and polymerisation of chlorothiophene and MTDCS, there are also small amounts of carbon, resinous products and gases (HCl and H_2) formed. Resinous and solid products remain on the walls of the synthesis reactor.

Raw stock: trichlorosilane (the 31-35 °C fraction with 78-79% of chlorine and 0.75-0.78% of hydrogen) and chlorothiophene with the main product content of not less than 95% (mass). The production process of methylthyenildichlorosilane (Fig. 15) comprises two main stages: the synthesis of MTDCS and the rectification of the condensate.

First, the working mixture of methyldichlorosilane and chlorothiophene.is prepared in agitator *1*. The components in the agitator are mixed by the supply of nitrogen to barbotage. Then, from agitator *1* with a dispensing pump (not shown in the diagram), the working mixture is fed into the lower part of the synthesis reactor 5 through the nozzle. At 540-590 °C the reactor synthesises MTDCS. It is permissible if the temperature in the first zone of the reactor (the heating zone) falls to 450-500 °C. The contact time is maintained in the range of 25-30 s.

Apart from MTDCS, the synthesis yields by-products, such as methyl-trichlorosilane (MTCS), thiophene, bis(methyldichlorosilyl)thiophene. There is also the thermal destruction and polymerisation of chlorothio-phene and MTDCS, which forms carbon, high-boling resinous products, HCl and hydrogen.

The reaction gases from the higher part of reactor 5 enter filter 6, then cooler 7, cooled by salt solution; after that, the abgases enter the abgase purification unit (not shown in the diagram). The condensate from the cooler self-flows into receptacle 8 and is poured into tanks 9 as it accumu-lates. The average composition of the condensate is (%):

MTCS	5-8	Chlorothiophene	3-5
Methyldichlorosilane (MDCS)	25-30	MTDCS	50-65
Thiophene	2-4.5	High-boiling products	4-6

The synthesis can be conducted continuously, at regular intervals adding the raw stock from the supply batch boxes into agitator 1.

The condensate is rectified in two stages: stage I is the distillation of the head fraction, which contains chlorosilanes, at atmospheric pressure; stage II is the separation of chlorothiophene at the residual pressure of 40-65 GPa and of MTDCS at the residual pressure of 13.3-20 GPa. MTDCS is separated at the rectification system which consists of tank 10, rectification tower 11, refluxer 12 and receptacles 13-17. The tower is filled with Ra-schig rings. Before the separation of the fractions, the rectification unit is checked for hermeticity with nitrogen under the pressure of 0.7 atm (0.07 MPa). Then the condensate is sent from tanks 9 into tank 10, and refluxer 12 is filled with water. The contents of the tank are slowly heated by send-ing a heat carrier (ditolylmethane, or DTM) into the tank jacket. The va-pours from tank 10 enter tower 11 and then refluxer 12, where they con-dense. When the reflux appears in the box, the tower operates in the "self-serving" mode for 3-4 hours without separating the distillate, until the temperature in the vapours reaches the constant 40 °C. After the constant operation mode is established, the towers separate the distillate.

The first fraction, which consists of MDCS and MTCS, is separated at atmospheric pressure and the temperature at the top of the column of 30-40 °C in the beginning of the separation and 70-72 °C at the end of the sepa-ration. The distillate is collected in receptacle 13. The composition of the first fraction is: 50-60% of MDCS and 50-40% of MTCS. As soon as fraction I accumulates, it is sent for repeated rectification to extract pure MDCS and MTCS.

After that, the tank and tower are cooled to 20-30 °C for the transition to the vacuum rectification stage. At 40-65 GPa in vacuum (residual pressure) one begins to separate intermediate fraction II into receptacle *14*. The temperature in the tank is 50-70 °C, and in the higher part of the tower it is 40-50 °C. This fraction is then sent for repeated rectification.

Fraction III, which consists of unreacted chlorothiophene, is separated at 40 GPa in vacuum in the beginning of the separation and at 13.3-20 GPa at the end. In this case, the temperature in the tank is 50-80 °C, and in the higher part of the tower it is 30-40 °C. The distillate is collected in receptacle *15*.

Fraction IV, the head MTDCS fraction, is separated into receptacle *16* in vacuum at 13.3-20 GPa and at the tank temperature of 70-110 °C and 40-75 °C in the higher part of the tower. This fraction is sent for repeated rectification.

When the MTDCS content in the distillate is not less than 95%, one begins the separation of fraction V, the main MTDCS fraction, in vacuum at 13.3-20 GPa into receptacle *17*. The temperature in the tank is 110-130 °C, and in the higher part of the tower it is 75-80 °C.

After rectification, DTM is no longer sent into tank jacket *10*, tank residue is cooled to 30-40 °C and by nitrogen pressure *(P = 0.7 atm)* is pressurised into collector 18, from where it is sent to bio-chemical cleaning (BCC).

Methylthyenildichlorosilane is a colourless liquid (the boiling point is 197 °C) with a specific organochlorosilane odour. It can be easily hydrolysed with water and humidity in air.

Technical methylthyenildichlorosilane should meet the following requirements:

Appearance.	Transparent colourless liquid with permissible metallic impurities due to steel containers.
MTDCS content (the boiling point is 190-200 °C), % (mass), not less than	98.0
Total impurity content, % (mass), not less than	2.0
including:	
MTCS	0.5
benzene	0.3
chlorothiophene	1.0
dichlorothiophene	1.0
MDCS	0.1
high-boiling fractions	0.5

Table 7. Physicochemical properties of some chlorine derivatives of unsaturated silicone compounds

Substance	Boiling point, °C	d_4^{20}	n_D^{20}
$CH_2=CHSiCl_3$	90.6	1.2426	1.4295
$CH_2=CHCH_2SiCl_3$	117.5	1.2011	1.4460
$C_4H_3SSiCl_3$	198.0	1.4555	1.5398
$CH_2=CH(CH_3)SiCl_2$	93.0	1.0868	1.4270
$CH_2=CH(C_2H_5)SiCl_2$	123.7	1.0664	1.4385
$CH_2=CH(C_6H_5)SiCl_2$	121.0 (at 48 GPa)	1.1960 (at 25 °C)	1.5335 (at 25 °C)
$CH_2=CHCH_2(CH_3)SiCl_2$	119-120	1.0758	1.4419
$CH_3(C_4H_3S)SiCl_2$	197.0	1.2754	1.5265

Methylthyenildichlorosilane is used to produce oligomethylthyenilsiloxane liquids with improved lubricating capacity.

Similarly to vinyltrichlorosilane, vinylmethyldichlorosilane and methylthyenildichlorosilane, HTC is used to produce allyltrichlorosilane, thyeniltrichlorosilane, vinylethyldichlorosilane, allylmethyldichlorosilane, vinylphenyldichlorosilane and other chlorine-containing silicone monomers with organic radicals attached to the silicon atom. Table 7 lists the physicochemical properties of some chlorine derivatives of unsaturated silicone compounds.

Preparation of phenyltrichlorosilane

The high-temperature condensation (HTC) technique proved to be very convenient and economical for the production of phenyltrichlorosilane and in recent years has become one of the main industrial techniques for its synthesis.

The synthesis of phenyltrichlorosilane by HTC technique is conducted according to the reaction:

$$HSiCl_3 + C_6H_5CI \xrightarrow[-HCI]{} C_6H_5SiCl_3 \qquad (2.61)$$

In this case, similarly to the vinyltrichlorosilane case, in the conditions of synthesis (600-640 °C) there is a secondary reaction of reduction, which forms silicon tetrachloride and benzene. However, if the process is conducted in the presence of an initiator (5% of diazomethane), it is possible to reduce the temperature down to 500-550 °C. Then the secondary reaction proceeds very slowly, which increases the yield of phenyltrichlorosilane by 1.5 times. To suppress the reduction process, one can also add benzene and silicon tetrachloride into the reactive mixture. It is also possible to boost the yield of phenyltrichlorosilane and increase the degree of the chlorobenzene transformation by using two subsequent reactors.

The technique of the dehydration of hydridechlorosilanes in their interaction with aromatic hydrocarbon

The synthesis of alkyl(aryl)chlorosilanes by dehydration is based on the interaction of hydridechlorosilanes with aromatic hydrocarbons:

$$HSiCI_3 + Ar\text{-}H \xrightarrow[-H_2]{} ArSiCI_3 \qquad (2.62)$$

$$RSiHCI_2 + AR\text{-}H \xrightarrow[-H_2]{} ARRSiCI_2 \qquad (2.63)$$

where R is alkyl or aryl.

The reaction is conducted in the autoclave apparatus at 240-300 °C and 1.5-20 MPa in the presence of a catalyst. The reaction can be catalysed by Lewis acids, for example, $AICI_3$, BCI_3 or $B(OH)_3$, amounting to 0.1-5%. The yield of target products is 20-40%, because the main reaction is accompanied by the secondary disproportioning of organochlorosilanes. This technique is convenient for the preparation of organochlorosilanes with various radicals attached to the silicon atom.

Preparation of methylphenyldichlorosilane

The synthesis of methylphenyldichlorosilane by dehydration is based on the interaction of methyldichlorosilane with benzene at heating under pressure:

$$CH_3SiHCI_2 + C_6H_6 \xrightarrow[-H_2]{} CH_3(C_6H_5)SiCI_2 \qquad (2.64)$$

The reaction is conducted in the presence of a catalyst, the solution of boric acid in methylphenyldichlorosilane with the mole ratio of parent reactants CH_3SiHCI_2:C_6H_6 of 1:3. The excess of benzene has a beneficial effect on the output of methylphenyldichlorosilane. It should be noted that in the synthesis conditions there is a disproportioning of methyldichlorosilane, which forms by-products, such as methyltrichlorosilane and dimethyldichlorosilane. These products can be separated in the process of the rectification of the condensate and added to the reactive mixture during the synthesis of methylphenyldichlorosilane to suppress the reaction of disproportioning.

Raw stock: methyldichlorosilane (not less than 98% of the main fraction, 61.3-62.5% of chlorine); benzene $(d_4^{20}$ =0.876-0.879); technical boric acid. Methylphenyldichlorosilane production (Fig.16) comprises three main stages: the preparation of equipment and reactive mixture; the synthesis of methylphenyldichlorosilane; the rectification of methylphenyldichlorosilane.

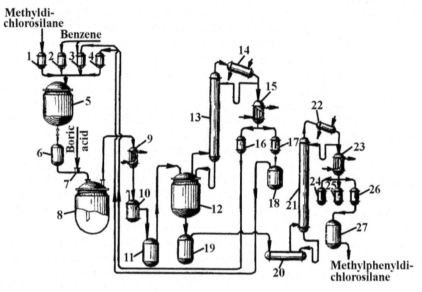

Fig. 16. Production diagram of methylphenyldichlorosilane: *1-4* - batch boxes; *5* –agitator; *6* - batch box; *7* - choke; *8* - autoclave; *9, 15, 23* -coolers; *10* - separator; *11, 18, 19, 27* - collectors; *12, 20* - rectification tower tanks; *13, 21* - rectification towers; *14, 22* - refluxers; *16, 17, 24-26* - receptacles.

Methylphenyldichlorosilane is synthesised under pressure in an autoclave reactor. The apparatus has chokes, a thermocouple pocket and a siphon for checking the degree of its filling. Before the beginning of the synthesis the autoclave should be checked for hermiticity. For this purpose, the operating pressure of 10-12 MPa is created in the apparatus and held for 10 minutes. During this the pressure in the autoclave should remain constant.

Agitator *5* is filled with methyldichlorosilane from batch box *1*, with benzene from batch boxes *2* and *3* and with recirculating methylchlorosilanes (after rectification) from batch box *4*. The reactive mixture is mixed in agitator *5* for 20-30 minutes; after that one determines the chlorine content and density. The mixture prepared in this way is then sent into batch box *6* and from there to the synthesis. Then autoclave *8* is electrically heated (it can be heated by sending vapour into the jacket) and fed part of reactive mixture from the batch box; from special choke 7 the reactor receives a solution of boric acid in methylphenyldichlorosilane (1.35-1.4% of the quantity of the reactive mixture). The autoclave is heated to 240±10 °C and is held at this temperature for approximately 30 minutes until constant pressure (12 MPa) is established.

Under these conditions the autoclave receives the reactive mixture from batch box 6. The vapours of the reaction products from the autoclave continuously enter cooler 9, which is cooled by salt solution (-15 °C), where they condense, and through separator 10 flow into collector 11. The reaction forms a condensate of the following average composition, (%):

Methyldichlorosilane (unreacted)	4-6	Benzene (unreacted)	50-55
Methyltrichlorosilane	2-5	Methylphenyldichlorosilane	18-25
Dimethyldichlorosilane	1-3	Tank residue	10-12

From collector 11 this condensate is sent into tank 12, which is heated by vapour (0.6 MPa). First, rectification tower 13 for 2 hours operates in the "self-serving" mode, and then starts separating the fractions.

Head fraction I, which is a mixture of methyldichlorosilane, methyltrichlorosilane, dimethyldichlorosilane and a small amount of benzene, is separated in the 36-78 °C range and collected in receptacle 16. Then this fraction can enter batch box 4. Fraction II (benzene) is distilled in the 78-82 °C range and collected in receptacle 11, and then poured into receptacle 18. It can be re-used in the synthesis (in this case benzene from collector 18 is sent into batch box 3). Tank residue, which after the distillation of the first two fractions is a concentrate with 50% of methylphenyldichlorosilane, is sent from tank 12 into collector 19 and from there into tank 20, heated with vapour (1.4 MPa).

For 1 hour vacuum rectification tower 21 operates in the "self-serving" mode, and then starts separating benzene, which is collected in collector 24 (from there it can be sent to the synthesis again into batch box 3). After the distillation of benzene residual pressure of 107 GPa is created in the rectification system; after the constant mode is established, the intermediate fraction is separated into receptacle 25. If the methylphenyldichlorosilane content in the intermediate fraction exceeds 5%, this fraction can be sent for repeated rectification in tank 20. After the intermediate fraction, the main fraction, methylphenyldichlorosilane, is separated into receptacle 26. The fraction with the density of 1.1750-1.1815 g/cm^3 and chlorine content of 36.9-37.8% is separated. The separation is conducted as long as reflux is extracted. From receptacle 26, technical methylphenyldichlorosilane flows into collector 27.

Similarly to methylphenyldichlorosilane, by this technique one can obtain phenylethyldichlorosilane and other organochlorosilanes with different radicals at the silicon atom.

The technique based on the bonding of hydridechlorosilanes with unsaturated hydrocarbons (hydrosilylation)

A convenient method for producing some silicone monomers is a reaction of bonding hydridechlorosilanes to unsaturated hydrocarbons. In recent years this method has been used in industry quite extensively. Its essence is in hydridechlorosilanes entering the reaction of hydrosilylation of unsaturated hydrocarbons in the presence of catalysts.

$$HSiCl_3 + CH_2=CHR \longrightarrow RCH_2CH_2SiCl_3 \qquad (2.65)$$

$$R'HSiCl_2 + CH_2=CHR \longrightarrow RCH_2CH_2SiR'Cl_2 \qquad (2.66)$$

where R is an organic radical or a functional group (e.g., CN, CI, COOH, NH_2), and R' is an organic radical.

This is a technique for producing silicone monomers with higher organic radicals or carbofunctional groups at the silicon atom. The reaction can be catalysed by a 0.1 M solution of chloroplatinic acid in isopropyl alcohol or tetrahydrofuran, and other platinum catalysts, as well as organic peroxides, main type catalysts (triethylamine, dimethylformamide) or cyclopentadienyltricarbonylmanganese. The reaction can be initiated with UV rays.

Preparation of nonyltrichlorosilane

Nonyltrichlorosilane can be synthesised by the hydrosilylation of propylene trimer with (nonen-1)trichlorosilane in the presence of a catalyst.

$$HSiCl_3 + CH_3-(CH_2)_6-CH=CH_2 \longrightarrow C_9H_{19}SiCl_3 \qquad (2.67)$$

Raw stock: trichlorosilane; propylene trimer (125-150 °C fraction, the density is 0.73-0.74 g/cm^3, $n_D^{20} = 1.420$-1.430).

Nonyltrichlorosilane can be obtained by the continuous technique (Fig. 17). Trichlorosilane from container 5 is sent by nitrogen into batch box 1, and the original olefine from container 7 into batch box 2.

If the reaction is catalysed by chloroplatinic acid, it is sent into batch box 7; if it is catalysed by main type catalysts (e.g. dimethylformamide), they are added into batch box 2 to avoid the formation of complex compounds with chlorosilanes.

From batch boxes *1* and *2* the original reactants in given quantities are sent into agitator *3*, where they are mixed with nitrogen, which is fed from container *6*, for 10-15 min. Before the installation is launched, reactor *8* is loaded to 2/3 of its volume with reactive mixture and electrically heated. The temperature is raised to 250-260 °C and after it is held for 5 hours, the reactive mixture from apparatus *3* is continuously fed through run-down box *4*. The products formed are separated from the lower part of the reactor at 2-2.1 MPa. The products are sent through cooler *10* are sent into separator *11* and collector *12*; from there they are sent to rectification.

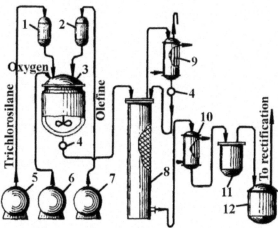

Fig. 17. Production diagram of nonyltrichlorosilane by continuous technique: *1, 2*- batch boxes; *3* - agitator; *4*- run-down boxes; *5-7* - containers; *8*- reactor; *9, 10* – coolers; *11* - separator; *12*- collector.

To improve the contact of reacting components with the catalyst, the reactor is filled with a carrier, pumice, which is pre-processed with hydrochloric acid to eliminate traces of iron, which deactivates the platinum catalyst.

Technical nonyltrichlorosilane is a transparent liquid, colourless or light brown. It should meet the following requirements: 95-98% of the fraction which boils out in the 195-240 °C range; 40-41% of hydrolysed chlorine: Nonyltrichlorosilane is used for producing film-forming substances, polymers for highly thermostable varnishes, waterproofing agents, etc.

Similarly to nonyltrichlorosilane, one can obtain other alkylchlorosilanes and other higher radicals at the silicon atom. For example, methylnonyldichlorosilane can be obtained from methyldichlorosilane and propylene trimer, hexyltrichlorosilane can be obtained from trichlorosilane and propylene dimer, isobutyltrichlorosilane can be produced from trichlorosilane and isobutene, isobutylmethyldichlorosilane can be produced from dichloromethylsilane and isobutene, etc.

Table 8. . Physicochemical properties of higher alkylchlorosilanes and alkylchlorosilanes with carbofunctional groups at the silicon atom

Substance	Boiling point, °C	d_4^{20}	n_D^{20}
$CH_3(iso-C_4H_9)SiCl_2$	45 (at 26.6 GPa)	—	1.4343 (at 25 °C)
iso -$C_4H_9SiCl_3$	141.0	1.4670	1.4358
$C_6H_{13}SiCl_3$	191.6	1.1070	1.4435
$C_7H_{15}SiCl_3$	210.7	1.0860	1.4462
$C_8H_{17}SiCl_3$	231-232 (at 970 GPa)	1.0744	1.4490
$C_9H_{19}SiCl_3$	116,2 (at 13.3 GPa)	1.0645	1.4498
$CH_3(C_9H_{19})SiCl_2$	115-117 (at 6.6 GPa)	0.9931	1.4548
$ClCH_2CH_2CH_2SiCl_3$	181.5 (at 1000 GPa)	1.3540	1.4638
$CH_3(ClCH_2CH_2CH_2)SiCl_2$	185.0	1.2045	1.4580
$CNCH_2CH_2SiCl_3$	84-86 (at 13.3 GPa) melting point 34-35 °C	-	-
$CH_3(CNCH_2CH_2)SiCl_2$	215.0 (at 1000 GPa)	1.2015	1.4550 (at 25 °C)

The hydrosilylation technique is also convenient for preparing organohalogensilanes with functional groups (ether, cyano, etc.) or halogens in the organic radical. E.g., attaching trichlorosilane to allylchloride yields γ-chloropropyltrichlorosilane:

$$HSiCl_3 + CH_2{=}CHCH_2Cl \longrightarrow ClCH_2CH_2CH_2SiCl_3 \quad (2.68)$$

and bonding trichlorosilane to acrylonitrile gives β-cyanethyltrichlorosilane:

$$HSiCl_3 + CH_2{=}CHCN \longrightarrow CNCH_2CH_2SiCl_3 \quad (2.69)$$

Table 8 lists some physicochemical properties of higher alkylchlorosilanes and alkylchlorosilanes with carbofunctional groups at the silicon atom.

2.2. Preparation of halogenated organochlorosilanes

Halogenated organochlorosilanes, especially chlorinated methyl- and phenylchlorosilanes of the common structure

$$[CI_nCH_{3-n}(CH_2)_m]_xSiR_yCI_{4-(x+y)}$$

$$n=1 \div 3, \quad x=1 \div 2, \quad y=0 \div 2, \quad R=CH_3, \ C_6H_5$$

$$[CI_nC_6H_{5-n}]_xSiR_yCI_{4-(x+y)}$$

$$n=1 \div 5, \quad x=y=1 \div 2, \quad R=CH_3, \ C_2H_5 \ \text{and others}$$

have been recently widely used to obtain silicone monomers with various carbofunctional groups at the silicon atom and to synthesise polyorganosiloxane with an improved mechanical durability and adhesion to various materials.

There are several techniques used to prepare chlorinated organochlorosilanes:

1. attachment of hydrogen chloride at the double link of unsaturated silicone compounds:

$$CH_2=CH(CH_2)_nSiCI_3 \xrightarrow{+HCI} CICH_2CH_2(CH_2)_nSiCI_3 \qquad n=0 \div 3. \quad (2.70)$$

2. organomagnesium synthesis:

$$CH_3SiCI_3 + CI_mRMgX \xrightarrow[-MgCIX]{} CH_3(CI_mR)SiCI_2 \quad (2.71)$$

3. catalytic dehydrocondensation of hydridehalogensilanes with halogen derivatives of hydrocarbons, catalysed with Lewis acids:

$$HSiCI_3 + C_6H_5CI \xrightarrow[-H_2]{} CIC_6H_4SiCI_3 \qquad (2.72)$$

4. the interaction of silicon halogenides with diazomethane, which occurs at -45 --55 °C in a solution of diethyl ether:

$$SiCI_4 + CH_2N_2 \xrightarrow[-N_2]{} CICH_2SiCI_3 \qquad (2.73)$$

However, in practice halogenated organochlorosilanes are most conveniently prepared by the reaction of direct halogenation, which in the simplicity of its experimental realisation and in its possibilities surpasses all the techniques stated above.

The introduction of functional groups (halogen, CN, COOH, NH$_2$, etc.) into the organic part of the molecule is a way for a controlled modification of silicone polymers. These groups increase the polarity of the molecule and the intermolecular forces of interaction between the chains, which has a positive effect on the mechanical and adhesive characteristics of the polymers.

The chlorination of silicone monomers compared to the chlorination of typical organic compounds has some peculiarities caused by the properties of silicon , which, unlike carbon (with the valency and coordination number of 4) has a valency of 4 and a coordination number of 6. The electronegativity of silicon is lower than that of carbon; by its electronegativity silicon is closer to group IV metals (see Table 1). This makes the energy of the Si-Hal bond higher than that of the C-Hal bond (see Table 2). At the same time, the higher polarity of the Si-Cl bond conditions its higher reactive capacity compared to the C-Cl bond. However, the reactive capacity of the Si-Cl bond drops as the chlorine atoms at the silicon atom are replaced with organic radicals, whereas similar carbon compounds have an inverse mechanism. All the characteristics of the silicon atom stated above affect the reactivity of organochlorosilanes in the process of their halogenation.

Free chlorine and sulfurylchlorideSO$_2$CI$_2$ can be used as the halogenating agent, in particular the chlorinating agent in the chlorination of organochlorosilanes.

2.2.1. Preparation of chlorinated methylchlorosilanes

The chlorination of methylchlorosilanes is realised by free chlorine in the liquid or gaseous phase. The reaction takes place by the radical chain mechanism; its initiators (substances capable of generating free radicals) can be peroxides (benzoyl, kumyl, etc.), the dinitrile of 2,2'-azobis(isobutyric) acid, fluorine, [60]Co γ-rays, as well as UV rays.

The chlorination of methylchlorosilanes with UV rays happens readily and forms a whole range of chlorinated methylchlorosilanes. For example, apart from chloromethyltrichlorosilane CICH$_2$SiCl$_3$, the reaction also forms dimethyltrichlorosilane   Cl$_2$CHSiCl$_3$ and   trimethyltrichlorosilane Cl$_3$CSiCl$_3$:

$$\text{CH}_3\text{SiCI}_3 + n\text{CI}_2 \xrightarrow[-n\text{HCI}]{h\nu} (\text{CH}_{3-n}\text{CI}_n)\text{SiCI}_3 \qquad (2.74)$$

$$n = 1 \div 3$$

The photochemical chlorination of methylchlorosilanes can be performed by the diagram given in Fig. 18. Reactor *6* is an enameled apparatus with a leaded steel lid. The lid has a filling hatch, lead pockets for thermocoupls (the temperature is measured both in liquid and in vapours), flanges for connecting the light cone and the backflow condenser, and an input for the bubbler. Light cone *5* is a thick-walled class hood of an explosion-proof lamp. The source of UV rays can be a spherical ultrahigh pressure mercury quartz lamp. Backflow condenser *8* is a lead tube encased in a steel tube; it is cooled by salt solution. Ring-shaped bubbler *7* is manufactured from lead; the holes in it are drilled in the lower part of the ring towards the reactor bottom.

Gaseous chlorine enters buffer container *1*, then flushing container *2* filled with copper sulfate, and then through tower *3* with Raschig rings, rotameter *4* and bubbler *7* enters reactor *6*. All the chlorine communication by the triple valve situated before the rotameter is connected with the line which sends nitrogen for eliminating hydrogen chloride in chlorination products. The hydrogen chloride formed escapes through backflow condenser *8*, tower *9* filled with calcium chloride and fire-resistant apparatus *10*, and enters the faolite absorption column.

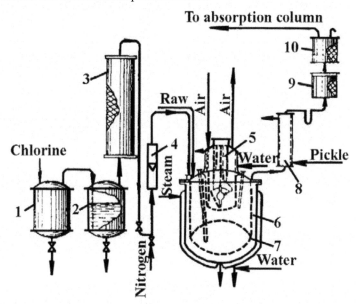

Fig. 18. Production diagram of the photochemical chlorination of methylchlorosilanes: *1, 2* - containers; *3* - tower filled with Raschig rings; *4* - rotameter; *5* – light cone; *6* - reactor; *7* - bubbler; *8* - backflow condenser; *9* - tower with CaCl$_2$; *10* - fire-resistant apparatus.

The contamination of methylchlorosilanes with products of steel corrosion (iron peroxides) sharply retards liquid phase chlorination. The yield of monochlorosubstitutes is reduced, whereas the yield of di- and polychlorosubstitutes and the quantity of the unreacted methylchlorosilane increase. To prevent the corrosion products from entering the reactor, the chlorine line (from the rotameter to bubbler 7) and the valve before the bubbler are manufactured from vinyplast. Besides, the line connecting condenser 8 with the further section of the diversion communication is given a U-shape.

For a satisfactory yield of the most valuable product, monochloride, the chlorination should be carried out at a large excess of methylchlorosilanes. However, the photochemical process has some considerable disadvantages, which make it harder to use in industry. The disadvantages include:

- the necessity of special equipment with quartz elements;
- the difficulty in manufacturing powerful chlorination apparatuses, since the force of light diminishes in proportion to the square distance from the source;
- the danger of self-inflammation and even explosion of the reacting substances, because the photocatalytic reaction of replacing hydrogen in the methyl group with chlorine is, as stated above, a chain radical process, which under certain conditions acquires an avalanche character. Self-inflammation or explosion can be caused by powerful light, contact with heated surfaces, etc.; that is why light sources should be encased in a special explosion-proof fitting, which considerably reduces the photochemical effect;
- the considerable reduction of the photochemical effect even at the slightest darkening of the reactive layer (due to tarring or contamination).

It is more practical to chlorinate organochlorosilanes with the help of initiators. The use of radical initiators (the dinitrile of 2,2'-azobis(isobutyric) acid, etc.) allows to carry out the chlorination reaction in the dark. This chlorination method is known in the literature as the dark technique.

Preparation of methyl(chloromethyl)dichlorosilane

Methyl(chloromethyl)dichlorosilane is prepared by chlorinating dimethyldichlorosilane with free chlorine in the presence of radical initiators, such as the dinitrile of 2,2'-azobis(isobutyric) acid.

$$(CH_3)_2SiCI_2 + CI_2 \xrightarrow[-HCI]{65\text{-}70^0C,\ Initiator} CH_3(CH_2CI)Si \qquad (2.75)$$

The process develops according to the common chain reaction scheme:

$$NC(CH_3)_2CN=NC(CH_3)_2CN \xrightarrow{-N_2} 2NC(CH_3)_2C \cdot \qquad (2.76)$$

$$NC(CH_3)_2C \cdot + CI_2 \longrightarrow NC(CH_3)_2CCI + CI \cdot \qquad (2.77)$$

$$(CH_3)_2SiCI_2 + CI \cdot \xrightarrow{-HCI} CH_3(\overset{\bullet}{C}H_2)SiCI_2 \qquad (2.78)$$

$$CH_3(\overset{\bullet}{C}H_2)SiCI_2 + CI_2 \longrightarrow CH_3(CH_2CI)SiCI_2 + CI \cdot \quad etc. \qquad (2.79)$$

The reaction of chlorination does not stop at this stage, which forms, apart from methyl(chloromethyl)dichlorosilane, products of a deeper chlorination, i.e. methyl(dichloromethyl)dichlorosilane and methyl(trichloromethyl)dichlorosilane:

$$CH_3(CH_2CI)SiCI_2 \xrightarrow[-HCI]{+CI_2} CH_3(CHCI_2)SiCI_2 \xrightarrow[-HCI]{+CI_2} CH_3(CCI_3)SiCI_2 \qquad (2.80)$$

The presence of the chlorine atom in the methyl group of dimethyldichlorosilane increases the speed of hydrogen replacement in it; that is why the chlorination of dimethyldichlorosilane with the preference of obtaining methyl(chloromethyl)dichlorosilane should be conducted superficially (to the conversion degree of 8-14%). The yield of methyl(chloromethyl)dichlorosilane will be 70-80% for the reacted dimethyldichlorosilane, and the unreacted dimethyldichlorosilane is returned into the cycle.

Raw stock: dimethyldichlorosilane (not less than 99% of the main fraction); liquid chlorine (not less than 99.5% of chlorine, not more than 0.06% of humidity); initiator is the dinitrile of 2,2'-azobis(isobutyric) acid (the melting point is 105-106°C, not more than 0.5% of humidity), used as a 2% solution in dimethyldichlorosilane.

The process comprises two main stages (Fig. 19): continuous chlorination of dimethyldichlorosilane (with simultaneous distillation of unreacted substance); vacuum rectification of chlorination products.

From container *18* dimethyldichlorosilane is sent into batch box *1* placed over apparatus *2*, where a 2% solution of the initiator in dimethyldichlorosilane is prepared. Apparatus 2 is an enameled flask with an agitator and a filling hatch. While the agitator operates, one sends there dimethyldichlorosilane and adds the initiator solution in the amount necessary to form a 2% solution. After 30 minutes of agitation at 20 °C the mixture of dimethyldichlorosilane and initiator is poured into intermediate container *17*, from where it is periodically pumped into pressure batch box *4*. After that, the mixture self-flows through rotameter *6* into chlorinator *5*. The chlorinator is a steel cylindrical apparatus with a heating jacket and a thermometer pocket; the lower part of the apparatus contains a distribution device which feeds chlorine. The temperature in the chlorinator is maintained within 65-70 °C and regulated with vapour sent into the jacket of the apparatus and with the speed at which chlorine is fed.

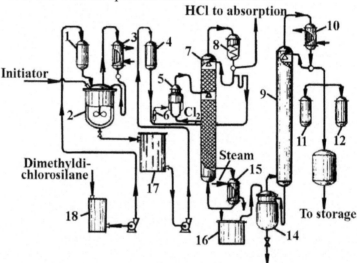

Fig. 19. Production diagram of methyl(chloromethyl)dichlorosilane: 1,4 - batch boxes; 2 - apparatus for preparing the initiator solution; 3 - backflow condenser; 5 - chlorinator; 6 - rotameter; 7, 9 - rectification towers; 8, 10 - refluxers; 11-13 - receptacles; 14- tank; 15 - boiler; 16-18- containers

The average exit composition of the chlorination products is the following: 3-3,5% of methyl(dichloromethyl)dichlorosilane, 11-14% of methyl(chloromethyl)dichlorosilane and 82.5-86% of unreacted dimethyldichlorosilane. The mixture is sent into the middle part of tower *7* for continuous rectification to distill the unreacted dimethyldichlorosilane and steam out the dissolved hydrogen chloride.

The unreacted dimethyldichlorosilane with hydrogen chloride is distilled when the temperature in the higher part of the tower is 70-72 °C. Dimethyldichlorosilane vapours condense in refluxer *8*, which is cooled with water. The condensate flows into the separation box, from where the most part is returned to reflux the tower, and the rest is sent through rotameter into the chlorinator, where it is mixed with the 2% initiator solution for repeated chlorination. After refluxer *8*, hydrogen chloride enters the backflow igurit condenser (not shown in the diagram), where it is purified from the impurity of dimethyldichlorosilane, and is sent for water absorption.

After partial separation of the unreacted dimethyldichlorosilane, the average composition of the chlorination products is the following: 20-44% of dimethyldichlorosilane, 45-67% of methyl(chloromethyl)dichlorosilane and 11-13% of polychlorides (a mixture of methyl(dichloromethyl)- and methyl(trichloromethyl)dichlorosilanes). They are sent into boiler *15*, which is a shell-and-tube heat exchanger, where the temperature of 115-120 °C is maintained with vapour (0.4 MPa).

From the boiler, the chlorination products continuously flow into container *16*. From container *16*, the mixture is periodically fed into tank *14* to distil methyl(chloromethyl)dichlorosilane in vacuum.

After the chlorinated products enter tank *14*, its jacket is filled with vapour (0.4-0.5 MPa); the tank temperature during the distillation is maintained at 100-120 °C. The distilled vapours are sent into rectification tower *9*, filled with Raschig rings. The rectification separates four fractions.

Fraction I, which mainly consists of dimethyldichlorosilane, is distilled at atmospheric pressure when the temperature at the top of the tower is below 72 °C. The vapours from the higher part of tower *9* condense in water-cooled refluxer *10*. The condensate is sent into the separation box, from where part of it is returned to reflux the tower, and the rest is collected in receptacle *11*. From the receptacle dimethyldichlorosilane with a density not exceeding 1.08 g/cm^3 can be used to prepare the initiator solution.

Fraction II, the mixture of dimethyldichlorosilane and methyl(chloromethyl)dichlorosilane, is separated into receptacle 12 72-98 °C at the tower top until the distillate density reaches 1.28 g/cm^3. After the separation of fraction II at 98 °C at the top of the tower, a vacuum pump is switched on to create vacuum, which is followed by the separation of fraction III. The rectification of dimethyldichlorosilane chlorination products at atmospheric pressure causes considerable tarring in the tank; the use of vacuum largely eliminates this phenomenon and facilitates the tank heating.

Fraction III, methyl(chloromethyl)dichlorosilane, is separated when the temperature of the tower top is 98-102 °C and residual pressure is 480-530

GPa. It is collected in receptacle 13 until the distillate density reaches 1.29 g/cm³.

Fraction IV, a mixture of methyl(chloromethyl)dichlorosilane and methyl(dichloromethyl)dichlorosilane is separated in a constantly increasing vacuum (the residual pressure of 350-300 GPa) in the 102-108 °C temperature range. The separation continues until the distillate density reaches 1.40 g/cm³. As soon as fraction IV accumulates, it can be returned for vacuum rectification.

As noted above, the transformation degree of dimethyldichlorosilane when chlorinated by this technique is rather low; however, it can be considerably increased (to 80%) without any significant drop in the methyl(chloromethyl)dichlorosilane yield, if dimethyldichlorosilane is chlorinated in the presence of the given initiator directly in the rectification tower. In this case the higher part of the tower can be filled with dimethyldichlorosilane with 2-2.5% of the initiator, and the middle part can be filled with gaseous chlorine. Chlorination products can be withdrawn from the reaction zone through the lower part of the tower, where they are separated from the unreacted dimethyldichlorosilane. The latter can be again returned into the reaction zone to be mixed with a concentrated solution of the initiator. The heating of the tower tank should be regulated so that the liquid in the top part, which serves as the chlorinator, is always in an emulsified state. Installing a head in the higher part of the tower and maintaining there an emulsifying mode make for a good contact of reactants.

The use of the top part of the packed tower as the chlorinator will also ensure good mass exchange between chlorination and rectification zones; this will allow to quickly withdraw chlorination products from the reaction zone and continuously send back the unreacted dimethyldichlorosilane as a result of the rectification of the mixture in the lower part of the tower. However, this chlorination technique can be possible only with chemical initiators, not UV rays, since in the latter case the effect of light will be screened by the head.

Methyl(chloromethyl)dichlorosilane is a colourless motile liquid (the boiling point is 120 °C) with a pungent odour. It fumes in air and can be easily hydrolysed with water releasing hydrogen chloride. It can be easily dissolved in organic solvents.

Technical methyl(chloromethyl)dichlorosilane should meet the following requirements:

Appearance	Transparent liquid
Density at 20 °C, g/cm³	1.280-1.290

Chlorine content, %:

silicon-bound	42.5-44
total	64-67

Methyl(chloromethyl)dichlorosilane is used to obtain silicone liquids, elastomers and polymers for varnishes, as well as monomers with various carbofunctional groups at the silicon atom.

Industrial chlorination of methyltrichlorosilane and trimethylchlorosilane is conducted by a similar technique. It should be borne in mind, however, that the chlorination speed of methyl groups greatly depends on the number of methyl radicals in the original methylchlorosilanes. Trimethylchlorosilane is the easiest to chlorinate, methyltrichlorosilane is the hardest. For example, the chlorination speed at the transition from dimethyldichlorosilane to trimethylchlorosilane increases 9-fold.

2.2.2. Preparation of chlorinated phenylchlorosilanes

Unlike the chlorination of methylchlorosilanes, the direct chlorination of phenylchlorosilanes is based on the mechanism of electrophylic substitution. The reaction easily occurs in the presence of catalytic quantities of Fe, $FeCl_3$, $AlCl_3$, PCl_5, $SbCl_3$ or I_2 and results in the replacement of hydrogen atoms in the aromatic nucleus with chlorine; moreover. depending on the temperature and mole ratio of reactants, it is possible to introduce various numbers of chlorine atoms into the phenylchlorosilane molecule with a 55-90% yield of the target products. Chlorination proceeds according to the following general scheme:

$$C_6H_5Si\diagdown \quad + nCl_2 \longrightarrow C_6H_{5-n}Cl_nSi\diagdown \quad + nHCl \tag{2.81}$$

In the absence of catalysts phenylchlorosilanes do not chlorinate even at increased temperature (150-200 °C). It should be kept in mind, however, that the replacement chlorination of phenylchlorosilanes depending on the conditions of the reaction (the presence of a catalyst and its composition, the effect time of chlorine, temperature) is accompanied by breaking up the $Si-C_{ar}$ bond:

$$\diagup\diagdown Si\text{-}C_6H_5 + HCl \longrightarrow \diagup\diagdown Si\text{-}Cl + C_6H_6 \tag{2.82}$$

This secondary reaction is especially noticeable in the presence of catalysts, metal halogenides, which are ranged by their dearylising effect as

$AlCl_3> FeCl_3>SbCl_5>SbCl_3$. The breaking-up of the Si—C_{ar} bond can be considerably lowered by conducting chlorination at a lower temperature and in a solvent medium (e.g. CCl_4). However, since during the chlorination of phenylchlorosilanes in the presence of $SbCl_3$ or Fe the Si—C_{ar} bond breaks up very slightly, solvents in this case are no longer necessary.

Preparation of chlorophenyltrichlorosilanes

The chlorination of phenyltrichlorosilane with chlorine on heating and in the presence of $SbCl_3$ or Fe proceeds according to the mechanism of electrophylic substitution of the hydrogen atom in the phenyl radical with chlorine:

$$C_6H_5SiCl_3 \xrightarrow[-HCl]{+Cl_2} C_6H_4ClSiCl_3 \xrightarrow[-HCl]{+Cl_2} C_6H_3Cl_2SiCl_3 \text{ etc.} \quad (2.83)$$

When Fe is used as a catalyst, chlorine interacts with the catalyst in the first place:

$$2Fe + 3Cl_2 \longrightarrow 2FeCl_3 \xrightarrow{+Cl_2} [FeCl_4]^- + Cl^+ \quad (2.84)$$

The chlorocation formed in this case detaches a proton from the phenyl radical in phenyltrichlorosilane and occupies its place:

$$C_6H_5SiCl_3 \xrightarrow[-H^+]{+ Cl^+} C_6H_4ClSiCl_3 \quad (2.85)$$

The proton in its turn interacts with the anion $[FeCl_4]^-$; thus forming hydrogen chloride and iron chloride:

$$H^+ + [FeCl_4]^- \longrightarrow HCl + FeCl_3 \quad (2.86)$$

The chlorination of phenyltrichlorosilane can also form products of a deeper chlorination:

$$C_6H_4ClSiCl_3 \xrightarrow[-H^+]{+ Cl^+} C_6H_3Cl_2SiCl_3 \text{ etc.} \quad (2.87)$$

Chlorination is accompanied by some by-processes.

1. The hydrogen chloride released in the reaction breaks up the Si—C_{ar} bond, which leads to the formation of chlorobenzenes and silicon tetrachloride:

$$C_6H_{5-n}CI_nSiCI_3 + HCI \xrightarrow{\text{FeCI}_3} C_6H_{6-n}CI_n + SiCI_4 \qquad (2.88)$$

2. Chlorobenzenes in the presence of $FeCl_3$ are in their turn subjected to further chlorination:

$$C_6H_{6-n}CI_n + CI_2 \xrightarrow[\text{-HCI}]{\text{FeCI}_3} C_6H_{5-n}CI_{n+1} \qquad (2.89)$$

3. When the concentration of the catalyst is not sufficient, chlorine attaches itself to the chlorinated phenyltrichlorosilane, forming polychlorocyclohexyltrichlorosilanes:

$$C_6H_{5-n}CI_nSiCI_3 + 3CI_2 \longrightarrow C_6H_{5-n}CI_{n+6}SiCI_3 \qquad (2.90)$$

Raw stock: phenyltrichlorosilane (not less than 99% of the main fraction); chlorine (not less than 99.5% of chlorine, not more than 0.06% of humidity); nitrogen (not less than 99.5% of nitrogen, not more than 0.5% of oxygen); iron chipping.

The production process of chlorinated phenyltrichlorosilanes (Fig. 20) comprises three main stages: the chlorination of phenyltrichlorosilane; the elimination of hydrogen chloride and unreacted chlorine with nitrogen; the separation of the reactive mixture.

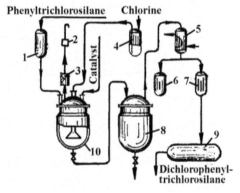

Fig. 20. Production diagram of chlorinated phenyltrichlorosilanes: *1* - batch box; *2* - fire-resistant apparatus; *3* - tower with $CaCl_2$; *4* - receiver; *5* - condenser; *6, 7* - collectors; *8* - distillation tank; *9* - container; *10* - chlorinator

Phenyltrichlorosilane is chlorinated in chlorinator 10, which is an enameled steel apparatus with a water vapour jacket and two bubblers (funnel-shaped) for feeding chlorine. From batch box *1* the chlorinator is loaded with phenyltrichlorosilane and with the catalyst, iron chipping, through the hatch. After that, the jacket of the chlorinator is filled with vapour and its contents are heated to 65-70 °C. Receiver 4 is filled with chlorine; after that, the chlorine is fed from there at the speed that would keep the chlorination temperature below 70 °C. This ensures intensive agitation of the reactive mass and good contact of chlorine with phenyltrichlorosilane.

The effluent gases (chlorine, hydrogen chloride) pass through tower 3 with $CaCl_2$ and fire-resistant apparatus 2 and are sent for neutralisation (not shown in the diagram). The temperature during the whole chlorination process is maintained within 65-70 °C, for which purpose the chlorinator jacket is filled with water. In 48 hours after the chlorine feeding begins, the process is stopped.

In order to eliminate hydrogen chloride and unreacted chlorine, after the process ends, the chlorinator is for 1 hour filled through bubblers with dehydrated nitrogen at the speed of 2 m^3/h. After that, the chlorinator is cooled to 40 °C, and the density of the reaction mass is determined. Chlorination is considered to be finished at a density of 1.530-1.550 g/cm^3. If the density is less, the process is continued.

The reactive mixture from the chlorinator is sent into vacuum distillation tank 8 for separation. A residual pressure of 50-70 GPa is created in the system and the tank jacket is filled with ditolylmethane or other heat carrier. After that, the fractions are separated.

Fraction I, a mixture of unreacted phenyltrichlorosilane and chlorophenyltrichlorosilane, is distilled below 190 °C (in liquid form) into collector 6. After fraction I is separated, the mixture in the tank is sampled to determine the content of hydrolysed chlorine and the density of the reactive mass. When the content of hydrolysed chlorine is not lower than 37.8% and the density is 1.482 g/cm^3 , fraction II is separated. To extract chlorophenyltrichlorosilane, fraction I is rectified.

Fraction II, which mainly consists of dichlorophenyltrichlorosilane, is separated into collector 7 below 215 °C (in liquid form). From there dichlorophenyltrichlorosilane is sent into container 9.

Dichlorophenyltrichlorosilane is a colourless transparent liquid (the boiling point is 269 °C) with a specific chloroanhydride odour. It dissolves well in common organic solvents, is easily hydrolysed with water and humidity in air. It fumes in air.

Technical dichlorophenyltrichlorosilane should meet the following requirements:

Appearance	Transparent liquid, colourless or light brown
Density at 20 °C, g/cm^3	1.482-1.550

Content, %:

of hydrolysed chlorine	37.8-39
of distillate at 240-270 °C	not less than 80

Dichlorophenyltrichlorosilane is used as raw stock in the manufacture of silicone oligomers and polymers.

2.2.3. Preparation of chlorinated methylphenyldichlorosilanes

When organochlorosilane molecules have at the same time alkyl and aryl groups, there is a possibility of controlled chlorination of these compounds. The determining factors are the conditions of the reaction, the type of the catalyst and the type of the chlorinating agent.

Thus, when methylphenyldichlorosilane is chlorinated in the presence of radical initiators (the dinitrile of 2,2'-azobis(isobutyric) acid, etc.), only the methal radical is chlorinated, while the phenyl radical is untouched.

$$CH_3(C_6H_5)SiCl_2 \xrightarrow[-HCl]{+Cl_2} CH_2Cl(C_6H_5)SiCl_2 \qquad (2.91)$$

The technology of methylphenyldichlorosilane chlorination is similar to that of methylchlorosilanes. The main difference is that the radical chlorination of methylphenyldichlorosilane occurs at a sufficient speed at higher temperatures (100-110°C), than in the case of methylchlorosilanes (60-70°C). At lower temperature (50-70 °C) the chlorination of methylphenyldichlorosilane is slow, and at higher temperatures (140-150 °C) the bond Si—C_{alk} is destroyed. This difference in the conditions for methylphenyldichlorosilane chlorination seems to be caused by spatial difficulties due to the presence of the phenyl radical.

The technological diagram of the production of phenyl(chloromethyl)dichlorosilane is similar to the diagram given in Fig. 19. In this process, similar to methylchlorosilane chlorination, in order to maximise the yield of phenyl(chloromethyl)dichlorosilane, the original methylphenyldichlorosilane should have a low conversion degree; otherwise, the process forms a great amount of products of a deeper chlorination.

On the contrary, when methylphenyldichlorosilane is chlorinated with chlorine in the presence of electrophilic catalysts (e.g. metals or their halogenides), only the phenyl radical is chlorinated, while the methal radical is untouched. In this case chlorination develops in the following way:

$$CH_3(C_6H_5)SiCI_2 \xrightarrow[-nHCI]{+nCI_2} CH_3(C_6H_{5-n}CI_n)SiCI_2 \qquad (2.92)$$

However, similarly to phenylchlorosilane chlorination, one cannot fully exclude the by-processes of bond breaking. The breaking-up occurs at Si—C_{ar} bonds:

$$CH_3(C_6H_5)SiCI_2 \xrightarrow{+CI_2} CH_3SiCI_3 + C_6H_5CI \qquad (2.93)$$

$$CH_3(C_6H_{5-n}CI_n)SiCI_2 \xrightarrow{+CI_2} CH_3SiCI_3 + C_6H_{5-n}CI_{n+1} \qquad (2.94)$$

As a result,. the chlorination products, apart from target methyl(chlorophenyl)dichlorosilanes, also contain methyltrichlorosilane, chlorobenzene and polychlorobenzenes.

Table 9. Physicochemical properties of chlorinated methyl-, phenyl- and methyl-phenylchlorosilanes

Substance	Boiling point, °C	Melting point, °C	d_4^{20}	n_D^{20}
$CH_2ClSiCl_3$	116.5 (at 1000 GPa)	-	1.4441	1.4535
$CHCl_2SiCl_3$	143.5 (at 995 GPa)	-	1.5518	1.4714
CCl_3SiCl_3	155-156	115-116	-	-
$CH_3(CH_2Cl)SiCl_2$	121.5-122	-	1.2858	1.4500
$CH_3(CHCl_2)SiCl_2$	148.5 (at 1000 GPa)	-	1.4116	1.4700
$CH_3(CCl_3)SiCl_2$	160-162	98	-	-
$(CH_3)_2(CH_2Cl)SiCl$	115	-	1.0865	1.4360
$C_6H_4ClSiCl_3$	102-108 (at 20 GPa)	-	1.4480	-
$C_6H_3Cl_2SiCl_3$	125 (at 10.7 GPa)	-	1.5528	1.5641
$C_6Cl_5SiCl_3$	146-147 (at 10.7 GPa)	59.5	-	-
$CH_2Cl(C_6H_5)SiCl_2$	108 (at 15 GPa)	-	1.3170	1.5365
$CH_3(C_6H_4Cl)SiCl_2$	231-232		1.3017	1.5360

The technological diagram of the production of methyl(chlorophenyl)dichlorosilanes is similar to the production diagram of chlorinated phenyltrichlorosilanes (see Fig. 20).

Phenyl(chloromethyl)dichlorosilanes and methyl(chlorophenyl)-dichlorosilanes are colourless, transparent, motile liquids, which fume in air. Like all organochlorosilanes, they are easily hydrolysed with water and humidity in air and dissolve well in organic solvents. Phenyl(chloromethyl)- and methyl(chlorophenyl)dichlorosilanes are raw stock for preparing various silicone liquids, elastomers, and polymers for varnishes.

The physicochemical properties of chlorinated methyl-, phenyl- and methylphenylchlorosilanes are given in Table 9.

2.3. Preparation of ethers of orthosilicon acid and their derivatives

Ethers of orthosilicon acid and their derivatives, tetraalkoxy(aroxy)silanes and alkyl(aryl)alkoxy(aroxy)silanes are a rather extensive class of silicone compounds. They are independently applied in various spheres of technology, but are particularly valuable as semi-products for preparing important silicone oligomers and polymers.

Of interest are also substituted ethers of orthosilicon acid with an amide group in the organic radical, which have the following general structure:

$$[R_2N(CH_2)_n\frac{\quad}{m}\overset{\displaystyle R'}{\underset{\displaystyle |}{Si}}_x(OR'')_{4-(m+x)}$$

R=H, CH$_3$, C$_2$H$_5$, C$_4$H$_9$, C$_6$H$_5$ and others; R'=R''=CH$_3$, C$_2$H$_5$, C$_4$H$_9$, C$_6$H$_5$ and others;

$$n=1\div5; \ m=x=1\div2; \ m\neq x.$$

as well as substituted ethers of orthosilicon acid with nitrogen in the ether group:

$$RSi\overset{\displaystyle OCH_2CH_2}{\underset{\displaystyle OCH_2CH_2}{\overline{\ \ OCH_2CH_2\ \ }}}N$$

R=CH$_3$, C$_2$H$_5$, C$_6$H$_5$, CH$_2$CI, OR, OCOR, H and others.

The main raw stock for the synthesis of tetraalkoxy- and tetraaroxysilanes is silicon tetrachloride; thus, below we consider the methods of its preparation.

2.3.1. Preparation of silicon tetrachloride

Silicon tetrachloride was first prepared by Berzelius in 1823 by subjecting silicon to chlorine at red heat. Further on, a range of other methods to prepare silicon tetrachloride have been suggested: heating a mixture of silica, carbonised starch and coal in chlorine flow (D.I.Mendeleev); subjecting silica to phosgene in the presence of soot as a catalyst at 700-1000 °C; subjecting ferrosilicon to hydrogen chloride at 500 °C.

The contemporary process of silicon tetrachloride production is based on research by Martin, who in 1914 was the first to obtain silicon tetrachloride by chlorinating ferrosilicon with gaseous chlorine:

$$2FeSi + 7Cl_2 \longrightarrow 2SiCl_4 + 2FeCl_3 \tag{2.95}$$

The chlorination of ferrosilicon was also used in the first USSR production of silicon tetrachloride, the credit for which goes to Academician K.A.Andrianov.

The mechanism of SiCl$_4$ formation is as follows. Silicon molecules, in which Si atoms are bonded by the forces of main valencies, are chlorinated under the influence of chlorine, with Si—Si bonds breaking up first. The chlorine atoms join the silicon atoms to form linear molecules of polychlorosilanes. Under continued influence of chlorine, the polychlorosilanes are broken up into lower-molecular chlorosilanes.

$$(2.96)$$

Further chlorination breaks up all Si—Si bonds and forms SiCl$_4$:

$$(2.97)$$

The silicides of other metals found in ferrosilicon are also chlorinated; however, since the boiling point of silicon tetrachloride is low (57.7 °C), it can be easily distilled from secondary metal chlorides.

In the process of $SiCl_4$ preparation the main raw stock is *ferrosilicon* , an alloy of iron and silicon, made in shaft electric furnaces. Electrothermal ferrosilicon is produced from quartzite and iron chipping; it is reduced with charcoal or coke (petroleum or metallurgical). The process is based on the endothermal reaction of silica reduction with carbon, which takes place at high temperature.

Our industry prepares several brands of ferrosilicon, the most important of which are Cu-45, Cu-75 and Cu-90:

Brand	Content, % (the rest is Fe)				
	Si	Mn	Cr	P	S
Si-45	40-47	0.80	0.50	0.05	0.04
Si-75	74-80	0.70	0.50	0.05	0.04
Si-90	87-95	0.50	0.20	0.04	0.04

The production of silicon tetrachloride uses Cu-75 and Cu-90 ferrosilicon. One can also use as raw stock evaporated chlorine (not less than 99.6% of Cl_2, not more than 0.02% of humidity) and lime milk (not less than 100 g of CaO per 1 litre). $SiCl_4$ can also be synthesised from a mixture consisting of 70% of crystal silicon Si-1 and 30% of ferrosilicon Cu-75.

The production of silicon tetrachloride (Fig. 21) comprises two main stages: the chlorination of ferrosilicon and the rectification of silicon tetrachloride.

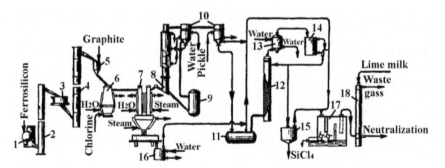

Fig. 21. Production diagram of silicon tetrachloride by ferrosilicon chlorination: *1* - jaw crusher; *2* - bucket elevator; *3* - grate; *4* - shaft hoist; *5* –bin; *6* - chlorinating furnace; *7* - condenser; *8* - scrubber; *9* - boiler; *10, 14* – condensers; *11* - distillation tank; *12* - rectification tower; *13* - refluxer; *15* -collector; *16*- apparatus for the destruction of solid chlorides; *17*- hydrolysis chamber; *18*- absorption column.

After the system is checked for hermeticity (chlorine pressure testing), chlorinating furnace 6 is loaded with 80-120 mm pieces of broken and pre-dried graphite from top bin 5 for steady supply and distribution of chlorine. The graphite layer is 600-700 mm high. Ferrosilicon is pre-crushed on jaw crusher *1* into 30-80 mm pieces and with bucket elevator *2* is sent into grate *3* with sieves of various sizes. Large pieces from the top sieve are sent back for another crushing, and small pieces, which passed the last sieve, are discarded; however, it happens only in case chlorination is conducted in a horizontal apparatus shown in Fig. 22. The crushed ferrosilicon is sent in shaft hoist *4* into filling bin *5*, from where it self-flows into the heater (when the furnace is started).

Heated to 300-400 °C, ferrosilicon is periodically loaded into furnace *6*, which is a steel vertical two-coned shaft apparatus with a water jacket. The cooling surface of the furnace is 7 m^2. The diameter of the wider section of the furnace is 900 mm, the diameter of the narrower section of the furnace is 480 mm. The total height of the furnace is 3300 mm. The lower cone has a smaller height (650 mm) and is lined with diabase tile in one layer. Chlorination is conducted with evaporated chlorine, which is fed into the lower section of the furnace through the distribution collector at a speed of 70-100 m^3/h.

When the vapours at the exit from the chlorinator reach 150 °C, the heating is switched off. Due to exothermicity the temperature in the reaction zone rises up to 200 °C. After that the jacket of the apparatus is filled with water, and the process continues normally. The temperature of SiCl$_4$ vapours and of metal chlorides at the exit from the chlorinator is kept below 700 °C (if the temperature rises higher, the chlorine supply is automatically stopped). During the process, the chlorinator is regularly (approximately every 2 hours) loaded with ferrosilicon. The raw stock level in the chlorinator is maintained at 250-350 mm higher than the graphite layer. After 8-12 operating days (which depends on the raw stock quality) the chlorinator is stopped to load off cinder.

When the system is stopped to be loaded with raw stock, it receives air with humidity, which breaks down silicon tetrachloride into hydrogen chloride and gel of silicon acid, as well as self-inflammable silanes. Calcium chloride (the melting point is 772 °C, the boiling point is 1600 °C) remains in the chlorinator and combines with pieces of unreacted ferrosilicon to form cinder, which is accumulated in the furnace and loaded off with graphite.

Apart from silicon tetrachloride, the process of chlorination forms a considerable amount of chlorides of iron and other metals; thus, the condensation of the chlorides formed is conducted in two stages.

The first stage, the condensation of solid chlorides ($FeCl_2$, $FeCl_3$, $AlCl_3$, etc.), takes place in condenser 7 (see Fig.21). $SiCl_4$ and metal chloride vapours enter the condenser from furnace 6 through the gas flue. The condenser consists of two vertical tubes with a common cone bin.

The tubes have rake agitators, which scrape the condensed chlorides off the walls. The tubes and the bin have jackets. The jacket of the first tube is filled with water for cooling; as a result, the main mass of solid chlorides condenses and deposits. Gaseous electrolytic chlorine is not usable for this process, because it contains impurities of carbon dioxide and oxygen, which oxydise silicon. The second tube and the cone bin of the condenser are warmed with vapour to 80-120 °C to avoid $SiCl_4$ vapours condensing in them. The condenser collects up to 90-95% of solid chlorides, which are then degraded with water in apparatus 16.

The second condensation stage, at which $SiCl_4$ vapours are finally purified from solid chlorides and other impurities, is conducted in scrubber 8, which is washed with silicon tetrachloride (the so-called "wet technique"). The wet technique purifies $SiCl_4$ almost completely. In the scrubber, which is a vertical tower with distribution plates, reaction gases are passed countercurrent to liquid silicon tetrachloride, which washes the gases, entraining all the solid particles. The suspension enters boiler 9 and is stripped there. After condensation $SiCl_4$ vapours are sent back to reflux the scrubber, and the solid chlorides and other particles accumulated in the boiler are periodically loaded off. After that, $SiCl_4$ vapours are condensed in tubular condensers 10, which are cooled with water and salt solution (-30 °C). The condensate flows into distilling tank 11.

The rectification of raw silicon tetrachloride is conducted in tower 12, which is a vertical steel tube filled with 50x50x5 mm ceramic rings. First, in order to eliminate the gaseous chlorine dissolved in the raw stock, coil condenser 14 is inversed and the mixture is heated until the temperature of the vapours after refluxer 13 reaches 55 °C. After that, the condenser is switched, and the main fraction ($SiCl_4$) is separated into collector 15. The separation of the finished product is stopped when the temperature of the vapours is 75 °C. The effluent gases from different stages of the process (condensation, rectification, destruction of solid chlorides and tank residue) are directed into chamber 17, which is made from brick and lined with diabase tile, to hydrolyse $SiCl_4$.vapours. To ensure the contact of gases with water, the chamber is separated into sections with rotating diffusers, which create a dense veil from water drops. The unabsorbed gases (mainly chlorine) are evacuated from the chamber with an air ejector and sent into tower 18, washed with lime milk. Acid slime waters are sent from the tower into neutralising boxes.

The chlorination of ferrosilicon is also possible in a horizontal apparatus shown in Fig. 22. This changes the construction of the whole condensation system, which is basically one unit with the chlorinator. Horizontal furnace *1* is a 600 mm tube; from one side this tube is covered with a water-cooled door, and from the other side it joins the first condensation tube *3*. The chlorinator has a filling device *2*, which is a small vertical tube with a gate on top, used for loading ferrosilicon. The furnace, the vertical tube and the chlorine-feeding choke have cooling jackets.

The condensation section of the apparatus consists of three condensers 3 located one above another. They are horizontal tubes, which are also fashioned with cooling jackets.

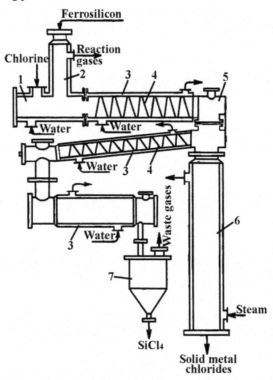

Fig. 22. Horizontal apparatus for the chlorination of ferrosilicon: *1* - furnace; *2* - filling device; *3* - condensers; *4* - screws; *5* - abscheider; *6* -tube for the withdrawal of solid chlorides; *7* – collector.

The first and second tubes have screws *4* with rakes, which continuously scrape the condensed chlorides off the walls. The end of the first tube is connected to a long abscheider *5* with tube *6*, which is connected with a branch pipe with the second condensation tube *3*. The abscheider is used to collect solid chlorides; its lower end is closed with a schieber (sliding vane) and fashioned with a pneumatic vibrator for loosening the deposit on the walls. Between the second and third tubes 3 there is a hydraulic gate for maintaining constant pressure in the system. The end section of the condensation system and the rectification apparatus do not differ from the technological diagram given in Fig. 21.

The temperature of the water which exits the jackets of furnace 1 and filling device 2 should not exceed 60-69 °C; the temperature of the water which exits the jackets of condenser 3, from the door and the chlorine choke should not exceed 50-55 °C. If these conditions are observed, it is possible to prevent the corrosion of the equipment.

In the temperature range below 500 °C the yield of silicon tetrachloride depends on the temperature of chlorination. If ferrosilicon is chlorinated at a temperature below 500 °C, e.g. at 300-350 °C, the chlorination products will contain alongside with $SiCl_4$ small amounts (less than 0.4%) of high-boiling silicon chlorides, hexachlorodisilane and octachlorotrisilane. At further reduction of temperature the content of polychlorosilanes grows. at 200 °C it reaches 0.4%, and at 180 °C it is already 5.7% of the condensate quantity.

The production of silicon tetrachloride by the technique described is ac-companied by the release of a great amount of heat in a small volume, which makes heat withdrawal more difficult. Thus, one has to use contact reactors low-productive reactors (2-2.5 tonnes of $SiCl_4$ daily). Because of high temperature in the reaction zone (до 1200 °C), there are by-processes: at a temperature above 1100 °C silicon tetrachloride interacts with free silicon, forming the compounds which energetically reduce $FeCl_3$ to $FeCl$ and Fe. That is why the solid substances deposited in condensers 7 (see Fig.21) are a mixture of $FeCl_3$, $FeCl_2$ and Fe, with $SiCl_4$ vapours and small amounts of silicon polychlorides adsorbing on their surface. These sub-stances inflame in air, that is why total hermeticity is required when they are unloaded. The technical and economical characteristics of the process considerably decrease due to a limited use of fine ferrosilicon formed dur-ing the crushing.

All these faults can be largely eliminated by chlorinating ferrosilicon in a medium of melted salts. In this case, chlorination should be started in a melt of sodium chloride (or an equimolar mixture of sodium and potassium chlorides). Then the iron chlorides which form alongside with $SiCl_4$, are held in the melt in the form of complexes with chlorides of alkaline metals ($NaFeCl_4$ and $KFeCl_4$). After the concentration of Fe in the melt reaches 45-50% (equivalent to $FeCl_3$), the new portions of ferrosilicon are supple-mented with a corresponding quantity of sodium chloride (or an equimolar mixture of sodium and potassium chlorides) to maintain the given concentration of iron chloride; the excess melt is withdrawn from the reactor.

The chlorination of ferrosilicon in salt melt is conducted in two stages, which take place rather quickly:

$$Si + 4FeCI_3 \rightleftharpoons 4FeCI_2 + SiCI_4 \qquad (2.98)$$

$$Fe + 2FeCI_3 \longrightarrow 3FeCI_2 \xrightarrow{1.5CI_2} 3FeCI_3 \qquad (2.99)$$

With this technique and the temperature of 600-1000 °C, the speed of the process is virtually independent of the temperature and the size of ferrosilicon particles (ranging from 0.25 to 3 mm).

Silicon tetrachloride is a motile, colourless or light yellow liquid (the boiling point is 57.7 °C) with a pungent odour. It is soluble in dichloroethane, gasoline and other organic solvents. It combines with alcohols to form ethers of orthosilicon acid. It strongly fumes in air, because humidity hydrolyses it and releases hydrogen chloride vapours.

Technical silicon tetrachloride should meet the following requirements:

Appearance   Transparent colourless or yellowish liquid

Density at 20 °C, g/cm^3 1.48-1.50

The fraction composition, determined at 0.1 MPa, °C

the onset of boiling, not lower than   55

the end of boiling, not higher than.   59

residue after distillation, %, not more than   2.5

Content, %, not more than

of iron.   0.001

of free chlorine   0.2

Silicon tetrachloride is a raw stock for preparing ethers of orthosilicon acid and is used in the production of silicone polymers used to obtain highly thermostable plastics and synthetic lubricants, as well as high-quality electroinsulation. Silicon tetrachloride is also used to prepare superfine silicon dioxide (Aerosil). A mixture of silicon tetrachloride and ammonia was used as early as during World War I as a fume-forming sub-stance.

2.3.2. Preparation of alkoxy(aroxy)silanes

Tetraalkoxy(aroxy)silanes, or ethers of orthosilicon acid H_4SiO_4.have the most practical applications among all the silicone compounds without Si—C bonds in their molecule.

The first ether of orthosilicon acid, *tetraethoxysilane*, was obtained by Ebelmen in 1844 by reacting silicon tetrachloride with ethyl alcohol.

$$SiCI_4 + 4C_2H_5OH \xrightarrow[-4HCI]{} Si(OC_2H_5)_4 \qquad (2.100)$$

Tetraethoxysilane can be obtained not only from $SiCl_4$, but also from $SiBr_4$ and SiF_4. Silicon tetraiodide SiI_4 is not advisable, since in this case the reaction breaks up the Si—OC_2H_5:bond.

$$SiI_4 + 2C_2H_5OH \longrightarrow SiO_2 + 2HI + 2C_2H_5I \qquad (2.101)$$

SiI_4 can yield tetraethoxysilane only by this reaction:

$$SiI_4 + 4(C_2H_5)_2O \longrightarrow Si(OC_2H_5)_4 + 4C_2H_5I \qquad (2.102)$$

Further   on,   a   range   of   other   methods   to   prepare tetraalkoxy(aroxy)silanes have been suggested:
1. the interaction of $SiCl_4$ with ethers of nitrous acid:

$$SiCI_4 + 4RONO \rightleftharpoons Si(OR)_4 + 4NOCI \qquad (2.103)$$

2. the effect of methyl alcohol on tetraisocyanatesilane:

$$4CH_3OH + Si(NCO)_4 \longrightarrow Si(OCH_3)_4 + 4HNCO \qquad (2.104)$$

3. the reaction of tetraaminosilanes with alcohols:

$$Si(NHR)_4 + 4R'OH \longrightarrow Si(OR')_4 + 4RNH_2 \qquad (2.105)$$

4. the exchange interaction of tetraaroxysilanes with aliphatic alcohols (or tetraalkoxysilanes with phenols):

$$(C_6H_5O)_4Si + 4 C_2H_5OH \underset{OH^-}{\overset{H^+}{\rightleftharpoons}} (C_2H_5O)_4Si + 4C_6H_5OH \qquad (2.106)$$

5. the effect of methyl alcohol on free silicon or on metal silicides in the presence of catalysts:

$$Si + 4CH_3OH \xrightarrow[-2H_2]{} Si(OCH_3)_4 \qquad (2.107)$$

$$SiMg_2 + 8CH_3OH \xrightarrow[-4H_2]{} Si(OCH_3)_4 + 2Mg(OCH_3)_2 \quad (2.108)$$

6. the interaction of alcohols with silicon sulfide:

$$SiS_2 + 4ROH \xrightarrow[-2H_2S]{} Si(OR)_4 \qquad (2.109)$$

7. the effect of metallic sodium on alkoxychlorosilanes:

$$2(RO)_2SiCI_2 + 4Na \longrightarrow Si(OR)_4 + 4NaCI + Si \qquad (2.110)$$

Nevertheless, only one technique is currently used in the production of tetraalkoxy(aroxy)silanes. It is based on the etherification of silicon tetra-chloride with alcohols or phenols and is the simplest and the most eco-nomical.

The etherification of silicon tetrachloride with anhydrous alcohols or phenols occurs stagewise and can be schematically written thus:

$$SiCI_4 \xrightarrow[-HCI]{+ROH} ROSiCI_3 \xrightarrow[-HCI]{+ROH} (RO)_2SiCI_2 \xrightarrow[-HCI]{+ROH} (RO)_3SiCI \xrightarrow[-HCI]{+ROH} (RO)_4Si \quad (2.111)$$

Hence, etherification always forms some amount of intermediate com-pounds, in the molecules of which silicon atoms are bound both with chlo-rine and with alkoxy(aroxy)groups. These compounds are called chloro-ethers of orthosilicon acid, or *alkoxy(aroxy)chlorosilanes*.

The process of tetrachlorosilane etherification is virtually irreversible; its course depends on the ratio of parent substances, temperature, agitation intensity and the order of component mixing. To maximise the yield of the target product, it is advisable to pour silicon tetrachloride into a small sur-plus of alcohol (not more than 5-10% of the theoretic amount). A large surplus is not advisable, because the main reaction may be accompanied by the secondary interaction of alcohol and hydrogen chloride with the re-lease of water. In the presence of hydrogen chloride water easily hydroly-ses ethers of orthosilicon acid; that is why a large surplus of alcohol will inevitably intensify the process. It is not recommended to use a surplus of silicon tetrachloride, because in this case there is another secondary reac-tion, the formation of chloroethers of orthosilicon acid:

$$SiCI_4 + (RO)_4Si \rightleftharpoons 2(RO)SiCI_2 \qquad (2.112)$$

The formation of the ethers of orthosilicon acid is completed more efficiently, if hydrogen chloride is withdrawn from the reactive mixture. For this purpose the mixture is heated or treated with dehydrated air or nitrogen.

Ethyl ether of orthosilicon acid (tetraethoxysilane) is most widely applied of all the ethers of orthosilicon acid because it is cheap, relatively easy to prepare and not toxic. Since the production of tetraethoxysilane uses the mixture of anhydrous and hydrous alcohol, etherification as a rule yields a mixture of tetraethoxysilane and ethylsilicates, the products of its partial hydrolysis and condensation. That is why we view below at the same time the technology to prepare tetraethoxysilane and ethylsilicate.

Ethylsilicate is a mixture of tetraethoxysilane and linear oligomers, mainly of the following composition:

$$C_2H_5O\left[\begin{array}{c}OC_2H_5\\|\\Si-O\\|\\OC_2H_5\end{array}\right]_n C_2H_5$$

Depending on the degree of polymerisation (n) and, consequently, on silicon content (equivalent to $SiCl_4$), ethylsilicate is produced in various brands: ethylsilicate-32 (30-34% of SiO_2), ethylsilicate-40 (38-42% of SiO_2) and ethylsilicate-50 (up to 50% of SiO_2).

Production of tetraethoxysilane and ethylsilicate-32

The production of tetraethoxysilane is based on the etherification of silicon tetrachloride with ethyl alcohol:

$$SiCl_4 \xrightarrow[HCl]{C_2H_5OH} C_2H_5OSiCl_3 \xrightarrow[HCl]{C_2H_5OH} (C_2H_5O)_2SiCl_2 \xrightarrow[HCl]{C_2H_5OH} \qquad (2.113)$$

$$(C_2H_5O)_3SiCl \xrightarrow[HCl]{C_2H_5OH} Si(C_2H_5O)_4$$

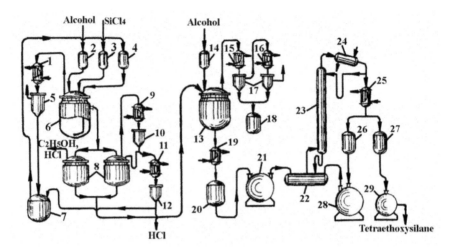

Fig. 23. Production diagram of tetraethoxysilane and ethylsilicate-32: *1, 9, 11, 15, 16, 19, 25* - coolers; *2-4, 14*- batch boxes; *5, 10, 12, 17* - phase separators; *6* - etherificator; *7, 18* - collectors; *8* - distillation tanks; *13* - vacuum distillation tank; *20* - settling box; *21, 28, 29* - depositories; *22* - rectification tower tank; *23* - rectification tower; *24* - refluxer; *26, 27* - receptacles.

This is accompanied by secondary processes, which form ethylsilicate in the long run.

$$C_2H_5OH \xrightarrow[H_2O]{HCl} C_2H_5Cl; \qquad nSiCl_4 \xrightarrow[2((n-1)HCl]{(n-1)H_2O} \sim \overset{\overset{\displaystyle Cl}{|}}{\underset{\underset{\displaystyle Cl}{|}}{Si}}-O-\overset{\overset{\displaystyle Cl}{|}}{\underset{\underset{\displaystyle Cl}{|}}{Si}}-Cl \qquad (2.114)$$

$$\sim \overset{\overset{\displaystyle Cl}{|}}{\underset{\underset{\displaystyle Cl}{|}}{Si}}-O-\overset{\overset{\displaystyle Cl}{|}}{\underset{\underset{\displaystyle Cl}{|}}{Si}}-Cl \xrightarrow[2((n-1)HCl]{2(n+1)C_2H_5OH} \sim \overset{\overset{\displaystyle OC_2H_5}{|}}{\underset{\underset{\displaystyle OC_2H_5}{|}}{Si}}-O-\overset{\overset{\displaystyle OC_2H_5}{|}}{\underset{\underset{\displaystyle OC_2H_5}{|}}{Si}}-OC_2H_5 \qquad (2.115)$$

$$\sim \overset{\overset{\displaystyle OC_2H_5}{|}}{\underset{\underset{\displaystyle OC_2H_5}{|}}{Si}}-O-\overset{\overset{\displaystyle OC_2H_5}{|}}{\underset{\underset{\displaystyle OC_2H_5}{|}}{Si}}-OC_2H_5 \xrightarrow[-C_2H_5OH]{H_2O} \sim \overset{\overset{\displaystyle OC_2H_5}{|}}{\underset{\underset{\displaystyle OC_2H_5}{|}}{Si}}-O-\overset{\overset{\displaystyle OC_2H_5}{|}}{\underset{\underset{\displaystyle OC_2H_5}{|}}{Si}}-OH \qquad (2.116)$$

Raw stock: silicon tetrachloride (55-59°C fraction, $d_4^{20} = 1.47 \div 1.50$, not more than 0.001% of Fe) and anhydrous ethyl alcohol (not less than 99.8% of 78.3-79 °C fraction, $d_4^{20} = 0.789-0.790$).

The production process of tetraethoxysilane and ethylsilicate-32 (Fig. 23) comprises three main stages: the etherification of silicon tetrachloride with ethyl alcohol; the two-stage distillation of excess alcohol and hydrogen chloride (at atmospheric pressure and in vacuum); the extraction of tetraethoxysilane.

After the raw stock is prepared and the equipment is checked for hermeticity, the etherification of silicon tetrachloride is started. The process is carried out in etherificator *6*, which is a cast iron enameled apparatus with a jacket. From batch boxes *2* and *3* the etherificator is simultaneously filled with anhydrous ethyl alcohol and silicon tetrachloride. Apart from anhydrous alcohol, the etherificator receives recirculating ethyl alcohol from batch box *4*. In certain volume ratios (usually from 1:2.2 to 1:2.3) silicon tetrachloride and alcohol enter through siphons the lower part of the etherificator. The temperature of the process (30-40°C) is maintained by regulating the supply of the components. The pressure in the apparatus should not exceed 0.015--0.016 MPa.

The hydrogen chloride liberated in the reaction and the vapours of the unreacted alcohol and silicon chloroethers carried with it enter cooler *1*. The condensed liquid products (recirculating ethyl alcohol and silicon chloroethers) flow through phase separator 5 into collector 7, whereas hydrogen chloride is sent to purification. The alcohol from collector *7* is sent into batch box 4 and into the etherificator.

The raw ethylsilicate formed is continually fed through an overflow pipe into one of distillation tanks *8*. Usually there are several tanks; while some are used for distillation, others are filled with etherification products. After the tanks are filled, the temperature is raised to 78-80 °C and during approximately 3 hours the alcohol vapours condensed in cooler *9* are directed through phase separator *10* back into the tank; i.e. the tank operates in the "self-serving" mode. This makes the etherification more complete. After that the temperature is gradually (at the speed of 5-10 grad/h) raised to 140 °C. The excess pressure in distillation tanks should not exceed 0.02 MPa.

The steam and gas mixture enters water-cooled cooler *9* and through phase separator *10* enters salt-cooled cooler *77*. The condensed alcohol is sent into collector *7*, and hydrogen chloride vapours are sent through phase separator *12* to be purified in bubble tanks (not shown in the diagram). After the excess alcohol and hydrogen chloride are distilled, the mixture is sampled at 140°C to determine hydrogen chloride content. If HCl content is below 0.5%, the reactive mass is cooled with water sent into the tank jacket. If hydrogen chloride content exceeds 0.5%, the distillation is continued.

After the process ends, the reactive mixture cooled down to 60 °C is sent from distillation tanks 8 into vacuum distillation tanks *13*. To make the distillation of hydrogen chloride complete, the tanks are filled with fresh ethyl alcohol from batch box *14*; after that, hydrogen chloride is distilled, first without heating and with the help of vacuum. Gradually the vacuum is increased and the residual pressure is brought to 480-370 GPa. When the distillation slows down, vapour heating is started.

The steam and gas mixture from vacuum distillation tanks *13* successively enters water-cooled cooler *15* and salt-cooled cooler *16*. Through phase separators *17* the condensate flows into collector *18*, and hydrogen chloride flows through the vacuum line into the alcohol trap and then into alkaline traps (not shown in the diagram). At 130 °C the pressure in tank *13* is brought to the atmospheric level and the hydrogen chloride content is determined. If it does not exceed 0.1%, the mixture of ethylsilicate and tetraethoxysilane is poured through cooler *19* into settling box *20*. If hydrogen chloride content exceeds 0.1%, the reactive mixture is cooled down to 60-70°C, the vacuum distillation tanks *13* are filled with fresh alcohol from batch box *14* and the distillation is continued. The mixture stands for about 24 hours in box *20* to separate mechanical impurities. After that, the mixture is poured into depository *21*, from where it is poured into containers or sent to rectification to extract pure tetraethoxysilane.

Ethylsilicate-32 is a mixture containing tetraethoxysilane and oligoethoxysiloxanes in various ratios. It is a transparent liquid, which is relatively easy to hydrolyse with water. The hydrolysis of ethylsilicate is accompanied by further condensation of hydrolysis products, including the formation of an amorphous substance $(SiO_2)_x$. In contact with naked flame burns in air. It dissolves well in benzene, toluene or xylene and completely dissolves in ethyl alcohol. The boiling point of ethylsilicate-32 is above 110 °C (at 0.1 MPa).

Technical ethylsilicate-32 should meet the following requirements:

Appearance	Transparent liquid
Density, g/cm^3., not more than	1.1
Viscosity at 20 °C, mm^2/s, not more than	1.5
Content, %:	
of fraction below 110°C, not more than	1.5
of hydrogen chloride, not more than	0.1
of silicon (equivalent to SiO_2)	31-34

Ethylsilicate-32 is widely used in various spheres of economy, e.g. as a cementing and impregnating agent; in the production of molds for precision molding, etc. For example, the cement obtained by mixing ethylsilicate with various fillers (quartz flour, cinder) solidifies in the cold. It is resistant to acids and weak alkali. Water only improves its mechanical properties. After baking at 300°C, the cement becomes resistant to concentrated alkali as well.

Ethylsilicate-32 can be used as an impregnating agent to reduce porosity and to waterproof various materials (brick, graphite, asbestos, leather, textile, plaster). The use of ethylsilicate-32 in the productions of precision molds is particularly important, since it reduces the amount of metals used, as well as the costs of processing the parts. Molds manufacted with ethylsilicate-32 can reproduce the given dimensions of a molded article accurate to 0.2 mm. Ethylsilicate can also be used as a basis for antifoaming agents. If added to solid polymers, it considerably increases their water resistance and adhesion to glass, metal and wood; if it is added to paints, the coatings become much more durable. Construction materials treated with a ethylsilicate-32 solution are considerably more long-lived. Ethylsilicate-32 can be also used as a modifier for carbon-chain elastomers.

The production of ethylsilicate-40 and ethylsilicate-50 is similar to that of ethylsilicate-32. The main difference is that if it is necessary to obtain more condensed products (by hydrolysing tetraethoxysilane with subsequent condensation of the products formed, which naturally increases silicon content in the polymer), one uses ethyl alcohol with an increased water content.

Ethylsilicate-40 and ethylsilicate-50 have generally the same application as ethylsilicate-32. However, they have some advantages compared to ethylsilicate-32. Thus, when used to produce precision molds, they noticeably reduce the consumption, decrease the drying time and increase the durability of the molds. Besides, they can be used to tan dried leather, to impregnate wood in order to make it resistant toward bacterial activity and to improve the quality of colophonic varnishes. Ethylsilicate-40 and ethylsilicate-50 can be used as binding agents in the production of dentures. A mixture of tetraethoxysilane and ethylsilicates can also be used in the production of silicone paints (gouache).

The production of pure tetraethoxysilane is basically put down to the rectification of the mixture of ethylsilicate and tetraethoxysilane. The mixture from depository 21 (see Fig.23) is poured into tank 22, the jacket of which is filled with vapour (0.9 MPa). The vapours of the product rise up, and after tower 23 enter refluxer 24, where they condense. Part of the distillate is used to to reflux the tower and the rest is sent through cooler 25 into receptacles 26 and 27. The first fraction, which boils out below 160 °C, is collected in receptacle 26; as it accumulates, it is poured into depository 28 and then re-loaded into tank 22. The second fraction (160-180 °C) is collected in receptacle 27 and is poured into depository 29 as it accumulates. This fraction is the end product, technical tetraethoxysilane.

Tetraethoxysilane is a transparent colourless liquid with a weak ether odour. The boiling temperature of the pure product is 166.5 °C. It is hydrolysed with water in the presence of catalytic quantities of acids or alkali. It dissolves completely in ethyl alcohol and dissolves well in benzene, toluene, xylene and other organic solvents.

Technical tetraethoxysilane should meet the following requirements:

Appearance	Transparent liquid
Colour according to the iodometric scale, mg I_2 not darker than	0.25
Content, %:	
of chlorine ion	Absent
of main substance, not less than	97
of the total of impurities, not more than	3
including ethylen alcohol, not more than	0.65

Tetraethoxysilane is the main raw stock in the production of valuable silicone products, e.g. heat-resistant and insulating varnishes, or oligoethylsiloxane liquids. The most important chemical property of tetraethoxysilane is its ability to replace the ethoxyl group with an organic radical under the effect of metalorganic compounds, i.e. to form substituted ethers of orthosilicon acid , which are the raw stock in the production of the silicone products mentioned above.

Similarly to tetraethoxysilane, one can also obtain other tetraalkoxysilanes; their physicochemical properties are given below (see Table 10).

Preparation of triethoxysilane

Triethoxysilane is prepared by the etherification of trichlorosilane with anhydrous ethyl alcohol. The reaction occurs stagewise; that is why the process forms various chloroethers, ethoxychlorosilanes.

$$HSiCl_3 \xrightarrow[HCI]{C_2H_5OH} (C_2H_5O)SiHCl_2 \xrightarrow[HCI]{C_2H_5OH} (C_2H_5O)_2SiHCI \xrightarrow[HCI]{C_2H_5OH} (C_2H_5O)_3SiH \quad (2.117)$$

Like tetraethoxysilane, triethoxysilane can be synthesised in a bubble etherificator (see app.6 in Fig. 23) below 60 °C, and rectified in a packed tower. This allows to separate two fractions: below 131 °C (unreacted alcohol with an impurity of triethoxysilane) and 131-135 °C (triethoxysilane).

Etherification in the given conditions due to the liberated hydrogen chloride has some by-processes. the triethoxysilane formed interacts with hydrogen chloride and forms chlorotriethoxysilane, which in its turn reacts with alcohol forming tetraethoxysilane:

$$(C_2H_5O)_3SiH \xrightarrow[H_2]{HCI} (C_2H_5O)_3SiCI \xrightarrow[HCI]{C_2H_5OH} Si(C_2H_5O)_4 \quad (2.118)$$

The following reaction is also possible:

$$(C_2H_5O)_3SiH + C_2H_5OH \xrightarrow[\quad H_2 \quad]{HCl} Si(C_2H_5O)_4 \qquad (2.119)$$

since ethyl alcohol in the presence of hydrogen chloride reacts with triethoxysilane at the Si-H bond.

To increase the yield of triethoxysilane, it is necessary to eliminate the hydrogen chloride formed from the reaction zone as soon as possible. It is hardly conceivable in periodical bubble reactors with a small phase contact surface. Because it is difficult to bring large amounts of heat for HCl desorption to the reactive mixture, packed towers do not allow for a continuous process either. The most convenient apparatus for the etherification of trichlorosilane is film-type, which allows for a continuous process.

The main parts of the apparatus (Fig.24) are reaction chamber 2, receiving chamber 4 and film desorber 3. Agitator 1, mounted into the top vinyl-plast flange of reaction chamber 2, has the shape of a U-tube with a hole in the bottom. Both chambers have heat-resistant rings, which allows one to observe the process visually. The flanges of the receiving chamber and the lower flange of the reaction chamber are manufactured from 1X18H9T steel and are protected with fluoroplast-4 from inside.

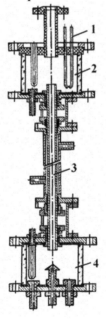

Fig. 24. Film apparatus for the synthesis of triethoxysilane: 1 - agitator; 2 - reaction chamber; 3 - film desorber; 4 - receiving chamber.

Desorber *3* is a film tower with a jacket of common steel (St. 3). The desorber is sealed with two gaskets; the sealing filler is powdered fluoroplast-4. The chokes in the lower flange of receiving chamber *4* are used to introduce nitrogen, to discharge raw triethoxysilane and to take samples. In film apparatuses the contact time of the reactive mixture with hydrogen chloride is reduced from 8-12 hours to 40-60 sec and makes for an intensive heat brought to the flowing liquid film.

Triethoxysilane is a colourless transparent liquid (the boiling point is 131.5°C) with a specific ether odour. It is stable in the absence of humidity; it does not dissolve in water, but is slowly hydrolysed. It can be easily dissolved in organic solvents. Technical triethoxysilane should meet the following requirements:

Appearance	Colourless transparent liquid
Content, %:	
of distillate at 131-135 °C, not less than	98
of hydrogen (active), not less than	0.58
of chlorine, not more than	0.05

Triethoxysilane is mostly used to obtain pure silicon, which is used in semiconductor technology, as well as in the production of aminopropyltriethoxysilane.

Preparation of alkyl(aryl)alkoxysilanes

The substituted ethers of orthosilicon acid, alkyl(aryl)alkoxysilanes or alkyl(aryl)aroxysilanes, have the common formula $R_nSi(OR')_{4-n}$,, where $n = 1 \div 3$, whereas R and R' are any organic radicals; the radicals situated at the silicon atom and in the ether group can be the same or different.

At present substituted ethers of orthosilicon acid are prepared industrially by two main techniques: the Grignard technique, based on the interaction of tetraalkoxysilanes with organomagnesium compounds:

$$Si(OR')_4 + nRMgCI \longrightarrow R_nSi(OR')_{4-n} + nMg(OR')CI \quad (2.120)$$

$$n = 1 \div 3$$

and the etherification of organochlorosilanes with alcohols:

$$R_nSiCI_{4-n} + (4-n)R'OH \xrightarrow[(4-n)HCI]{} R_nSi(OR')_{4-n} \quad (2.121)$$

$$n = 1 \div 3$$

The first technique is currently used only in the production of oligoethylsiloxanes and polyphenylethylsiloxanes, which does not require the cumbersome and labour-consuming stage when individual alkyl- or arylalkoxysilanes have to be purified and extracted; the second technique is used to obtain individual alkyl- or arylalkoxysilanes.

Most chemical properties typical of tetraalkoxy(aroxy)silanes also characterise substituted ethers of orthosilicon acid, with the only difference that the reactivity of the ether group is greatly affected by the type and number of organic radicals directly at the silicon atom.

When organochlorosilanes are etherified with alcohol, the reactive mixture always contains not only the target product, but also unreacted alcohol, hydrogen chloride and products of partial substitution, the chloroethers of orthosilicon acid. If the conditions of the reaction are changed (e.g. the duration of the synthesis or the quantity of the parent components), the qualitative composition of the mixture remains the same. The only changes occur in the quantitative ratio of the components; moreover, the ratio of the product of the mole quantities of alcohol and chloroethers to the product of the quantity of complete substitution products and hydrogen chloride at one and the same temperature remains virtually constant.

This demonstrates that the reactive mixture during the synthesis of alkyl- or arylalkoxysilanes has an equilibrium

$$\diagdown \!\!\!\!\diagup \!\!\! \text{Si}-\text{Cl} + \text{ROH} \underset{k_2}{\overset{k_1}{\rightleftharpoons}} \diagdown \!\!\!\!\diagup \!\!\! \text{Si}-\text{OR} + \text{HCI} \tag{2.122}$$

with an equilibrium constant (K), which equals the ratio of the speed constants of the reverse and forward reactions: $K=k_2/k_1$. In the process of etherification the speed of the forward reaction (k_1) is always higher than that of the reverse reaction (k_2); that is why the mixing of parent reactants yields mostly the target product, and the reaction is accompanied by an intense liberation of hydrogen chloride. This contributes to the displacement of the equilibrium towards the formation of the target product. It should be remembered, however, that although the etherification of silicon tetrachloride is virtually irreversible, the replacement of chlorine atoms in $SiCl_4$ with organic radicals will greatly increase the value of K. Thus, in case of organochlorosilane etherification, we shall find a sharp increase of the reverse reaction, the interaction of ethers with hydrogen chloride. That is why the relative amount of chloroethers at equilibrium will be significantly large for alkoxyorganosilane synthesis than for tetraalkoxy(aroxy)silane synthesis. It follows that optimal yields of substituted ethers of orthosilicon acid should be gained if parent components are sent to react simultaneously and at an as large speed as possible; one should also quickly eliminate the hydrogen chloride formed in the reaction.

Preparation of methylphenyldimethoxysilane

Methylphenyldimethoxysilane is prepared by the etherification of methyl-phenyldichlorosilane with methyl alcohol in the presence of carbamide, the acceptor of hydrogen chloride released during the reaction.

$$CH_3(C_6H_5)SiCI_2 + 2CH_3OH + CO(NH_2)_2 \longrightarrow \qquad (2.123)$$
$$\longrightarrow CH_3(C_6H_5)Si(OCH_3)_2 + CO(NH_2)_2 \cdot 2HCI$$

It is advisable to carry out the reaction in a toluene solution, which dissolves methylphenyldimethoxysilane, with an excess of methyl alcohol, because the by-product, muriatic carbamide, easily dissolves in an excess of methyl alcohol and thus easily separates from target methylphenyldimethoxysilane.

Raw stock: methylphenyldichlorosilane (not less than 99% of the main fraction); methyl alcohol (not less than 0.5% of humidity); technical carbamide (urea), which is a colourless crystal substance with the melting point of 132-135 °C (not less than 0.2% of humidity); oil or coal toluene (the boiling point is 109.5-111 °C, d_4^{20} - 0.865±0.002); activated coal (not less than 0.02% of humidity).

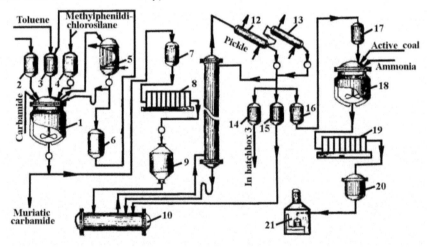

Fig. 25. Production diagram of methylphenyldimethoxysilane: *1, 18* - reactors; *2-4, 17* - batch boxes; *5, 13* - coolers; *6, 7, 14-16* - receptacles; *8, 19--* druck filters; *9, 20* - collectors; *10* - tank; *11* - rectification tower; *12* - refluxer; *21* - draft.

The production of methylphenyldimethoxysilane comprises the following main stages (Fig. 25): reception and preparation of raw stock and equipment; synthesis; distillation of toluene methanol mixture and filtering

of raw material; rectification and purification of methylphenyldimethoxysilane.

Reactor *1* is loaded from batch boxes *2* and *3* with methanol (poison!)
and toluene; carbamide is dried in the draft (not less than 6 hours at 100-
120 °C) and loaded through a filling hatch at agitation. The mixture in the
reactor is agitated and at 20-25 °C methylphenyldichlorosilane is introducted from batch box *4*. Cooler *5* during the introduction of methylphenyldichlorosilane works in the backflow mode; the temperature in the
reactor at that time is maintained below 35°C by sending water into the reactor jacket. The released hydrogen chloride is bound by carbamide. After
all methylphenyldichlorosilane has been introduced, the reactive mixture is
agitated for 4-12 hours at 30-35 °C until pH is 3.5÷6.5. After that the agitator is switched off and the reactive mixture is settled for 4-6 hours.

If pH is less than 3.5, the reactive mixture is kept for 1-2 hours more at
30-35 °C and then checked again. If pH is still below 3.5, one adds 1-2%
of carbamide into the reactive mixture and holds it for 1-2 hours.

The bottom layer, the methanol solution of muriatic carbamide, is
poured off. The top layer, the toluene solution of methylphenyldimethoxysilane, is left in the reactor, backflow cooler *5* is switched in the direct
operation mode and the reactor jacket is filled with vapour. The contents of
the apparatus are heated and at the tank temperature below 150 °C the mixture of methyl alcohol and toluene is distilled into receptacle *6* until the
distillation stops (the distillation speed is regulated by sending vapour into
the reactor jacket *7*). The toluene methanol mixture (97% of toluene and
3% of methyl alcohol) is sent from receptacle *6* into batch box *3*, from
where it enters reactor *7*. After the toluene methanol mixture is distilled
completely, the contents of reactor *7* are cooled with water sent into the
jacket down to 30 °C and sent by nitrogen flow into receptacle *7*.

Raw methylphenyldimethoxysilane is separated from muriatic carbamide in druck filter 8, which has been filled with coarse calico, filter paper and glass cloth and compressed with nitrogen (0.3 MPa). After filtering, the raw material under the nitrogen pressure of 0.15-0.3 MPa is sent
into collector *9*. When the filtering is finished, the pressure in receptacle *7*
is reduced to atmospheric; the druck filter is cleaned from the cake and re-
filled.

The raw methylphenyldimethoxysilane from collector *9* and the intermediate fraction from receptacle *15* self-flow into tank *10*. After the loading is finished, the heater of the tank is filled with vapour. After the reflux
appears in the box, the tower operates in the "self-serving" mode for 1 or 2
hours; then the rest of methyl alcohol and toluene are distilled. The vapours pass through tower *11*, refluxer *12* and cooler *13*; the condensate enters receptacle *14*. Methyl alcohol and toluene are distilled when the tem-

perature in the higher part of the tower is below 115 °C. From receptacle *14*, the return toluene containing some methyl alcohol under nitrogen pressure is returned into batch box *3*. After the separation of toluene (containing some methyl alcohol) the contents of tank *10* are cooled with water sent into the heater down to 80-100 °C; a vacuum pump is switched on to create vacuum in the system (the residual pressure in the higher part of the tower is 65-200 GPa) and the contents of the tank are heated by sending vapour into the heater. When the reflux appears, tower *11* operates in the "self-serving" mode for 1-2 hours After that the intermediate fraction is collected in receptacle *15* when the temperature in the higher part of the tower is below 135 °C.

The distillate in tower *11* is periodically sampled. At first the analysis is carried out until the density reaches 1.002 g/cm^3, after that one determines the main substance content. When the main substance content is at least 93% and that of the head fraction does not exceed 1-2%, the separation of the intermediate fraction is completed.

After that, the target fraction, methylphenyldimethoxysilane, is collected into receptacle *16* when the temperature in the higher part of tower *11* is below 140 °C. When the methylphenyldimethoxysilane content in the target fraction is less than 93%, the separation is completed. After that, the tank residue is cooled by sending water into heater 70 down to 30-60 °C. The residue can be used in the production of low polydispersity oligomethylphenylsiloxanes.

The methylphenyldimethoxysilane is sent from receptacle *16* by nitrogen flow (0.07 MPa) into batch box *11*, from where the product flows into reactor *18* to be purified with active coal.

First druck filter *8* is filled with coarse calico, filter paper and glass cloth and compressed with nitrogen (0.3 MPa). The pressure should not fall for 10 minutes; after that it is reduced to atmospheric. Then reactor *18* is filled from batch box *17* with self-flowing methylphenyldimethoxysilane and newly dried active coal (1-2% of the product weight). After that, the agitator is switched on and the reactor contents are heated to 60-70 °C by sending vapour into the jacket. At this temperature the mixture is agitated for 2-4 hours and then analysed.

In case of negative results (pH<4) methylphenyldimethoxysilane is treated with ammonia for 10-20 sec; after that it is agitated for 1 or 2 hours and analysed again. If pH is still below 4, the ammonia treatment is repeated.

At pH 4÷6.5 the product is filtered through druck filter *19* into collector *20* under the nitrogen pressure of 0.15-0.3 MPa. The dispensing of methylphenyldimethoxysilane should be carried out in draft *21*. The workers should use rubber gloves.

Methylphenyldimethoxysilane is a colourless transparent liquid (the boiling point is 199-200 °C). Technical methylphenyldimethoxysilane should meet the following requirements:

Appearance	Colourless transparent liquid; no mechanical impurities
Density at 20 °C, g/cm^3	1.003-1.010
n_D^{20}	1.4790-1.4800
pH of the aqueous extract	4-7
Fraction content, % (vol.):	
with the boiling point of below 200°C, not more than	1
with the boiling point of 200-204°C, not more than	93

Methylphenyldimethoxysilane is used as a stabiliser (antistructuring additive) in the production of rubber compounds based on silicone elastomers and highly active fillers. Introducing up to 10% (weight) of methylphenyldimethoxysilane into a rubber mixture improves the physicochemical properties of vulcanised rubbers and helps to preserve the technological characteristics of the compounds in storage.

Table 10. Physicochemical properties of the most important tetraalkoxysilanes and alkylalkoxysilanes

Substance	Boiling point, °C	d_4^{20}	n_D^{20}
$Si(OCH_3)_4$	121-122	1.0232	1.3683
$Si(OC_2H_5)_4$	166,5	0.8330 (at 17 °C)	1.3852
$Si(OC_3H_7-n)_4$	225-227	0.9180	1.4019
$Si(OC_4H_9-n)_4$	173 (at 26.6 GPa)	0.9130 (at 25 °C)	1.1431
$Si(OC_4H_9-iso)_4$	256-260	0.9530 (at 15 °C)	-
$Si(OCH_2CH=CH_2)_4$	115-116 (at 16 GPa)	0.9842 (at 17 °C)	1.4329
$CH_3Si(OC_2H_5)_3$	151	0.9380	1.3869
$(CH_3)_2Si(OC_2H_5)_2$	111	0.8900	1.3839
$(CH_3)_3SiOC_2H_5$	75	0.7573	1.3741
$C_2H_5Si(OC_2H_5)_3$	159	0.9407	1.3853
$(C_2H_5)_2Si(OC_2H_5)_2$	155	0.8752 (at 0 °C)	-
$(C_2H_5)_3SiOC_2H_5$	153	0.8414	-
$CH_3(C_6H_5)Si(OCH_3)_2$	199 (at 1000 GPa)	0.9934	1.4694

Similarly to methylphenyldimethoxysilane, one can also obtain other alkyl(aryl)alkoxysilanes; their physicochemical properties are given in Table 10.

2.3.3. Preparation of aromatic ethers of orthosilicon acid and their derivatives

Due to their high boiling points, aromatic ethers of orthosilicon acid and their derivatives are of great interest to industry and can be used as heat carriers. Tetraphenoxysilane and its derivatives are the most important among them.

Preparation of tetraphenoxysilane and 1,3-bis(triphenoxysiloxy)benzene

Tetraphenoxysilane is obtained by the reaction of silicon tetrachloride with phenol:

$$SiCl_4 + 4\ C_6H_5OH \xrightarrow[-HCl]{} Si(OC_6H_5)_4 \qquad (2.124)$$

It is advisable to carry out the reaction in a solvent (toluene) medium with a small amount [0.2-0.5% (weight)] of a catalyst (dimethylformamide), which allows to reduce the temperature down to 120 °C and decrease the time of synthesis 6-fold.

1,3-bis(triphenoxysiloxy)benzene can be obtained by the re-etherification of tetraphenoxysilane with resorcinol at 220-300 °C:

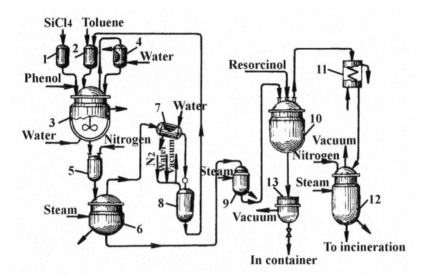

Fig.   26.   Production   diagram   of   tetraphenoxysilane   and   1,3-bis(triphenoxysiloxy)benzene: *1, 2* - batch boxes; *3* - etherificator; *4, 11* - back-flow condensers; *5, 8, 9* - collectors; *6* -distillation tank ; *7* - direct condenser; *10* - re-etherificator; *12* - receptacle; *13* - nutsch filter.

Raw stock: silicon tetrachloride (55-59°C fraction, not more than 0.2% of free chlorine, not more than 0.001% of Fe); phenol (the melting point is 40.9-41 °C);   resorcinol (the melting point is 110-111 °C);   dimethylfor-mamide (152-154 °C fraction, d_4^{20}= 0.945±0.005);   coal toluene (109.5-111 °C fraction , d_4^{20}= 0.865±0.002).

Preparation of tetraphenoxysilane and 1,3-bis(triphenoxysiloxy)benzene (Fig. 26) comprises two main stages: the production of tetraphenoxysilane and the re-etherification of tetraphenoxysilane with resorcinol.

From the melting batch box, etherificator *3* receives a necessary amount of phenol. From batch box *2* it also receives toluene, and through the fill-ing hatch, dimethylformamide [0.2-0.3% (weight) of the phenol quantity]. After that, silicon tetrachloride is sent from batch box *7* under a layer of phenol solution in toluene; at that time the temperature is maintained at 7-10 °C (due to the endothermicity of the process and the changing speed of SiCl₄ supply). After introducing the necessary amount of SiCl₄ the tem-perature is gradually (at the speed of 20 degrees an hour) raised to 120 °C and the mixture is agitated at this temperature for 2 hours.

The reactive mixture is cooled for 1 hour to 60 °C and poured into col-lector *5* and then into distillation tank *6*. After the reactive mixture is loaded, the tank is heated with vapour, cooler *7* is filled with water, and toluene is distilled into collector *5*. The distillation is continued up to 170

°C. After that the reactive mixture is cooled to 60 °C and a residual pressure of 50-80 GPa is created in the tank for the complete distillation of toluene and excess phenol. The distillation is continued up to 180 °C; after that, the water is no longer supplied to the cooler. The distilled toluene and phenol mixture is sent as it accumulates from collector *8* into batch box *2*.

Pure *tetraphenoxysilane* is a colourless crystal product (the melting point is 47-48°C, the boiling point is 415-420°C). It can be easily dissolved in organic solvents.

The residue from tank 6, tetraphenoxysilane, flows at 60 °C into collector *9* or directly into apparatus *10* for re-etherification; this apparatus is also filled with a calculated amount of resorcinol. The reaction occurs at 220-300 °C. First, apparatus 10 is heated for 2 hours to 220 °C; then, at the speed of 20 degrees an hour the temperature is increased to 300 °C. The reactive mixture is kept at this temperature for 2 hours. The vapours of liberated phenol are directed into condenser *11*, which is heated with hot water (70-80 °C), and collected in receptacle *12*.

After 2 hours of standing at 300 °C the reactive mixture in apparatus *10* is cooled to 100 °C; a residual pressure of 13-20 GPa is created to distil the residual phenol. The distillation is carried out until the temperature is 290-300 °C. After that the reactive mixture is cooled to 60 °C and sent to nutsch filter *13* to filter the product, 1,3-bis(triphenoxysiloxy)benzene.

Table 11. Physicochemical properties of tetraaroxysilanes and their derivatives

Substance	Boiling point, °C	Melting point, °C
$Si(OC_6H_5)_4$	415-420	47-48
$Si(OC_6H_4CH_3-o)_4$	338-342 (at 0.26 GPa)	140
$Si(OC_6H_4CH_3-p)_4$	347-351 (at 0.26 GPa)	83
$C_6H_5Si(OC_6H_5)3$	248 (at 17.3 GPa)	40
$(C_6H_5)_2Si(OC_6H_5)_2$	-	70-71
$C_6H_4[(OC_6H_5)_3Si]_2$	240 (at 13.3 GPa)	-

1,3-bis(triphenoxysiloxy)benzene is a liquid ranging in colour from light yellow to light brown. Technical 1,3-bis(triphenoxysiloxy)benzene should meet the following requirements:

Density at 20 °C, g/cm^3	1.195-1.210
Viscosity at 20 °C, mm^2/s	70-120
Temperature, °C	
of boiling	240 (at 13.3 GPa)
of flash	235
Mechanical impurities	Absent

1,3-bis(triphenoxysiloxy)benzene can be used as a heat carrier at operating temperatures below 350 °C. Unlike individual tetraphenoxy- or phenyltriphenoxysilane, this product is liquid, which considerably facilitates its use.

Other tetraaroxysilanes and their derivatives can be obtained similarly to tetraphenoxysilane and 1,3-bis(triphenoxysiloxy)benzene. The physico-chemical properties of some tetraaroxysilanes and their derivatives are given in Table 11.

2.3.4. Preparation of substituted ethers of orthosilicon acid with an aminogroup in the organic radical

Substituted ethers of orthosilicon acid with an aminogroup in the organic radical are a rather extensive class of silicone compounds, which is finding more and more applications. These substances also proved to be active hardeners for organic and silicone polymers, stabilisers for polyolefines and binding agents for fiberglass plastics, as well as modifiers for organic and silicone products.

Preparation of methyl(phenylaminomethyl)diethoxysilane

Methyl(phenylaminomethyl)diethoxysilane (AM-2) is prepared by the etherification of methyl(chloromethyl)dichlorosilane with anhydrous ethyl alcohol and subsequent amidation of the etherified product with aniline.

The etherification stage entails only the substitution of chlorine atoms which are situated directly at the silicon atom, forming methyl(chloromethyl)diethoxysilane (the chlorine in the chloromethyl radical is not affected):

$$CH_3(ClCH_2)SiCl_2 + 2C_2H_5OH \xrightarrow[2HCl]{} CH_3(ClCH_2)Si(OC_2H_5)_2 \quad (2.126)$$

The chlorine in the chloroalkyl radical can be substituted only under the influence of alcoholates of alkali metals. That ensures the substitution of the chlorine atoms both at the silicon atom and in the chloroalkyl radical.

$$CH_3(ClCH_2)SiCl_2 \xrightarrow[3NaCl]{3RONa} CH_3(ROCH_2)Si(OR)_2 \quad (2.127)$$

At the stage of methyl(phenylaminomethyl)diethoxysilane amidation with aniline, the chlorine atoms in the chloromethyl radical are replaced with a phenylamine group, which forms methyl(phenylaminomethyl)diethoxysilane.

$$CH_3(ClCH_2)Si(OC_2H_5)_2 + 2\ C_6H_5NH_2 \longrightarrow \qquad (2.128)$$

$$\longrightarrow CH_3(C_6H_5NHCH_2)Si(OC_2H_5)_2 + C_6H_5NH_2 \bullet HCl$$

Raw stock: methyl(chloromethyl)dichlorosilane (119-123 °C fraction, 42.5-44% of hydrolysed chlorine., d_4^{20} = 1.280-5-1.290); anhydrous ethyl alcohol (not less than 99.8% of C_2H_5OH); aniline (the boiling point is 181-184 °C, d_4^{20} = 1.022, not less than 0.5% of humidity); toluene (109.5-111 °C fraction, d_4^{20} = 0.865+0.002, not less than 0.3% of humidity).

The production process of methyl(phenylaminomethyl)diethoxysilane (Fig. 27) comprises two main stages: the etherification of methyl(chloromethyl)dichlorosilane with anhydrous ethyl alcohol and amidation of methyl(chloromethyl)diethoxysilane with aniline.

Methyl(chloromethyl)diethoxysilane is synthesised in etherificator *1*, which is an enameled steel apparatus with a removable cover, spherical bottom and jacket for heating (with a heat carrier or vapour). From batch box *2* the etherificator receives a necessary amount of anhydrous ethyl alcohol; after that cooler *4* is filled with water.

The lead bubbler of the etherificator is filled with dehydrated nitrogen; then the original methyl(chloromethyl)dichlorosilane is slowly introduced from batch box *3*. After a necessary amount of methyl(chloromethyl)dichlorosilane has been introduced, the jacket of the etherificator is filled with a heat carrier and within 2 or 3 hours the reactive mixture is heated to 80-85 °C. The mixture is held at this temperature for 6-8 hours; the condensate is always returned through cooler *4* into the etherificator.

During the introduction of methyl(chloromethyl)dichlorosilane, as well as during the heating and holding of the reactive mixture there is a possibility of a profuse discharge of hydrogen chloride, which raises the temperature and pressure in the etherificator.

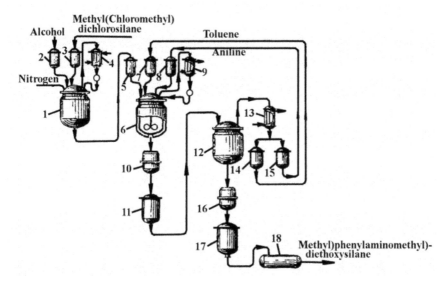

Fig. 27. Production diagram of methyl(phenylaminomethyl)diethoxysilane: *1* - etherificator; *2, 3, 5, 7, 8* - batch boxers; *4, 9, 13* - coolers; *6* - reactor; *10, 16* - nutsch filters; *11, 17*- collector boxes; *12*- distillation tank; *14, 15* - receptacles; *18* - depository

That is why one should pay close attention to the speed of methyl(chloromethyl)dichlorosilane and to the temperature in the etherificator, and constantly regulate these parameters. After 6-8 hours of standing the etherificator is cooled to about 40°C, and the content of hydrolysed chlorine is determined. The reaction is considered finished when the content of hydrolysed chlorine does not exceed 3%.

The obtained methyl(chloromethyl)diethoxysilane is sent by nitrogen flow into batch box *5*, from where it is sent for amidation into reactor *6*, which is an enameled cylindrical apparatus with a jacket and an agitator. First, the reactor is loaded with a necessary amount of methyl(chloromethyl)diethoxysilane from batch box *5*; then cooler *9* is filled with water, the agitator is switched on and anilinee is introduced from batch box *8*.

We should note that the reactor receives an excess of aniline, so that the amidation of methyl(chloromethyl)diethoxysilane is accompanied by the binding of released hydrogen chloride. When introducing aniline, one should keep a close watch on the temperature in the reactor, since the process takes place with the liberation of heat. After aniline has been introduced, the reactive mixture is heated to 135-140 °C within 2-3 hours and agitated at this temperature for 8-8.5 hours. After that reactor *6* is cooled down to 80 °C, receives toluene to dilute the reactive mixture, and is cooled to 30 °C.

The product of amidation, methyl(phenylaminomethyl)diethoxysilane, is separated from muriatic aniline in nutsch filter *10*. Before the filtering, the filtering material on the filter grate is thoroughly packed (several layers of cloth and filter paper). The reactive mixture is poured from reactor *6* onto the filter, vacuum is created in collector box *11* and the suction is maintained until the layer of muriatic aniline on the filter dries. After that, muriatic aniline is additionally washed with toluene for a more thorough extraction of methyl(phenylaminomethyl)diethoxysilane. Part of muriatic aniline may pass through the filtering material; therefore, the filtrate in collector box *11* is kept for 2 days.

Then, the solution of methyl(phenylaminomethyl)diethoxysilane in the mixture of toluene and aniline is sent by nitrogen flow into distillation tank *12*. There a residual pressure of 25-35 GPa is created, cooler *13* is filled with water, the agitator is switched on and the tank is slowly heated. The distilled toluene enters receptacle *14* through run-down box *4*; the distillation continues up to 100°C. The toluene in the receptacle is sampled to determine the density (at the density of 0.854-0.870 g/cm³ the distillation is considered complete). The distilled toluene from receptacle *14* is sent in vacuum into batch box *7* and used for another process of amidation.

After the distillation of toluene begins the distillation of aniline. The distillation of aniline is conducted up to 160-165°C in the tank (under a residual pressure of 30 GPa) and collected in receptacle *15*. From there aniline is sent in vacuum into batch box *8* and used for another process of amidation. The target product which remains in tank *12*, methyl(phenylaminomethyl)diethoxysilane, is cooled down to 30-50 °C, filtered in nutsch filter *16* and collected in box *17*. After 2-3 days of standing, methyl(phenylaminomethyl)diethoxysilane is sent from box *17* into depository *18*.

Methyl(phenylaminomethyl)diethoxysilane is a trasparent liquid ranging in colour from light yellow to light brown (the boiling point is 152-153 °C at 20 GPa). It can be easily dissolved in organic solvents; is hydrolysed in the presence of water.

Technical methyl(phenylaminomethyl)diethoxysilane should meet the following requirements:

Appearance	Transparent liquid with a specific amine odour*
Density at 20 °C, g/cm³, not more than	1.035
Viscosity at 20 °C, mm²/s, not more than	10
n_D^{20}, not more than	1.5100
The fraction composition under a residual	

pressure of 16 GPa, %:

below 115 °C, not more than	5
115-130 °C, not more than	70
above 130 °C, not more than	25

*After 1 month of storing some sediment ($C_6H_5NH_2 \cdot HCl$) in the amount below 1% (weight).can be tolerated.

Methyl(phenylaminomethyl)diethoxysilane is used as a modifying agent for some polymethylphenylsiloxane varnishes and polyolefines, as well as to obtain binding agents for fiberglass plastics.

Preparation of diethylaminomethyltriethoxysilane

Diethylaminomethyltriethoxysilane (ADE-3) is synthesised in two stages. At the first stage, the etherification of chloromethyltrichlorosilane with an-hydrous ethyl alcohol forms chloromethyltriethoxysilane.

$$ClCH_2SiCl_3 + 3C_2H_5OH \xrightarrow[3HCl]{} ClCH_2Si(OC_2H_5)_3 \qquad (2.129)$$

At the second stage, chloromethyltriethoxysilane is subjected to the amidation with diethylamine, which forms diethylaminomethyltriethoxysi-lane:

$$ClCH_2Si(OC_2H_5)_3 \xrightarrow[(C_2H_5)_2NH \bullet HCl]{2(C_2H_5)_2NH} (C_2H_5)_2NCH_2Si(OC_2H_5)_3 \qquad (2.130)$$

Raw stock: chloromethyltrichlorosilane (116-119°C fraction , 56.8-58.8% of hydrolysed chlorine, d_4^{20}= 1.440-5-1.470); anhydrous ethyl alcohol (not less than 99.8% of C_2H_5OH); diethylamine (d_4^{20}=0.7056, d_4^{20}=1.3873); toluene {d_4^{20} = 0.865±0.002, there should be no humidity).

Diethylaminomethyltriethoxysilane production (Fig.28) comprises three main stages: the etherification of chloromethyltrichlorosilane with anhy-drous ethyl alcohol; the amidation of chloromethyltriethoxysilane with di-ethylamine; the vacuum distillation of the amidation product with the ex-traction of diethylaminomethyltriethoxysilane.

The etherification of chlorophenyltrichlorosilane is carried out in steel enameled apparatus *1* with a bubbler and a jacket (for heating or cooling). First, the etherificator receives from batch box *2* with vacuum a necessary amount of anhydrous ethyl alcohol, and toluene from batch box *3*. After that, backflow condenser *5* is filled with water, and chloromethyltrichloro-silane is loaded from batch box *4* into the etherificator at such speed that the temperature in the apparatus does not exceed 45-50°C. After the whole of chloromethyltrichlorosilane has been introduced, the reactive mixture is heated to 80 °C. The mixture is held at 80-85 °C for 7 hours; the conden-sate is returned through cooler *5* into the etherificator. Then the condensate is no longer returned into the etherificator, cooler 5 is switched in the direct operation mode, the etherificator jacket is filled with a heat carrier or vapour, and the alcohol-toluene mixture is distilled into receptacle 6; the distillation is continued up to 115 °C.

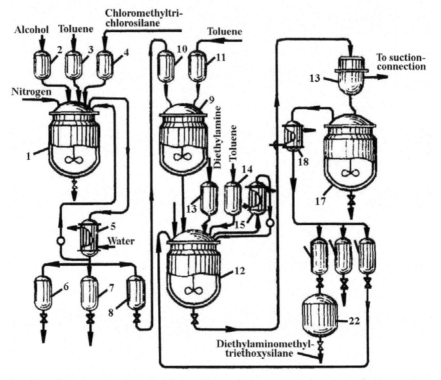

Fig. 28. Production diagram of diethylaminomethyltriethoxysilane: *1* - etherificator; *2, -4, 10, 11, 13, 14* - batch boxers; *5, 15, 18* - coolers; *6-8, 19-21* - receptacles; *9* - agitator; *12* - reactor; *16* - nutsch filter; *17*- distillation tank; *-22* - collector.

The residual pressure of 52 GPa is created in the system and the etherificator is slowly heated. The first fraction to be distilled into receptacle 7 is the below 105° C with a content of hydrolysed chlorine below 1%; the second is the target fraction, chloromethyltrichlorosilane (below 130 °C), which is distilled into receptacle 8. When the target fraction is being distilled, dehydrated nitrogen is passed through the bubbler of the etherificator at 0.01 MPa.

The first distillates of the target fraction should be sampled for the content of hydrolysed chlorine and the refraction index. The separation of the target fraction should begin when the content of hydrolysed chlorine is below 1% and $n_D^{20} = 1.4145$; the end of the distillation is determined by an abrupt decrease in the quantity of the condensate if the temperature is raised.

The amidation of chloromethyltrichlorosilane is carried out in steel enameled reactor *12* with an agitator and a jacket. In agitator *9*, a mixture of chloromethyltriethoxysilane and toluene is prepared. Then, apparatus *12* is loaded with diethylamine from batch box *13* and with toluene from batch box *14*; backflow condenser *15* is filled with water and the reactor jacket is filled with vapour. The mixture in the apparatus is heated to 35 °C and at agitation one starts to introduce ether and toluene mixture from agitator *9* at such speed that the temperature of the reactor does not exceed 50 °C. At the same time, the apparatus is constantly fed with dehydrated nitrogen. After the whole mixture has been introduced, the reactive mixture is kept at 75-85 °C for 15-18 hours.

After standing in reactor *12*, the mixture is cooled there down to 30 °C and filtered in nutsch filter *16* from diethylaminochloride. The filtrate is sent into tank *17* for distillation, and the filter cake is washed with toluene to eliminate amidation products as completely as possible. After the filtrate has been loaded, cooler *18* is filled with water, and the tank agitator is switched on. A residual pressure of 40-55 GPa is created in the system and the tank jacket is filled with a heat carrier or vapour. First, receptacle 20 receives toluene (below 60-65 °C); after separating toluene, amidation products are distilled into fractions. Receptacle 21 receives the intermediate fraction (below 106 °C); the distillation is monitored by the refraction index. At $n_D^{20} = 1.4210 \div 1.4230$ the target fraction, diethylaminomethyltriethoxysilane, is separated into receptacle *19*. The distillation is continued up to 140 °C. As it accumulates, the intermediate fraction from receptacle *21* is sent into apparatus *12* for repeated amidation, and the ready product, diethylaminomethyltriethoxysilane, is sent after additional filtering (in case there is a filter cake) from receptacle *19* into collector *22*.

Diethylaminomethyltriethoxysilane is a colourless transparent liquid (the boiling point is 196-198 °C) with a specific amine odour. It can be easily dissolved in organic solvents. In the presence of light it slowly breaks down.

Technical diethylaminomethyltriethoxysilane should meet the following requirements:

Appearance	Transparent colourless or yellowish liquid with a specific amine odour*
Density at 20 °C, g/cm^3	0.900-0.918
n_D^{20}	1.4155-1.4180
Content, %:	

of silicon (equivalent to SiO_2)	23.0-25.5
of chlorine, not more than	0.5
of nitrogen	4.5-5.7
Solubility in distilled water (the quantity of dis- solved product, %)	5
Stability of filtered 5%-solutions, h, not less than	24

*When storing, $(C_2H_5)_2NH \cdot HCl$ flakes can be tolerated.

Diethylaminomethyltriethoxysilane is used as a hardener for organic and silicone polymers.

Similarly to methyl(phenylaminomethyl)diethoxysilane and diethylaminomethyltriethoxysilane, one can also obtain other substituted ethers of orthosilicon acid with an aminogroup in the organic radical, e.g. α-aminomethyltrialkoxysilanes, as well as 1-aminohexamethylene-6-aminomethyltriethoxysilane.

Preparation of 1-aminohexamethylene-6-aminomethyltriethoxysilane

1-aminohexamethylene-6-aminomethyltriethoxysilane (AHM-3) is prepared by the etherification of chloromethyltrichlorosilane with anhydrous ethyl alcohol and subsequent amidation of the etherified product with hexamethylenediamine in toluene medium.

When chloromethyltrichlorosilane is etherified with ethyl alcohol, the chlorine atoms which are situated directly at the silicon atom are substituted with ethoxyl groups.

$$ClCH_2SiCl_3 + 3C_2H_5OH \xrightarrow[3HCl]{} ClCH_2Si(OC_2H_5)_3 \qquad (2.131)$$

After that, the amidation of the formed chloromethyltriethoxysilane with an excess of hexamethylenediamine yields 1-aminohexamethylene-6-aminomethyltriethoxysilane:

$$ClCH_2Si(OC_2H_5)_3 + 2\ H_2N(CH_2)_6NH_2 \longrightarrow \qquad (2.132)$$

$$\longrightarrow H_2N(CH_2)_6NHCH_2Si(OC_2H_5)_3 + H_2N(CH_2)_6NH_2 \cdot HCl$$

Raw stock: chloromethyltrichlorosilane [fraction with the boiling point of 116-120 °C not less than 85% (vol.), hydrolysed chlorine 56.8-58.8%, $d_4^{20} = 1.440 \div 1.470$ g/cm^3]; anhydrous ethyl alcohol (not less than 99.8% of C_2H_5OH); hexamethylenediamine (colourless crystal substance with the melting point of 39-42 °C, $d_4^{20} == 0.825 \div 0.828$ g/cm^3); toluene $(d_4^{20} = 0.865 \pm 0.002$ g/cm^3, there should be no humidity).

AHM-3 production comprises the following main stages: the etherification of chloromethyltrichlorosilane with anhydrous ethyl alcohol; the extraction of chloromethyltriethoxysilane; the amidation of chloromethyl-triethoxysilane with hexamethylenediamine in the toluene medium; the vacuum distillation of toluene and filtration of AHM-3.

The technological diagram of the production of 1-aminohexamethylene-6-aminomethyltriethoxysilane is similar to the production diagram of ADE-3 (see Fig. 28).

The first stage, the etherification of chloromethyltrichlorosilane with anhydrous ethyl alcohol, is usually carried out at 80-85 °C for 6-8 hours until the content of hydrolysed chlorine does not exceed 2%. The chloro-methyltriethoxysilane (target fraction) formed is distilled at 125-145 °C *(P* = 52 GPa). When the target fraction is being distilled, dehydrated nitrogen is passed through the bubbler of the etherificator at 0.01 MPa.

The second stage, the amidation of chloromethyltriethoxysilane with hexamethylenediamine solution in toluene is conducted at 90-120 °C for 3.5-5 hours. The 1-aminohexamethylene-6-aminomethyltriethoxysilane (raw AHM-3) formed is filtered in the nutsch filter and purified from the cake of hydrogen chloride hexamethylenediamine. Then toluene is distilled from raw AHM-3. Toluene is distilled in vacuum *(P* = 52+160 GPa) up to 145 °C (liquid). The distillation is considered complete when the density of the product ranges from 0.945 to -0.965 g/cm^3 and n_D^{20} =1.440-1.446. The   prepared   product,   1-aminohexamethylene-6-aminomethyltriethoxysilane, is sent into the collector of the prepared product after additional filtering.

1-Aminohexamethylene-6-aminomethyltriethoxysilane is a transparent slightly yellowish liquid with a specific amine odour; it is not toxic. It can be easily dissolved in organic solvents.

Technical   1-aminohexamethylene-6-aminomethyltriethoxysilane should meet the following requirements:

Appearance	Transparent liquid with colour ranging from light green to dark brown Mechanical impurities are absent Opalescence is tolerated.
Density at 20 °C, g/cm^3	0.945-0.965
Refraction index at 20 °C	1.440-1.446
Content, %:	
of silicon (equivalent to SiO$_2$)	19.0-22.0
of chlorine (total)	not more than 0.5

AHM-3 is used for dressing glass cloth, as well as a component of a glass fiber sizer.

γ-Aminopropyltrialkoxysilanes can be prepared by the reaction of hy-drosilylation of allylamine with trialkoxysilanes; e.g. this is the industrial method for obtaining γ-aminopropyltriethoxysilane.

Preparation of γ-aminopropyltriethoxysilane

γ-Aminopropyltriethoxysilane (AHM-9) is prepared by the hydrosilylation of allylamine with triethoxysilane in the presence of a catalyst, a 0.2-0.4 N. solution of chloroplatinic acid in a mixture of isopropyl alcohol and allyl-glycidil ether:

$$(C_2H_5O)_3SiH + CH_2=CHCH_2NH_2 \longrightarrow H_2N(CH_2)_3Si(OC_2H_5)_3 \quad (2.133)$$

It should be noted that under such conditions hydrosilylation occurs not only according to Farmer's rule, but also by Markovnikov's rule, forming β-aminoisopropyltriethoxysilane:

$$(C_2H_5O)_3Si-\underset{\underset{CH_3}{|}}{C}HCH_2NH_2$$

and consequently forming a mixture of isomers.

Besides, the synthesis of rectification of the obtained products may be accompanied by certain by-processes, when the unreacted triethoxysilane is disproportioned to tetraethoxysilane and silane (SiH_4).

Raw stock: triethoxysilane (not less than 98% of the 131-135 °C fraction, not less than 0.58% of H_{act}); allylamine (not less than 99% of the 53-54 °C fraction, not more than 0.4% of humidity); isopropyl alcohol (not less than 99.85% of the 82.3-83 °C fraction); allylglycidil ether (the density is 0.9695-0.9750 g/cm^3, n_D^{20} = 1.4325÷1.4355);

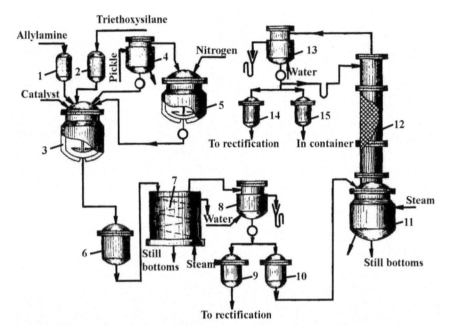

Fig. 29. Production diagram of γ-аминопропилтриэтоксисилана: *1, 2* - batch boxes; *3, 5* - reactors; *4, 8, 13* - refluxers; *6, 9, 10, 14, 15* - collectors; *7, 11* - tanks; *12*- rectification tower

The solution of chloroplatinic acid in a mixture of isopropyl alcohol and allylglycidil ether (the concentration of 0.2-0.4 N.).

In the production of γ-аминопропилтриэтоксисилана by the hydrosilylation technique (Fig. 29) reactor *3* receives allylamine from batch box 1 by nitrogen (pressure below 0.07 MPa) and triethoxysilane from batch box *2*. The mixture of allylamine and triethoxysilane in the weight ratio of 1:3 is agitated for 15-20 minutes; then apparatus 3 is loaded through a hatch with a necessary amount (0.5-0.8% of the weight of the reactive mixture) of the catalyst, which consists of 0.2-0.4 N. solution of chloroplatinic acid in isopropyl alcohol and allylglycidil ether in the ratio of (1÷1,2):1. Refluxer *4* and reactor *5* are filled with a salt solution (-15 °C), and the jacket of reactor *3* is filled with vapour under the pressure of 0.6 MPa. When after refluxer *4* there is reflux in the run-down box, the heating of reactor *3* is regulated so that the reflux is not too turbulent. The temperature in apparatus *3* is gradually increased to 120-125 °C and is maintained at this level during the synthesis.

Since the synthesis is accompanied by the by-process of triethoxysilane disproportioning, which forms monoethoxysilane and other hydrogen-containing products which are flammable in air, the synthesis should be

carried out in nitrogen and periodically checked in case the mixture inflames at the end of the air vent (placed above apparatus 5).

The synthesis in reactor *3* is conducted at 120-125 °C and constant agitation for 3-6 hours. After that, the apparatus is cooled down to 25-30 °C
and sampled to determined the content of the main substance. If the content of the main substance is below 60%, the reactor is heated again to
120-125 °C, and the reactive mixture is agitated for another 2-4 hours. If
the main substance content is 60% or more, the reactive mixture is poured
into collector *6*.

The pre-distillate fraction, which accumulates in reactor *5* after several
syntheses, is sampled to determine the content of unreacted allylamine and
triethoxysilane, and then loaded into reactor *3*. In this case the loading of
raw stock for synthesis is re-calculated depending on the quantity and
composition of the pre-distillate.

The vacuum distillation of γ-aminopropyltriethoxysilane is carried out
in tank *7*, which is a steel apparatus with a coil and a jacket welded to the
bottom. The coil and jacket are fed with vapour, and refluxer *8*, with water
for cooling. A residual pressure of 80-105 GPa is created in the system and
the tank is vacuum-filled with raw γ-aminopropyltriethoxysilane from collector *6*. When there is reflux in the run-down box after the refluxer, fraction I pre-distillate) is separated into collector *9*. The distillation continues
till the temperature in tank 7 is 135 °C; after that the tank is cooled to 90-
100 °C and a residual pressure of 3-10 GPa is created in the system. At the
given residual pressure and tank temperature not higher than 180 °C, fraction II (raw γ-aminopropyltriethoxysilane) can be separated into collector
10. After the separation of fraction II, the tank is cooled down to 30-40 °C
and the tank residue is poured into containers.

From collector *10*, the raw substance is sent to vacuum rectification into
tank *11*, the jacket of which is filled with vapour under 1 MPa. Rectification is carried out in tray towers *12* (or towers filled with Raschig rings)
which are heated with 0.6 MPa vapour (the diagram shows one). A residual pressure of 3-7 GPa is created in the rectification system. When there
is reflux in the run-down box after water-cooled refluxer *13*, tower *12* operates in the "self-serving" mode until constant temperature is established
at the tower top; after that, the pre-distillate is separated into collector *14*.
This collector is periodically sampled for tetraethoxysilane content. When
the content of tetraethoxysilane is below 3%, the target product, γ-
aminopropyltriethoxysilane, is collected into collector *15*. During the distillation the temperature in the tank is maintained at 140-200 °C, and in the
top part of the tower, 90-110 °C. After the separation of the target fraction,

tank *11* is cooled down to 30-40 °C and the tank residue is poured into containers.

The pre-distillates of the vacuum distillation from collector 9 and of vacuum rectification from collector 14 are distilled in a vacuum rectification system similar to the one described above. It is done in order to extract the unreacted allylamine and triethoxysilane, as well as tetraethoxysilane and the target product.

After the vacuum distillation and rectification tank residue consists of condensed products, e.g. oligomers of the following composition:

$$(C_2H_5O)_2RSiO-[(C_2H_5O)RSiO]n-Si(OC_2H_5)_2R$$

$$R=(CH_2)_3NH_2 \text{ or } (CH_2)_3NH(OC_2H_5)$$

They can be used as waterproofing compounds.

γ-Aminopropyltriethoxysilane is a liquid, which actually consists of a mixture of γ- *and* β-isomers with the boiling point of 79-80 °C (at 4 GPa). It can be dissolved in organic solvents, alcohols and is rather easily hydrolysed, forming a gel.

Table 12. Physicochemical properties of substituted ethers of orthosilicon acid with an aminogroup in the organic radical

Substance	Boiling point, °C	d_4^{20}	n_D^{20}
$(NH_2CH_2)Si(OC_2H_5)_3$	93 (at 34.6 GPa)	0.9550 (at 25 °C)	1.4080 (at 25 °C)
$NH_2(CH_2)_3Si(OC_2H_5)_3$	68 (at 4 GPa)	0,9506	1.4225
$NH_2(CH_2)_4Si(OC_2H_5)_3$	123-124 (at 20 GPa)	0.9340 (at 25 °C)	1.4222 (at 25 °C)
$(CH_3NHCH_2)Si(OC_2H_5)_3$	95 (at 26.6 GPa)	0.9560	1.4082
$(C_2H_5)_2NCH_2Si(OC_2H_5)_3$	196-198	0.9100	1.4167
$CH_3[C_2H_5(C_6H5)NCH_2]Si(OCH_3)_2$	151-154 (at 30.66 GPa)	1.0240	1.5131
$CH_3(NH_2CH_2)Si(OC_2H_5)_2$	65.7 (at 32 GPa)	0.9140- 0.9160 (at 25 °C)	1.4120- 1.4126 (at °C)
$CH_3[NH_2CH_2)_3]Si(OC_2H_5)_2$	85-88 (at 10.66 GPa)	0.9162	1.4272
$CH_3[NH_2(CH_2)_4]Si(OC_2H_5)_2$	115-116 (at 38.6 GPa)	0.9125 (at 25 °C)	1.4320

CH$_3$(C$_2$H$_5$NHCH$_2$)Si(OC$_2$H$_5$)$_2$	180-181 (at 1009 гПа)	0.8870	1.4120
CH$_3$(C$_6$H$_5$NHCH$_2$)Si(OC$_2$H$_5$)$_2$	152-153 (at 20.1 GPa)	1.0020	1.4975

Technical γ-aminopropyltriethoxysilane should meet the following requirements:

Appearance	Liquid without mechanical impurities
Density at 20 °C, g/cm^3	0.945-0.957
Colour according to the iodometric scale, not darker than	0.25
Content, % (weight):	
γ-aminopropyl- and β-aminoisopropyl-triethoxysilanes, not less than	95
tetraethoxysilane, not more than	3

γ-Aminopropyltriethoxysilane is used for dressing glass cloth. Besides, it can be also used as a modifier for rubbers based on carbon-chain elastomers.

The physicochemical properties of important substituted ethers of ortho-silicon acid with an aminogroup in the organic radical are given in Table 12.

2.3.5. Preparation of substituted ethers of orthosilicon acid with nitrogen in the ether group

Substituted ethers of orthosilicon acid with nitrogen in the ether group are triptychlike silicon compounds:

$$X-Si(OCH_2CH_2)_3N \qquad \text{X - alkyl or aryl, H, OR, OCOR et al.}$$

These compounds have received the general name of silatranes; in them silicon atoms are connected directly with nitrogen atoms:

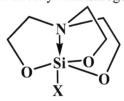

Silatranes are increasingly used in agriculture, medicine and other spheres of economy. Credit for the development of the silatrane chemistry and technology in Russia goes to Academician of the Russian Academy of Sciences, M.G.Voronkov et al.

Silatranes can be synthesised in several ways.

Re-etherification of alkyl- or aryltriethoxysilanes with triethanolamine:

$$RSi(OC_2H_5)_3 + (HOCH_2CH_2)_3N \xrightarrow[3C_2H_5OH]{} R-\overset{\displaystyle\lceil\qquad\qquad\rceil}{Si(OCH_2CH_2)_3N} \qquad (2.134)$$

If alkyl or aryl radicals at the silicon atom should be replaced with alkoxyl or aroxyl, tetraethoxysilane is re-etherified with triethanolamine in the presence of alcohols or phenols.

$$Si(OC_2H_5)_4 + (HOCH_2CH_2)_3N + ROH \xrightarrow[4C_2H_5OH]{NaOH} RO-\overset{\displaystyle\lceil\qquad\qquad\rceil}{Si(OCH_2CH_2)_3N} \qquad (2.135)$$

Acyloxysilatranes are obtained by the re-etherification of tetraethoxysilane with triethanolamine in the presence of appropriate organic acids.

The re-etherification of boratrane with triethoxysilane in the presence of catalysts (aluminum alcoholates) in boiling xylene can yield hydridsilatrane:

$$\overset{\displaystyle\lceil\qquad\qquad\rceil}{B(OCH_2CH_2)_3N} + SiH(OC_2H_5)_3 \xrightarrow[B(OC_2H_5)_3]{} H-\overset{\displaystyle\lceil\qquad\qquad\rceil}{Si(OCH_2CH_2)_3N} \qquad (2.136)$$

This reaction can be carried out without a catalyst, but in that case the process takes considerably longer.

2. The decomposition of the products of hydrolithic condensation of organotrichlorosilanes; the liberated water is eliminated from the reaction area by constant azeotropuc distillation with a solvent (e.g. xylene).

$$1/n[RSiO_{1,5-x}(OH)_x]_n + n(HOCH_2CH_2)_3N \xrightarrow[n(1,5+x)H_2O]{KOH} R-\overset{\displaystyle\lceil\qquad\qquad\rceil}{Si(OCH_2CH_2)_3N} \qquad (2.137)$$

Oligohydridorganosiloxanes are used to form organosilatranes similarly:

$$1/n[RSi(H)O-]_n + n(HOCH_2CH_2)_3N \xrightarrow[nH_2O;\ nH_2]{KOH} R-\overset{\displaystyle\lceil\qquad\qquad\rceil}{Si(OCH_2CH_2)_3N} \qquad (2.138)$$

First, the hydrogen at the silicon atom reacts with the hydroxyl group of triethanolamine (liberating molecular hydrogen), and only after that the siloxane link breaks up, forming organosilatranes.

3. The dehydrocondensation of hydridsilatrane with alcohols, phenols or organic acids can yield corresponding alkoxy-, aroxy- or acyloxysilatranes (in the presence of alkali or $ZnCI_2$):

$$H-Si(OCH_2CH_2)_3N \quad \xrightarrow[\substack{H_2}]{ROH} \quad RO-Si(OCH_2CH_2)_3N \quad (2.139)$$

$$\xrightarrow[\substack{H_2}]{RCOOH} \quad RCOO-Si(OCH_2CH_2)_3N$$

Method 1, the reaction of re-etherification, seems the most convenient.

Among the most widely used silatranes are chloromethyl- and ethoxysi-latranes.

Preparation of chloromethylsilatrane

Chloromethylsilatrane (Mival) is obtained by the etherification of chloro-methyltrichlorosilane with anhydrous ethyl alcohol and subsequent re-etherification of the obtained chloromethyltriethoxysilane with trietha-nolamine.

The first stage, the etherification of chloromethyltrichlorosilane, occurs stagewise, with a gradual substitution of chlorine atoms at the silicon atom with ethoxyl groups:

$$ClCH_2SiCl_3 \xrightarrow[\substack{HCl}]{C_2H_5OH} ClCH_2Si(OC_2H_5)Cl_2 \xrightarrow[\substack{HCl}]{C_2H_5OH} \quad (2.140)$$

$$\longrightarrow ClCH_2Si(OC_2H_5)_2Cl \xrightarrow[\substack{HCl}]{C_2H_5OH} ClCH_2Si(OC_2H_5)_3$$

The nucleophilic attack of the $C_2H_5O^-$ anion is directed at the silicon atom (rather than at the carbon atom) due to the following distribution of the electron density in the chloromethyltrichlorosilane molecule:

$$\underset{Cl}{\overset{\delta^-}{\longleftarrow}} \overset{\delta^+}{CH_2-Si} \underset{Cl}{\overset{Cl}{\Longleftrightarrow}}$$

This explains why the chlorine atom at the silicon atom is not touched during etherification.

In this reaction the substitution of the first chlorine atom with an eth-oxygroup occurs relatively easily, whereas that of the subsequent chlorine atoms is much more complicated due to steric hindrances. That is why, for a more complete etherification it is necessary to heat the reactive mixture to 80-85°C; if the reaction is conducted in a solvent (light petroleum or toluene), to the boiling point of the solvent.

The second stage, the re-etherification of chloromethyltriethoxysilane with triethanolamine, is actually a process of ester condensation; since the ester group of triethanolamine has a low carbonyl activity, one should employ strong bases, e.g. alcoholates of alkali metals, as the condensing agent. Consequently, re-etherification can be generalised in the following way:

$$C_2H_5ONa \rightleftharpoons C_2H_5O^- + Na^+ \qquad (2.141)$$

$$3C_2H_5O^- + N(CH_2CH_2OH)_3 \xrightarrow{3C_2H_5OH} N(\overset{-}{C}CH_2OH)_3 \rightleftharpoons \qquad (2.142)$$
$$\rightleftharpoons N(CH_2CH_2\overset{-}{O}H)_3$$

$$\overset{\delta^+}{ClCH_2}-Si(OC_2H_5)_3 + N(CH_2CH_2\overset{-}{O})_3 \xrightarrow{3C_2H_5OH} \qquad (2.143)$$

$$\rightleftharpoons ClCH_2-Si\underset{OCH_2CH_2}{\overset{OCH_2CH_2}{\underset{\displaystyle}{\diagdown}}}N$$

Since all these stages are balanced, the output of the target chloromethylsilatrane can be increased by conducting the process with anhydrous alcoholate rather than in the excess of alcohol.

Raw stock: chloromethyltrichlorosilane (not less than 98% of the 114-118°C fraction); triethanolamine (not less than 99% of the 186-190 °C fraction); anhydrous ethyl alcohol (not less than 99.6% of C_2H_5OH); chloroform (medical); light petroleum (not less than 98% of the 40-70 °C fraction); potassium hydroxide (technical product).

Chloromethylsilatrane production (Fig.30) comprises the following main stages: the etherification of chloromethyltrichlorosilane and extraction of pure chloromethyltriethoxysilane; the re-etherification chloromethyltriethoxysilane with triethanolamine; the re-crystallisation and dehydration of chloromethylsilatrane.

The etherification of chloromethyltrichlorosilane with anhydrous ethyl alcohol is carried out in an etherificator by the technique described for obtaining methyl(phenylaminomethyl)diethoxysilane and according to the diagram described in Fig. 27.

The etherification can be carried out in a solvent (light petroleum) at the boiling point of the solvent and in the presence of a hydrogen chloride acceptor (carbamide or any amine).

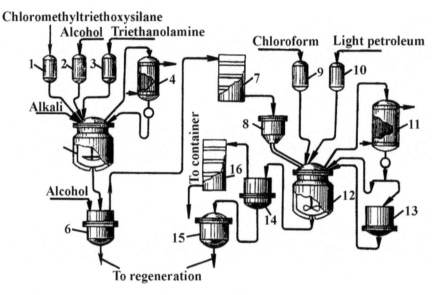

Fig. 30. Production diagram of chloromethylsilatrane: 1-3; 8-10 - batch boxes; *4, 11* - coolers; *5* - reactor; *6, 13, 14* -nutsch filters; *7, 16* - shelf drafts; *12* - crystalliser; *15* - collector:

The synthesis of chloromethylsilatrane is carried out in steel enameled reactor *5* with an agitator and a water vapour jacket. First the reactor is loaded with ethyl alcohol and freshly distilled triethanolamine from batch boxes; a calculated amount of potassium hydroxide is loaded through a hatch. The mixture is intensively agitated, the temperature raised till KOH dissolves completely, as shown by the appearance of reflux in the run-down box situated after cooler *4*. Then chloromethyltriethoxysilane is fed at such speed that the reactive mixture boils uniformly. After the reaction is finished, the mixture is cooled to 15°C, sent to nutsch filter 6 and filtered through   coarse calico. The technical chloromethylsilatrane in the nutsch filter is washed with ethyl alcohol twice, thoroughly pressed and dried in draft 7 at a temperature below 100 °C till its weight is constant.

The re-crystallisation of the technical chloromethylsilatrane is carried out in enameled apparatus *12* with an agitator, a water vapour jacket and cooler *11*. The crystalliser receives chloromethylsilatrane from batch box *8* with a screw batcher, and chloroform, from batch box *9*. These components are loaded into the crystalliser in the weight ratio of 1:10. The mixture in the crystalliser is boiled at agitation until chloromethylsilatrane dissolves completely. Then the hot chloroform solution of the product is filtered in nutsch filter *13* through coarse calico and re-loaded into crystalliser *12*. The crystalliser is filled with light petroleum from batch box *10* at room temperature (to precipitate chloromethylsilatrane from the chloroform solution); after that, the jacket of the crystalliser is filled with salt solution; the precipitated sediment is held at 5°C for 2-3 hours for the growth and complete deposition of crystals. From apparatus *12* crystal chloromethylsilatrane in the solution of chloroform and light petroleum mixture is sent to nutsch filter *14*, where the crystals of chloromethylsilatrane are pressed to extract the mother solution. The mother solution is sent into collector *15* and then to regenerate chloroform and light petroleum; the recrystallised chloromethylsilatrane is sent into shelf draft 16 to be dried till its weight is constant.

Pure chloromethylsilatrane is a white crystal substance with a bitttersweet taste. It is soluble in chloroform and dimethylformamide, less soluble in water and acetone, virtually insoluble in alcohols and diethyl ether.

Technical chloromethylsilatrane should meet the following requirements:

Appearance	White crystals
Content, %:	
of silicon	12.2-13.4
of nitrogen	6-6.6
of chlorine	15.5-16.5
Solubility in chloroform	Complete

Chloromethylsilatrane is used in medicine (to heal wounds and burns, to treat dermatitises and atherosclerosis, to stimulate hair growth and prevent hair loss). It is used in agriculture to increase the yield of many crops.

Other organosilatranes can be obtained similarly to chloromethylsilatrane. All organosilatranes are solid substances, which crystallise in the form of large beautiful crystals mostly of needlelike shape. They also have a definite melting point. As a rule, they dissolve well in polar organic solvents; methyl- and ethylsilatranes are also soluble in water.

2.4. Preparation of acyloxyorganosilanes

Tetraacetoxysilane and alkyl(aryl)acetoxysilanes of the common structure $R_nSi(OCOCH_3)_{4-n}$ where R are various organic radicals, and n ranges from 0 to 3, are prepared by two main techniques: the acetylation of organochlorosilanes with acetic anhydride:

$$R_nSiCl_{4-n} + (4 - n)(CH_3CO)_2O \xrightarrow[(4-n)CH_3COCl]{} R_nSi(OCOCH_3)_{4-n} \quad (2.144)$$

and the acetylation of organochlorosilanes with metal acetates:

$$R_nSiCl_{4-n} + (4 - n)CH_3COOK \xrightarrow[(4-n)KCl]{} R_nSi(OCOCH_3)_{4-n} \quad (2.145)$$

Preparation of triacetoxymethylsilane with acetic anhydride

The synthesis of triacetoxymethylsilane with acetic anhydride takes place in one stage, with simultaneous distillation of the by-product, acetylchloride. It is all the more accessible, since acetylchloride boils at 55 °C, methyltrichlorosilane boils at 66.1 °C, acetic anhydride, at 140 °C and triacetoxymethylsilane, at 84-90 °C (at 13.3 GPa). The reaction proceeds thus:

$$CH_3SiCl_3 + 3(CH_3CO)_2O \rightarrow CH_3Si(OCOCH_3)_3 + 3CH_3COCl \quad (2.146)$$

Fig. 31 shows a diagram of the preparation of triacetoxymethylsilane with acetic anhydride. The acetylation of methyltrichlorosilane can be carried out in reactor 6, a steel enameled cylindrical apparatus with an agitatorand, a water vapour jacket and rectification tower *3* filled with Raschig rings. The reactor is loaded with necessary amounts of methyltrichlorosilane and acetic anhydride from the batch boxes, the agitator is switched on and the jacket is filled with vapour. The process ends with the complete distillation of the fraction which boils below 58 °C. The reactor is still filled with triacetoxymethylsilane with an impurity of unreacted acetic anhydride. The product is collected in receptacle *7*.

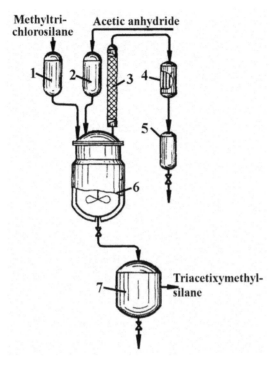

Fig. 31. Production diagram of triacetoxymethylsilane: *1, 2* - batch boxes; *3* – rectification tower; *4* - cooler; *5* - receptacle; *6* -reactor; *7*-collector

Similarly to triacetoxymethylsilane, acetylation of organochlorosilanes with acetic anhydride can also yield other alkyl(aryl)acetoxysilanes.

E.g., by joint acetylation of methyl- and phenyltrichlorosilanes with phenyltrichlorosilane one can obtain a mixture of triacetoxymethyl- and phenylsilanes, which after methacrylation forms a mixture of silanes with methacrylethyl groups at the silicon atom, apart from the acetate ones.

Preparation and methacrylation of the mixture of triacetoxymethyl- and triacetoxyphenylsilanes

The preparation of the mixture of methyl- and phenylsilanes with methacrylethyl and acetate groups at the silicon atom, comprises two main stages:

1. The acetylation of the mixture of methyl- and phenyltrichlorosilanes with acetic anhydride:

$$CH_3SiCl_3 + C_6H_5SiCl_3 + 6(CH_3CO)_2O \longrightarrow \qquad (2.147)$$

$$\longrightarrow CH_3Si(OCOCH_3)_3 + C_6H_5Si(OCOCH_3)_3 + 6CH_3COCl$$

The acetylation of the mixture of methyl- and phenyltrichlorosilanes with acetic anhydride:

Partial methacrylation of the mixture of triacetoxymethyl- and triacetoxyphenylsilanes with glycolmethacrylate (MEG):

$$2CH_3Si(OCOCH_3)_3 + 2C_6H_5Si(OCOCH_3)_3 + 2HOCH_2CH_2OCO\overset{\underset{\mathrm{CH_3}}{|}}{C}=CH_2\longrightarrow \qquad (2.148)$$

$$\longrightarrow CH_3Si(OCOCH_3)_2(OCH_2CH_2OCO\overset{\underset{\mathrm{CH_3}}{|}}{C}=CH_2 + CH_3Si(OCOCH_3) +$$

$$+ C_6H_5Si(OCOCH_3)_2(OCH_2CH_2OCO\overset{\underset{\mathrm{CH_3}}{|}}{C}=CH_2 + C_6H_5Si(OCOCH_3)_3 + 2CH_3COOH$$

This stage is accompanied by the distillation of the liberated acetic acid.

Products of partial methacrylation are then sent to hydrolytic cocondensation to obtain a branched oligomer.

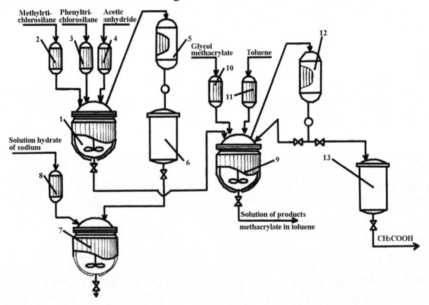

Fig. 32. . Production and methacrylation diagram of triacetoxymethyl- and triacetoxyphenylsilanes: 1,9- synthesis reactors; 2-4, 8, 10, 11 - batch boxes; 5, 12- coolers; 6, 13- receptacles; 7 – neutraliser

Raw stock: methyltrichlorosilane [not less than 99.6% of the main substance, hydrolysed chlorine content 69-71.5% (weight)]; phenyltrichlorosilane [not less than 99% (weight) of the 196-202°C fraction, hydrolysed chlorine content 49-50.5% (weight)]; acetic anhydride (not less than 97% of the main substance, $d_4^{20=}$ 1.076÷1.082); glycolmethacrylate (the boiling point is 68-73 °C at P=2.6 GPa, d_4^{20} = 1.07÷1.08).

Fig. 32 shows a production and methacrylation diagram of triacetoxy-methyl- and triacetoxyphenylsilanes.

Reactor *1* is loaded with a necessary amount of methyltrichlorosilane, phenyltrichlorosilane and acetic anhydride from batch boxes *2, 3* and *4*. Direct condenser *5* is switched on; then the reactive mixture is agitated with the agitator for 15 minutes; after that the reactor contents are heated to 60 °C for 30 minutes. At 50-60 °C acetylchloride is vapour-distilled; it enters receptacle 6 through the run-down box; the reactor contents are heated to 85-90 °C Further distillation of acetylchloride is conducted in vacuum under 550-480 GPa and at

the vapour temperature of 35-50 °C (50-70 °C in the reactor). After ace-tylchloride has been completely distilled, it is poured into neutraliser *7*, which has been loaded with a 40% solution of hydrate of sodium from batch box *8*. Reactor 1 is sampled to determine the content of the residual chlorine ion (not more than 2%) and acetoxygroups (68-75%); if the re-sults are satisfactory, the acetoxysilane mixture is pressurised with nitro-gen from reactor *1* into reactor *9*.

Reactor 9 is used to methacrylate the acetoxysilane mixture. First, back-flow condenser *12* is switched on; the agitator is switched on and the jacket of the apparatus is filled with vapour; then the contents of reactor *9* are heated to 60°C. At this temperature glycolmethacrylate self-flows from batch box *10* at such speed that the temperature in the reactor does not exceed 65°C. After the glycolmethacrylate has been loaded, the con-tents of reactor 9 are agitated at 60-65°C for one more hour; then con-denser 12 is switched into the direct operation mode, vacuum is created (the residual pressure is 130-80 GPa), and acetic acid is distilled into re-ceptacle 13 at 40-50 °C (in vapour); reactor 9 is periodically sampled to determine the acid number (a.n.). If the analysis is satisfactory (a.n. = 450÷500 mg/g), reactor *9* is loaded with toluene from batch box *11*.

Acetic acid (the boiling point is 118 °C), which is collected in receptacle 13, contains small amounts of unreacted methyltriacetoxysilane and gly-colmethacrylate. In order to be utilised, it should be neutralised with iron hydroxide and distilled.

The obtained solution of the products of acetoxysilane methacrylation in toluene are then sent to hydrolytic cocondensation to obtain a branched oligomer.

It is not advisable to use this technique to prepare tetraacetoxysilane and acetoxytrimethylsilane, because this requires silicon tetrachloride and

trimethylchlorosilane, the boiling points of which (57.7 and 57.3 °C correspondingly) are close to that of acetylchloride (55 °C), thus forming mixtures which are hard to separate.

To obtain tetraacetoxysilane and acetoxytrimethylsilane, as well as pure
alkyl(aryl)acetoxysilanes, the acetylation of organochlorosilanes should
make use of Na or K acetates.

Preparation of triacetoxymethylsilane with potassium acetate

The synthesis of triacetoxymethylsilane with potassium acetate occurs in
the following way:

$$CH_3SiCl_3 + 3CH_3COOK \xrightarrow[3KCl]{} CH_3Si(OCOCH_3)_3 \qquad (2.149)$$

Raw stock: methyltrichlorosilane (not less than 99.6% of the main substance); potassium acetate (not less than 85% of the main substance); oil
or coal toluene $[d_4^{20} = 0.865\pm0.002$, not less than 95% (vol.) of the 109.5-
111 °C fraction].

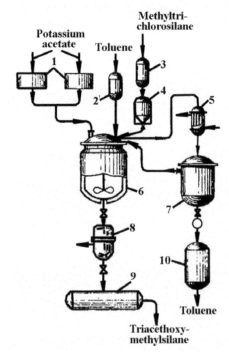

Fig. 33. Diagram of triacetoxymethylsilane production with potassium acetate: *1* -
vacuum drafts; *2-4* -batch boxers; *5* - cooler; *6* - reactor; *7, 9, 10* - collecors; *8*
nutsch filter.

The production process of vinyltrichlorosilane (Fig. 33) comprises three main stages: the dehydration of potassium acetate; the acetylation of methyltrichlorosilane; the filtering of triacetoxymethylsilane.

Potassium acetate has the form of crystals melting in air, which contain crystalline hydrate water.

Since methyltrichlorosilane should be acetylated in an anhydrous medium, potassium acetate is dried in vacuum drafts *1*. Before drying the salt is ground on special grounders and then loaded on trays into vacuum drafts; the salt is held there at 100-120 °C and a residual pressure of 520-660 GPa. The drafts can be warmed with vapour (0.3 MPa). The drying process takes at least 12 hours.

The final dehydration of potassium acetate is carried out in reactor *6* (until the total elimination of moisture). For that purpose the washed and dried reactor is filled with toluene from batch box *2*; then potassium acetate is sent at agitation through an overflow pipe. The potassium acetate has already been dried in vacuum drafts and weighed. The jacket of the reactor is filled with vapour (0.3 MPa), and cooler *5*, with water. The dehydration is carried out by distilling the azeotropic mixture of toluene and water at 100-115 °C. The distilled mixture is sent into cooler *5* to condense. The condensate is collected in receptacle *7*. The speed of the distillation of the azeotropic mixture is regulated by changing the vapour supply in the reactor jacket. After 200-250 l of the azeotropic mixture have been distilled, toluene is returned from collector *7* into the reactor. The distilled water is collected in the lower part of collector *7*, and the separated toluene is sent back into the reactor.

During the distillation of the azeotropic mixture the condensate (toluene) is sampled to determine moisture content. When the analysis is positive (absence of moisture by $CuSO_4$), the sample is additionally analysed by the gasometric technique; if the moisture content in toluene is not more than 0.02%, the dehydration of potassium acetate is considered complete. To account for the amount of dehydrated potassium acetate, the quantity of water poured off from collector *7* is measured; by the difference of the weights of the loaded potassium acetate and the separated water one determines the precise amount of anhydrous salt. Toluene from collector *7* (after a thorough separation and its weighing) is poured into collector *10*, and then sent to regeneration and returned into the production process.

The acetylation of methyltrichlorosilane is conducted in a toluene medium in reactor 6, which is an enameled apparatus with an agitator and a water vapour jacket Methyltrichlorosilane from batch box 3 is poured into weight batch box 4, where a necessary amount is measured and then analysed to determine the chloroion content. From batch box 4, methyltrichlorosilane is sent into reactor 6 through a siphon under the toluene layer at agitation. The temperature during acetylation is maintained below 45 °C by regulating the water supply into the reactor jacket and the speed of methyltrichlorosilane supply. After the whole methyltrichlorosilane has been introduced, the reactive mixture is kept at 40-45 °C for 2-2.5 hours (at agitation).

Apart from triacetoxymethylsilane, the acetylation also yields potassium chloride, which is separated from the target product in nutsch filter 8 at 0.07 MPa. The toluene solution of triacetoxymethylsilane, separated from potassium chloride, is collected into collector 9 and then sent to the vacuum distillation of toluene, which takes place in a conventional distillation tank (not shown in the diagram).

Triacetoxymethylsilane is a colourless transparent liquid (the boiling point is 84-90 °C at 13.3 GPa). It can be easily hydrolysed with water and dissolves well in common organic solvents.

Similarly to triacetoxymethylsilane, acetylation of chlorosilanes with metal acetates can also yield tetraacetoxysilane and other alkyl(aryl)acetoxysilanes. The physicochemical properties of important acyloxyorganosilanes are given in Table 13. The practical value of acetoxysilanes is that their hydrolysis, unlike the hydrolysis of organochlorosilanes, forms weak acetic acid, rather than hydrogen chloride. That is why acetoxysilanes can be used to waterproof various materials (textiles, paper, etc.). Alkyl(aryl)acetoxysilanes are also used to obtain some silicone varnishes and as hardeners for low-molecular silicone elastomers.

Table 13. Physicochemical properties of tetraacyloxysilanes and acyloxyorganosilanes

Substance	Boiling point, °C	Melting point, °C	d_4^{20}	n_D^{20}
$Si(OCOCH_3)_4$	148 (at 6-8 GPa)	110	—	—
$CH_3Si(OCOCH_3)_3$	84-90 (at 13.3 GPa)	-	1.1750	1.4083
$(CH_3)_2Si(OCOCH_3)_2$	165 (at 1000 GPa)	-	1.0540	1.4030
$(CH_3)_3Si(OCOCH_3)$	102.5-103 (at 985 GPa)	-	1.8914	1.3890
$Si(OCOC_2H_5)_4$	-	55-56	—	-
$C_2H_5Si(OCOCH_3)_3$	107.5-108.5 (at 10.6 GPa)	-	1.1428	1.4123
$(C_2H_5)_2Si(OCOCH_3)_2$	192-193	-	1.0240	1.4152
$(C_2H_5)_3Si(OCOCH_3)$	173.4	-	0.8926	1.4190
$(C_6H_5)_2Si(OCOCH_3)_2$	176-178 (at 4 GPa)	-	-	—
$(C_6H_5)_3Si(OCOCH_3)$	-	97	-	-

2.5. Preparation of hydroxyorganosilanes

There are several techniques for obtaining hydroxyorganosilanes. These techniques generally amount to the hydrolysis of halogen-, alkoxy-, acyloxy-, amino- and hydride substituted organosilanes with water, as well as aqueous solutions of soda or alkali. The hydrolysis proceeds thus:

$$R_nSiX_{4-n} \xrightarrow[(4-n)\ HX]{(4-n)\ H_2O} R_nSi(OH)_{4-n} \qquad (2.150)$$

X=Hal, OR, OCOR, NH_2 or H, n=1÷3

Apart from the given common technique, one can prepare hydroxyorganosilanes by the breakdown of the organochlorosilane mixture (with an aqueous alkali solution) obtained by the organomagnesium synthesis:

$$R_nSiCl_{4-n} + mR'MgCl \xrightarrow[mMgCl_2]{} R_nR'_mSiCl_{4-(m+n)} \qquad (2.151)$$

$$R_nR'_mSiCl_{4-(m+n)} + [4-(m+n)]H_2O \xrightarrow[4-(m+n)HCl]{NaOH} R_nR'_mSi(OH)_{4-(n+m)} \qquad (2.152)$$

However, in this case a mixture of mono-, di- and trihydroxyorganosilanes, which is difficult to separate, is formed.

With the help of organomagnesium substances it is possible to obtain hydroxyorganosilanes from α,ω-dihydroxydiorganosiloxanes:

$$HO(SiR_2O)H + nR'MgCl \xrightarrow[Mg(OH)Cl]{} nR_2R'SiOH \qquad (2.153)$$

In this case the Grignard agent serves as a reactant and an agent that breaks the Si—O—Si bonds.

The synthesis of hydroxyorganosilanes based on the use of organomagnesium compounds, is convenient mostly for the preparation of mono- and difunctional hydroxyorganosilanes with different radicals at the silicon atom.

Finally, hydroxyorganosilanes can be obtained by breaking up the Si—C bond in tetraorganosilanes with sulfuric or hydrochloric acid:

$$R_3SiC_6H_5 + H_2SO_4 \longrightarrow R_3SiOH + C_6H_5SO_3H \qquad (2.154)$$

$$R_3SiC_6H_5 + HCl \xrightarrow[C_6H_6]{} R_3SiCl \xrightarrow[HCl]{H_2O} R_3SiOH \qquad (2.155)$$

The second reaction also occurs under the effect of an aluminum chloride solution. In these reactions, it is easier to break away larger radicals from the silicon atom.

Of these techniques to obtain hydroxyorganosilanes, the most widely used is the water hydrolysis of organochlorosilanes.

However, it is rather problematic to extract pure hydroxyorganosilanes due to their considerable mobility and the condensation ability of hydroxyl groups at the silicon atoms. The stability of hydroxyorganosilanes is noticeably affected by the number and type of the organic radicals at the silicon atom. Mono- and dihydroxyorganosilanes with lower organic radicals at the silicon atom can be extracted in the pure form in case certain precautions are observed (the hydrolysis with an aqueous alkali solution or ammonia flow), on the other hand, trihydroxyorganosilanes with lower organic radicals cannot be separated at all. At the same time, hydroxytriphenyl- and dihydroxydiphenylsilanes are easily formed in the hydrolysis of corresponding phenylchlorosilanes even without a hydrogen chloride acceptor. It happens due to the presence of large phenyl radicals, which screen the hydroxyl groups. For the same reason, it has been possible to obtain and separate only such trihydroxyorganosilanes as trihydroxyphenyl-, trihydroxydichlorophenyl- and trihydroxypentachlorophenylsilanes.

Thus, the hydrolysis of organochlorosilanes is governed by two general regularities:

1. the reduction in the number of the organic radicals at the silicon atom decreases the stability of hydroxyorganosilanes;
2. the enlargement of these organic radicals increases the stability of hydroxyorganosilanes.

A natural question arises: why is the condensation of hydroxyl groups at the silicon atoms so easy? Hydroxyorganosilane should be viewed as acids.

E.g., orthosilicon acid $Si(OH)_4$ is a very unstable substance. It can be fixed only at -20 °C, but it cannot be separated (even at very low temperatures), because its condensation, which liberates a water molecule, forms siloxane bonds. When hydroxyl groups are partially substituted with organic radicals, the hydroxyl groups become a little more stable, because the electron density at the silicon atom is re-distributed. This decreases the acid characteristics of these substances and thus decreases their ability to condense. However, the acid characteristics of trihydroxyorganosilanes are still considerable.

The condensation of hydroxyorganosilanes does not occur by the intramolecular mechanism, like the corresponding carbon compounds:

$$RC(OH)_3 \xrightarrow[H_2O]{} RCOOH \qquad\qquad (2.156)$$

but rather by the intermolecular mechanism, forming silanoacids:

$$2RSi(OH)_3 \xrightarrow[H_2O]{} R-\underset{\underset{OH}{|}}{\overset{\overset{OH}{|}}{Si}}-O-\underset{\underset{OH}{|}}{\overset{\overset{OH}{|}}{Si}}-R \qquad \text{etc.} \qquad (2.157)$$

This is explained by the fact that a silicon atom is much larger (the co-valence radius of a Si atom is 0.117 nm) than a carbon atom (the covalence radius of a C atom is 0.077 nm); therefore, the distances between the hydroxyl groups at the silicon atom are rather large and impair intramolecular condensation (such a reaction requires a considerable deformation of valence angles).

Hydroxyorganosilanes can be easily dissolved in organic solvents; only dihydroxydimethylsilane dissolves well in water. In principle, the enlargement of the organic radicals at the silicon atom decreases the solubility of hydroxyorganosilanes. For example, the water solubility of hydroxyorganosilanes decreases in the following sequence:

$(CH_3)_2Si(OH)_2 > (C_2H_5)_2Si(OH)_2 > CH_3(C_6H_5)Si(OH)_2,$
(71%)   (10%)   (0.43%)

whereas dihydroxydiphenylsilane is virtually water-insoluble.

Preparation of dihydroxydiphenylsilane

Dihydroxydiphenylsilane is synthesised by the water hydrolysis of diphenyldichlorosilane in petrol.

$$(C_6H_5)_2SiCl_2 \xrightarrow[2HCl]{2H_2O} (C_6H_5)_2Si(OH)_2 \qquad (2.158)$$

However, the hydrolysis in these conditions is accompanied by the by-processes in which the products condense:

$$n\,(C_6H_5)_2Si(OH)_2 \xrightarrow[nH_2O]{} [(C_6H_5)_2SiO]_n \qquad n=3 \text{ or } 4 \qquad (2.159)$$

$$n(C_6H_5)_2Si(OH)_2 \xrightarrow[(n-2)H_2O]{} HO-\underset{\underset{C_6H_5}{|}}{\overset{\overset{C_6H_5}{|}}{Si}}-O\left[\underset{\underset{C_6H_5}{|}}{\overset{\overset{C_6H_5}{|}}{Si}}-O\right]_n\underset{\underset{C_6H_5}{|}}{\overset{\overset{C_6H_5}{|}}{Si}}-OH \qquad (2.160)$$

$$(C_6H_5)_2SiCl_2 \xrightarrow[HCl]{H_2O} (C_6H_5)_2\underset{\underset{OH}{|}}{Si}Cl \qquad (2.161)$$

$$2\,(C_6H_5)_2\underset{\underset{OH}{|}}{Si}Cl \xrightarrow{H_2O} \underset{\underset{C_6H_5}{|}}{\overset{\overset{C_6H_5}{|}}{Cl-Si}}-O-\underset{\underset{C_6H_5}{|}}{\overset{\overset{C_6H_5}{|}}{Si}}-Cl \xrightarrow[2HCl]{2H_2O} \underset{\underset{C_6H_5}{|}}{\overset{\overset{C_6H_5}{|}}{HO-Si}}-O-\underset{\underset{C_6H_5}{|}}{\overset{\overset{C_6H_5}{|}}{Si}}-OH \quad etc. \qquad (2.162)$$

These by-processes reduce the yield of the target product and contaminate it with impurities of oillike cyclic and linear oligomers.

To decrease by-processes and increase the yield of the target product, dihydroxydiphenylsilane, diphenyldichlorosilane is hydrolyses in the presence of butil alcohol. Replacing chlorine atoms at the silicon atom, butoxyl groups block some places where hydroxyl groups may form, thus impairing the by-processes of condensation.

The hydrolysis of diphenyldichlorosilane in the presence of butil alcohol occurs thus:

$$(C_6H_5)_2SiCl_2 \xrightarrow[HCl]{C_4H_9OH} (C_6H_5)_2Si(OC_4H_9)Cl \xrightarrow[HCl]{H_2O} \qquad (2.163)$$

$$\longrightarrow (C_6H_5)_2Si(OC_4H_9)OH \xrightarrow[C_4H_9OH]{H_2O} (C_6H_5)_2Si(OH)_2$$

Naturally, even in this case it is not possible to completely avoid condensation by-processes, e.g.:

$$2(C_6H_5)_2Si(OC_4H_9)Cl + (C_6H_5)_2Si(OH)_2 \xrightarrow[HCl;\, 2C_4H_9OH]{H_2O} \qquad (2.164)$$

$$\longrightarrow \underset{\underset{C_6H_5}{|}}{\overset{\overset{C_6H_5}{|}}{HO-Si}}-O-\underset{\underset{C_6H_5}{|}}{\overset{\overset{C_6H_5}{|}}{Si}}-O-\underset{\underset{C_6H_5}{|}}{\overset{\overset{C_6H_5}{|}}{Si}}-OH$$

Raw stock: diphenyldichlorosilane (not less than 98% of the main substance); butil alcohol (the boiling point is 115-118 °C, $d_4^{20}==$ 0.810±0.002); petrol solvent (60-90 °C, d_4^{20} = 0.693-0.730); toluene (the boiling point is 109-111 °C, d_4^{20} = 0.865±0.002).

The basic diagram of dihydroxydiphenylsilane production is given in Fig. 34. First, a solution of diphenyldichlorosilane in petrol is prepared in agitator 3 . The synthesis of dihydroxydiphenylsilane is carried out in enameled reactor 7 with an agitator and a jacket. In this apparatus a mixture of petrol, butanol and water is prepared and agitated. After the mixture has been prepared, the reactor jacket is filled with salt solution, the contents of the apparatus are cooled down to +1 - -5°C, and a petrol solution of diphenyldichlorosilane is introduced from agitator 3 at constant agitation under the nitrogen pressure of 0.07 MPa. As the solution is introduced, reactor 7 is filled with nitrogen through the bubbler under the liquid layer at a speed of 170-850 l/min. The hydrolysis temperature should range from -7 to +20 °C. After the solution has been introduced, the reactive mixture is agitated for 30 minutes; then the agitator is switched off, the supply of nitrogen is stopped and the reactive mixture is settled.

The hydrolysis forms two phases: the solid phase, dihydroxydiphenylsilane, and the liquid phase, the aqueous solution of butanol and petrol.

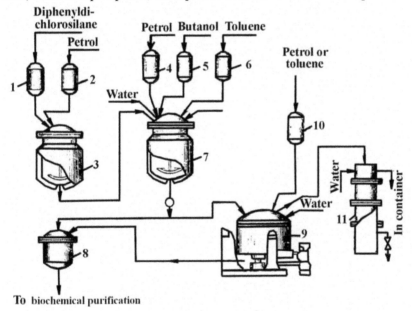

Fig. 34. Production diagram of dihydroxydiphenylsilane: 1, 2, 4-6, 10- batch boxes; 3 - agitator; 7- reactor; 8 - collector; 9- centrifuge; 11 -shelf draft

The liquid phase is poured into receptacle 8. Then, to wash dihydroxydiphenylsilane, the reactor is filled with a necessary amount of water, the mixture is agitated and after settling the acid waters are poured into collector 8. The target product is sampled to determine its solubility in acetone. If the solubility is complete, reactor 7 is filled with water again, the aqueous solution of the product is poured into centrifuge 9, which is a vertical suspended three-tower self-installing machine.

If the solubility of dihydroxydiphenylsilane in acetone is incomplete, re-actor *7* is filled with toluene from batch box *6*, and the mixture is washed with toluene until the solubility is complete. After that the product and toluene are poured into the centrifuge to be washed with water until they give a neutral reaction. From centrifuge *9*, the acid waters are sent into col-lector *8*.

The mixture in centrifuge *9* is sampled to determine the solubility of di-hydroxydiphenylsilane in acetone. If the solubility is complete, the product is washed with petrol, which is fed from batch box *10* If the solubility is incomplete, the product is washed in the centrifuge, first with toluene (un-til the solubility is complete), and then with petrol.

Table 14. Physicochemical properties of hydroxyorganosilanes

Substance	Boiling point, °C	Melting point, °C	d_4^{20}	n_D^{20}
$(CH_3)_3SiOH$	97	–	0.8112	1.3883
$(C_2H_5)_3SiOH$	64-65 (at 18.7 GPa)	-	0.8630	1.4328
$(C_6H_5)_3SiOH$	-	153	-	-
$(CH_3)_2Si(OH)_2$	-	100.5	-	-
$(C_2H_5)_2Si(OH)_2$	140 (diff.)	96	-	-
$(C_6H_5)_2Si(OH)_2$	-	131-132	-	-
$(C_6H_5)Si(OH)_3$	-	180	-	-
$(C_6H_3Cl_2)Si(OH)_3$	-	188	-	-

After washing, dihydroxydiphenylsilane is pressed in the centrifuge and loaded into vacuum shelf rack *11* with four shelves and a heating jacket. The drying is conducted at a temperature not higher than 115 °C (the tem-perature is regulated by hot water or vapour). During the drying, the con-tent of hydroxyl groups in the products is periodically checked. The drying is stopped, when the hydroxyl group content is 14.5-16.5%, and the prod-uct is loaded into containers.

Dihydroxydiphenylsilane is a white crystal product with a phenol odour. It dissolves well in ethers, chloroform, poorly in benzene and naphtha, and does not dissolve in water.

Technical dihydroxydiphenylsilane should meet the following require-ments:

Appearance	Crystal powder; white; some foreign inclusions can be tolerated
Solubility in acetone	Complete
pH of the aqueous extract	6-7
Hydroxyl group content, %	14.5-16.5
The onset of melting temperature, °C, not lower than	140

Dihydroxydiphenylsilane is used as a stabiliser in the processing of silicone rubber compounds and as a raw stock in the production of branched, spirocyclic or cyclolinear oligoorganodiphenylsiloxanes.

Similarly to dihydroxydiphenylsilane, one can also obtain other hydroxyorganosilanes; their properties are given in Table 14.

3. Chemistry and technology of silicone oligomers

At present, industry produces a wide range of silicone and elementosilicone high-molecular compounds of three classes: polyorganosiloxanes, polyorganosilazanes and polyelementorganosiloxanes.

Polyorganosiloxanes are high-molecular compounds with the backbone consisting of alternating silicon and oxygen atoms with different groups or organic radicals at the silicon atom.

The production of polyorganosiloxanes is accompanied by some processes:

1. the hydrolytic condensation or cocondensation of organohalogensilanes or alkoxyorganosilanes;
2. the stagewise polycondensation of the products of the hydrolytic condensation of organohalogensilanes or alkoxyorganosilanes;
3. the catalytic polymerisation or regrouping of the cyclic products of the hydrolytic condensation;
4. the polycondensation or copolycondensation of the products of the hydrolytic condensation with various organic compounds.

The mechanism of the hydrolytic condensation of organohalogensilanes is as follows. Under the influence of water on alkyl- or arylhalogensilane the halogen atom near the silicon atom is replaced with the hydroxyl group. This produces intermediate products, *hydroxyorganosilanes*, which can be easily condensed, liberating water and forming a siloxane bond.

R_3SiCl-type monofunctional compounds are hydrolysed with water to form *hydroxytriorganosilanes,* which are easily condensed, liberating water and forming hexaorganodisiloxanes:

$$2R_3SiCl \xrightarrow[2HCl]{2H_2O} 2[R_3SiOH] \underset{H_2O}{\rightleftharpoons} R_3Si-O-SiR_3 \qquad (3.1)$$

Depending on the nature of the organic radical and the reaction conditions, the equilibrium can be shifted in one or the other direction. Hexaorganodisiloxanes as such are of little practical interest; thus, the hydrolysis of monofunctional triorganochlorosilanes has limited value. On the other hand, the use of monofunctional compounds for the hydrolytic cocondensation with di- and trifunctional alkyl(aryl)chlorosilanes yields polymers with different preset chain lengths depending on the mole ratios of the components, e.g.:

$$2R_3SiCl + nR_2SiCl_2 \xrightarrow[2(n+1)\,HCl]{(n+1)\,H_2O} R_3Si{\left[OSiR_2\right]}_nOSiR_3 \qquad (3.2)$$

The hydrolytic condensation of difunctional R_2SiCl_2-type organohalogensilanes forms linear and cyclic compounds depending on the pH of the medium:

$$nR_2SiCl_2 \xrightarrow[2nHCl]{2nH_2O} n[R_2Si(OH)_2] \xrightarrow[nH_2O]{} HO{\left[SiR_2O\right]}_nH + [R_2SiO]_n \qquad (3.3)$$

At pH > 7 it forms mainly linear oligomers,. and cyclic oligomers at pH < 7. The organic radical also considerably affects the process. The catalytic polymerisation of the organocyclosiloxanes forms linear polyorganosiloxanes.

Depending on the reaction conditions, the solvent used and the length of the radical at the Si atom, the hydrolytic condensation of trifunctional halogen-containing compounds drastically changes the structure, composition and properties of polyorganosiloxanes formed as a result of hydrolytic condensation and polycondensation. This leads to the formation of branched (I), ladder (II) or cross-linked molecular structures in the polymer.

$$nRSiCl_3 + 3nH_2O \xrightarrow[3nHCl]{} n[RSi(OH)_3] \qquad (3.4)$$

(3.5)

The hydrolytic condensation of catalytic polymerisation of di- and trifunctional organohalogensilanes forms cyclolinear polyorganosiloxanes:

$$nR_2SiCl_2 + mR'SiCl_3 + 5(n+m)H_2O \xrightarrow[5(n+m)HCl;\ H_2O]{}$$

$$\longrightarrow \left[H \!-\!\! \left(O \underset{\underset{O}{\overset{Si}{|}}}{} \ \cdots \ O\!-\!SiR_2 \right)_{mn} \!\!\! OH \right] \tag{3.6}$$

At present polyorganosiloxanes are manufactured in the form of: 1) oligomers with linear or cyclic chains (silicone liquids); 2) polymers with linear chains (silicone elastomers); 3) polymers with cyclolinear, ladder and branched chains.

Polyorganosilazanes are high-molecular compounds with the backbone consisting of alternating silicon and nitrogen atoms with different surrounding groups or organic radicals at the silicon atom. Polyorganosilazanes can be synthesised by the ammonolysis or coammonolysis of organohalogensilanes with subsequent polycondensation or polymerisation of the products formed in the ammonolysis. For example, the ammonolysis of diorganodichlorosilanes yields linear diorganocyclosilazanes:

$$nR_2SiCl_2 + 3nNH_3 \xrightarrow[2nNH_4Cl]{} [R_2SiNH]_n \tag{3.7}$$

The ion polymerisation of these substances forms polycyclic organosilazanes (unlike the polymerisation of organocyclosiloxanes):

$$(3.8)$$

etc.

Polyorganosilazanes are manufactured as linear or cyclic oligomers and as branched or cyclolinear.

Polyelementorganosiloxanes are high-molecular compounds with the backbone consisting of alternating silicon, oxygen atoms and another atom (aluminum, iron, titanium) with different surrounding groups or organic radicals at the silicon atom. *Polyelementorganosiloxanes* can be obtained either by the exchange decomposition of organosilanolates of alkali metals with salts of other metals (aluminum or iron):

$$3nRSi(OH)_2ONa \xrightarrow[nNaCl]{nMeCl_3} \begin{bmatrix} \cdots \end{bmatrix}_n \quad (3.9)$$

or by the heterofunctional condensation of the products of organotrihalogensilane hydrolytic condensation with tetraalkyl derivatives of metals (e.g., titanium).

$$m[RSiO_{0,5x}(OH)_{2-x}]_n \xrightarrow[mR'OH]{mMe(OR')_4} \begin{bmatrix} \cdots \end{bmatrix}_m \quad (3.10)$$

Polyelementorganosiloxane molecules have a branched, ladder or star-shaped structure.

3.1. Preparation of linear oligoorganosiloxanes

Oligoorganosiloxanes are divided into linear (I) and cyclic (II) oligomers:

$$R_3SiO \!\!+\!\! SiR_2O \!\!+\!\!_n SiR_3; \qquad [R_2SiO]_n$$

$$\text{I} \qquad\qquad\qquad \text{II}$$

where R represents various organic radicals.

The properties of oligoorganosiloxanes, as well as those of polyorganosiloxanes (linear and branched) largely depend on the type and structure of their organic radicals. Thus, the thermal-oxidative stability of oligoorganosiloxanes with the aliphatic radical decreases as the number of C atoms in the radicals rises, but increases if there are aryl radicals at the Si atom. For example, oligomethylsiloxanes are more resistant to thermal-oxidative destruction than oligoethylsiloxanes, whereas oligomethylphen-ylsiloxanes are more resistant than oligomethyl- and oligoethylsiloxanes. However, if there are aryl radicals at the Si atom, oligomer viscosity has a greater dependence on temperature, and their congelation point increases.

At present, industry produces a wide range of linear oligoorganosilox-anes with the same and different organic groups at the silicon atom ; how-ever, the most widely used are oligomethylsiloxanes (PMS), oligoethylsi-loxanes (PES), oligomethylphenylsiloxanes (PFMS) and their copolymers.

3.1.1. Preparation of oligomethylsiloxanes

Oligomethylsiloxanes are a mixture of liquid products, mainly of the linear structure:

$$(CH_3)_3Si \left[O-\underset{\underset{CH_3}{|}}{\overset{\overset{CH_3}{|}}{Si}} \right]_n OSi(CH_3)_3;$$

$$n = 3 \div 700$$

They differ in viscosity, which has a wide range (from 0.65 to $1 \cdot 10^6$ mm^2/s) and depends on the degree of polymerisation n.

These liquids can be obtained by the hydrolytic cocondensation of mono- and dichlorine-containing methylsilanes

$$2(CH_3)_3SiCl + n\,(CH_3)_2SiCl_2 \xrightarrow[2(n+1)HCl]{(n+1)H_2O} (CH_3)_3Si \left[O-\underset{\underset{CH_3}{|}}{\overset{\overset{CH_3}{|}}{Si}} \right]_n OSi(CH_3)_3 \qquad (3.11)$$

or by the separate hydrolytic condensation of these compounds and sub-sequent catalytic rearrangement of the products formed. The second tech-nique of oligomethylsiloxane preparation comprises three main stages: the hydrolytic condensation of dimethyldichlorosilane, forming cyclic di-methylsiloxanes:

$$n(CH_3)_2SiCl_2 + n\,H_2O \xrightarrow[2nHCl]{} [(CH_3)_2SiO]_n \qquad (3.12)$$

the hydrolytic condensation of trimethylchlorosilane, forming hexame-thyldisiloxane:

$$2(CH_3)_3SiCl \xrightarrow[2HCl]{H_2O} (CH_3)_3Si\text{-}O\text{-}Si(CH_3)_3 \qquad (3.13)$$

and the catalytic rearrangement of the obtained products, which breaks the Si—O—Si bonds and forms linear oligomer homologues:

$$m[(CH_3)_2SiO]_n + (CH_3)_3Si\text{-}O\text{-}Si(CH_3)_3 \longrightarrow (CH_3)_3Si\left[O\text{-}\underset{\underset{CH_3}{|}}{\overset{\overset{CH_3}{|}}{Si}}\right]_{mn}OSi(CH_3)_3 \qquad (3.14)$$

$$n=3 \div 700$$

The process can be catalysed wih mineral acids, metal sulphates, activated silica-alumina clays or zeolites.

Raw stock: dimethyldichlorosilane [not less than 58% (weight) of chlorine, the density of 1.06-1.07 g/cm³]; trimethylchlorosilane [32.1-33.2% (weight) of chlorine, the density of 0.856-0.859 g/cm³]; kil clay treated with sulfuric or hydrochloric acid.

The process comprises the following stages (Fig. 35): the hydrolytic condensation of dimethyldichlorosilane; the distillation of cyclic dimethylsiloxanes; the hydrolytic condensation of trimethylchlorosilane; the catalytic rearrangement of the hydrolytic condensation products; the vacuum distillation of the rearrangement products.

The hydrolytic condensation of dimethyldichlorosilane is carried out in apparatus 4 with a jacket, an agitator, and backflow condenser 6. First, the apparatus is loaded with a necessary amount of water; then the agitator is switched on and the reactor receives dimethyldichlorosilane at such speed that the hydrolytic condensation temperature does not exceed 40 °C. The reactive mixture from the hydrolyser enters separator 11, where hydrochloric acid and the products of hydrolytic condensation (a mixture of cyclic dimethylsiloxanes) are continuously separated. The hydrochloric acid from the lower part of the separator enters a special collector (not shown in the diagram), and the mixture of cyclic dimethylsiloxanes from the top of the separator is sent into collector 10 and from there into apparatus 7, where it is neutralised with soda.

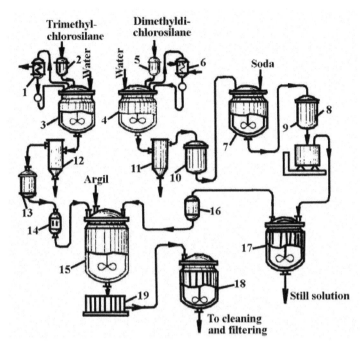

Fig. 35. Production diagram of oligomethylsiloxanes by separate hydrolytic condensation: *1,6-* backflow condensers; *2, 5, 9, 14, 16* - batch boxes; *3, 4* - hydrolysers; *7-* neutraliser; *8, 10, 13* - collectors; *11, 12* - separators; *15* - reactor; *17, 18* - distillation tanks; *19* - pressure filter.

After neutralising, cyclic dimethylsiloxanes are sent into collector *8* and from there through weight batch box *9* into distillation tank *17*. The distillation of the light fractions, which are low-molecular cyclic dimethylsiloxanes, is conducted at a temperature below 230°C in vapours and at 260-280 °C in the tank at a residual pressure of 5.2--9.3 GPa. The distilled product enters batch box *16* and then is subjected to catalytic rearrangement.

The liquid which remains in tank *17* is a nondistillable product, oligomethylsiloxane with a viscosity of 180-220 mm²/s at 25 °C. After purifying with active carbon and filtering, the liquid is bottled and labeled.

The distillation of the hydrolysate obtained in the hydrolytic condensation of dimethyldichlorosilane forms tank residue, which is a colourless liquid without a scent (the solidification point is below 60 °C; the flash point is not lower than 250 °C). This liquid, just like all oligomethylsiloxanes, is completely nontoxic and inert. It can be used as an anti-foaming agent in aqueous-alcoholic media and in mineral oils.

The hydrolytic condensation of trimethylchlorosilane is conducted in apparatus *3,* similarly to that of dimethyldichlorosilane. The reactive mixture enters separator *12*, where hydrochloric acid and the products of hydrolytic condensation (hexamethyldisiloxane) are continuously separated. Hexamethyldisiloxane from the top part of the separator enters collector *13* and is sent through batch box *14* to the rearrangement stage.

Catalytic rearrangement is conducted in reactor *15*, which is a vertical cylindrical apparatus with a spheric bottom and an anchor agitator. The apparatus is fed from batch boxes *14* and *16* with hexamethyldisiloxane and cyclic dimethylsiloxanes in the amounts corresponding to the specified polymerisation stage or the viscosity of the target products. After loading the parent components, the agitator is switched on and the reactor is filled with active clay (7-8% of the amount of the loaded components) through a hatch. The reactor is heated with vapour sent into the jacket to 90-95 °C and is maintained at this level during the process, until a constant viscosity of the product is established. After that, the product is filtered in pressure filter *19*.

If the viscosity of the rearrangement product does not meet the necessary requirements, it should be supplemented with a corresponding component (cyclic dimethylsiloxanes or hexamethyldisiloxane) to achieve a necessary viscosity.

From pressure filter the rearranged product enters vacuum distillation tank *18*. Distillation is conducted at a residual pressure of 1.3-6.6 GPa. In this case mainly low viscosity oligomethylsiloxanes are distilled (5-55 mm^2/s viscosity at 25 °C), and high viscosity liquids remain in the tank. The tank is cooled, and the liquids are collected in a special collector. After the distillation, oligomethylsiloxanes are purified with active carbon from the remaining acid in a special apparatus at 80-90 °C. After filtering is completed, they are bottled and labelled.

The hydrolytic condensation of dimethyldichlorosilane can be achieved by the continuous technique in the flow circuit: pump - heat exchanger - hydrolyser. In the continuous process (Fig.36) circulation pump *3* is fed with water through a regulating gate or rotameter at a certain speed; at the same time, it is filled with dimethyldichlorosilane, which self-flows from collector *1* through a regulating gate or rotameter and shell-and-tube heat exchanger *2* cooled with salt solution (-15 °C). The volume ratio of water and dimethyldichlorosilane is (1.5÷2):1. After that, the reactive mixture is sent through heat exchanger *4* cooled with salt solution (-15 °C) into hydrolyser *5*, and then through a siphon to circulation. The reaction in the flow circuit creates dimethyldichlorosilane hydrolysate and hydrogen chloride, which dissolves in water forming *27%* hydrochloric acid.

Dimethyldichlorosilane

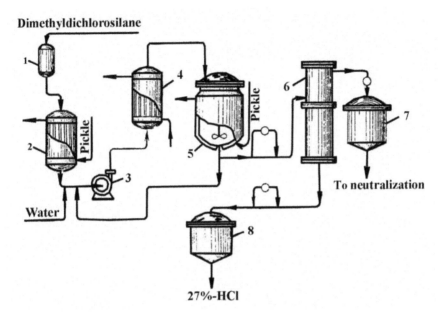

27%-HCl

Fig. 36. Diagram of continuous hydrolytic condensation of dimethyldichlorosilane:1 - batch box; *2, 4* - heat exchangers; *3* - circulation pump; 5 - hydrolyser; *6* - separator; *7,8*- receptacles

The temperature during hydrolytic condensation does not exceed 50 °C; it is maintained with salt solution (-15 °C) fed into heat exchanger *4* and hydrolyser jacket *5*. The mixture of hydrolysate and hydrochloric acid is continuously sent through the hydraulic gate into the middle part of apparatus *6*, where it is separated. The top layer, the product of dimethyldichlorosilane hydrolytic condensation, continuously enters receptacle *7*. From there it is vacuum-loaded into the reactor (not shown in the diagram) to be neutralised with a 15% solution of soda ash. The bottom layer, 27% hydrochloric acid, is poured into receptacle *8*.

Oligomethylsiloxanes are colourless or light yellow liquids, which have no odour. Their density ranges from 0.90 to 0.98 g/cm^3 at 20 °C, the solidification point is below -60 °C. All are soluble in benzene, toluene, alcohols and chlorinated hydrocarbons; the liquids with a viscosity below 10 mm^2/s) are also soluble in acetone and dioxane. Oligomethylsiloxanes are inert and nontoxic.

Some physicochemical properties and applications of oligomethylsiloxanes are given in Table 15. As seen from the table, PMS-300, PMS-400 and PMS-500 oligomethylsiloxane brands can be used as a basis in the production of mineral butters, anti-foaming and anti-adhesion emulsions.

Table 15. Important physicochemical properties of oligomethylsiloxanes and their applications

Brand	Density at 20 °C, g/cm³	Viscosity at 20 °C, mm²/s	Temperature, °C	
			flash, not lower than	solidification, not higher than
PMS-5	-	4.5-5.5	115	-60
PMS-6	0.95	5.6-6.6	130	-60
PMS-10	0.94	9-11	170	-60
PMS-15	0.95	13.5-16.5	200	-60
PMS-20	0.96	18-22	200	-60
PMS-25	0.96	22.5-27.5	200	-60
PMS-30	-	27.6-33	200	-60
PMS-40	0.97	36-44	200	-60
PMS-60	-	57-63	300	-60
PMS-70	_	66.5-73.5	300	-60
PMS-150	_	142.5-157.5	300	-60
PMS-50	0.97	45-55	220	-60
PMS-100	0.98	95-105	300	-60
PMS-200	_	190-210	300	-60
PMS-300	0.98	285-315	300	-60
PMS-400	-	389-420	300	-60
PMS-500	-	475-525	300	-60
PMS-700	-	665-735	300	-60
PMS-1000	-	950-1050	300	-60

Note 1. The refraction index n_D^{20} for all liquids ranges from 1.3900 to 1.4040.
Note 2. The thermal conductivity coefficient for all liquids is 0.33--2.7 KJ/(m·h·K). Specific thermal capacity for all liquids is on the average 1.25-1.59 KJ/(kg K).

3.1.2. Performance characteristics and applications of oligodimethylsiloxanes

The performance characteristics of oligodimethylsiloxanes are given in references (Skorokhodov 1978). However, a review of these properties in combination with general information about their applications and the no-menclature of technical oligodimethylsiloxane materials seems appropriate and may be useful for many specialists.

Taking this into account, this monograph gives information about the performance characteristics, applications and nomenclature of these materials in the form of a table (Table 15.1) (Sobolevsky 1985). The temperature performance characteristics are necessary to determine the acceptable temperature performance range for choosing oligodimethylsiloxanes for different applications.

Oligodimethylsiloxanes (PMS) and oligodimethyl(methyl)siloxanes (PMS-b) ("b" stands for "branched structure") have a wide range of properties depending on the composition, structure and molecular weight. The main characteristic which determines their fields of application is kinematic viscosity. Its value is found in the brand of PMS and PMS-b liquids.

The oligomers of this oligoorganosiloxane group, given in Table 15.1, are classified according to their applications in the following way.

PMS-1 - PMS-2 liquids are used as cooling and damping liquids in various devices below 60 °C; PMS-10 - PMS-1000 are used as damping and hydraulic liquids in devices and mechanisms, as well as dispersion media for plastic lubricants, mineral butters and pastes, which have performed well in the stop valves of gas pipelines.

Liquids with a viscosity higher than 10 000 mm^2/s are used as damping liquids in devices, dampers of torsional vibrations in diesel locomotives and as dispersion media for high viscosity plastic lubricants.

Liquids with a branched molecular structure (PMS-1p - PMS-3p) are used in devices and mechanisms as cooling and damping liquids to -100°C, and PMS-10p - PMS-400p as dispersion media in low temperature oils, lubricants and damping liquids.

All these applications are mainly based on PMS liquids with a small dependence of viscosity on temperature, low temperatures of glass transition and solidification (melting) and a relatively high thermal and thermal-oxidative stability (up to 180-200°C).

Because PMS have low surface tensions (18-20 mH/m); they are also widely used as anti-foaming additives in mineral oils and other nonpolar organic media of the $10~^4$% order (weight). Small quantities of PMS liquids eliminate bubbles and give a good flowability to varnishes and coatings, as well as better polishing abilities to polishing liquids.

Table 15.1. Temperature performance characteristics of technical oligodimethylsiloxanes and oligodimethyl(methyl)siloxanes

Brand of the oligomer	Temperature, °C							Temperatures of explosive limits of vapours with air	
	crystal melting	glass transition	flash	inflammation	self-inflammation	start of thermal destruction	start of oxidation	lower	higher
PMS-1	-	-	32	43	-	-	-	-	-
PMS-1.5	-	-	40	71	340	-	-	32	144
PMS-2	-	-	56	79	-	-	-	-	-
PMS-2 -PMS-10	-70	-135	85-170	100-160	305-330	-	-	60-128	240-256
PMS-15 - PMS-50	43-58	-125	200-220	271-274	360-378	330	180	200-208	292-295
PMS-100 - PMS-1000	-35-40	-125-127	300	315	390	350	180	214	297
PMS-10000- PMS-100000	-50-55	-	200	315	400	370	180	270	306
PMS-1000000	-28	-128	200	-	460	-	-	-	-
PMS-1p	-110*	-	30	-	280	-	-	15	147
PMS-1.5b	-	-	50	-	340	-	-	32	144
PMS-2b- PMS-3b	-	-	70-85	-	305	-	-	60	240
PMS-10b- PMS-50b	-120*	-125	170-220	-	400	-	-	-	-
PMS-100b- PMS-400b	-120	-125							

These properties are also connected with low surface tension of oligodimethylsiloxanes.

PMS liquids have shown good results as the immobile phase in gasliquid chromatography (SE-30, DC 200(220, 410,550), OV-1, etc.).

The data on the volume properties of PMS liquids (i.e. coefficient of volumetric expansion, relative volume variation, coefficient of isothermal compressibility) are essential for the performance characteristics of oligodimethylsiloxanes in hydraulic systems, hydraulic shocks and dampers: they allow one to determine the working characteristics of these systems with some brands of PMS liquids at different temperatures and pressures.

The dielectric properties of oligodimethylsiloxanes and their dependence on temperature point to good dielectric characteristics of PMS liquids. Taking into consideration that PMS do not form conductive carbon particles in case of electric breakdown or sparking, it is obvious why they are used as liquid dielectrics in transformers and other electric devices. There is a law in Japan and the USA which forbids the use of nonflammable yet toxic pentachlorodiphenyl and leaves room for oligodimethylsiloxanes. This has stimulated a much more active production of PMS liquids in Japan, Germany and other countries.

PMS liquids have mediocre lubricating capacities for steel-on-steel friction, although some combinations of friction couples, e.g. bronze-on-steel, brass-on-plastic, perform satisfactorily if lubricated hydrodynamically with PMS liquids. The use of PMS as liquid media in oils and lubricants compounded with various antiwear additives and dispersions, creates favourable conditions for selective adsorption of additives on friction surfaces and helps to form an antiwear film which greatly increases the pressures and slip velocities.

PMS liquids are corrosion-inert substances. Under normal conditions and heated to 100-150 °C they do not cause corrosion and for a long period of time do not change in airflow when in contact with aluminum and magnesium alloys, bronzes, carbon and doped steels, as well as titanium alloys. PMS liquids do not change their properties under 100 °C in air for 200 hours in contact with the above-listed alloys as well as with beryllium, bismuth, cadmium, Invar alloy, brass, copper, melchior, solder, lead, silver. The stability of the properties of PMS liquids in these conditions is usually accompanied by the absence of metal and alloy corrosion, although the colour of the metal surface may slightly change.

At 65-100 °C PMS liquids do not change their viscosity and do not cause any significant swelling or washing in many polymers and polymer materials (The applications of PMS liquids see in Skorokhodov 1978).

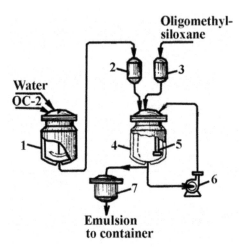

Fig. 37. Production diagram of antifoaming silicone emulsion: *1* - reactor agitator; *2, 3* - batch boxes; *4* - reactor emulsifier; *5* - hydrodynamic changer; *6* - pump; *7*- collector

Preparation of emulsions

Recently, emulsions based on oligomethylsiloxane liquids have been widely used as antifoaming agents to suppress or prevent foam in aqueous, nonaqueous and low-aqueous solutions, and as separating lubricants in pressure molds for tyre production. They are also used in foundry work to manufacture rods and in the production of elastomers and plastics.

Preparation of the antifoaming silicone emulsion. Antifoaming silicone emulsion (OS-2) is a 30% aqueous emulsion of oligomethylsiloxane liquid (the viscosity is below 500 mm^2/s) stabilised with the diethyleneglycol ether of higher fatty alcohols C_{14}—C_{18}.

To obtain antifoaming silicone emulsion (Fig. 37), agitator reactor *1* is filled with metered water and loaded with OS-2 through a hatch at agitation. The amount of water taken corresponds to a 3% solution of OS-2. The mixture in the apparatus is heated to 60-70 °C and agitated at this temperature for 30 minutes. The obtained solution is cooled to 15-20 °C and held at agitation for 24 hours. The ready solution is loaded at a residual pressure of 400-520 GPa into batch box *2*. Steel enameled reactor emulsifier *4*, which has a hydrodynamic changer and a pump, is first loaded from batch box *2* with a 3% solution of OS-2; then, through pump *6*, the lower part of the reactor receives oligomethylsiloxane liquid from batch box *3*. The solution is sent with pump *6* into hydrodynamic changer *5* under 0.2-0.3 MPa. After all the oligomethylsiloxane liquid has been introduced, the solution is emulsified for 1 hour until stable emulsion is obtained. The finished emulsion is poured into receptacle *7*. Silicone anti-foaming agent should be stored in dry enclosed rooms at a temperature not higher than 30 °C.

Antifoaming silicone emulsion is a homogeneous liquid with colour ranging from white to light brown; it is nonflammable and nontoxic. It should meet the following requirements:

Stability, hours, not lower than   24
Solid residue content, not less than   25
pH, not lower than   6
Nonvolatile substance content, % (weight), not less than   25
Antifoaming ability (the speed of foam elimination), ml/s, not less than   0.5

Silicone is used to suppress foam or to preventing foam in aqueous solutions of anionic and nonionic surface-active substances.

Other silicone antifoaming agents can be based on oligomethylsiloxanes of various viscosity and are prepared similarly. Instead of OS-2 solution, some emulsion brands use the Solvar solution or solutions of emulsifiers OP-7 and OP-10.

Preparation of the separating silicone emulsion. Separating silicone emulsion is a 70% aqueous emulsion of oligomethylsiloxane liquid (with a viscosity below 400 mm^2/s), emulsified with OP-7 or OP-10.

The separating silicone emulsion is prepared (Fig.38) by sending distilled water into reactor *3* from collector *2*. The reactor jacket is filled with vapour; the water is heated to 70 °C or lower. The agitator is switched on and the reactor receives a required amount of emulsifier (OP-7 or OP-10) through a hatch; the contents of the reactor are agitated until the solution becomes homogeneous. The emulsifier solution heated to 70 °C is pumped under nitrogen pressure (0.07 MPa) through filter *5* and homogeniser *6* to wash the homogeniser. To prepare the emulsion mixture, the emulsifier solution is again sent from reactor *7* by nitrogen flow (0.07 MPa) into reactor *3*; it is filled at agitation from batch box *4* with a necessary amount of oligomethylsiloxane. The reactor contents are agitated for at least 25-30 minutes to distribute the oligomethylsiloxane liquid in the emulsifier solution at a temperature below 40 °C.

As soon as it is prepared, the mixture from reactor *3* is sent by nitrogen flow (0.07 MPa) at agitation through filter *5* into homogeniser *6*. The mixture is emulsified due to the high pressure in the homogenising head of apparatus *6*. Maximum pressure is 30 MPa. The resulting shock forces, stream cutting, as well as cavitation make for a superfine emulsion.

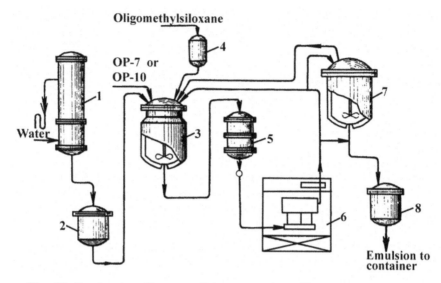

Fig. 38. Production diagram of the separating silicone emulsion:

1 - distiller; 2, 8 - collectors; 3, 7 - reactors; 4 - batch box; 5 - filter; 6 - homogeniser

At first, before the pressure reaches 10-25 MPa, subquality emulsion is returned into reactor 3. After the operating pressure (30 MPa) has been established, quality emulsion is pumped into collector 8. The duration of homogenisation depends on the amount of emulsion mixture and the pressure in apparatus 6. The finished product from collector 8 is poured into readied and weighed containers.

The separating silicone emulsion is a homogeneous white liquid. It should meet the following requirements:

Silicon content, %	25-28
Surface tension, N/m, not more than	40-1
pH of the aqueous solution	6-7.5
Stability (at 1:4 dilution) in 5 hours, %, not less than	11.5

Silicone emulsion is used in foundry work as a separating agent for manufacturing rods based on furane resins at hot mounting.

Other silicone separating emulsions based on oligomethylsiloxanes are prepared similarly. They can be used as separating lubricants in pressure molds for tyre production. as well as in the production of elastomers and plastics.

As stated above, oligomethylsiloxanes can be obtained not only by the separate hydrolytic condensation of original methylchlorosilanes, but also by the joint hydrolytic condensation of mono- and dihalogenmethylsilanes.

Moreover, apart from dimethyldichlorosilane, the second component of high-viscosity oligomethylsiloxanes can be tank residue, which remains after the distillation of methylchlorosilanes.

Preparation of high-viscosity oligomethylsiloxanes

High viscosity (more than 1000 mm^2/s) oligomethylsiloxanes can be prepared by the hydrolytic cocondensation of trimethylchlorosilane and tank residue after methylchlorosilane distillation with subsequent partial polycondensation of the obtained products.

Cohydrolysis and partial condensation proceed thus:

$$2(CH_3)_3SiCI + mCH_3SiR_{3-n}CI_n + (m+1)H_2O \xrightarrow[2(m+1)HCI;\ H_2O]{}$$

$$\longrightarrow (CH_3)_3SiO\text{-}[Si(CH_3)_nR_{3-n}O]_m\text{-}Si(CH_3)_3 \qquad (3.15)$$

R=CH3, H or a fragment of the molecule of a partially condensed product.

In this process the product is partially condensed during hydrolytic cocondensation and during solvent distillation.

Raw stock: tank residue after distilling methylchlorosilanes (40-55% of chlorine, not more than 3% of dimethyldichlorosilane); trimethylchlorosilane (33-45% of chlorine); oil or coke toluene (not less than 98% of the 109.5-111 °C fraction).

The production process comprises two main stages (Fig. 39): hydrolytic cocondensation; solvent distillation and filtering of the finished product.

The reactive mixture is prepared in agitator 5, a cylindrical apparatus with an anchor or gate agitator. Tank residue from batch box 1 self-flows through weight batch box 4 into the apparatus, which is also filled with trimethylchlorosilane (14-16% of the quantity of the loaded tank residue). During the introduction of the components the agitator in the reactor is switched on. Then, the apparatus receives toluene from batch box 3 (approximately 1 volume per 1 volume of methylchlorosilanes). The mixture is agitated for 15-30 minutes and sent into weight batch box 6.

Hydrolytic cocondensation is carried out in enameled hydrolyser 7 with a water vapour jacket and an agitator. The hydrolyser is filled with a required amount of water; then, under the water layer, the reactive mixture is sent through a siphon from batch box 6. The temperature during hydrolytic cocondensation should not exceed 50-55 °C. Most of the released hydrogen chloride dissolves in the water in the apparatus, and the excess of HCl passes through backflow condenser 8 and enters tower 9, where is absorbed by water. The vapours of water and reaction products, which are entrained by hydrogen chloride and condensed in the backflow condenser, are sent back into the hydrolyser. After all the reactive mixture has been introduced, it is agitated for 30 minutes.

The toluene solution of oligomethylsiloxanes, which is formed by hydrolytic cocondensation, is held in the hydrolyser for 1.5-2 hours to separate water. The lower layer (water) is poured through a run-down box into collector *10* and sent to biochemical purification; the organic layer (toluene solution of oligomethylsiloxanes) is washed with water until it gives a neutral reaction. Acid flush waters are also poured through a run-down box into collector *10* and then sent to biochemical purification.

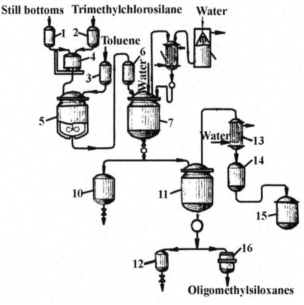

Fig. 39. Production diagram of high-viscosity oligomethylsiloxanes: *1-4, 6-* batch boxes; *5-* agitator; *7-* hydrolyser; *8, 13* - condensers; *9-* absorption column; *10, 12* - collectors; *11* - vacuum distillation tank; *14* - receptacle; *15* - settling box; *16* - nutsch filter

The neutral product from the hydrolyser is sent by nitrogen flow (or in vacuum) to distil the solvent in vacuum distillation tank 11, which is an enameled apparatus with an agitator and a water vapour jacket. Before distillation, the tank jacket is filled with vapour and the product is heated to 80-100 °C. The product is additionally purified from traces of moisture ("clarified") by 2-4 hours of settling; the clarified product should be transparent. The water with impurities of the product is settled in the lower part of the tank and poured into collector *12* for a prolonged settling and water separation; the clarified product is distilled to obtain toluene.

Toluene vapours are sent into cooler *13* to condense. The condensate is collected in receptacle *14*. First, toluene is distilled at atmospheric pressure and 110-120 °C, then, vacuum is gradually created in the system. The final distillation is carried out at 110-120 °C and the residual pressure of 200 GPa, until the toluene content in the product is not more than 1%. When the analysis is positive, the distillation of toluene is stopped and the product is cooled to 50-60 °C by sending water into the tank jacket. The distilled toluene from receptacle *14* self-flows into settling box *15*, where it is separated from water and mechanical impurities. Purified toluene can be re-used in the production of oligomethylsiloxanes.

The cooled liquid from tank *11* is sent through the lower choke to nutsch filter *16* with a coarse calico filter and after filtering is poured into containers.

High-viscosity oligomethylsiloxanes are a mixture of α,ω-bis(trimethylsiloxy)dimethylsiloxanes. They are transparent liquids (or yellowish brown) without mechanical impurities. They should meet the following requirements: the viscosity at 20 °C is more than 1000 mm^2/s, the flash point is not lower than 200 °C, the solidification point is below -50 °C.

Table 16 contains important physicochemical properties of high-viscosity oligomethylsiloxanes (PMS) and their applications.

Similarly to the technique described above, the hydrolytic cocondensation of trimethylchlorosilane and various difunctional organochlorosilanes with subsequent partial polycondensation of the products can be used to obtain a great variety of α,ω-bis(trimethylsiloxy)diorganosiloxanes.

For example, α,ω-bis(trimethylsiloxy)dimethyl(methyldichlorophenyl)siloxanes can be obtained by the hydrolytic cocondensation of trimethylchlorosilane, dimethyldichlorosilane and methyldichlorophenyldichlorosilane with subsequent partial polycondensation:

$$(CH_3)_3Si-[OSi(CH_3)_2]_n-[OSi(CH_3)(C_6H_3Cl_2]_m-OSi(CH_3)_3$$

Table 16. Important physicochemical properties of high-viscosity oligomethylsiloxanes and their applications

Brand	Density at 20 °C, g/cm^3	Viscosity* at 20 °C, MPa	Temperature, °C		Applications
			flash, not lower than	solidification, not higher than	
PMS-2500	0.98	2500	200	-60	Antifoaming additives to nonviscous oils

PMS-5000	0.98	5000	200	-50	Damping liquids in gages, vibration indicators and dampers of torsional vibrations
PMS-10000	0.98	10000	200	-50	
PMS-15000	0.98	15000	200	-50	
PMS-20000	0.98	20000	200	-50	
PMS-30000	0.98	30000	200	-50	
PMS-50000	0.98	50000	200	-50	
PMS-75000	0.98	75000	200	-50	
PMS-100000	0.98	100000	200	-50	
PMS-200000	0.98	200000	200	-50	
PMS-250000	0.98	250000	200	-50	
*A 10% fluctuation is tolerated.					

The hydrolytic cocondensation of trimethylchlorosilane, dimethyldichlorosilane and isobutylmethyldichlorosilane with subsequent partial polycondensation of the products can be used to obtain oligomers with methyl and isobutyl groups at the Si atom:

$$(CH_3)_3Si\text{-}[OSi(CH_3)_2]_n\text{-}[OSi(CH_3)(i\text{-}C_4H_9)]_m\text{-}OSi(CH_3)_3$$

The hydrolytic cocondensation of trimethylchlorosilane, dimethyldichlorosilane and methyl-β-cyanoethyldichlorosilane with subsequent partial polycondensation of the products can be used to obtain oligomers with methyl and cyanoethyl groups at the Si atom:

$$(CH_3)_3Si\text{-}[OSi(CH_3)_2]_n\text{-}[OSi(CH_3)(CH_2CH_2CN)]_m\text{-}OSi(CH_3)_3$$

It should be kept in mind that alongside with the product of joint condensation, hydrolytic cocondensation also forms the products of separate condensation, i.e. low molecular cyclic diorganosiloxanes, oligoorganosiloxanes with end hydroxyl groups and hexamethyldisiloxane. Thus, to increase the yield of target products it is advisable to include the stage of catalytic regrouping with 90-95% sulfuric acid, active kil clay or bentonite. The regrouping with sulfuric acid should be conducted at 20 °C for 5-7 hours, and the process using clay or bentonite, at 90 °C for 2-4 hours.

The modified oligomers mentioned above are nontoxic liquids, which boil above 200-300 °C at 1.3-4 GPa. Their viscosity ranges from 40 to 300

mm^2/s, and the solidification point is much lower than that of oligomethyl-siloxanes and ranges from -90 to -125 °C.

The applications of these liquids are quite varied. E.g., oligomers with isobutylmethylsiloxygroups are used as antifoaming agents and low temperature dielectrics; oligomers with methyl-p-cyanethylsiloxygroups are used as antifoaming agents in nonaqueous media as solvent-stable lubricants and oils, as the immobile phase in gas-liquid chromatography, as polishing agents and additives for varnishes and paints, as well as dielectrics with high dielectric permeability.

3.1.3. Preparation of oligoethylsiloxanes

Oligoethylsiloxanes are a mixture of liquid products, mainly of the linear structure:

$$(C_2H_5)_3Si \left[O-\underset{\underset{C_2H_5}{|}}{\overset{\overset{C_2H_5}{|}}{Si}} \right]_n OSi(C_2H_5)_3$$

They are liquids with a viscosity from 1.5 to $1 \cdot 10^6$ mm^2/s, which changes depending on the conditions for condensation and the ratio of parent reactants. These oligomers can also include impurities of cyclic compounds $[(C_2H_5)_2SiO]_{n,}$; high-boiling fractions may contain branched oligomers.

Oligoethylsiloxanes are obtained by the hydrolytic condensation of ethylethoxysilanes synthesised by Grignard reaction, or by the hydrolytic condensation of diethyldichlorosilane with subsequent catalytic regrouping of the hydrolysis products.

A. The production of oligoethylsiloxanes by the hydrolytic condensation of ethylethoxysilanes is described below.

1. The synthesis of ethylethoxysilanes takes place when ethylchloride interacts with tetraethoxysilane in the presence of magnesium in toluene medium:

$$C_2H_5CI + Mg \longrightarrow C_2H_5MgCI \qquad (3.16)$$

$$Si(OC_2H_5)_4 \xrightarrow[C_2H_5OMgCI]{C_2H_5MgCI} C_2H_5Si(OC_2H_5)_3 \xrightarrow[C_2H_5OMgCI]{C_2H_5MgCI}$$

$$\longrightarrow (C_2H_5)_2Si(OC_2H_5)_2 \xrightarrow[C_2H_5OMgCI]{C_2H_5MgCI} (C_2H_5)_3SiOC_2H_5 \qquad (3.17)$$

It is accompanied by a secondary reaction

$$C_2H_5MgCI + C_2H_5CI \longrightarrow C_4H_{10} + MgCI_2 \qquad (3.18)$$

and other side processes caused by the presence of the traces of moisture in the parent reactants:

$$C_2H_5MgCI + H_2O \longrightarrow C_2H_6 + Mg(OH)CI \qquad (3.19)$$

$$Mg + 2H_2O \longrightarrow Mg(OH)_2 + H_2 \qquad (3.20)$$

The degree of tetraethoxysilane ethylation depends on the ratio of the components; therefore, depending on the ratio of ethyl magnesium chloride and tetraethoxysilane, one can obtain a mixture of ethylethoxysilanes with a preferential formation of a certain ethylethoxysilane. For example, the 2:1 ratio shifts the reaction towards the preferential formation of diethyldiethoxysilane.

2. The hydrolytic condensation of ethylethoxysilanes occurs in the excess of water and in the presence of sulfuric acid, which is a catalyst and a reactant that interacts with the by-product of Grignard reaction, ethyl magnesium chloride, to form magnesium chloride and ethyl alcohol. The process proceeds thus:

$$2(C_2H_5)_3SiOC_2H_5 + n(C_2H_5)_2Si(OC_2H_5)_2 + (n+1)H_2O \xrightarrow[2(n+1)C_2H_5OH]{HCI}$$

$$(C_2H_5)_3Si \left[O-\underset{\underset{C_2H_5}{|}}{\overset{\overset{C_2H_5}{|}}{Si}} \right]_n OSi(C_2H_5)_3 \qquad (3.21)$$

The main reaction is accompanied by some by-processes:

$$n(C_2H_5)_2Si(OC_2H_5)_2 \xrightarrow[2nC_2H_5OH]{nH_2O,\ HCI} [(C_2H_5)_2SiO]_n \qquad (3.22)$$

$$2(C_2H_5)_3SiOC_2H_5 \xrightarrow[2C_2H_5OH]{H_2O,\ HCI} (C_2H_5)_3Si-O-Si(C_2H_5)_3 \qquad (3.23)$$

$$n(C_2H_5)_2SiCI_2 \xrightarrow[2nC_2H_5OH]{(n+1)H_2O,\ HCI} HO-[Si(C_2H_5)_2O]_n-H \qquad (3.24)$$

$$(C_2H_5)_3SiOC_2H_5 + n(C_2H_5)_2Si(OC_2H_5)_2 \xrightarrow[2nC_2H_5OH]{(n+1)H_2O,\ HCI}$$

$$\longrightarrow (C_2H_5)_3SiO\text{-}[Si(C_2H_5)_2O]_{n\text{-}1}\text{-}Si(C_2H_5)_2OH \qquad (3.25)$$

Small impurities of ethyltriethoxysilane form a certain amount of branched oligomers.

3. The yield of the main product (I) can be increased by the catalytic regrouping of the products of the hydrolytic condensation of ethylethoxysilanes. For this purpose, the products of hydrolytic condensation are treated with sulfuric acid or various types of clay (kil, askanite, bentonite) activated with mineral acids. It is accompanied by the following processes:

$$(C_2H_5)_3Si\text{-}O\text{-}Si(C_2H_5)_3 + m[(C_2H_5)_2SiO]_n \underset{}{\overset{H^+}{\rightleftharpoons}}$$

$$\rightleftharpoons (C_2H_5)_3SiO\text{-}[Si(C_2H_5)_2O]_{mn}\text{-}Si(C_2H_5)_3 \qquad (3.26)$$

$$HO\text{-}[Si(C_2H_5)_2O]_n\text{-}H + (C_2H_5)_3Si\text{-}O\text{-}Si(C_2H_5)_3 \underset{H_2O}{\overset{H^+}{\rightleftharpoons}} (C_2H_5)_3SiO\text{-}[Si(C_2H_5)_2O]_n\text{-}Si(C_2H_5)_3 \quad (3.27)$$

$$(C_2H_5)_3SiO\text{-}[Si(C_2H_5)_2O]_{n\text{-}1}\text{-}Si(C_2H_5)_2OH \underset{H_2O}{\overset{H^+}{\rightleftharpoons}} (C_2H_5)_3SiO\text{-}[Si(C_2H_5)_2O]_{2n}\text{-}Si(C_2H_5)_3 \quad (3.28)$$

B. The production of oligoethylsiloxanes by the hydrolytic condensation of diethyldichlorosilane is described below.

1. The hydrolytic condensation of diethyldichlorosilane is conducted in the excess of water:

$$n(C_2H_5)_2SiCl_2 \xrightarrow[2nHCl]{nH_2O} [(C_2H_5)_2SiO]_n \qquad (3.29)$$

$$n(C_2H_5)_2SiCl_2 \xrightarrow[2nHCl]{(n+1)H_2O} HO\text{-}[Si(C_2H_5)_2O]_n\text{-}H \qquad (3.30)$$

2. The catalytic regrouping of the products of the hydrolytic condensation of diethyldichlorosilane is conducted in the following way. They are mixed with 15-20% of ethylsiloxane liquid (synthesised by the hydrolysis of the reactive 3:1 mixture of ethyl magnesium chloride and tetraethoxysilane and containing mostly hexaethyldisiloxane) and treated with activated kil clay or sulfuric acid (the reaction is similar to the one shown above).

Raw stock: tetraethoxysilane (160-180 °C fraction, $d_4^{20}=0.930\div0.940$); ethylchloride (the boiling point is 12.5 °C, $d_4^{20} == 0.916\div0.920$); toluene (the boiling point is 109.5-111°C, $d_4^{20} = 0.865\pm0.002$); magnesium; 25-30% hydrochloric acid.

Oligoethylsiloxane production comprises three main stages: the synthesis of ethylethoxysilanes; the hydrolytic condensation of ethylethoxysilanes or diethyldichlorosilane; the catalytic regrouping of the products of hydrolytic condensation.

In large-scale production it is more economical to conduct the hydrolytic condensation of ethylethoxysilanes and the catalytic regrouping of the products by the continuous technique. In this case ethylethoxysilanes are synthesised not with magnesium chipping, but with pelleted magnesium, which contains at least 97% of active magnesium.

In the production of oligoethylsiloxanes by the continuous technique (Fig. 40) agitator 5 receives from batch boxes toluene, ethylchloride and tetraethoxysilane. The mixture is agitated for 30 minutes and checked for correct dosage.

Ethylethoxysilanes (ethyl paste) are synthesised in continuous reactor 6 with backflow condenser 7 cooled with salt solution.

Reactor 6 is a vertical cylindrical apparatus consisting of five sections (rings) with a blade agitator ($n = 22$ rotations/min). Rings I and II have the diameter of 400 mm, rings III and IV, 500 mm, and ring V, a separator, has the diameter of 1500 mm. Each ring has a water vapour jacket and a thermocouple. Rings II, III and IV have indicators of magnesium level. The allowable pressure in the reactor is 0.07 MPa, in the jackets 0.3 MPa.

The synthesis is carried out in the excess of pelleted magnesium. Depending on the ratio of the parent components, one obtains ethyl paste of various brands from 2.2 to 3.0 (the numbers denote the ratio of ethylchloride and tetraethoxysilane in the reactive mixture). First, reactor 6 is heated to 110-120 °C in nitrogen flow (to remove moisture traces) for 8 hours; then it is cooled to 80°C, and part of the reactive mixture is pumped into the lower part of the ring.

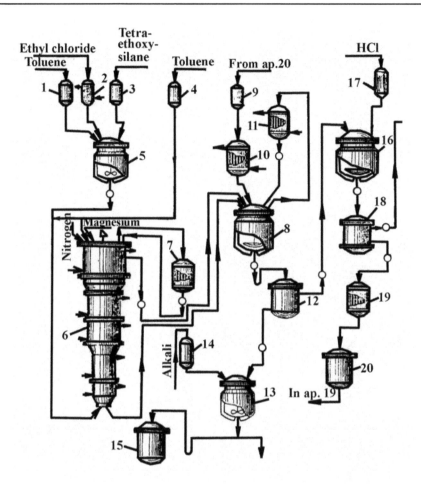

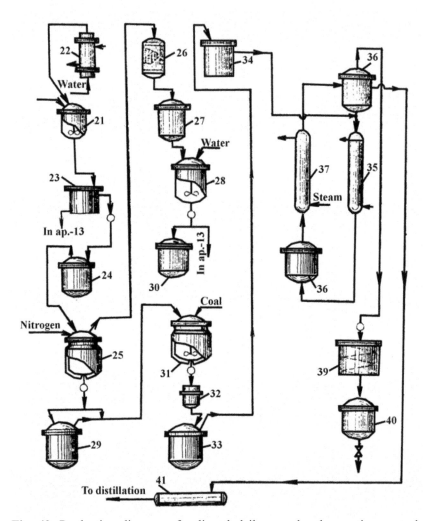

Fig. 40. Production diagram of oligoethylsiloxanes by the continuous technique: *1-4, 9, 14, 17* - batch boxes; *5, 21* - agitators; *6, 36*- reactors; *7, 10, 11, 19, 26* - coolers; *8* - forehydrolyser; *12, 18, 23* - settling boxes; *13* - neutraliser; *15, 20, 24, 29, 40* - collectors; *16* - hydrolyser; *22* - heat exchanger; *25* - distillation tank; *27, 30, 33* - receptacles; *28* - flusher; *31* - purification apparatus; *32* - nutsch filter; *34* - pressure container; *35* - circulation tube; *37* - bubble tube; *38* - separator; *39* - trap; *41* - container

At 50-60 °C the reactor is loaded with pelleted magnesium and the mixture is held until the temperature in ring I rises spontaneously to 100±10 °C. After the "activation" of the reaction, the jacket of ring I is filled with water; the lower part of the apparatus is continuously and gradually filled with the reactive mixture. Each 10 minutes pelleted magnesium is added from the top.

After that (at the same time as magnesium) the reactive mixture is loaded faster and the jackets of rings II—IV and the separator are filled with water to keep the temperature in the following range: ring I - 90±5 °C; ring II - 85±5 °C; rings III and IV - 95±5 °C; separator - 90±5 °C. Magnesium content in the rings should be the following: 35-45% for the top of ring III, 25-30% for the bottom of ring IV and 10-12% for the top of ring IV. In this operation mode the ethylethoxysilanes (ethyl paste) formed are continuously withdrawn from the separator to hydrolytic condensation in forehydrolyser 8.

To prevent water vapours and HCl from hydrolyser 16 from entering the separator of the reactor , the reactor is filled with nitrogen (0.02 MPa) and the synthesis is conducted with the air gate closed. To dilute ethyl paste (in case it thickens), the separator is fed toluene from batch box 4.

If it is necessary to stop the reactor, the flow of magnesium and reactive mixture is stopped; the lower part of the apparatus is filled with toluene (until the reaction is completely terminated), the reactor is cooled to 80 °C; the indicators of magnesium content in the rings are switched off; the liquid contents of the reactor are loaded into forehydrolyser 8 through the dumping line; the unloading line from the reactor to the hydrolyser is disconnected and the remaining paste with magnesium is sent through a choke into closeable pots. The agitator is switched off; the lower lid of the reactor is opened and the reactor is unloaded and cleaned.

The hydrolytic condensation of ethyl paste is carried out in the following way. Ethyl paste continuously self-flows from the top part of the separator through a siphon into forehydrolyser 8; recirculating hydrochloric acid is sent there from batch box 9 through cooler 10. The jacket of the forehydrolyser is filled with water. The vapours condensed in backflow condenser 11.

The products of hydrolytic condensation enter settling box 12 through a siphon and a hydraulic gate. The bottom layer, the acidic aqueous-alcoholic solution of $MgCl_2$, is sent through a hydraulic gate to be neutralised in apparatus 13; the top layer, the acidic toluene solution of the products of partial condensation, is sent through a siphon to further hydrolytic condensation in hydrolyser 16. This hydrolyser is also filled through a siphon with 15% hydrochloric acid. The jacket of the apparatus is filled with vapour (0.3 MPa).

From the hydrolyser the products of condensation enter settling box 18 through a siphon and a hydraulic gate. The bottom layer, hydrochloric acid, self-flows through cooler 19 into collector 20; it is then pumped into batch box 9 to be used at the forehydrolysis stage. The top layer, the acidic toluene solution of condensation products, is first sent into agitator *21*, where it is washed with water heated in heat exchanger *22*, and then into settling box *23*. When the toluene solution of condensation products is washed and separated from flush waters, it self-flows into collector *24*, from where it is pumped to toluene distillation into tank *25*. From settling box *23*, the flush waters are sent into neutraliser *13*.

In the neutraliser, the acidic aqueous-alcoholic solution of magnesium chloride is treated with a 20% alkali solution. The contents of the neutraliser are agitated for 15 minutes and sampled for acidity. At 6-8 pH the solution is pumped into a tower (not shown in the diagram) to extract the alcohol-toluene fraction from the neutral solution of magnesium chloride; the 25% solution of magnesium chloride is continuously collected into collector *15* through a hydraulic gate as a marketable product. It can be used as operating fluid in cooling systems.

In tank *25* the products of hydrolytic condensation are distilled from toluene. Cooler *26* is filled with water, and the tank jacket is filled with water vapour. The contents of the tank are heated to 80-90 °C and held at this temperature for 1 hour. The separated water and the intermediate layer are poured off into the intermediate container (not shown in the diagram); then toluene is distilled. First, the temperature in the tank at atmospheric pressure reaches 130 °C; then, the tank is cooled to 70-90 °C and a residual pressure of 0.04-0.06 MPa is created in the system. Further distillation is conducted in the tank to 150 °C. The toluene vapours condensed in cooler *26* are collected in receptacle *27* and sent by compressed nitrogen flow (0.07 MPa) into flusher *28* as they accumulate. The flusher is filled with water, and the mixture is agitated for 10 minutes; after that the agitator is switched off and the mixture is settled for 2 hours. The bottom layer, aqueous-alcoholic solution, is poured into neutraliser *13*, and the top layer, washed toluene, is sampled for moisture content. If moisture content does not exceed 0.06%, toluene is poured into receptacle *30*, sent to azeotropic drying (until the moisture content does not exceed 0.02%) and re-used in reactive mixtures.

After toluene distillation, the contents of tank *25* are sampled for toluene content, which should not exceed 1.5%. When the analysis is satisfactory, the technical product is cooled to 60 °C with water sent into the tank jacket, loaded into collector *29* by compressed nitrogen flow (0.07 MPa) and pumped into apparatus *31* for purification. The purification is done with active coal or askanite. Apparatus 31 is loaded with 1-2% of coal or 4-6% of askanite (as against the quantity of the loaded product of hydrolytic condensation). After loading the adsorbent, the contents of the apparatus are heated to 120-140 °C and agitated at this temperature for 3-4 hours.

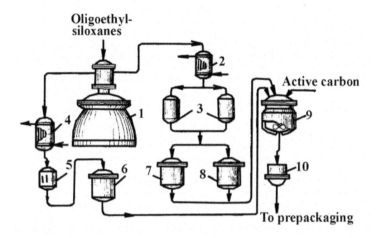

Fig. 41. Diagram of oligoethylsiloxane distillation and completion *1* - vacuum distillation tank; *2, 4* - coolers; *3* - receptacles; *5* - batch box; *6-8* - collectors; *9* - completion and purification apparatus; *10* - nutsch filter

After that pH and colour are tested. If the analysis is satisfactory, the hot mixture is filtered through coarse calico and paper in nutsch filter *32* into receptacle *33* and pumped into pressure container *34* for continuous catalytic regrouping.

The apparatus for continuous catalytic regrouping provides for three subsequent stages (the diagram shows one). The purified products of hydrolytic condensation self-flow from pressure container *34* to the first stage of the apparatus (circulation tube *35*, reactor *36*, bubble tube *37* and separator *38*). After the first stage, the liquid products of hydrolytic condensation self-flow to the second and then third stage into the corresponding apparatuses (not shown in the diagram).

The circulation and bubble tubes are vertical apparatuses with vapour jackets. The reactors are vertical cylindrical apparatuses with spherical welded bottoms and removable lids (with cylindrical "baskets" with cationite KU-23 or askanite. The reactors are fashioned with capacitance moisture indicators. To moisturise cationite or ascanite, the apparatus is filled with live steam. The separators are vertical cylindrical apparatuses with conical bottoms and removable spherical lids. The reactors and separators operate under 0.02 MPa.

The jackets of the circulation and bubble tubes are filled with vapour; the inner part of the bubble tubes is filled with air (1.5-5 m^3/h) and heated to 100-115 °C. Circulation is made possible by the difference in the density of the liquid in the circulation and bubble tubes created by air in them.

The discharge air with vapours of by-products is separated in separators from oligoethylsiloxanes. The condensed vapours accumulate in trap *39*, which is cooled with water in a coil. As the condensate fills the trap, it is poured into collector *40*. At the last stage oligoethylsiloxane liquid is sampled through the sampler of separator *38* to determine its viscosity and ethoxyl groups content. If the analysis is satisfactory, the finished product, oligoethylsiloxanes, is poured into container 41 and sent to high-vacuum distillation.

The undistilled oligomethylsiloxanes obtained after regrouping are oil-like transparent colourless or light yellow liquids without odour. Their boiling point is not lower than 80 °C (at 1.3-4 GPa); the solidification point is below -70 °C. They can be easily dissolved in organic solvents; but are not soluble in water; they are nontoxic.

These liquids are semiproducts used to prepare oligoethylsiloxanes of various viscosity (by fraction distillation and completion). Oligoethylsiloxane combine nicely with organic oils and are also used to prepare heat resistant oils and lubricants.

Distillation and completion of technical oligoethylsiloxanes In order to extract pure oligoethylsiloxanes the technical products obtained by the listed methods are distilled into fractions.

The basic diagram of the distillation and completion of oligoethylsiloxanes is given in Fig. 41. The mixture of oligoethylsiloxanes is loaded from container *41* (see Fig. 40) into electrically heated vacuum distillation tank *7*. After the liquid has been loaded, the tank jacket is filled with water for cooling, the system is vacuumised, the electric heating is switched on and volatile products are distilled at a residual pressure of 520-800 GPa to 80 °C (in vapour). After that, a higher vacuum is created in the system, and the liquid is distilled into fractions under a residual pressure not exceeding 8 GPa. The vapours are condensed in cooler *2*; the distillate is collected in receptacles *3*.

At the given residual pressure the distillation is carried out in the following temperature mode (°C, in vapour):

Fraction I	below 110	Fraction III	150-185
Fraction II	110-150	Fraction IV	185-250

It is also possible to distil fractions I, II and III below 185 °C in vapour (fraction Ia) or fractions I and II below 150 °C in vapour (fraction Ib). Fractions I-IV self-flow or are sent by vacuum from receptacles *3* into collector *7*, and fractions Ia and Ib into collector *8*. After the distillation is completed, the electric heating of the tank is switched off, the vacuum pump is stopped and tank *1* is filled with nitrogen (0.07 MPa). When the pressure in the system is atmospheric, the tank residue which boils above 250 °C (at 1.3-6.6 GPa) is sent by vacuum through cooler *4* into batch box

5. The cooled tank residue is sent from batch box *5* by vacuum or compressed air into collector *6* and then sent to complete liquids PES-5, 132-24 and 132-25 in corresponding apparatuses *9*.

Fractions I-IV (unpurified liquids) from collector *7* and fractions Ia and Ib from collector *8* are also sent to the completion stage of marketable liquids or to the stage of catalytic regrouping to obtain higher-boiling oligomers.

The completion of oligoethylsiloxane liquids, which is conducted in apparatus *9*, involves the preparation of marketable liquids from the corresponding fractions by mixing, further treatment with active coal (to eliminate residual acidity), decolouring and subsequent filtering in nutsch filter *10*.

Liquids PES-1, PES-2, PES-3 and PES-4 are completed from fractions I, II, III and IV separately or mixed; liquids PES-5, 132-24 and 132-25 are completed from tank residue after distillation into fractions; damping liquids I-IX are completed from corresponding fractions and tank residue; hydraulic liquids 132-10 and 132-10D are completed from the products of the catalytic regrouping of fractions Ia and Ib with mineral oil MVP; the electric insulation liquid 132-12D is completed from the products of the catalytic regrouping of fractions II and III with an addition of fraction IV.

Lubricating oils 132-08, 132-20 and 132-21 are completed from fraction IV with an addition of mineral oil MS-14. Products KRP-1 and KRP-2 are obtained by the thermal-oxidative condensation of tank residue below 250 °C.

3.1.4. Performance characteristics and applications of oligoethylsiloxanes

Oligoethylsiloxanes and products based on them are as a rule transparent colourless nontoxic liquids. They can be used at operating temperatures ranging from $-60-70^0$ C to $+180^0$ C. The applications of these liquids are quite varied. They are used as coolants and heat carriers in hydraulic systems, as liquids for diffusion pumps, damping liquids and bodies for instrument oils and lubricants.

Table 17. Properties of oligoethylsiloxane-based products

Characteristics	PES-1	PES-2	PES-3	PES-4	PES-5	I	II	III	IV	V	VI	VII	VIII	IX
Appearance	Transparent liquids					Transparent liquids from colourless to light yellow								
Colour according to the iodometric scale, not more than	0.25	0.25	0.25	0.25	0.25	-	-	-	-	-	-	-	-	-
Viscosity at 20 °C, mm²/s	1.5-4.5	6-12	14-17	42-48	200-500	10	20	65	100	200	250	350	500	1000
Boiling point at 1.3-6.6 GPa, °C, not less than	80-110	110-150	150-185	185-250	>250	-	-	-	-	-	-	-	-	-
Flash point, °C, not lower than	-	110	125	170	260	110	110	150	170	250	250	250	250	250
Solidification point, °C	<-70	<-70	<-70	<-70	<-70	-	-	-	-	-	-	-	-	-
Density at 20 °C, g/cm³	0.86-0.94	0.93-0.95	0.95-0.97	0.97-1.18	0.99-1.02	-	-	-	-	-	-	-	-	-
Ethoxyl groups, not more than	0.35	0.35	0.35	0.35	0.35	0.35	0.35	0.35	0.35	0.35	0.35	0.35	0.35	0.35
Content, %: Silicon	≤19.6	24.3-27.1	25.2-28	26-27.1	26.4-28	24-28	25-28	26-28	26-28	26-28	26-28	27-29	27-29	27-35
pH of the aqueous extract	6-7	6-7	6-7	6-7	6-7	6-7	6-7	6-7	6-7	6-7	6-7	6-7	6-7	6-7
Solubility	Can be easily dissolved in aromatic hydrocarbons and ethers, combine with mineral oils, are not soluble in water					Can be easily dissolved in aromatic hydrocarbons and ethers, combine with mineral oils, are not soluble in water								

Table 17 (cont.)

Characteristics	132-24	132-25	KRP-1	KRP-2	Electric insulation liquid 132-122D		Hydrolytic liquids		Lubricating oils		
					grade I	grade II	132-10	132-102	132-08	132-20	132-21
Appearance	Transparent liquids from colourless to light yellow				Transparent liquids				Transparent light brown liquids		
Colour according to the iodometric scale, not more than	-				0.25	0.25	7.0	7.0	-	-	-
Viscosity at 20°C, mm²/s	200-300 (at 60°C not more than 250)	190-290	10-20	8-18	70-140	70-140	20-33 (at 50°C not less than 0.1; at -55°C not more than 11)		47-55 (at 50°C not more than 22)	65-75 (at 50°C not more than 60)	45-50 (at 50°C not more than 22)
Boiling point at 1.3-6.6 GPa, °C, not less than	-				≥165 (at 2.6-5.2 GPa)		-	≥230 (at 10-13 GPa)			
Flash point, °C, not lower than	265	260	203		150	150	130	130	173	175	170
Solidification point, °C	-70	-70			-60	-60	-70	-70	-70	-70	-70
Density at 20°C, g/cm³	0.95-1.05	0.95-1.05			0.96-1	0.96-1	0.89-0.96	0.89-0.96	0.92-0.94	0.93-0.95	0.94-0.96
Content, %:											
Ethoxyl groups, not more than	0.2				-	-	-	-	-	-	-
Silicon	-	-	27.5-30.5	27.5-30.5	-	-	19-28	19-28	21-24	18-21	21-25
pH of the aqueous extract	-				6-7	6-7	6-7	6-7	6-7	6-7	6-7
Solubility	Can be easily dissolved in ethers and other organic solvents.				Can be easily dissolved in ethers and other solvents.		Can be easily dissolved in ethers and other organic solvents.		Are easily dissolved in aromatic hydrocarbons.		

Notes. 1. For PES-1 - PES-5 nD20 varies from 1.4350 to 1.4470; average thermal capacity at 20 °C varies from 1.64 to 1.86 KJ/(kg·K); the thermal conductivity coefficient at 20 °C changes from 0.13 to 0.16 Wt/(m·K). 2. For liquid 132-12D (grade I): dielectric permeability at 15-35 °C and 1000 Hz varies from 2.4 to 2.8; specific volume electric resistance is not less than 2.5·1014 Ohm cm at 15-35 °C and not less than 1·1014 Ohm cm at 98-1020C; electric density at 15-350C and 50 Hz is at least 18 KV/mm; dielectric loss tangent is 0.0003 at 1000 Hz and 15-350C and 0.0008 at 98-1020C. For liquid 132-12D (grade II): specific volume electric resistance is not less than 2.5·1013 Ohm cm at 15-35 °C and at least 1·1013 Ohm cm at 98-1020C. 3. Oils 132-08, 132-20 and 132-21 do not corrode metals. 4. Liquids 132-24 and 132-25 do not corrode metals for 3 hours at 1000C.

Industry manufactures a large range of oligodiethylsiloxanes. They can be easily dissolved in common organic solvents and, unlike other oligoorganosiloxanes, can be fully combined with mineral oils, which accounts for their use as bodies for oils and lubricants.

Because methyl substituents are replaced in oligoorganosiloxanes with ethyl substituents, the freedom of atom and group rotation around ≡Si-O- and ≡Si-C≡ bonds is limited. This makes the chains more rigid and thus prevents the formation of spiral conformations of siloxane chains and full intramolecular compensation of the dipoles of the polar ≡Si-O- bonds. The same effect is caused by the branched structure of some oligodiethylsiloxanes. As a result, the intermolecular compensaiton of the dipoles in the siloxane bonds becomes more probable; consequently, the energy of intermolecular interaction also grows, which improves physical characteristics. On the other hand, ethyl substituents and branches interfere with the dense packing of chains; thus, the distances between chains in oligodiethylsiloxanes are larger and intermolecular interaction is weaker. The combination of these factors forms the specific set of physical properties which characterises oligodiethylsiloxanes: higher solubility, lower freezing and glass transition points compared to oligodimethylsiloxanes.

The properties of oligoethylsiloxane-based products are listed in Table 17. Owing to their performance characteristics, such as motor volatility, operating fraction and tendency to form varnish, oligomers 132-24 and 132-25 are used as bodies for oils and lubricants. The maximum allowable performance temperature of these oligomers in oils and lubricants is within the 100-150^0 C range. The outlined information demonstrates the following applications for oligoethylsiloxane-based products:

- PES-1 is a heat carrier for low temperatures (coolant).
- PES-2 is a heat carrier and operating fluid in hydraulic systems.
- PES-3, ditto and additive for polishers.
- PES-4 is a liquid in gages and body for low-temperature oils.

- PES-5 is a heat carrier and additive for polishing and cleaning systems; lubricant for case molding and for pressure molding of plastic and rubber products.
- Damping liquids I-IX are used in oscillograph vibrators in the -60 - +100°C temperature range.
- Liquids 132-24 and 132-35 are used to lubricate steel-on-steel and steel-on-rubber friction surfaces; body for plastic lubricants.
- KRP-1 is a filler for semiconductor devices.
- KRP-2 is used to produce sealants for air spark plugs, which can operate at 250°C.
- 132-12D are used for saturating and filling condensers and other devices operating in the -60 - +100°C temperature range.
- 132-10 and 132-D are used for hydraulic and electric isolated systems operating in the -70 - +100°C temperature range.
- Lubricating oils 132-08. 132-20 and 132-21 are used for bearings in various devices and in friction units of the machines operating in the -70 - +100°C temperature range; as a body for plastic lubricants.

3.1.5. Preparation of oligomethylphenylsiloxanes

Oligomethylphenylsiloxanes are a mixture of oligomer homologues with the following common formula:

$$(CH_3)_3Si-O\left[\begin{array}{c} CH_3 \\ | \\ Si-O \\ | \\ C_6H_5 \end{array}\right]_n Si(CH_3)_3$$

$$n = 1 \div 12$$

The viscosity of industrial oligomethylphenylsiloxanes, as well as that of oligomethylsiloxanes, varies from 4 to 1000 mm²/s. Similarly to oligomethylsiloxanes, oligomethylphenylsiloxanes are obtained by: 1) the joint hydrolytic condensation of trimethylchlorosilane and methylphenyldichlorosilane with subsequent catalytic regrouping and partial polycondensation of the condensation products; the separate hydrolytic condensation of trimethylchlorosilane and methylphenyldichlorosilane with subsequent catalytic regrouping of the condensation products.

Preparation of oligomethylphenylsiloxanes The preparation of oligomethylphenylsiloxanes by the hydrolytic cocondensation of trimethylchlorosilane and methylphenyldichlorosilane comprises two stages: the cohydrolysis and partial condensation of the obtained products; the catalytic regrouping of the products.

The hydrolytic cocondensation proceeds thus:

$$2(CH_3)_3SiCI + n\ CH_3(C_6H_5)SiCI_2 \xrightarrow[2(n+1)H_2O]{(n+1)H_2O} (CH_3)_3Si\text{-}O\underset{\underset{\displaystyle C_6H_5}{|}}{\overset{\overset{\displaystyle CH_3}{|}}{Si}}\text{-}O\text{-}Si(CH_3)_3 \qquad (3.31)$$

Depending on the ratio of the components, one obtains liquids with various characteristics. Hydrolytic cocondensation is accompanied by by-processes, such as the separate hydrolytic condensation of the parent substances:

$$2(CH_3)_3SiCI \xrightarrow[2HCI]{H_2O} (CH_3)_3Si\text{-}O\text{-}Si(CH_3)_3 \qquad (3.32)$$

$$mCH_3(C_6H_5)SiCI_2 + m\ H_2O \xrightarrow[2mHCI]{} [(CH_3)(C_6H_5)SiO]_m \qquad (3.33)$$

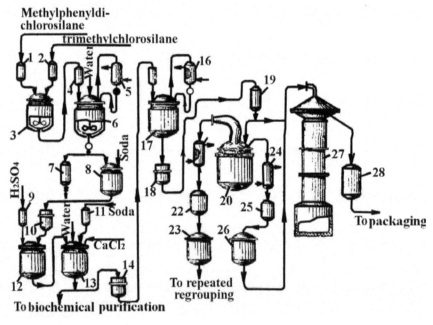

Fig. 42. Production diagram of oligomethylphenylsiloxanes: *1, 2, 4, 9, 11, 15, 19* - batch boxes; *3* - agitator; *5, 16, 21, 24* - coolers; *6* - hydrolyser; *7, 23, 26, 28* - collectors; *8* - settling box; *10, 14, 18* - nutsch filters; *12* - apparatus for catalytic regrouping; *13* - neutraliser; *17* - purification apparatus; *20*- distillation tank; *22, 25* - receptacles; *27* - distillation tank

and the formation of hydroxyl-containing methylphenylsiloxanes:

$$(CH_3)_3SiCl + mCH_3(C_6H_5)SiCl_2 \xrightarrow[2(m+1)HCl]{(m+1)H_2O} (CH_3)_3Si-O\left[\!\begin{array}{c} CH_3 \\ | \\ Si-O \\ | \\ C_6H_5 \end{array}\!\right]_m\!\!\!H \qquad (3.34)$$

Catalytic regrouping in the presence of acids (H_2SO_4) or alkali can transform all these by-products into desired oligomers:

$$[(CH_3)(C_6H_5)SiO]_m + (CH_3)_3Si-O-Si(CH_3)_3 \longrightarrow (CH_3)_3Si-O\left[\!\begin{array}{c} CH_3 \\ | \\ Si-O \\ | \\ C_6H_5 \end{array}\!\right]_m\!\!\!Si(CH_3)_3 \qquad (3.35)$$

$$2(CH_3)_3Si-[OSi(CH_3)(C_6H_5)]_m-OH \xrightarrow[H_2O]{} 2(CH_3)_3Si-[OSi(CH_3)(C_6H_5)]_{2m}-OSi(CH_3)_3 \qquad (3.36)$$

Raw stock for the production of oligomethylphenylsiloxanes is: trimethylchlorosilane (not less than 93% of 56-58 °C fraction, 32.1-33.2% of chlorine, density 0.856-0.859% g/cm^3); methylphenyldichlorosilane (the boiling point is 196-204 °C, 36.9-37.8% of chlorine), sulfuric acid (up to 90% concentration).

The production process (Fig. 42) comprises three main stages: hydrolytic cocondensation; catalytic regrouping; vacuum distillation of the purified regrouping product.

The necessary quantities of methylphenyldichlorosilane and trimethylchlorosilane are sent from batch boxes into agitator 3. After the organochlorosilanes have been loaded, the agitator is switched on, the mixture is agitated for 30 minutes and then sent into weight batch box 4. Hydrolyser 6 is filled with a necessary quantity of water and the jacket is filled with vapour; the water in the hydrolyser is heated to 60 °C. At 60-70 °C the organochlorosilane mixture is sent from weight batch box 4 at such speed that the temperature in the hydrolyser does not exceed 70 °C. After the mixture has been introduced, the mixture in the hydrolyser is agitated for 1 hour at the same temperature. Then, the agitator is stopped and the mixture is held for 30 minutes.

The settled bottom layer (18-20% hydrochloric acid) is poured through a run-down box into collector 7, and the top layer, the product of hydrolytic cocondensation, is sent into settling box 8. After 2 hours of settling in the apparatus, the top layer is separated from the remaining acid; at agitation and 20-25 °C the apparatus is gradually, within 30 minutes, filled through a hatch with a a calculated amount of dry soda ash. The mass is mixed for 1-2 hours. After that, the medium is sampled for acidity. If the reaction is acid, some soda (as calculated) is added into the apparatus; after an additional 0.5 hours of mixing the mixture is sampled again. The neutral product of hydrolytic cocondensation is filtered in nutsch filter *10*.

If the neutral product contains moisture (i.e. it is cloudy), apparatus *8* is loaded before filtering with burnt calcium chloride. The mixture is agitated for 4-5 hours and filtered.

The dried product is sent from the nutsch filter for catalytic regrouping in apparatus 12, which is filled with 90% sulfuric acid (5% of the product quantity) at agitation. The mixture is agitated at 20 °C for approximately 6 hours. The product of regrouping is sent from apparatus *12* by nitrogen flow (or in vacuum) into neutraliser *13*, it is also filled with water for flushing. The agitator is switched on; after 0.5-1 hours of agitation the apparatus is filled with a certain amount of 20% soda solution from batch box *11*. The mixture is agitated for 0.5 hours and after 1 hour of settling the bottom (acid) layer is poured off. The top layer, the product of regrouping, is sampled for acidity and loaded with soda ash in the amount necessary for neutralisation. The mixture is agitated for 1-2 hours and sampled for acidity. The process of neutralisation is finished at pH 5-7.

Apparatus 13 is loaded with burnt calcium chloride (1% of the product weight) through the top hatch; the mixture is agitated for 4-5 hours. Then the mixture is filtered in nutsch filter *14*; from there the filtrate is sent into apparatus *17* through batch box *15* to be purified with active coal. At agitation the apparatus is loaded with dry active coal (2% of the product amount). The jacket is filled with vapour, the contents of the apparatus are heated to 70-80 °C and are agitated at this temperature for 2 hours. The acidity is sampled. When pH of the aqueous extract is 6-7, the hot purified product (70-80 °C) is filtered in nutsch filter *18*. The filtered product is collected in receptacle *19*.

Vacuum distillation of the purified product of regrouping To prepare low-viscosity oligomethylphenylsiloxanes (4-12 mm^2/s at 50 °C), the distillation is conducted in two stages. First, volatile fractions are distilled in tank *30*, which is made of doped steel, fashioned with a sectional electric heating, coolers *21* and *24* and receptacles *22* and *25* to collect the distillate and tank residue.

The tank is loaded with the purified product of regrouping from batch box *19*. A residual pressure of 330-200 GPa is created in the system. The lower heating section is switched on, and volatile substances are distilled at 120 °C. The electric heating is switched off, and the contents of the apparatus are cooled to 50 °C at the same residual pressure. The residual pressure is vacuum-pumped to 2.6-4 GPa; the volatile products are collected in receptacle 22 at a temperature up to 120 °C. Tank residue is cooled to 50-60 °C in cooler *24*, loaded into receptacle *25* and then sent by vacuum into receptacle *26*. The volatile products are sent into collector *23* from receptacle *22* and then to another regrouping with sulfuric acid.

Tank residue from collector 26 is sent for a deeper vacuum distillation into tank 27 with a corrugated brass head. The lid of the tank has a water-cooled condenser; the tank is heated with a three-sectioned electric heater. Distillation is conducted at a residual pressure of 0.13-0.26 GPa. At 50 °C the viscosity of the fraction is sampled. The distillation is stopped when the viscosity reaches 16-17 mm^2/s.

To prepare high-viscosity oligomethylphenylsiloxanes (450-1000 mm^2/s at 20 °C), the distillation of the product from batch box *19* is conducted in one stages in tank 27 at a residual pressure from 0.26 GPa to 360°C (liquid). At 20 °C the viscosity of the fraction is sampled. The distillation is stopped when the viscosity of tank residue reaches 450 mm^2/s. The product is sent to the completion stage.

Oligomethylphenylsiloxanes are nontoxic liquids (the boiling point is above 200 °C at 1.3-4 GPa), colourless and odourless. They have superior thermal, thermal-oxidative and radiation stability. They have found a wide application in various spheres of technology, e.g. as heat resistant lubricants, as operating fluids in diffusion pumps, as heat carriers and dielectrics.

Preparation of low-dispersion oligomethylphenylsiloxanes

The synthesis of low-dispersion oligomethylphenylsiloxanes is based on the separate hydrolytic condensation of trimethylchlorosilane and methylphenyldichlorosilane. It comprises three stages: the hydrolytic condensation of trimethylchlorosilane and subsequent distillation of pure hexamethyldisiloxane; the hydrolytic condensation of methylphenyldichlorosilane and subsequent depolymerisation of the formed methylphenylcyclosiloxanes; the catalytic regrouping of the mixture of hexamethyldisiloxane and methylphenylcyclotrisiloxane.

The hydrolytic condensation of trimethylchlorosilane in the excess of water forms hexamethyldisiloxane as the main product:

$$2(CH_3)_3SiCl \xrightarrow[\text{2HCl}]{\text{H}_2\text{O}} (CH_3)_3Si\text{-}O\text{-}Si(CH_3)_3 \qquad (3.37)$$

and trimethylsilanol as a by-product, which condenses when treated with sulfuric acid, also forming hexamethyldisiloxane:

$$2(CH_3)_3SiCl \xrightarrow[\text{2HCl}]{\text{H}_2\text{O}} 2(CH_3)_3SiOH \xrightarrow[\text{H}_2\text{O}]{} (CH_3)_3Si\text{-}O\text{-}Si(CH_3)_3 \qquad (3.38)$$

The hydrolytic condensation of methylphenyldichlorosilane in the excess of water forms both methylphenylcyclosiloxanes and α,ω-dihydroxymethylphenylsiloxanes:

$$nCH_3(C_6H_5)SiCl_2 \xrightarrow[2nHCl]{nH_2O} HO\left[\begin{array}{c} CH_3 \\ | \\ Si-O \\ | \\ C_6H_5 \end{array}\right]_n H + [(CH_3)(C_6H_5)SiO]_n \qquad (3.39)$$

To extract pure methylphenylcyclotrisiloxane, the products of hydrolytic condensation can be treated with aqueous solutions of alkali metal hydroxides, e.g. LiOH:

$$[(CH_3)(C_6H_5)SiO]_n \xrightarrow{LiOH} LiO\left[\begin{array}{c} CH_3 \\ | \\ Si-O \\ | \\ C_6H_5 \end{array}\right]_n H \qquad (3.40)$$

$$HO\left[\begin{array}{c} CH_3 \\ | \\ Si-O \\ | \\ C_6H_5 \end{array}\right]_n H \xrightarrow[2H_2O]{2LiOH} LiO\left[\begin{array}{c} CH_3 \\ | \\ Si-O \\ | \\ C_6H_5 \end{array}\right]_n Li \qquad (3.41)$$

Methylphenylcyclosiloxanes depolymerise under the influence of lithium siloxanolates and heating to form an equilibrium system:

$$[(CH_3)(C_6H_5)SiO]_n \rightleftharpoons [(CH_3)(C_6H_5)SiO]_x + [(CH_3)(C_6H_5)SiO]_y \qquad (3.42)$$

$$n=x+y; \ x=3, \ y=4$$

During the continuous separation of methylphenylcyclotrisiloxane in the rectification tower the equilibrium shifts towards the formation of a trimer cycle.

The catalytic regrouping of the mixture of hexamethyldisiloxane and methylphenylcyclotrisiloxane in the presence of hydrochloric acid finally forms the target product, oligomethylphenylsiloxane with the degree of polymerisation n = 3:

$$(CH_3)_3Si-O-Si(CH_3)_3 + [(CH_3)(C_6H_5)SiO]_3 \xrightarrow{HCl} (CH_3)_3SiO\left[\begin{array}{c} CH_3 \\ | \\ Si-O \\ | \\ C_6H_5 \end{array}\right]_3 Si(CH_3)_3 \qquad (3.43)$$

Raw stock: trimethylchlorosilane (32.1-33.2% of chlorine, d_4^{20}=0.856-0.859); methylphenyldichlorosilane (36.9-37.8% of chlorine, d_4^{20}=

1.175÷1.182); sulfuric acid (not lower than 94% concentration); lithium hydroxide (10% solution).

After the raw stock and equipment have been prepared, hydrolyser 4 (Fig. 43) is loaded with with a calculated amount of water. The agitator is switched on, and trimethylchlorosilane is sent from batch box *1* under the water layer at such speed that the temperature in the hydrolyser does not exceed 30 °C.

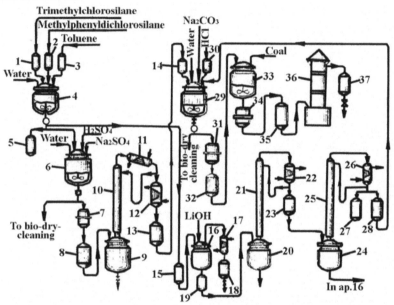

Fig. 43. Production diagram of low-dispersion oligomethylphenylsiloxanes: *1-3, 14, 30*- batch boxes; *4*- hydrolyser; *5, 8, 15, 19, 32, 35* - collectors; *6*- neutraliser; *7, 31, 34* - nutsch filters; *9, 16, 20, 24* - distillation tanks; *10, 21, 25* - rectification tanks; *11* - refluxer; *12, 17, 22, 26* - coolers; *13, 18, 23, 27, 28, 37* - receptacles; *29* - apparatus for catalytic regrouping; *33* - purification apparatus; *36* - vacuum apparatus

After the trimethylchlorosilane has been loaded, the mixture is agitated for 1 more hour. Then the agitator is switched off. The mixture splits. The bottom layer (hydrochloric acid) is poured into collector *5*, and the top layer (the product of hydrolysis) is sent into apparatus *6*, which is also filled with 15% (weight) of sulfuric acid. The mixture is agitated at room temperature for 1 hour. Then, the apparatus is filled with water to dilute the acid, and the mixture is split. The bottom layer is poured off; the product remaining in apparatus *6* is flushed with water, neutralised, dried and sent to nutsch filter *7*. The filtrate is sent into collector *8* and then to tank *9* to extract pure hexamethyldisiloxane. The target fraction, hexamethyldisi-loxane, is separated from tower *10* into receptacle 13 at ~100 °C and from there into batch box *14*.

Hydrolyser *4* is filled with a calculated amount of water, the agitator is switched on, and methylphenyldichlorosilane is sent from batch box *2* under the water layer at such speed that the temperature in the apparatus does not exceed 80 °C. After the methylphenyldichlorosilane has been loaded, the mixture is agitated for 1 hour. To improve the splitting of the reactive mixture, the apparatus is filled with toluene from batch box *3*. The mixture is agitated for 1 more hour; after that the bottom layer (hydrochloric acid) is poured into collector *5*, and the toluene solution of the products of hydrolytic condensation is flushed with water at high temperature (70-90 °C) until it gives a neutral reaction.

The neutral solution is poured into collector *15*, and from there into tank *16*. The tank is also filled with a calculated amount of a 10% solution of lithium hydroxide. Vapour heating is started and toluene is distilled, first under atmospheric pressure, then in vacuum. The distillation is finished when the distillate no longer enters receptacle *18*, which is achieved approximately at 180 °C and a residual pressure of 52-80 GPa. From the receptacle, toluene is sent into batch box *3* for repeated use. The products of the hydrolytic condensation of methylphenyldichlorosilane, methylphenyl-cyclosiloxanes, are sent from tank *16* into collector *19* and then into tank *20* to decompose.

After the methylphenylcyclosiloxanes have been loaded, a residual pressure of 3-13 GPa is created in tank *20*, and electric heating is switched on. Decomposition is usually conducted at 300-350 °C. The products of decomposition (low-molecular cyclic substances) are collected in receptacle *23*. After that the products enter distillation tank *24*. A residual pressure of 3-13 GPa is created in the tank; electric heating is switched on; the light fractions are distilled at 240-260 °C and collected in receptacle *27*. The distillation of the light fractions can be monitored by the refraction index: the distillation is stopped at $n_D^{20} = 1.5350 \div 1.5360$.

After that, the target fraction, methylphenylcyclotrisiloxane, is collected into receptacle *28;* when $n_D^{20} = 1.5420$, the separation is stopped. Methyl-phenylcyclotrisiloxane is poured from receptacle *28* at a temperature not lower than 100 °C into reactor *29*. The distillation residue (methylphenyl-cyclotetrasiloxane) is loaded from tank *24* into tank *16* for repeated use or put into containers.

The catalytic regrouping of the mixture of hexamethyldisiloxane and methylphenylcyclotrisiloxane is conducted in reactor *29*. The apparatus is loaded from batch box 14 with hexamethyldisiloxane, and from receptacle

28, with methylphenylcyclotrisiloxane. The mixture in the reactor is mixed, heated to 50 °C to dissolve cyclic trimer and filled with a required amount of hydrochloric acid from batch box *30*. The reaction is carried out at 45-55 °C; the process is monitored by the refraction index and chromatographic data.

After catalytic regrouping the reactive mixture is split. The bottom layer, diluted hydrochloric acid, is sent to biochemical purification, and the product of regrouping is flushed with a solution of sodium chloride, neutralised with dry soda ash, filtered in nutsch filter *31* and collected in collector *32*. From there the products of regrouping are loaded into apparatus *33*; this apparatus is filled with dry active coal through a hatch. The suspension in the apparatus is heated to 70 °C and is held at this temperature for several hours. If the sample is neutral, the suspension is poured into nutsch filter *34*, and the filtrate is collected in collector *35*.

The regrouping products are distilled in vacuum apparatus 36 at the residual pressure of 0.13 GPa. The light fractions are separated at the speed of about 0.5 l/h. After separating the light fractions, the speed of separation should be reduced to 0.2 l/h; the intermediate fractions are separated at the same speed. The distillation is monitored by the refraction index, viscosity and chromatographic data. The target fraction can be separated at the speed of 5-6 l/h; the separation should be finished when the quantity of the distillate abruptly drops.

Low-dispersion oligomethylphenylsiloxanes are colourless nontoxic liquids. They can be easily dissolved in organic solvents, but are not soluble in water. They are used as operating fluids in high-vacuum pumps.

The most important physicochemical properties and applications of some oligomethylphenylsiloxanes are given in Table 18.

Table 18. Important physicochemical properties of oligomethylphenylsiloxanes and their applications

Brand	Viscosity at 50 °C, mm^2/s	Boiling point, °C	Flash point, °C, not lower than	Solidi-fication point, °C	Applications
PFMS-1	3.6-4.6	65-75(at 1.3$10^{-2}$Pa)	-	-	Operating fluids in high-vacuum pumps (the maximum residual pressure is 2.5 10^{-2} GPa)
PFMS-2/51	14-17	95-120 (at 1.3$10^{-2}$ Pa)	-	-	
PFMS-3	6.6-9	94-112 (at 1.3$10^{-2}$ Pa)	-	-	

Table 18. (cont.)

PFMS-4	600-1000	290 (at 1.3-4 GPa)	300	-20	Heat carrier; body of high-temperature consistent greases and oils; high-temperature lubricant
PFMS-6	45-110(at 100 °C)	330 (at 1.3-4 GPa)	360	+10	High-temperature immobile phase in gas-liquid chromatography; High-temperature heat carrier

Note. The density is standardised for PFMS-4 (1.100 g/cm^3 at 20 °C) and for PFMS-6 (1.150 g/cm^3 at 20 °C).

3.1.6. Performance characteristics and applications of oligomethylphenylsiloxanes

The main dissimilarity of oligomethylphenylsiloxanes from oligoalkylsiloxanes is their higher thermal-oxidative and thermal stability. This characteristic determines their wide application in technology. The introduction of phenyl groups into oligoorganosiloxanes considerably increases the degree of intermolecular interactions due to the greater rigidity of molecular chains, limited rotation of atoms and groups around ≡Si-O- and ≡Si-C≡, as well as to specific intermolecular interactions caused by the presence of aromatic nuclei in these oligomers. We should mention the three main practical uses for oligomethylphenylsiloxanes:

- high-vacuum oils for diffusion pumps;
- heat carriers for high and low temperatures;
- dispersion media for heat resistant oils and lubricants.

All oligomethylphenylsiloxanes are transparent colourless or yellowish liquids. The most viscous oligomers range from light yellow to light brown.

In all these spheres the temperature performance characteristics of oligomethylphenylsiloxanes have the greatest importance. The maximum allowable performance temperatures for oligomethylphenylsiloxanes range from -20- -100 до 200-350^{0}C depending on their composition, degree of polymerisation and presence of phenyl substituents in molecules.

Oligomethylphenylsiloxanes are characterised by improved thermal stability, low glass transition temperatures, low vapour pressure and compatibility with organic media. This set of properties determines the fields of application which can be classified by oligomer brands in the following way:

Liquids FM-5, FM-6, FM-5,6 AP are used as: dispersion media in low-temperature oils and lubricants; cooling heat carriers. They are also used in light-loaded high-speed ball bearings in freon refrigerating systems.

Oligomethylphenylsiloxane liquids are used as heat resistant and low-temperature media in oils and lubricants, which can be used in a wide temperature range and high vacuum. They are used as heat carriers and liquids for hydraulic systems; PFMS-2/5l, FM-1, FM-2 liquids are used in diffusion vacuum pumps with ultimate vacuum from 133.322 nPa to 13.332 μPa; PFMS-4 are used as high-temperature and low-flammable heat carriers, dielectrics, operating and dispersion media for lubricants, oils, immobile phases in gas-liquid chromatography. The use of oligomethylphenylsiloxanes in consistent greases comprises thermostable, vacuum antifriction lubricants, instrument oils, as well as electric contact lubricants, jointing pastes and extreme pressure lubricants.

Diffusion oils for high-vacuum pumps (ultimate vacuum up to 13.3 μPa) are the most valuable and high-quality materials based on oligomethylphenylsiloxanes. Their vacuum properties cannot be rivalled by any other classes of chemical compounds. First and foremost, one should take into account that high-vacuum liquids for diffusion pumps should be resistant to oxidation at high temperatures.

Of great importance nowadays are oligodimethyl(methylthyenil)siloxanes (PMTS), liquids with an improved lubricating capacity and resistance to oil and petrol.

3.1.7. Preparation of oligodimethyl(methylthyenil)siloxanes

Oligodimethyl(methylthyenil)siloxanes are a mixture of oligomer homologue liquids with the following common formula:

$$(CH_3)_3SiO \left[\begin{array}{c} CH_3 \\ | \\ Si-O \\ | \\ CH_3 \end{array} \right]_n \left[\begin{array}{c} CH_3 \\ | \\ Si-O \\ | \\ C_4H_3S \end{array} \right]_m Si(CH_3)_3$$

where the average value of $n = 14$, $m = 2$ (ПМТС-1); $n = 20$, $m = 3$ (ПМТС-2); $n = = 10$, $m = 4$ (ПМТС-3).

Oligodimethyl(methylthyenil)siloxane liquids are obtained by the joint hydrolytic condensation of dimethyldichlorosilane, methylthyenildichlorosilane (MTDCS) and trimethylchlorosilane taken in certain mole ratios for each brand with subsequent treatment of the products of the hydrolytic co-condensation with sulfuric acid and separation of the low-boiling products from the products of catalytic regrouping.

Raw stock: dimethyldichlorosilane [not less than 58% (weight) of chlorine, the density of 1.06-1.07 g/cm^3]; MTDCS [not less than 98% (weight) of the 190-200 °C fraction]; trimethylchlorosilane [32.1-33.2% (weight) of chlorine, the density of 0.856-0.859 g/cm^3]; sulfuric acid (up to 70% concentration).

The hydrolytic cocondensation proceeds thus:

$$2(CH_3)_3SiCI + n\ CH_3)_2SiCI_2 + mCH_3(C_4H_3S)SiCI_2 \xrightarrow[2(n+m+1)HCI]{(n+m+1)H_2O}$$

$$\longrightarrow (CH_3)_3SiO\text{-}[Si(CH_3)_2O]_n\text{-}[Si(CH_3)(C_4H_3S)O]_m\text{-}Si(CH_3)_3 \qquad (3.44)$$

Depending on the ratio of the parent components, one obtains liquids with different values of n and m.

Moreover, in the process of ПМТС synthesis there are secondary reactions, which form cyclohexanes, hexamethyldisiloxane and hydroxy-containing oligodimethyl(methylthyenil)siloxanes.

$$n(CH_3)_2SiCI_2 \xrightarrow[2nHCI]{nH_2O} [(CH_3)_2SiO]_n \qquad (3.45)$$

$$n(CH_3)_2SiCI_2 + mCH_3(C_4H_3S)SiCI_2 \xrightarrow[2(n+m)HCI]{(n+m)H_2O} [(CH_3)_2SiO]_n\text{-}[CH_3(C_4H_3S)SiO]_m \qquad (3.46)$$

$$2(CH_3)_3SiCI \xrightarrow[2HCI]{H_2O} (CH_3)_3Si\text{-}O\text{-}Si(CH_3)_3 \qquad (3.47)$$

$$(CH_3)_3SiCI + n\ CH_3)_2SiCI_2 + mCH_3(C_4H_3S)SiCI_2 \xrightarrow[2(n+m+1)HCI]{(n+m+1)H_2O}$$

$$\longrightarrow (CH_3)_3SiO \left[\begin{array}{c} CH_3 \\ | \\ Si-O \\ | \\ CH_3 \end{array} \right]_n \left[\begin{array}{c} CH_3 \\ | \\ Si-O \\ | \\ C_4H_3S \end{array} \right]_m H \qquad (3.48)$$

Treating the products of hydrolytic cocondensation with 70% H_2SO_4 presupposes an additional condensation by the remainder hydroxyl groups in the products obtained according to scheme 5, as well as the regrouping of the products obtained according to 3.45-3.48.

The production process of ПМТС (Fig. 44) comprises the following main stages: the hydrolytic cocondensation of the corresponding organochlorosilanes, the catalytic regrouping of the products of hydrolytic cocondensation, the distillation of the low-boiling products of hydrolytic cocondensation and the purification of the ready product with active coal.

Agitator *1* is loaded with corresponding organochlorosilanes from batch boxes *2-4* at 10-30 °C. The mixture is agitated for 30 min and pressurised with nitrogen *(P=0.7 atm)* into weight batch box *5*. Hydrolyser *6* is loaded with a metered amount of water at at a temperature not exceeding 20 °C the organochlorosilane mixture from weight batch box *5* at such speed that the temperature does not rise. The contents of the apparatus are agitated for 1 hour; the agitator is stopped and the mixture is settled. Within 2 hours the reactive mixture splits in two layers. The bottom muriatic layer is poured into container *7* and then sent to biochemical purification. The top layer, the product of hydrolytic cocondensation, is poured into neutraliser *8*. At agitation and a temperature of 18-30 °C the neutraliser with the product of hydrolytic cocondensation also receives soda ash in the amount of up to 4% (weight) of the reactive mixture. The agitation is carried out for 2 hours; then the mixture is sampled to determine pH.

If the reaction is acid (pH<6), the neutraliser receives an additional amount of soda and after 0.5 hours of agitation pH is tested again.

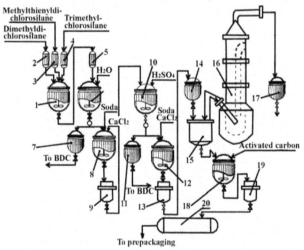

Fig. 44. Production diagram of oligodimethyl(methylthyenil)siloxanes: *1* - agitator; *2-5* - batch boxes; *6* - hydrolyser; *7,11* - intermediate containers for the muriatic layer; *8, 12* - neutralisers; *9, 13, 19* - nutsch filters; *10* - apparatus for catalytic regrouping; *14* - collector of the product of catalytic regrouping; *15* - loader-unloader; *16* - distillation vacuum apparatus; *17* - receptacle for light fractions; *18* - purification apparatus; *20* - container for the finished product

If the neutral product contains moisture (i.e. it is cloudy), the neutraliser is loaded with sodium sulphate or calcium chloride up to 1% (weight) and agitated for 6-10 hours. After that, the product is filtered in nutsch filter 9.

The catalytic regrouping (CR) of the products of hydrolytic cocondensation is conducted in enameled apparatus 10. The apparatus is loaded from nutsch filter 9 with the product of hydrolytic cocondensation. After adding 5% (weight) of 70% sulfuric acid, the mixture is agitated for approximately 4 hours. The CR temperature should be 20-25 °C. When the agitator is switched off, the mixture is settled for 2 hours. The bottom layer, acid, is flushed with water, poured into container 11 and sent to biochemical cleaning (BCC); the top layer is sampled to determine the viscosity of the CR product. For ПМТC-1 $\eta= 5\div6$ sSt [$(5+6)\cdot10^{-6}$м^2/c], for ПМТC-2$\eta = 11\div16$ sSt, and for ПМТC-3 $\eta=6\div8$ sSt. From apparatus 10 the product is sent to further neutralisation into neutraliser 12. To neutralise the CR product, neutraliser 12 is filled at agitation through a hatch with portions of 4% (weight) of soda ash. The agitation is continued until pH$\approx$7.

If the neutral CR product contains moisture (i.e. it is cloudy), the reactive mixture in apparatus 12 is supplemented with $CaCl_2$ or Na_2SO_4 in the amount of 3-4% of the product weight. After that, the product is agitated for 6-10 hours.

The neutral CR product is filtered in nutsch filter 13, and pressurised into intermediate collector 14.

Low-boiling products are separated from the CR product in distillation vacuum apparatus 16, which is a vertical cylindrical apparatus with a removable top cover which also serves as a condenser. The apparatus has special electric heating.

The CR product is sent from collector 14 into loader-unloader 15 and then into apparatus 16; the condenser cover is filled with water and the whole system is vacuumised until the residual pressure is below 1.3 GPa. The low-boiling fractions below the vapour temperature of 220-235 °C (depending on the brand of the liquid) are collected in receptacle 17. After the fraction has been separated, the distillation is stopped by turning off the electric heating of apparatus 16. Tank residue, which is the target product, ПМТC-1, ПМТC-2 or ПМТC-3 liquid, is cooled in the tank of the apparatus down to 100 °C in a vacuum of 1.3-2.6 GPa and then loaded into receptacle 15.

The finished products are purified in apparatus 18 with active coal, which is loaded in the amount of 3-4% of the weight of the finished product sent to purification. The mixture is agitated at 80-90°C; then it is sampled to determine its pH. If pH is 6-7, coal is filtered out of the liquid in nutsch filter *19*. From there the purified finished product is sent into container *20*.

Oligodimethyl(methylthyenil)siloxanes are virtually nontoxic liquids (the boiling point >250 °C at residual $P = 1.3 \div 2.6$ GPa); they do not possess cumulative characteristics.

The main technical requirements to PMTC liquids:

	PMTC-1	PMTC-2	PMTC-3
Appearance	Homogeneous transparent liquids from colourless to light yellow		
Mechanical impurities	Absent		
Kinematic viscosity, sSt:			
at 20 °C	21-25	35-45	20-40
at -60 °C	≤ 700	800-1300	≤2500
Temperature, °C			
congelation point	-80	-80	Not higher than -70
flash point	≥250	≥230	≥250
Sulfur content, % (weight), not less than	3.0	3.0	5.0
pH of the aqueous extract	6-7	6-7	6-7
Corrosion index in the testing of metal plates at 100 °C for 100 hours, mg/cm^2:			
for KhGSA steel	-	-	not >0.02
for duralumin D-16	-	-	the same
for bronze BRAZh	9-4	-	the same

PMTC liquids are used in hydraulic actuators of membrane compressors and as operating liquids in hydraulic systems.

3.2. Preparation of branched oligoorganosiloxanes

Branched oligoorganosiloxanes are increasingly used in industry due to their specific properties. For example, branched oligomethylsiloxanes have much lower solidification temperatures than linear oligomethylsiloxanes. Branched oligoorganosiloxanes with phenyl radicals at the silicon atom and end hydroxyl or alkoxyl groups have proved to be efficient modifiers for various organic polymers and rubbers made of carbo- and heterochain elastomers. Oligoorganosiloxanes with lateral alkoxyoxyalkylene or alkoxyoxyalkylenpropyl groups in their branches have high surface activity and are efficient foam regulators and stabilisers, which are widely used in the production of foamed polyurethanes.

Due to this, the production of branched oligoorganosiloxanes is steadily increasing.

3.2.1. Preparation of branched oligomethylsiloxanes

Low-viscosity branched oligomethylsiloxanes are prepared by the hydrolytic condensation of a mixture of methyltrichloro-, dimethyldichloro- and trimethylchlorosilanes, subsequent catalytic regrouping and the separation of the target products by rectification.

The process of hydrolytic condensation and subsequent catalytic regrouping proceeds according to the following general scheme:

$$m(CH_3)_3SiCI_2 + mCH_3SiCI_3 + (m+2)(CH_3)_3SiCI \xrightarrow[2(n+m+1)HCI;\ H_2O]{(n+m+2)H_2O}$$

$$\longrightarrow (CH_3)_3SiO \left[\begin{array}{c} CH_3 \\ | \\ Si-O \\ | \\ CH_3 \end{array} \right]_n \left[\begin{array}{c} CH_3 \\ | \\ Si-O \\ | \\ \left(O-Si(SH_3)_2 \\ | \\ OSi(CH_3)_3 \right)_x \end{array} \right]_m Si(CH_3)_3 \qquad (3.49)$$

The number of silicon atoms in these oligomers ranges from 9 to 18.

The rectification of the products of hydrolytic cocondensation and catalytic regrouping forms individual oligomethylsiloxane liquids PMS-1b, PMS-1,5b, PMS-2b, PMS-2,5b and PMS-3b, as well as higher-viscosity oligomethylsiloxanes, which differ in the number of siloxane elements in the chain.

The process comprises the following stages (Fig. 45): the preparation of the reactive mixture; hydrolytic condensation; the neutralisation and drying of the cocondensation products; catalytic regrouping and filtering; rectification.

Reactor *4* is loaded with a necessary amount of methyltrichlorosilane, dimethylchlorosilane and trimethylchlorosilane from batch boxes *7*, *2* and *3*. The mixture is agitated for 30 minutes at the ambient temperature and sent into weight batch box *5* by nitrogen flow (the pressure does not exceed 0.07 MPa). Reactor *4* is totally cleaned from the reactive mixture and filled with water (2.1 weights to 1 weight of the methylchlorosilane mixture). The reactive mixture is sent from weight batch box *5* at a speed that does not increase the temperature in the reactor above 30 °C. After the mixture has been introduced, the mixture is agitated for 1 hour at the same temperature. After that, the agitator is stopped and the reactive mixture is held at 30 °C for not more than 1 hour. The bottom muriatic layer is sent through a run-down box to biochemical purification. The product of hydrolytic cocondensation is poured into collector *6*.

The product is neutralised in filter *8* and dried in filter *9*. Filter *8* is loaded through a hatch with a necessary amount of soda ash; then, acid hydrolysate is pumped from collector *6* by pump *7* into filter *8* and sent back into collector *6*. The pumping takes about 2 hours; after that, the pH of the product is tested.

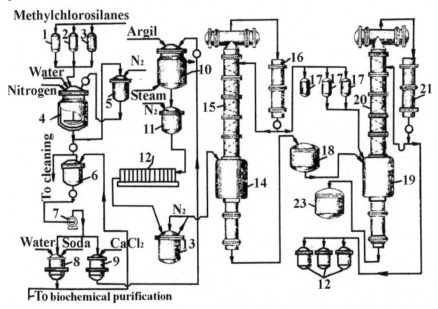

Fig. 45. Production diagram of branched oligomethylsiloxanes:*1-3* - batch boxes; *4* - reactor; *5* - weight batch box; *6, 11, 13, 18, 23* - collectors; *7* - pump; *8, 9* - nutsch filters; *10* - apparatus for catalytic regrouping; *12* - pressure filter; *14, 19* - distillation tanks; *15, 20* - rectification towers; *16, 21* - heaters; *17, 22* - receptacles

If the product is neutral, it is pumped from collector *6* by pump *7* through filter *9*, which is filled with calcium chloride (for drying), into apparatus *10* for catalytic regrouping. After neutralisation, soda ash in filter *8* is dissolved in warm water and sent to biochemical purification.

The neutralised and dried products of hydrolytic cocondensation are heated to 50 °C in apparatus *10* with vapour (0.3 MPa) sent into the jacket. The apparatus is loaded through a hatch with kil clay (8% of the quantity of the loaded cocondensation product) or cationite KU-23 (1-3%) and agitated at 80 °C for 2-3 hours. Then it is poured into receptacle *11* and by nitrogen flow (0.3 MPa) is sent to pressure filter *12*, where clay (or cationite) is filtered off. The filtrate is collected into collector *13*, from where it is pressurised with nitrogen (to 0.3 MPa) into tank *14* for rectification.

The product of hydrolytic cocondensation is a polydisperse mixture of oligomethylsiloxanes of the following composition (%):

Hexamethyldisiloxane (HMDS)	11-18	PMS-2,5b	9-12
PMS-1b	10-16	PMS-3b	5-12
PMS-1,5b	15-17	PMS-5b - PMS-10b	15-25
PMS-2b	12-17	Oligocyclosiloxanes	2-5

All the components have different boiling points; that is why the rectification of the oligomers comprises many stages. At the first stage the product of synthesis is separated into three fractions and tank residue, where the following components are concentrated: HMDS, PMS-lb, PMS-1,5b, PMS-2b and PMS-3b. The concentrates obtained at the first stage are accumulated and separated at further rectification stages.

I rectification stage. Tank 14 is loaded by nitrogen flow (the pressure is below 0.3 MPa) with the product of hydrolytic condensation in the amount not exceeding 65% of the volume; the contents of the tank are heated to the boiling point. When the temperature in the tank reaches 250 °C, rectification tower *15* for 36 hours operates in the "self-serving" mode, and then separates the fractions.

The rectification tower is packed and periodic. It has compensation heating and a built-in water-cooled condenser. Tank *14* and tower *15* are heated with a heat carrier.

The vapours of the products formed by the boiling liquid in tank *14* rise through tower *15*. After the tower the vapours saturated with low-boiling products enter the built-in condenser. Part of the condensate from the top of the tower is sent back through vapour-heated (1 MPa) heater *16* to reflux the tower; the other part is collected in receptacle *17*. The rectification takes place under atmospheric pressure. It allows to separate two fractions: HMDS (80-100 °C), concentrated PMS-1b (100-153 °C) and concentrated PMS-1,5b (153-194 °C). After the distillation the tank residue is cooled and loaded off into collector *18*.

HMDS is used in the synthesis of linear polymethylsiloxane liquids and as a solvent in the synthesis of hexamethyldisilazane; the concentrated PMS-lb, PMS-1,5b and tank residue are subjected to further rectification.

II rectification stage. The concentrated PMS-1b is separated in tower *20*. The concentrate in receptacle *17* is loaded into tank *19* and heated to 150 °C. When the temperature in the tank reaches this level, tower *20* operates for 24 hours in the "self-serving" mode, and then the fractions are separated. After the tower the vapours saturated with low-boiling products enter the built-in water-cooled condenser. Part of the condensate from the top part of the tower is sent back through vapour heater *21* to reflux the tower; part is collected in receptacle *22*. Concentrated PMS-1b is rectified under atmospheric pressure and separated into three fractions: the head fraction (80-100 °C), intermediate (100-152 °C) and concentrated PMS-1b (152-153 °C). After the distillation the heating is switched off, the tank residue is cooled and loaded off into collector *23*.

The head fraction is sent into the tank of tower *20*, the intermediate fraction is added to concentrated PMS-1b for further rectification, and the tank residue is added to concentrated PMS-1,5b.

III rectification stage. Concentrated PMS-1b is separated in the rectification apparatus similar to the one described above. It is distilled under atmospheric pressure into two fractions: the head fraction (153-193 °C) and PMS-1,5b (193-195 °C). The head fraction is added to concentrated PMS-1b for further rectification, the tank residue is added to the tank residue of stage I.

IV rectification stage. The tank residue of rectification stage I is separated in the apparatus similar to the one described above under a residual pressure of 26-53 GPa. The following fractions are separated: the head fraction (below 132 °C), PMS-2b (132-133 °C), intermediate fraction I (133-157 °C), PMS-2,5b (157-158 °C), intermediate fraction II (158-181 °C) and PMS-3b (181-182 °C). The head fraction is added to concentrated PMS-1,5b for further rectification, intermediate fractions I and II are added to the tank residue of stage I for further rectification. The tank residue of stage IV can be used to extract PMS-5b, PMS-6b and higher-viscosity liquids.

Branched oligomethylsiloxane liquids should meet the following technical requirements:

	PMS-1b	PMS-1,5b	PMS-2b	PMS-2,5b	PMS-3b
Appearance	Transparent colourless liquids without mechanical impurities				
Density at 20 °C, g/cm^3	0.82	0.85	0.87	0.89	0.90
Viscosity, mm^2/s					
at 20 °C	1-1.1	1.5-1.7	2.2-2.5	2.8-3.2	3.5-3.9
at -50 °C	≤4	-	≤12	≤16	≤21
at -70 °C	-	≤13	-	-	-
Flash point (in the open pot), °C, not lower than	30	50	70	80	85
Solidification point, °C, not more than	-100	-110	-100	-100	-100
Corroding effect on metals at 100 °C	Absent				
pH of the aqueous extract	6-7	6-7	6-7	6-7	6-7

Content, %:

of the main fraction	≥95	≥95	≥95	≥95	≥94
of dimethylcyclosiloxanes	≤5	≤2	≤5	≤5	≤6
of water	Not standardised	≤0.008	Not standardised		

Branched oligomethylsiloxanes PMS-1b - PMS-3b are soluble in aromatic hydrocarbons, do not corrode construction materials and alloys, are nontoxic. If heated above 200 °C, they form moderately toxic volatile substances, which excite the nervous system and cause conjunctivitis and inflammation of upper airways. The temperature of self-inflammation of these liquids exceeds 340 °C. Their coefficient of volumetric expansion ranges from 0.0013 to 0.0009 cm^3/grad.

Oligomethylsiloxanes PMS-1 - PMS-3b are used as heat carriers in temperature control systems operating in the -100 to +100 °C range in a loop, and as damping liquids. Some physicochemical properties of oligomethylsiloxanes are given in Table 15.1.

3.2.2. Preparation of branched methyloligodiphenylsiloxanes

Methyloligodiphenylsiloxanes are synthesised in two stages. The first stage consists of the catalytic condensation of dihydroxydiphenylsilane in the presence of a catalyst, an aqueous ammonia solution at 100-120 °C:

$$n(C_6H_5)_2Si(OH)_2 \xrightarrow[(n-1)H_2O]{} HO-\left[\begin{array}{c} C_6H_5 \\ | \\ Si-O \\ | \\ C_6H_5 \end{array}\right]_n H \qquad (3.50)$$

$$n=3 \div 5$$

The second stage consists of the heterofunctional condensation of α,ω-dihydroxydiphenylsiloxane obtained at the first stage with triacetoxymethylsilane:

$$HO-\left[\begin{array}{c} C_6H_5 \\ | \\ Si-O \\ | \\ C_6H_5 \end{array}\right]_n H \xrightarrow[CH_3COOH]{CH_3Si(OCOCH_3)_3} CH_3-Si \begin{array}{l} [O-Si(C_6H_5)_2]_n\text{-}OH \\ -[O-Si(C_6H_5)_2]_n\text{-}OH \\ [O-Si(C_6H_5)_2]_n\text{-}OH \end{array} \qquad (3.51)$$

Raw stock: dihydroxydiphenylsilane (14.5-16.5% of hydroxyl groups, pH of the aqueous extract is 6-7); triacetoxymethylsilane (toluene solution, the semi-product of polyphenylsilsesquioxane, which is obtained in the acetylation of phenyltrichlorosilane;aqueous ammonia (reagent grade, at least 25% of ammonia, not more than 0.002% of carbonates, not more than 0.0001% of chlorides), B brand of coal toluene (at least 95% (vol.) of the

109.9-111 °C fraction) or oil toluene (at least 98% (vol.) of the 109.9-111 °C fraction).

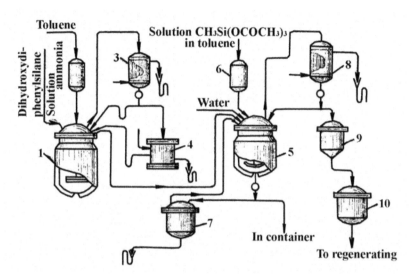

Fig. 46. Production diagram of branched methyloligodiphenylsiloxane: *1, 5* - reactors; *2, 6* - batch boxes; *3, 8* - coolers; *4* - Florentine flask; *7, 10* - collectors; *9*- receptacle

The production of silicon tetrachloride (Fig. 46) comprises the following main stages: the condensation of dihydroxydiphenylsilane and water distillation; the heterofunctional condensation of the product of stage I with tri-acetoxymethylsilane; the flushing of the obtained product with water and the distillation of the solvent.

Dihydroxydiphenylsilane is condensed in toluene in the presence of 25% ammonia solution. The process is carried out in reactor *1*, which is a steel apparatus with a water vapour jacket and a gate agitator. The reactor is loaded through a hatch with a corresponding amount of dihydroxydiphenylsilane and 25% ammonia solution (2% of the weight of dihydroxydiphenylsilane). The required amount of toluene self-flows into the reactor from batch box *2*. The agitator is switched on, the mixture in the apparatus is heated to 100-120 °C and agitated at this temperature for 1 hour. Cooler *3* works in the self-serving mode. The reactor is warmed by vapour (0.3 MPa) fed into the jacket. Then the cooler is switched into the direct operation mode and the mixture of toluene and released water is distilled at a temperature below 120 °C; the mixture is collected in Florentine flask *4*. The process is monitored by the amount of distilled water. The condensation is completed when the amount of water corresponding to the content of hydroxyl groups in the product does not exceed 7%.

The content of OH-groups can be calculated by the formula:

$$\%OH= \frac{X \cdot OH_x - 188{,}9 \,(Y-0{,}015X)}{X-Y+0{,}015X}$$

where X is the amount of dihydroxydiphenylsilane, kg; OH_X is the content of OH-groups in the original dihydroxydiphenylsilane, %; 188.9 is the coefficient; Y is the amount of distilled water, kg; 0.015x is the amount of water introduced with aqueous ammonia.

The speed of the distillation of the azeotropic mixture (water + toluene) is regulated by supplying water into the jacket of reactor 1. The distilled water is collected in the lower part of Florentine flask 4, and the separated toluene is sent back into the reactor. To calculate the number of hydroxyl groups, the water poured out of the Florentine flask is measured and sent to biochemical purification.

The reaction of heterofunctional condensation is carried out in reactor 5, which is a steel enameled apparatus with a water vapour jacket and an anchor agitator. Reactor 5 is loaded at a residual pressure of 730±200 GPa with α,ω-dihydroxydiphenylsiloxane, the product of dihydroxydiphenylsilane condensation; a calculated amount of the toluene solution of triacetoxymethylsilane self-flows at agitationfrom batch box 6 .

The quantity of triacetoxymethylsilane is calculated by the content of hydroxyl groups in α,ω-dihydroxydiphenylsiloxane according to the formula:

X=0.02YZ

where X is the amount of triacetoxymethylsilane, kg; Y is the amount of dihydroxydiphenylsilane, kg; Z is the content of OH-groups in α,ω-dihydroxydiphenylsiloxane, %; 0.02 is the coefficient.

After the whole of triacetoxymethylsilane has been introduced, the reactive mixture is heated to 120 °C and cooled to 40 °C by filling the jacket of reactor 5 with water. The obtained product is washed until the flush waters give a neutral reaction. The flush waters enter collector 7 and from there are sent through a hydraulic gate to biochemical purification.

The separation of toluene from the product of heterofunctional condensation is carried out in two stages. First, toluene is distilled at atmospheric pressure to 50% of the volume of the reactive mixture; then, the mixture is "clarified" (purified from traces of moisture) by settling at 80-100 °C for 1-2 hours. The settled water is poured off through the bottom drain of reactor 5 into collector 7; the product in the reactor is heated to 100-120 °C. At this temperature and a residual pressure of 730±200 GPa toluene is further distilled until the content of volatile substances is less than 3%. The vapours of distilled toluene condense in water cooler *8*. The condensate is collected in receptacle *9* and from there poured into collector *10*. From that collector, raw toluene is sent to regeneration and is re-used in manufacture. After toluene has been distilled, the jacket of reactor *5* is filled with water, the product is cooled to 60 °C and poured into special containers.

The branched oligomer, *methyltris(hydroxydiphenylsiloxy)silane,* should meet the following technical requirements:

Appearance   Resinous product with colour ranging from light yellow to dark brown

Content, %:
of volatile substances   <3
of hydroxyl groups   0.5-3.5
Note. If the stored product crystallises, it should be held before use for 3 hours at 130-150 °C.

Branched methyloligodiphenylsiloxanes are used as modifiers for silicone and organic rubbers to improve their physicomechanical characteristics.

3.2.3. Preparation of oligophenylethoxysiloxanes

Oligophenylethoxysiloxanes are a mixture of branched (A) and ladder (B) oligomer homologues:

$$[C_6H_5SiO_{(3-x)/2}(OC_2H_5)_x]_n$$

A

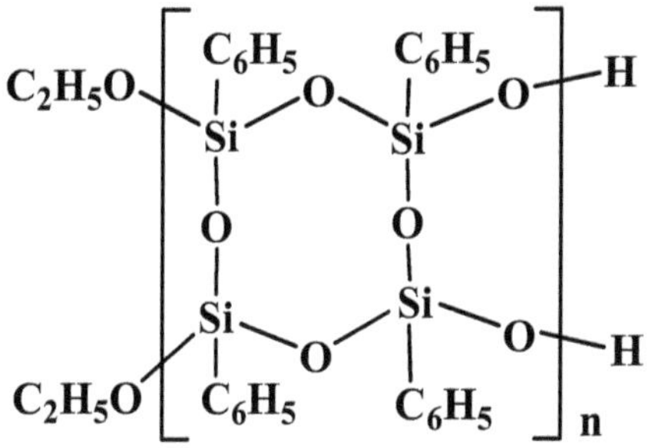

B (n from 2 up to 4)

Oligophenylethoxysiloxanes are obtained by the joint process of etherification and hydrolytic condensation of phenyltrichlorosilane. The degree of oligomer polymerisation depends on the amount of water and alcohol in the reaction. Apart from oligomers, the reaction forms products of partial hydrolytic condensation, such as 1,3-diphenyltetraethoxydisiloxane.

Raw stock: phenyltrichlorosilane [at least 99% (weight) of the 196-202 °C fraction, not more than 1% (weight) of impurities, including not more than 0.3% (weight) of diphenyl and not more than 0.002% (weight) of Si-bound hydrogen]; ethyl alcohol [at least 96.2% (vol.) of the main substance].

The process comprises the following stages (Fig. 47): preparing a mixture of ethyl alcohol and water; etherification of phenyltrichlorosilane and simultaneous hydrolytic condensation; desorption of hydrogen chloride; distillation of excess alcohol and filtering.

The mixture of ethyl alcohol and water is prepared in agitator 2. Alcohol is sent from batch box 1, and a calculated amount of water is sent through a meter. The mixture is agitated for 30 minutes and checked for density and alcohol content. The mixture can also be prepared with recirculating alcohol, sent into the agitator from collector 11.

The etherification of phenyltrichlorosilane (accompanied by hydrolytic condensation) is carried out in enameled reactor 4 with an agitator and a water vapour jacket. Phenyltrichlorosilane is sent from batch box 3 and the agitator is switched on. The mixture of ethyl alcohol and water is sent from batch box 2 through two siphons at such speed that the temperature in the reactor does not exceed 35 °C. When the temperature increases, the jacket is filled with water. After the mixture of alcohol and water has been introduced, the contents of the reactor are agitated at a temperature not exceeding 35 °C for 2 hours. To speed the desorption of the released hydrogen chloride, in the beginning reactor 4 is filled with nitrogen through through a bubbler at a speed of 4-12 m³/h. Hydrogen chloride mixed with nitrogen is sent into water cooler 5, then through gas separator 6 into cooler 7 cooled with salt solution (-15 °C), through gas separator 8 into cooler 9 cooled with salt solution (-40 °C), and through gas separator 10 into absorber 12. After this purification hydrogen chloride is absorbed with water in the absorber and in the form of 30% hydrochloric acid enters container 13. Recirculating ethyl alcohol is collected from the bottom of gas separators 6, 8 and 10 into collector 11.

Coolers 5 and 7 are block graphite apparatuses with the heat exchange surface of 5.4 m²; cooler 9 is a steel enameled apparatus of the "vessel in a vessel" type with the heat exchange surface of 4 m².

After 2 hours of holding, the jacket of reactor 4 is filled with vapour (to heat the reactive mixture to 70 °C) for 1 hour; at a residual pressure of 200-520 GPa, temperature not exceeding 70 °C and at agitation the released alcohol is distilled. Alcohol vapours are sent through cooler 5 and 7, gas separators 6 and 8 into collector 11. After the excess alcohol has been distilled, the contents of the reactor are sampled to determine the viscosity and hydrogen chloride content in the obtained oligophenylethoxysiloxane. If the analysis is positive, the product is cooled in the reactor to 60 °C and pumped (the pump is not shown in the diagram) into pressure filter 14 to filter off mechanical impurities. After that, it is poured into containers.

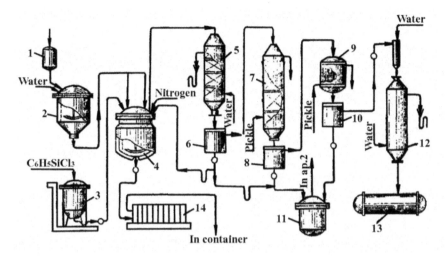

Fig. 47. Production diagram of oligophenylethoxysiloxanes: *1, 3* - batch boxes; *2* - agitator; *4* - reactor; *5, 7, 9* - coolers; *6, 8, 10* - gas separators; *11* - collector; *12* - absorber; *13* - container; *14* - pressure filter

Depending on the ratio of the parent components, one obtains two brands of oligophenylethoxysiloxanes, PES-50 and PES-80.

Pure *oligophenylethoxysiloxanes* are transparent, viscous, colourless liquids with low toxicity (mild narcotic action). They dissolve in polar and nonpolar solvents.

Technical oligophenylethoxysiloxanes should meet the following requirements:

	PES-50	PES-80
Appearance	Transparent viscous liquids, from colourless to light brown	
Density, g/cm^3	1.09-1.14	1.185
n_D^{20}	1.4920-1.5090	1.5175-1.5350
Viscosity of 50% toluene solution at 20 °C, mm^2/s	1.6-2.2	3.0-4.2
Content, %:		
of phenyltriethoxysilane	≤ 20	≤ 7
of 1,3-diphenyltetraethoxydisiloxane	≤ 15	≤ 5
of hydrogen chloride	≤0.04	≤0.04
of silicon	14.5-16.5	17-19
of ethoxygroups	27-40	12-17.5

Note: Opalescence is tolerated.

Oligophenylethoxysiloxanes are used as modifiers for various polymers to improve their weather resistance and other technical characteristics, as well as to increase the heat resistance of coatings. E.g., PES-50 is used to modify polyethers, acrylic and epoxy polymers; PES-80 is used to modify alkyd and urea-formaldehyde resins. Besides, PES-80 is used as an additive in paints and enamels (to improve their flow properties, gloss and colour), as well as in concrete mixes (to improve the water resistance and durability of concrete works).

3.2.4. Preparation of branched oligoorganosiloxanes with alkoxyoxyalkylene groups in the lateral chain

Oligoorganosiloxanes with alkoxyoxyalkylene fragments in their branches have recently been widely used as surface active agents, which regulate foam forming and foam stabilising in the production of various foamed polyurethanes.

   These oligoorganosiloxanes are prepared by the catalytic polymerisation of diorganocyclosiloxanes with organotriethoxysilanes with subsequent re-etherification of branched oligoorganosiloxane with monoalkyl ethers of oligoalkyleneglycols or hydrolytic cocondensation of hydrideorgano- and diorganodichlorosilanes with subsequent hydrosilylation of allylalkyl ethers of oligoalkyleneglycols with oligohydrideorganosiloxanes.

Preparation of tris{ω-butoxyoligo[(propyleneoxy)(ethyleneoxy)-(dimethylsiloxy)]}ethylsilane

Tris{ω-butoxyoligo[(propyleneoxy)(ethyleneoxy)(dimethylsiloxy)]}-ethylsilane is a liquid mixture of oligomer homologues with the following common composition:

$$C_2H_5-Si \begin{cases} [OSi(CH_3)_2]_{x'} -[OCH_2CH_2]_{y'} -[OCH_2CH_2CH_2]_{z'} -OC_4H_9 \\ [OSi(CH_3)_2]_{x''} -[OCH_2CH_2]_{y''} -[OCH_2CH_2CH_2]_{z''} -OC_4H_9 \\ [OSi(CH_3)_2]_{x'''} -[OCH_2CH_2]_{y'''} -[OCH_2CH_2CH_2]_{z'''} -OC_4H_9 \end{cases}$$

x', x" and x'" – from 12 to 24
y', y" and y'" – from 15 to 20
z', z" and z'" – from 10 to 15
The oligomer is synthesised in two stages.

1. The first stage is the anion polymerisation of dimethylcyclosiloxanes with ethyltriethoxysilane:

$$m[(CH_3)_2SiO]_n + C_2H_5Si(OC_2H_5)_3 \xrightarrow{KOH} C_2H_5-Si \begin{cases} [OSi(CH_3)_2]_{x'}-OC_2H_5 \\ [OSi(CH_3)_2]_{x''}-OC_2H_5 \\ [OSi(CH_3)_2]_{x'''}-OC_2H_5 \end{cases} \quad (3.52)$$

mn=x'+x"+x"'=12÷24; mole weight ≈1500

2. The obtained branched oligoorganosiloxane (A) is re-etherified with the monobutyl ether of oligooxypropyleneoxyethyleneglycol:

$$A + 3H-[OCH_2CH_2]_y-[OCH_2CH_2CH_2]_z-OC_4H_9 \xrightarrow[-3C_2H_5OH]{}$$

$$\longrightarrow C_2H_5-Si \begin{cases} [OSi(CH_3)_2]_{x'}-[OCH_2CH_2]_{y'}-[OCH_2CH_2CH_2]_{z'}-OC_4H_9 \\ [OSi(CH_3)_2]_{x''}-[OCH_2CH_2]_{y''}-[OCH_2CH_2CH_2]_{z''}-OC_4H_9 \\ [OSi(CH_3)_2]_{x'''}-[OCH_2CH_2]_{y'''}-[OCH_2CH_2CH_2]_{z'''}-OC_4H_9 \end{cases} \quad (3.53)$$

The reaction is catalysed with trifluoroacetic acid and solid potassium hydroxide.

Raw stock: dimethylcyclosiloxanes (the products of the depolymerisation of linear polydimethylsiloxanes (60-85% of tetramer, 13-20% of pentamer, 0.5% of hexamer and up to 5% of trimer, $d_4^{20} = 0.95+0.96$); ethyltriethoxysilane (at least 97% of the 158-160 °C fraction); the monobutyl ether of oligooxypropyleneoxyethyleneglycol (Laprol 1604-2-50, type A (1-1.2% of OH groups, not more than 0.1% of moisture); coal toluene of B grade [at least 95% (vol.) of the 109.9-111 °C fraction]; trifluoroacetic acid (the boiling point is 72.4 °C); potassium hydroxide (chemically pure); frozen acetic acid; monoethanolamine, a viscous hygroscopic liquid (the boiling point is 171.1 °C at 1013 GPa, the density is 1.017 g/cm³), or pure sodium bicarbonate - colourless crystals (the melting point is 160 °C, the density is 2.16-2.22 g/cm³).

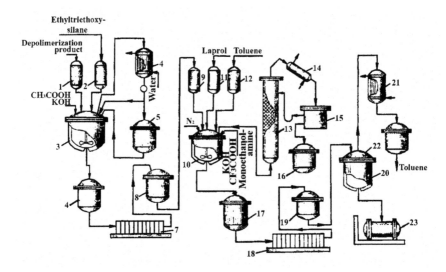

Fig. 48. Production diagram of tris{ω-butoxyoligo[(propyleneoxy)(ethyleneoxy)-(dimethylsiloxy)]}ethylsilane: *1, 2, 9, 11, 12* - batch boxes; *3, 10* - reactors; *4, 21* - coolers; *5, 6, 8, 16, 17, 19, 22* - receptacles; *7, 18* - pressure filters; *13* - packed tower; *14* - refluxer; *15* - Florentine flask; *20* - distillation tank; *23* – container

The process comprises the following main stages (Fig. 48): the synthesis of silicone branched oligomer; the synthesis of tris{ω-butoxyoligo[(propyleneoxy)(ethyleneoxy)(dimethylsiloxy)]}ethylsilane, the subsequent distillation of the solvent and the filtering of the product.

Reactor 3 is loaded with dimethylcyclosiloxane (depolymerisate) from batch box 1. The depolymerisate is dried at 50-80 °C for 1-3 hours (until the moisture content is lower than 0.01%). The reactor is cooled down to 40 °C , filled with pre-ground potassium hydroxide (0.12% of the reactive mixture) through a hatch, and the contents of the reactor are agitated for 30 minutes. After that, at agitation the reactor receives a calculated amount of ethyltriethoxysilane from batch box 2. The reactor is filled with vapour and ethyltriethoxysilane is heated to 120 °C and held at this temperature for 2 hours, then tested for kinematic viscosity. The viscosity of the oligomer should be 12-16 mm^2/s. It is sampled every 30 minutes. If viscosity increases slowly, the temperature can be raised to 150 °C. If the analysis is satisfactory, the contents of the reactor are cooled down to 40 °C by sending water in the jacket, agitated and neutralised with frozen acetic acid, which is fed into the reactor through a hatch for 45-90 minutes. Reactor 3 has a water cooler 4, which during the synthesis operates as a backflow condenser. After the reactive mixture has been neutralised, the cooler is switched into the direct mode, the jacket of the reactor is filled with vapour and a vacuum is created in the system. At a temperature in the tank not exceeding 150 °C, 80-100 °C in vapours and a residual pressure of 0.085-0.09 GPa the unreacted products are distilled and collected in receptacle *5*. After the volatile substances have been separated, the oligomer is tested for viscosity, which should amount to 16-25 mm^2/s. Silicon content in the synthesised oligomer should be 33-35%, and the content of ethoxyl groups should be 6.5-9%. The volatile products from receptacle 5 can be re-used in the synthesis of the oligomer.

If the analysis is satisfactory, the oligomer is cooled to 40 °C by sending water into the reactor jacket, loaded into receptacle *6* and under the pressure of dehydrated nitrogen (0.3 GPa) filtered through pressure filter *7* into receptacle *8*, from where it is sent into weight batch box *9*.

The target oligomer is synthesised in reactor *10*. For this purpose, Laprol is loaded from batch box *11*, and toluene is loaded from batch box *12*. The agitator is switched on, the temperature in the reactor is increased to 110-130 °C (to 85-110 °C in vapours) by sending vapour into the jacket and at this temperature toluene is subjected to azeotropic drying. The vapour of the azeotropic mixture (toluene + water) rises up packed tower *13* and condenses in refluxer *14*. The condensate splits in Florentine flask *15*. Toluene from the top part of the apparatus is sent back (through a side choke) to reflux tower *13*, and toluene-containing water is collected in receptacle *16*. Thus the toluene solution of Laprol is dehydrated until moisture content is not more than 0.01%.

After the drying, the contents of reactor *10* are cooled to 40 °C by sending water into the jacket. From weight batch box *9*, a calculated amount of silicone oligomer is loaded; through the hatch the reactor receives trifluoroacetic acid (0.12% of the amount of the reactive mixture) and potassium hydroxide (0.04%). The reactive mixture is agitated and heated to 120-140 °C; at this temperature re-etherification takes place. The released ethyl alcohol is withdrawn out of the system in the form of azeotropic mixture with toluene. The vapours of the azeotropic mixture rise up tower *13* and condense in refluxer *14*. From there, part of the condensate in the form of reflux is returned into the tower, and the rest is collected in receptacle *16*.

If within 6-10 hours after the beginning of the synthesis the reactive mixture does not become transparent, reactor *10* should be loaded with an additional amount of trifluoroacetic acid (50% of the main catalyst load) and blown with anhydrous nitrogen for 30 minutes.

When the synthesis ends, the reactive mixture is cooled down to 80-100 °C and reactor *10* is loaded with monoethanolamine or sodium bicarbonate to neutralise trifluoroacetic acid. The mixture is neutralised for 2-4 hours and sampled to determine pH. If its pH is 6.5-7-8.5, the mixture is cooled down to 30-40 °C, under nitrogen pressure loaded into receptacle *17* and sent to filtering.

If the toluene solution of the target product is neutralised with sodium bicarbonate, the solution is filtered in pressure filter *18* through through coarse calico and paper under the pressure of dehydrated nitrogen of 0.3 GPa.

The filtered toluene solution of the target product is collected in receptacle *19* and sent into tank *20* to distill toluene. A residual pressure of 0.07-0.09 GPa is created in the tank; the contents are heated to 120-135 °C. The distillation is carried out until toluene content is 0.6% or lower and the viscosity is 600-1100 mm^2/s, if the product completely dissolves in water. The distilled toluene vapours are condensed in cooler *21* and collected in receptacle *22*. Toluene is subsequently re-used in reactor *10*.

The target product is cooled in tank *20* to a temperature not exceeding 80 °C and filtered through brass mesh into scale-mounted container *23*.

Tris{ω-butoxyoligo[(propyleneoxy)(ethyleneoxy)(dimethylsiloxy)]}-ethylsilane is a nontoxic liquid which dissolves well in water. Technical oligomer product should meet the following requirements:

Appearance	Transparent brown liquid
Density at 25 °C, g/cm^3	1.0200-1.0400
Viscosity at 20 °C, mm^2/s	600-1100
Solubility in water	Complete
pH of the aqueous extract	6.5-8.5
Content, %:	
of silicon	6.5-11.5
of toluene	< 0.6
Average polyurethane shrinkage, mm, not more than	1
Average foam rising height, % of the standard sample, at least	90
Mechanical impurities	Absent

Tris{ω-butoxyoligo[(propyleneoxy)(ethyleneoxy)(dimethylsiloxy)]}-ethylsilane is used as a surface active additive, which regulates foam formation in the production of polyurethanes.

Preparation of branched oligo(ω-alkoxyalkyleneoxypropyl-organosiloxy)dimethylsiloxanes with end trimethylsiloxy groups

Branched oligoorganosiloxanes with alkoxyalkyleneoxypropylorganosiloxane elements in their branches and with end trimethylsiloxy groups are liquids consisting of oligomer homologues with the following common composition:

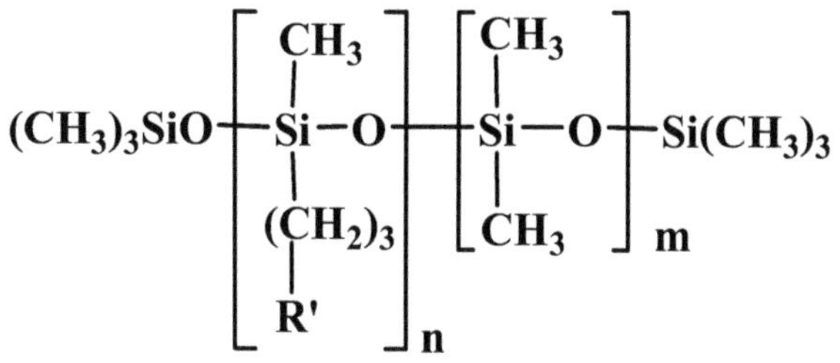

$$(CH_3)_3SiO \left[\begin{array}{c} CH_3 \\ | \\ Si-O \\ | \\ (CH_2)_3 \\ | \\ R' \end{array} \right]_n \left[\begin{array}{c} CH_3 \\ | \\ Si-O \\ | \\ CH_3 \end{array} \right]_m Si(CH_3)_3$$

R'= $(OCH_2CH_2)_x$-$(OCH_2CH_2CH_2)_y$-OC_4H_9 or $(OCH_2CH_2)_x$-OCH_3

Oligo(ω-butoxypropyleneoxyethyleneoxypropylethylsiloxy)-dimethylsiloxane with end trimethylsiloxy groups (oligomer I)

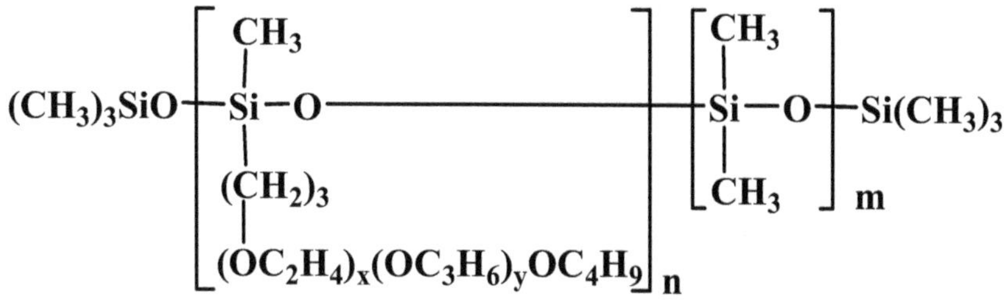

$$(CH_3)_3SiO \left[\begin{array}{c} CH_3 \\ | \\ Si-O \\ | \\ (CH_2)_3 \\ | \\ (OC_2H_4)_x(OC_3H_6)_yOC_4H_9 \end{array} \right]_n \left[\begin{array}{c} CH_3 \\ | \\ Si-O \\ | \\ CH_3 \end{array} \right]_m Si(CH_3)_3$$

Oligomer I n=3÷4; m=28÷32; x=15÷20; y=12÷15 is obtained by the hydrolytic cocondensation of ethyldichloro-, dimethyldichloro- and trimethylchlorosilanes with the subsequent hydrosilylation of the allylbutyl ether of oligooxypropyleneoxyethyleneglycol with the product of hydrolytic cocondensation, α,ω-bis(trimethylsiloxy)hydridethylsiloxy-dimethylsiloxane. The hydrolytic cocondensation of ethyldichloro-, dimethyldichloro- and trimethylchlorosilanes is conducted according to the following reaction:

$$nC_2H_5SiHCI_2 + m(CH_3)_2SiCI_2 + 2(CH_3)_3SiCI \xrightarrow[2(n+m+1)HCI;\ nH_2O]{2(n+m+1)H_2O}$$

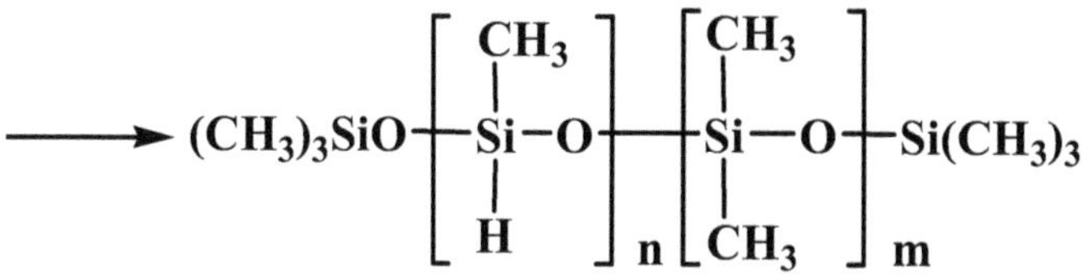

$$\longrightarrow (CH_3)_3SiO \left[\begin{array}{c} CH_3 \\ | \\ Si-O \\ | \\ H \end{array} \right]_n \left[\begin{array}{c} CH_3 \\ | \\ Si-O \\ | \\ CH_3 \end{array} \right]_m Si(CH_3)_3 \qquad (3.54)$$

However, the hydrolytic cocondensation is complicated and also involves reactions of the separate hydrolytic condensation of monomers:

$$nC_2H_5SiHCI_2 + nH_2O \xrightarrow[2nHCI]{} [C_2H_5Si(H)O]_n \qquad (3.55)$$

$$m(CH_3)_2SiCI_2 + mH_2O \xrightarrow[2mHCI]{} [(CH_3)_2SiO]_n \qquad (3.56)$$

$$2(CH_3)_3SiCI + H_2O \xrightarrow[-2HCI]{} (CH_3)_3Si\text{-}O\text{-}Si(CH_3)_3 \quad (3.57)$$

Therefore, to increase the yield of the main product, it is followed by the catalytic regrouping of the products of the separate hydrolytic condensation:

$$[C_2H_5Si(H)O]_n + [(CH_3)_2SiO]_m + (CH_3)_3Si\text{-}O\text{-}Si(CH_3)_3 \longrightarrow$$

$$\longrightarrow (CH_3)_3SiO\left[\begin{array}{c} C_2H_5 \\ | \\ Si\text{-}O \\ | \\ H \end{array}\right]_n \left[\begin{array}{c} CH_3 \\ | \\ Si\text{-}O \\ | \\ CH_3 \end{array}\right]_m Si(CH_3)_3 \quad (3.58)$$

The process can be catalysed wih sulfuric acid and its salts, various brands of activated clays or cationites.

At the final production stage of oligomer I the allylbutyl ether of oligooxypropyleneoxyethyleneglycol is hydrosilylated with a silicone oligomer:

$$(CH_3)_3SiO\left[\begin{array}{c} C_2H_5 \\ | \\ Si\text{-}O \\ | \\ H \end{array}\right]_n \left[\begin{array}{c} CH_3 \\ | \\ Si\text{-}O \\ | \\ CH_3 \end{array}\right]_m Si(CH_3)_3 +$$

$$+ nCH_2=CH\text{-}CH_2\text{-}(OCH_2CH_2)_x(OCH_2CH_2CH_2)_y\text{-}OC_4H_9 \xrightarrow{H_2PtCI_6} \text{Oligomer I} \quad (3.59)$$

Oligo(ω-methoxyethyleneoxypropylmethylsiloxy)dimethylsiloxane with end trimethylsiloxy groups (oligomer II) is synthesised

$$(CH_3)_3SiO\left[\begin{array}{c} CH_3 \\ | \\ Si\text{-}O \\ | \\ (CH_2)_3 \\ | \\ (OC_2H_4)_xOCH_3 \end{array}\right]_n \left[\begin{array}{c} CH_3 \\ | \\ Si\text{-}O \\ | \\ CH_3 \end{array}\right]_m Si(CH_3)_3$$

Oligomer II n=7÷9; m=4÷6; x=8÷10 is synthesised by the hydrolytic cocondensation of a mixture of methyldichloro-, dimethyldichloro- and trimethylchlorosilanes:

$$nCH_3SiHCI_2 + m(CH_3)_2SiCI_2 + 2(CH_3)_3SiCI \xrightarrow[2(n+m+1)HCI]{2(n+m+1)H_2O}$$

$$\longrightarrow (CH_3)_3SiO \left[\begin{array}{c} CH_3 \\ | \\ Si-O \\ | \\ H \end{array} \right]_n \left[\begin{array}{c} CH_3 \\ | \\ Si-O \\ | \\ CH_3 \end{array} \right]_m Si(CH_3)_3 \qquad (3.60)$$

Then, similarly to oligomer I, the formed products of the separate hydrolytic condensation are subjected to catalytic regrouping:

$$[CH_3Si(H)O]_n + [(CH_3)_2SiO]_m + (CH_3)_3Si\text{-}O\text{-}Si(CH_3)_3 \longrightarrow$$

$$\longrightarrow (CH_3)_3SiO \left[\begin{array}{c} CH_3 \\ | \\ Si-O \\ | \\ H \end{array} \right]_n \left[\begin{array}{c} CH_3 \\ | \\ Si-O \\ | \\ CH_3 \end{array} \right]_m Si(CH_3)_3 \qquad (3.61)$$

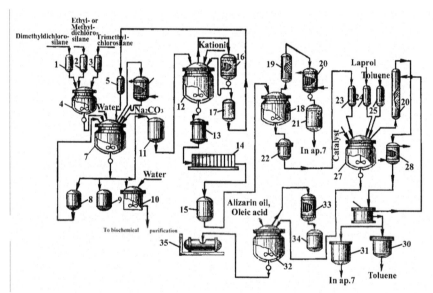

Fig. 49. Preparation of branched oligo(ω-alkoxyalkyleneoxypropylorganosiloxy)dimethylsiloxanes with end trimethylsiloxy groups 1-3,5,23 -batch boxes; 4 - agitator; 6,16,20,28,33 - coolers; 7 - hydrolyser; 8,9,13 - receptacles; 10 - neutraliser; 11,15, 17,21,22,30,31,34 - collectors; 12,17 – pressure filter; 18,32 – distillation tanks; 19,26 – packed towers; 29 – Florentine flask; 35 – container

And, finally, the allylmethyl ether of oligooxyethyleneglycol is hydrosilylated with a silicone oligomer:

$$\longrightarrow (CH_3)_3SiO\left[\begin{matrix}CH_3\\|\\Si-O\\|\\H\end{matrix}\right]_n\left[\begin{matrix}CH_3\\|\\Si-O\\|\\CH_3\end{matrix}\right]_m Si(CH_3)_3$$

$$+ \ nCH_2=CH-CH_2-(OCH_2CH_2)_x-OCH_3 \xrightarrow{H_2PtCl_6} \text{Oligomer II} \quad (3.62)$$

Raw stock for oligomer I: ethyldichlorosilane (at least 99% of the main fraction); dimethyldichlorosilane (at least 99.8% of the main fraction); trimethylchlorosilane (at least 99.6% of the main fraction); the allylbutyl ether of oligooxypropyleneoxyethyleneglycol (Laprol 1601-2-50, B type), yellow or brown transparent liquid (1% of OH groups, not more than 0.1% of moisture); chloroplatinic acid (brown-red or orange crystals); distilled and dehydrated tetrahydrofuran (the boiling point is 65-67 °C); cationite KU-23-40/100 or KU-23-30/100 (grey spherical granules); sodium bicarbonate (pure); monoethanolamine (pure).

Raw stock for oligomer II: methyldichlorosilane (at least 99% of the main fraction); dimethyldichlorosilane; trimethylchlorosilane; chloroplatinic acid; tetrahydrofuran; cationite KU-23-40/100 or KU-23-30/100; sodium bicarbonate (with the same characteristics as for oligomer I); the allylmethyl ether of oligooxyethyleneglycol (Laprol 501-2-100, B type), oily nonvolatile liquid (7-9% of OH groups, not more than 0.15% of moisture); methylenechloride (the boiling point is 40 °C).

The production of oligomer I comprises the following main stages: the preparation of the reactive mixture; the hydrolytic cocondensation of organochlorosilanes; the catalytic regrouping of the products of separate hydrolytic condensation and the distillation of the volatile synthesis products; the distillation of the solvent and the filtering of oligomer I.

The production of oligomer II comprises the following stages: the preparation of the reactive mixture; the hydrolytic cocondensation of organochlorosilanes and the distillation of the solvent; the catalytic regrouping of the products of separate hydrolytic condensation and the distillation of the volatile products; the synthesis; the distillation of the solvent and the preparation of oligomer II.

To prepare branched oligo(ω-alkoxyalkyleneoxypropylorganosiloxy)-dimethylsiloxanes with end trimethylsiloxy groups (Fig. 49), the reactive mixtures for the hydrolytic cocondensation of organochlorosilanes are mixed in agitator 4 with an anchor agitator and a cooling jacket. The agitator is loaded with calculated amounts of dimethyldichlorosilane, ethyldichlorosilane (for the synthesis of oligomer I) or methyldichlorosilane (for the synthesis of oligomer II) and trimethylchlorosilane. The agitator is switched on and the contents of the apparatus are agitated for 30 minutes. The mixture is sampled to determine its composition; if the analysis is positive, it is sent to cocondensation into hydrolyser *7* with an agitator and a jacket.

Production of oligomer I. Hydrolyser *7* is loaded with a calculated amount of water; then, at agitation and 40 °C it receives the reactive mixture from agitator *4*. The temperature is regulated by changing the speed of mixture supply and sending water into the jacket of the hydrolyser. After the whole mixture has been introduced, the contents of the hydrolyser are heated to 60-65 °C and kept for 2 hours. After that, the reactive mixture is cooled to 30 °C, the agitation is stopped and the mixture is settled.

The bottom layer (25-30% hydrochloric acid) is poured through a run-down box into receptacle *9* and used in the production of oligoethylsiloxane liquids. The product of hydrolytic cocondensation in the hydrolyser is mixed with a calculated amount of sodium bicarbonate dissolved in warm water. The mixture is agitated for 4-5 hours; then the hydrolyser is flushed with a required quantity of water. The mixture is agitated for 0.5-1 hours, the agitator is stopped and the bottom aqueous layer is settled. The water layer is poured into neutraliser 10 and at pH 6-8 sent through a hydraulic gate to biochemical purification. The product of hydrolytic cocondensation in the hydrolyser is heated to 40 °C and settled to separate water. The water is poured into the neutraliser, and the ready product is pumped into collector *11*.

The catalytic regrouping for oligomer I is carried out in reactor *12* with an agitator and a water vapour jacket. The product of hydrolytic cocondensation is sent from collector *11* by nitrogen flow into reactor *12*, heated to 80-90 °C and held at this temperature for 3-5 hours with an open hatch (for drying). The humidity of the product after drying should not exceed 0.015%. After drying, the reactor is filled through a hatch with ion-exchange resin (cationite KU-23) in the amount of 2% of the weight of the product of hydrolytic cocondensation. The mixture is agitated at 60-100 °C for 2-10 hours. Cooler *16* at this time works in the self-serving mode. Every hour the mixture is sampled to determine the viscosity of the oligomer. When the viscosity is 20-35 mm^2/s, the contents of the reactor are loaded into receptacle *13* and filtered through pressure filter *14* into collector *15*.

The product of catalytic regrouping is loaded from collector *15* into tank *18*, which has an agitator, a water vapour jacket and packed tower *19*. A residual pressure of 0.08-0.09 GPa is created in the tank; the contents are heated to 120-150 °C and the volatile products are distilled, until the viscosity is 35-55 mm^2/s and active hydrogen content in the oligomer (H_{act}) reaches 0.13-0.25%. The volatile products after cooler *20* are collected into collector *21* and are sent into hydrolyser *7* as they accumulate for repeated use in the synthesis. After the distillation the contents of tank *18* are cooled to   40   °C   and   the   finished   oligomer,   α,ω-bis(trimethylsiloxy)hydridethylsiloxydimethylsiloxane, is sent by nitrogen flow into collector *22* and on into weight batch box *23*.

Oligomer I is prepared in reactor *27* with an agitator, a water vapour jacket and packed tower *26*. The reactor is loaded with Laprol and toluene. The mixture is heated to 110-130 °C at agitation and subjected to azeotropic distillation (drying). The vapour of the azeotropic toluene + water mixture pass through tower *26* and cooler *28* and enter Florentine flask *29*, where the mixture of toluene and water splits. The top layer (toluene) is returned to reflux tower *26*, and the bottom layer (moist toluene) is sent into collector *30*. After drying, the contents of reactor *27* are cooled to a temperature not higher than 50 °C and the reactor is loaded with a calculated amount of the oligomer from weight batch box *23*.

The calculation is based on the iodine number of Laprol and the active hydrogen content in the oligomer:

$$\text{Oligomer quantity} = \frac{IN \cdot L}{254 \cdot H_{act}} 0.95 (kg)$$

where IN is the iodine number of Laprol, (mg I_2/g); L is the quantity of the loaded Laprol, kg; 254 is the gram-equivalent of iodine; 0.95 is the co-efficient which expressed the reduction of the loaded amount of the oligomer by 5% compared to the calculated amount (the calculated amount is reduced by 5% to guarantee the complete absence of reacted products).

Reactor *27* is loaded through a hatch with a catalyst (1% solution of H_2PtCl_6 in tetrahydrofuran) in the quantity of 0.45% of the reactive mixture. The mixture is heated to 130-145 °C and held for 40-60 hours. At this time toluene and volatile products are distilled. Toluene rises up tower *26*, is cooled and partially sent back to reflux the tower; the main amount is sent into collector *31* to be used in later synthesis. During the holding, the appearance of oligomer I is checked every 3-4 hours (it should be transparent). When the holding is over, the mixture is cooled in the reactor to 80-90 °C and neutralised with monoethanolamine. After the neutralisation, oligomer I is sampled.

If pH is 6.5-8.5, the toluene solution of oligomer I is sent into distillation tank *32*, which has an agitator and a water vapour jacket. A residual pressure of 0.08-0.09 GPa is created in the tank; the contents are heated to 100-125 °C. At this temperature the solvent is distilled until the content of volatile substances is not more than 1%. Toluene vapours are condensed in cooler *33* and collected in receptacle *34*. After the distillation, oligomer I in the tank is cooled to a temperature not exceeding 80 °C and sampled to determine its viscosity and pH. If the characteristics meet the standard, oligomer I is filtered through brass mesh into scale-mounted container *35*. If pH of the product does not correspond to the standard, it is neutralised with monoethanolamine in tank *32*.

Production of oligomer II. The reactive mixture is prepared and subjected to hydrolytic cocondensation in hydrolyser *7*. First, the apparatus is loaded with a calculated amount of water and then with methylchloride from batch box *5* (the cooling agent). The mixture in the apparatus is agitated for 30 minutes, the jacket of the apparatus is filled with salt solution and cooled down to 10-15 °C . The reactive mixture is introduced from batch box *4*; the temperature in the hydrolyser is maintained at 10-17 °C. After introducing the reactive mixture the supply of salt solution into the jacket is stopped and the contents of the hydrolyser are held for 3 hours at agitation. The temperature in the reactor rises spontaneously to 20 °C. The agitation is stopped and the reactive mixture is held for 1 hour. It splits: the product of hydrolytic cocondensation forms the bottom layer, and the solution of hydrochloric acid remains on top. The bottom layer is poured into receptacle *8* and pumped back into hydrolyser *7* to be flushed. The top layer (25-30% hydrochloric acid) is poured into receptacle *9* and used in the production of oligoethylsiloxane liquids.

To flush the product of hydrolytic cocondensation, the hydrolyser is loaded with a calculated amount of water, the contents of the apparatus are agitated for 30 minutes and settled for 1 hour. The lower layer, the flushed solution of the product, is sent into collector *11*, and the upper layer (flush waters) is sent into neutraliser *10* and at pH 6-8 is sent to biochemical purification.

The distillation of the solvent and catalytic regrouping are conducted in reactor *12*. For this purpose, from collector *11* the product of hydrolytic cocondensation is sent by nitrogen flow into reactor 12, the agitator is switched on and the reactor is heated to 36-40 °C by sending vapour into the jacket. At this temperature most of the solvent is distilled within 3-4 hours. Cooler *16* is switched into the direct operation mode. After cooling, methylenechloride vapours are collected into collector 17, poured into batch box 5 and re-ised in the process of hydrolytic cocondensation.

In 3-4 hours, when the distillation speed noticeably decreases, the temperature in the reactor is gradually (in 5-8 hours) raised to 90-100 °C. Methylenechloride and water are completely distilled. After that, the product of hydrolytic co-condensation is cooled down to 80 °C and tested for viscosity (2-4 mm^2/s). The reactor is loaded through a hatch with cationite KU-23 in the amount of 2% of the product weight. The contents of the reactor are agitated at 80-90 °C for 8-24 hours, sampling them every hour for viscosity. If the viscosity is 4-7 mm^2/s, the contents of the reactor are unloaded into receptacle *13* and filtered through pressure filter *14* into collector *15*.

The product of catalytic regrouping is sent from collector *15* by nitrogen flow into tank *18* to distil volatile products (under the same conditions as in the production of oligomer I). The distillation is monitored by the viscosity of the reactive mixture. If the viscosity is 7-12 mm^2/s and active hydrogen content is 0.77-0.8% , the distillation of volatiles (collected into collector *21*) is stopped, the contents of tank *18* are cooled down to a temperature below 40 °C and the ready silicone oligomer is sent by nitrogen flow into collector *22* and on into weight batch box *23*.

Hydrosylilation is conducted in reactor *27,* which is filled with Laprol and toluene. The azeotropic drying of the toluene Laprol solution is carried out just like in the production of oligomer I. After drying the contents of the reactor are cooled down to a temperature not exceeding 50 °C, a calculated amount of silicone oligomer is loaded from weight batch box *23* (the loading calculation is given on page 253), the catalyst, 1% tetrahydrofurane solution of H_2PtCl_6 (24% of the reactive mixture) is loaded through a hatch Hydrosylilation and subsequent distillation of toluene is conducted just just like in the production of oligomer I. After the distillation the contents of tank *32* are cooled down to a temperature not exceeding 80 °C and sampled for viscosity and pH. If the characteristics meet the standards, oligomer II is finished by adding alizarin oil and oleic acid in the following ratio: 65% of the hydrosylilation product, 20% of alizarin oil and 15% of oleic acid. The contents of apparatus *32* are agitated for an hour, and the finished product is poured into containers. Oligomers I and II are nontoxic transparent liquids; they do not form explosive mixtures with air. They disintegrate at the boiling temperature.

Oligomers I and II are nontoxic transparent liquids; they do not form explosive mixtures with air. They disintegrated at the boiling temperature.

Technical oligomers I and II should meet the following requirements:

	Oligomer I	Oligomer II
Appearance	transparent liquid, colour ranging from yellow to brown; for oligomer II, a sediment is tolerated	
Viscosity at 20 °C, mm^2/s	800-3000	200-500
Acid number, mg KOH/g, not more than	-	70
Solubility in water	Complete (opalescence is tolerated)	Complete
pH of the aqueous extract	6-8	6-8
Content, %:		
of toluene	≤ 1.0	-

of silicon   7.5-10   -
Average polyurethane shrinkage, 1   1
mm, not more than
Average foam rising height, % of 90   95
the standard sample, at least
Mechanical impurities   Absent

Oligomer I is used as a regulator and stabiliser in the production of elastic polyurethanes based on polyethers, rigid and semirigid polyurethanes; oligomer II is used in the production of elastic polyurethanes based on esters.

3.2.5. Preparation of branched oligomethyl(phenyl)siloxanes with metacrylethoxyl groups in the lateral chain

Oligomethyl(phenyl)metacrylethoxysiloxane (MAS) is a mixture of oligomer homologues in a solution of toluene and butanol with the following common composition:

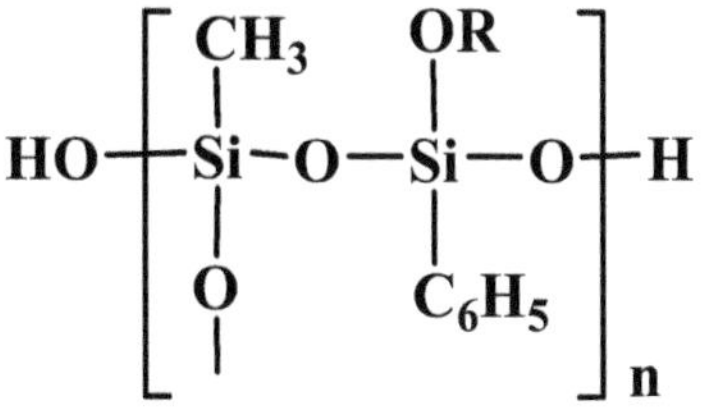

$$HO\left[\begin{array}{cc} CH_3 & OR \\ | & | \\ -Si-O-Si-O- \\ | & | \\ O & C_6H_5 \\ | \end{array}\right]_n H$$

where $R=-CH_2CH_2OCOC(CH_3)=CH_2$; $n=0.5\div3$.

MAS is synthesised by the hydrolytic cocondensation of metacrylated triacetoxymethyl- and triacetoxyphenylsilanes:

$nCH_3Si(OCOCH_3)_2OR + nCH_3Si(OCOCH_3)_3 + nC_6H_5Si(OCOCH_3)_2OR + nC_6H_5Si(OCOCH_3)_3 +$

$$+ (11n+1)H_2O + HO\left[\begin{array}{cc} CH_3 & OR \\ | & | \\ -Si-O-Si-O- \\ | & | \\ O & C_6H_5 \\ | \end{array}\right]_n H + 10nCH_3COOH + nHOCH_2CH_2OCOC(CH_3)=CH_2 +$$

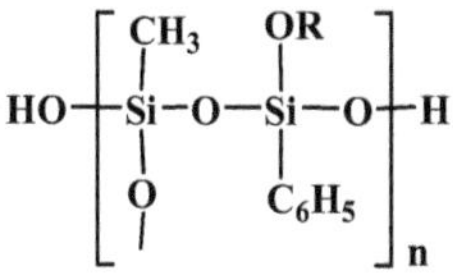

$$+ (10n-1)H_2O \qquad (3.63)$$

where $R = —CH_2CH_2OCOC(CH_3)=CH_2$.

MAS production comprises two main stages: 1) the hydrolytic cocondensation of the metacrylation products of the acetoxysilane mixture and the flushing of the products from acetic acid with an impurity (-0.5%) of glycolmethacrylate; 2) the partial distillation of the solvents from the product of hydrolytic cocondensation and the filtering of the finished product.

The production diagram of oligomethyl(phenyl)metacrylethoxysiloxanes is given in Fig. 50. Reactor *1* is filled with the help of pressurised nitrogen with the toluene solution of the metacrylation products from batch box *2*; backflow condenser *4* and an agitator are switched on. The reactor is filled with a calculated amount of water at such speed that the temperature in the reactor does not exceed 35-40 °C. At a temperature not higher than 35 °C the reactive mixture is agitated for 1 hour; reactor *1* receives the remaining self-flowing water and butanol from batch box *3*; the contents of the reactor are agitated for 30 minutes. The agitator is stopped and the reactive mixture is split for 20-30 minutes. The lower, water acidic layer from reactor *1* is poured into neutraliser *5*, which has been loaded with a -30-40% solution of NaOH from batch box *6*. The obtained product is washed until the flush waters give a neutral reaction.

After the product in reactor *1* has been flushed, the agitator is switched on, condenser *4* is switched into the direct mode, a vacuum is created in the system (the residual pressure is 550-480 GPa), the jacket of the apparatus is filled with vapour and the solvents (butanol and toluene mixture) are partially distilled up to 65 °C in vapour into receptacle *7*. The vapour is switched off, the jacket of the reactor is filled with water, the reactive mixture is cooled to 25-30 °C, the agitator is stopped and the vacuum is discharged. From reactor *1* the reaction product is sent to ultracentrifuge *8*. From there the finished MAS is sent into container *9*.

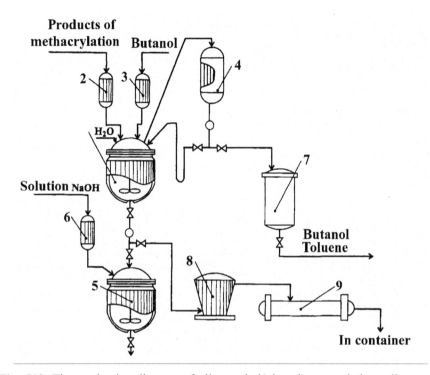

Fig. 510. The production diagram of oligomethyl(phenyl)metacrylethoxysiloxane (MAS): *1* - reactor; *2,3, 6* - batch boxes; *4* - condenser; *5* - neutraliser; *7* - receptacle, *8* - ultracentrifuge; *9* - container for the finished product

Oligomethyl(phenyl)metacrylethoxysiloxane (MAS) is a transparent homogeneous mixture in the solution of toluene and butanol (1:1). It has a low toxicity.

Technical oligomer MAS should meet the following requirements:

Appearance	Transparent colourless or light yellow liquid without visible mechanical impurities
pH of the aqueous extract	6.5-7.5
Content, %:	
of nonvolatile substances	≥ 50
of metacrylethoxyl groups	≤ 2.5
of silicon	≤ 2.5

Oligomethyl(phenyl)metacrylethoxysiloxane (MAS) is used to glue some nonmetallic materials with metals.

3.3. Preparation of oligoorganosiloxanes with reactive groups

Many silicone compounds have water-repellent (waterproofing) properties. Usually materials are waterproofed with silicone oligomers with various reactive groups, which interact with the treated surface chemically or physically. Effective waterproofers are silicone oligomers with ≡Si-H, ≡Si-OH or ≡Si-Ona groups. They differ in their chemical composition and application.

Oligomers with ≡Si—H groups, or so-called *oligohydrideorganosiloxanes*, are not soluble in water and are used as aqueous emulsions or solutions in organic solvents. On the other hand, oligomers with ≡Si—ONa groups dissolve in water. They are obtained by the etherification of alkyltrichlorosilanes or the hydrolytic cocondensation of alkyltrichlorosilanes mixed with the tank residue from alkylchlorosilane distillation and the subsequent saponification of the products with caustic soda. This forms the following compounds:

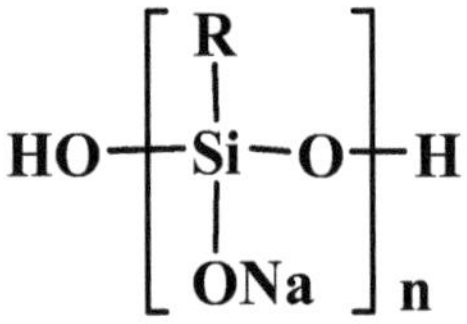

$$HO \left[\begin{array}{c} R \\ | \\ Si-O \\ | \\ ONa \end{array} \right]_n H$$

They are called *sodium organosiliconates*. By dehydrating these substances one can obtain solid products of the following composition:

[RSi(ONa)O]$_n$

3.3.1. Preparation of oligohydrideorganosiloxanes

Oligohydrideorganosiloxanes are products with the following structure of the element:

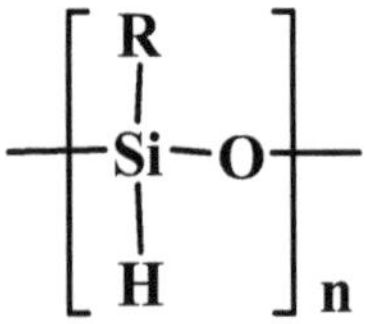

$$\left[\begin{array}{c} R \\ | \\ Si-O \\ | \\ H \end{array} \right]_n$$

R=CH$_3$ or C$_2$H$_5$, n=15÷20

The production of *oligohydridemethylsiloxane* comprises three main stages: the hydrolytic condensation of methyldichlorosilane; the neutralisation of the product of hydrolytic condensation; the settling and centrifuging of the finished product. The hydrolytic condensation of methyldichlorosilane proceeds thus:

$$nCH_3SiHCI_2 \xrightarrow[2HCI]{nH_2O} \left[\begin{array}{c} CH_3 \\ | \\ Si-O \\ | \\ H \end{array} \right]_n \qquad (3.64)$$

Hydrochloric acid is neutralised (after the acid aqueous layer is separated from the hydrolysate) wuth soda.

Raw stock: methyldichlorosilane (the 40-44° C fraction, 60.5-63% of chlorine, d_4^{20}=1.080-5-1.117); butil alcohol (the boiling point is 115-118 °C, d_4^{20}= 0.808-0.812); rectified ethyl alcohol; soda ash.

Enameled hydrolyser 1 (Fig. 51) with an agitator and a jacket is filled with a necessary amount of water, butil alcohol or hydrosite.

Hydrosite is an aqueous toluene solution of ethyl and butyl alcohols. It is formed at the stage of hydrolytic condensation in the production of polyphenyldiethylsiloxane varnishes. Before use hydrosite is neutralised with alkali solution.

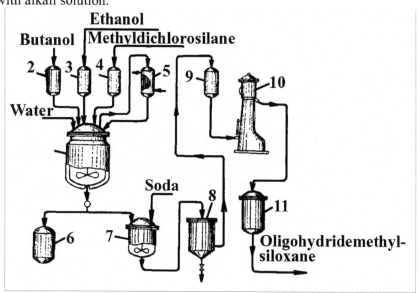

Fig. 511. Production diagram of oligohydridemethylsiloxane: *1* - hydrolyser; *2 -4* – batch boxes; *5* - cooler; *6, 11* - collectors; *7* - neutraliser; *8* - settling box; *9* - pressure container; *10* – ultracentrifuge

The contents of the apparatus are agitated for 15-20 minutes and cooled down to 0-1 °C by filling the jacket with salt solution. At this temperature the hydrolyser receives methyldichlorosilane from weight batch box *4*. The temperature of 10 °C is maintained in the hydrolyser by regulating the speed at which methyldichlorosilane and salt solution are fed. After the whole methyldichlorosilane has been introduced, the contents of the hydrolyser are agitated for 15 minutes, the agitator is stopped and the mixture is settled. The bottom layer (18-20% hydrochloric acid with an impurity of ethyl and butyl alcohols) is poured through a run-down box into collector *6*, and the top layer, the product of the hydrolytic condensation of methyldichlorosilane (hydrolysate), is sent into neutraliser *7*, which is an enameled apparatus with an agitator.

After the hydrolysate has been supplied, the neutraliser is filled at agitation and at 18-20 °C through a hatch with a calculated amount of soda ash. The mixture is agitated for 30 minutes and sampled for acidity. The end of neutralisation is determined by the pH of the aqueous extract. If the reaction is acid (pH < 6), the neutraliser receives another 1% of soda and after 0.5 hours of agitation pH is tested again.

When pH is 6.0-7.0, the finished product, oligohydridemethylsiloxane, is poured at agitation into settling box *8*. Usually after 12 hours the salts mostly settle to the bottom of the settling box. From the settling box oligohydridemethylsiloxane is sent into pressure container *i* and then into ultracentrifuge *10* to separate the salts completely. The liquid from the centrifuge (the finished product) is poured into collector *11*, sent into containers and labeled.

Oligohydridemethylsiloxane can also be produced by the continuous technique. In this case hydrolytic condensation should be carried out in a the flow circuit (pump - heat exchanger - hydrolyser) (see Fig. 36), in a shell-and-tube heat exchanger (see Fig. 58) or in common countercurrent spray towers with agitators. The neutralisation of the product of hydrolytic condensation should be conducted in sectional apparatus with agitators, which also use the countercurrent principle.

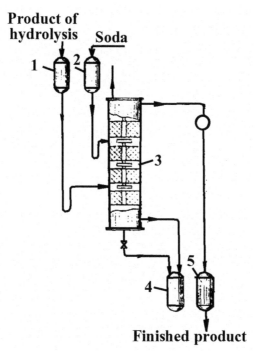

Product of hydrolysis   Soda

Finished product

Fig. 12. Diagram of the continuous neutralisation of the product of the hydrolytic condensation of ethyldichlorosilane: *1, 2* - batch boxes; *3* - neutraliser; *4, 5* – collectors for complete salt separation. The liquid from the centrifuge (the finished product) is poured into collector *11*, sent into containers and labeled

The diagram of continuous neutralisation is given in Fig. 52. The lower section of neutraliser *3* is continuously filled with the product of condensation from weight batch box *1*; the top section receives a soda solution from weight batch box *2*. The neutralisation is conducted at room temperature. The neutralised product from the top part of apparatus *3* is sent through a run-down box into collector *5*, and the discharge soda solution is sent from the lower part of the neutraliser into collector *4*.

Oligohydridemethylsiloxane is a colourless or yellowish liquid without mechanical impurities (the viscosity is 5-100 mm^2/s at 20 °C, 1.5-1.8% of active hydrogen, $d_4^{20} = 1.009 \div 1.024$). It is corrosion-inert, does not release harmful vapours or gases, dissolves well in aromatic and chlorinated hydrocarbons and is easily gelated under the influence of amines, aminoalcohols, strong acids and alkali. Oligohydridemethylsiloxane can be used in the form of 0.5% solutions in various organic solvents and in the form of a 0.5-10% aqueous emulsion. It is used for waterproofing glass cloth, metals, construction materials, etc.

3.3.2. Performance characteristics of oligoorganohydridesiloxanes

Oligohydrideethylsiloxane $[(C_2H_5)Si(H)O]_n$ is produced similarly. It is also a colourless or yellowish liquid (1.3-1.43% of active hydrogen, $d_4^{20} =$ 0.995-5-1.003, the viscosity is 45-200 mm^2/s at 20 °C), which dissolves well in aromatic and chlorinated hydrocarbons. It does not mix with water and forms an emulsion. It is corrosion-inactive and does not release harmful vapours or gases. Oligohydrideethylsiloxane is used in the form of diluted solutions in organic solvents or as an aqueous emulsion to waterproof cloth, paper, cardboard, gypsum, concrete, asbestos cement, brick, ceramics, glass, metal and other materials.

Oligoethylhydridesiloxane GKZh-94 is the most universal waterproofing agent. It is widely used in civil engineering, textile and light industries. GKZh-94 is obtained by the hydrolytic condensation of ethyldichlorosilane at low temperatures (0-30° C) in the ethanol and butanol medium:

$$nC_2H_5SiHCl_2 + H_2O \rightarrow [C_2H_5SiHO]_n + 2nHCI \quad (3.65)$$

The hydrolysate is separated from HCI dissolved in water; washed, neutralised and dried. The dried product is marketable liquid GKZh-94. This technology is based on the fact that GKZh-94 cannot be heated to distil volatile products due to gel formation. Well-washed, neutralised and dried liquid under normal conditions can be stored over a long period of time retaining its properties.

Liquid GKZh-94 is a mixture of oligomer homologues and analogues with cyclic and linear molecules with hydroxyl groups at the ends of the chain. Each silicon atom in the molecules has the $\equiv$Si-H bond, which determines the reactivity of GKZh-94 in the process of waterproofing. The active hydrogen content in the marketable product ranges from 1.3 to 1.42% (weight), which approaches the theoretical level. The liquid is insoluble in water and dissolves well in aromatic and aliphatic hydrocarbons.

Oligomethylhydridesiloxane (GKZh-94M), which as a waterproofing agent is not as universal as GKZh-94, has greater reactivity and other specific properties.

The main spheres of alkylhydridesiloxane application can be separated into two groups: in civil engineering and in the light and textile industries. In both spheres the substances yield considerable technical and economic results, but their ways of application are different.

In order to describe the reactions of waterproofing materials with alkylhydridesiloxanes, let us consider the interaction of the $\equiv$Si-H bond with different functional groups. If a negative ligand is introduced, the silicon atom with its larger volume receives a positive charge due to polarisation, and the hydrogen atom receives a negative charge $\equiv Si^{+\delta}\text{-}H^{-\delta}$. $\equiv$Si-H groups in siloxane and their polarity contribute to the following reactions:

$$\equiv Si\text{-}H + HOH \xrightarrow{\ OH^-\ } \equiv Si\text{-}OH + H_2 \qquad (3.66)$$

$$\equiv Si\text{-}H + HOR \xrightarrow{\ RONa\ } \equiv Si\text{-}OR + H_2 \qquad (3.67)$$

$$\equiv Si\text{-}H + NH_2R \xrightarrow{\ NH_2R\ } \equiv Si\text{-}NHR + H_2 \qquad (3.68)$$

$$\equiv Si\text{-}H + HOOCR \xrightarrow{\ Al(OR)_3\ } \equiv Si\text{-}OCOR + H_2 \qquad (3.69)$$

$$2\equiv Si\text{-}H + HOH \longrightarrow \equiv Si\text{-}O\text{-}Si\equiv + H_2 \qquad (3.70)$$

$$\equiv Si\text{-}H + HO\text{-}Cell \longrightarrow \equiv Si\text{-}O\text{-}Cell + H_2 \quad \text{(Cell-cellulose)} \qquad (3.71)$$

$$\equiv Si\text{-}H + [O] + H\text{-}Si\equiv \longrightarrow \equiv Si\text{-}O\text{-}Si\equiv + H_2O \qquad (3.72)$$

These reactions can bring us to the conclusion that oligoalkylhydridesiloxanes will interact by forming a chemical bond with the surfaces of cellulose, leather, synthetic and artificial fibres, hydroxyl groups of ceramic and metal-oxide materials. It should be kept in mind, however, that these and other reactions at the $\equiv$Si-H bond require the use of catalysts for waterproofing. The catalysts should catalyse not only the interaction of the $\equiv$Si-H bond with the functional groups on the surface of the material, but also the polymerisation of alkylhydridesiloxane, forming a polymer film on the surface.

These catalysts are the alkoxyderivatives of aluminum, titanium, lead, tin. This study shows that the interaction of oligoalkylhydridesiloxane with tetrabutoxytitanium first forms coordination bonds due to the donor-acceptor interaction of the titanium atom with the oxygen atoms in oligosiloxane. This is followed by the polymerisation of oligosiloxane, which forms a spatially cross-linked polymer.

When oligoalkylhydridesiloxane interacts with tetrabutoxytitanium, it releases hydrogen, which is supported by the chromatographic analysis of the gas. The quantity of the released gas depends on the ratio of the components and is at the maximum when the ratio of GKZh-94 and $Ti(OR)_4$ is equimolar.

The release of hydrogen in equimolar quantities points to simultaneous reactions of the 3.66, 3.67, 3.70, 3.71 and 3.72 type and the reaction of polycondensation at hydroxyl groups formed according to the reaction of the 3.66 type. The role of the active catalysts of the interaction of the $\equiv$Si-H bond of oligoalkylhydridesiloxanes with the functional groups of the material surface and the polymerisation of oligomers can be also played by aminoalkylethoxysilanes and aminoalkylsiloxanes. A typical representative of these catalysts is γ-aminopropyltriethoxysilane (AHM-9). It can interact with the oligomer at the ≡Si-H bond according to a 3.68 reaction; the three alkoxygroups at the silicon atom cause the spatial cross-linking of the oligomer due to their hydrolysis and subsequent polycondensation.

Oligoalkylhydridesiloxanes dissolve well in fatty and aromatic hydrocarbons. They do not dissolve in lower alcohols and acetone and do not mix with water; however, they can form stable aqueous emulsions if emulsifiers are used. Many surface active substances can be emulsifiers used for the preparation of aqueous emulsions of oligoalkylhydridesiloxanes; however, not all emulsifiers can be used to obtain the waterproofing effect. The best emulsifiers for GKZh-94 and GKZh-94M liquids are the derivatives of polyvinyl alcohol. Emulsions with this surfactant ensure a high waterproofing effect; they are stable in storage and virtually neutral (pH=6). Waterproofing is carried out with 2-5% solutions of oligoalkylhydridesiloxanes in solvents or a GKZh-94 emulsion diluted to the same concentration. The catalysts are introduced into the solutions of waterproofing agents or into diluted emulsions.

The waterproofing of materials is also due to the ability of oligoorganosiloxanes to be easily sorbed by the surface of the treated material and to form homogeneous thin and durable films with high adhesion to the surface. In this case the waterproofing effect is connected with the fact that when materials are treated, there is a protective film formed on their surface; its hydrophobic behavior is caused by a certain orientation of silicone molecules: the organic radicals face the environment, whereas the siloxane groups face the surface of the material.

Considering the physical aspect of waterproofing, we should note that all particles of the waterproofed material that have been in touch with the waterproofing agent are coated with a very thin film of silicon oligomer. This is what makes a product waterproof. The size of pores and particles, the appearance and texture of the surfaces of the material remain virtually the same.

The homogeneity and uniformity of the hydrophobic film formed on the surface of the fibres has been supported by an electron microscope study of the replicas of the surface of the waterproofed cotton fibres. The same result has been given by the study of the film on polyester fibres. Cellulose fibre, treated with waterproofing GKZh-94 catalysed with AHM-9 dissolved in a copper ammonia solution leaves a film which repeats the form of the fibre.

Aqueous emulsions of oligoalkylhydridesiloxanes GKZh-94 and GKZh-94M with copper acetate as a catalyst have proved to be the most acceptable in the hydrophobic treatment of capron cloth. They help to preserve the original valuable properties of capron fibre and to achieve a high and stable hydrophobic effect. Judging by the properties given to the fibre, this treatment is superior to the treatment with any other preparations.

The aqueous emulsions of oligoalkylhydridesiloxanes are catalysed with water-soluble acetates of copper, lead, zinc, nickel and mercury. The catalyst is introduced into the aqueous emulsion immediately before the treatment; the temperature of thermal treatment does not exceed 100° C and is combined with drying the capron cloth. By their hydrophobic effect the catalysts can be arranged thus: Cu>Pb>Ni>Li>Hg.

Copper acetate as a catalyst is superior to other metal salts; this can be explained by its ability to form complexes both with oligoalkylhydridesiloxane and with the fibre due to coordination linking. The reactivity of the complex creates favourable conditions to form a branched cross-linked structure of the polyorganosiloxane film.

According to IR spectroscopy, if a cloth is waterproofed with GKZh-94 emulsion catalysed with copper acetate, the intensity of the absorption 2180 cm^{-1}band, which corresponds to the oscillations of the $\equiv$Si-H bond, is considerably lower than without a catalyst. These results are consistent with the higher hydrophobic properties of a cloth waterproofed with GKZh-94 emulsion in the presence of copper acetate.

Due to their specific characteristics, oligoalkylhydridesiloxanes are widely used in the textile industry. After silicone treatment textiles acquire a high hydrophobicity, which is resistant towards various physicochemical stresses, do not change appearance and remain air- and vapour-permeable. Certain silicone substances considerably reduce the shrinkage and creasing of materials with cellulose and hydrocellulose fibres, improve their resistance towards abrasion, tearing, bending and other wear factors.

Silicone substances used in the textile industry fall into two groups according to their function: finishing agents and textile auxiliaries. Silicone finishing agents are widely used to improve the wear resistance of cloth. Oligoalkylhydridesiloxanes have the greatest practical value as finishing agents.

Regardless of their fibre composition and the type of the manufactured item, cloths for waterproof clothing should have a high and stable hydrophobicity, at the same time preserving their hygienic properties, wear resistance, form stability and attractive appearance.

To obtain a stable effect of hydrophobicity, the cloth should be thoroughly cleaned and prepared, because natural fats and size interfere with the uniform distribution of the finishing film. Silicone film on fibre reduces its sorption, penetration of moisture into the fibre and thus reduces the amount of moisture sorbed by the fibre.

As shown by the practical application of oligoalkylhydridesiloxanes, surface treatment of clothing cellulose textiles yields certain effects, the most important of which are shrink resistance, crease resistance and hydrophobicity.

Oligohydrideethylsiloxane $[(C_2H_5)Si(H)O]_n$ is produced similarly. It is also a colourless or yellowish liquid (1.3-1.43% of active hydrogen, $d_4^{20} =$ 1.003-5-1.003, the viscosity is 45-200 mm²/s at 20 °C), which dissolves well in aromatic and chlorinated hydrocarbons. It does not mix with water and forms an emulsion. It is corrosion-inactive and does not release harmful vapours or gases.

Oligohydrideethylsiloxane is used in the form of diluted solutions in organic solvents or as an aqueous emulsion to waterproof cloth, paper, cardboard, gypsum, concrete, asbestos cement, brick, ceramics, glass, metal and other materials.

3.3.3. Preparation of sodium oligoorganosiliconates

Sodium oligoorganosiliconates are aqueous-alcoholic solutions of the following composition:

$$HO\left[\begin{array}{c} R \\ | \\ Si-O \\ | \\ ONa \end{array}\right]_n H$$

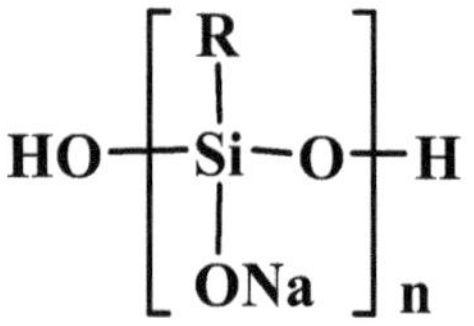

$R = CH_3$ or C_2H_5

or solutions of the mixture of oligomer homologues of the following composition:

$$HO\left[\begin{array}{ccc} R & R & R \\ | & | & | \\ Si-O-Si-O-Si-O \\ | & | & | \\ ONa & R' & O- \end{array}\right]_n H$$

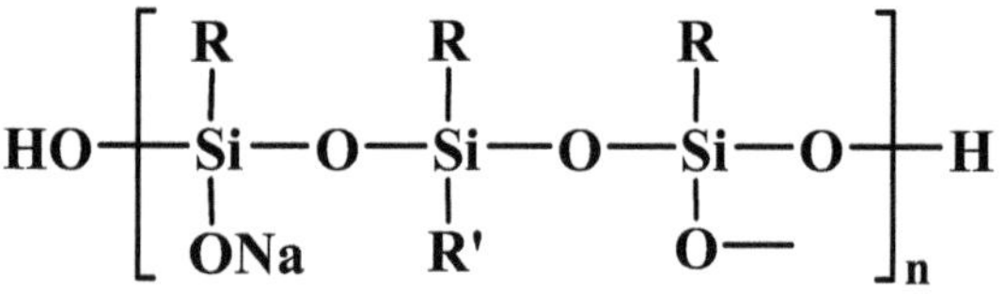

$R=CH_3$ or C_2H_5

$R=CH_3$ or C_2H_5; $R'=CH_3$, C_2H_5, H, OH or a fragment of the molecule of the product of the partial hydrolytic condensation of organochlorosilanes.

The structure of sodium oligoorganosiliconates first of all depends on the type of the original raw stock.

Preparation of alkyltrichlorosilane-based sodium oligoalkylsiliconates

Sodium oligoalkylsiliconates of the following common composition:

$$HO\left[\begin{matrix} R \\ | \\ Si-O \\ | \\ ONa \end{matrix}\right]_n H$$

are obtained by the etherification of alkyltrichlorosilanes with ethyl alcohol and subsequent saponification of alkyltriethoxysilanes with caustic soda.

Alkyltrichlorosilanes are etherified when the weight ratio of alkyltrichlorosilane and ethyl alcohol is 1:(3.05÷3.1); the etherification is accompanied by gradual replacement of chlorine atoms in alkyltrichlorosilane with ethoxyl groups:

$$R_3SiCl_3 \xrightarrow[HCl]{C_2H_5OH} RSiCl_2(OC_2H_5) \xrightarrow[HCl]{C_2H_5OH} RSiCl(OC_2H_5)_2 \xrightarrow[HCl]{C_2H_5OH}$$

$$\xrightarrow{\qquad} RSi(OC_2H_5)_3 \qquad (3.73)$$

If alcohol is lacking, the reaction can form only chloroethers, e.g.

$$R_3SiCl_3 + 2C_2H_5OH \xrightarrow[2HCl]{} RSiCl(OC_2H_5)_2 \qquad (3.74)$$

and an excess of alcohol can cause a secondary reaction:

$$ROH + HCl \longrightarrow RCl + H_2O \qquad (3.75)$$

The released water will contribute to the hydrolysis of alkyltriethoxysilanes:

$$RSi(OC_2H_5)_3 \xrightarrow[C_2H_5OH]{H_2O} RSi(OC_2H_5)_2OH \qquad (3.76)$$

which naturally reduces their yield.

The formed alkyltriethoxysilanes are saponified with a 42% NaOH solution:

$$nRSi(OC_2H_5)_3 + nNaOH \xrightarrow[3nC_2H_5OH]{nH_2O} HO\left[\begin{matrix} R \\ | \\ Si-O \\ | \\ ONa \end{matrix}\right]_n H \qquad (3.77)$$

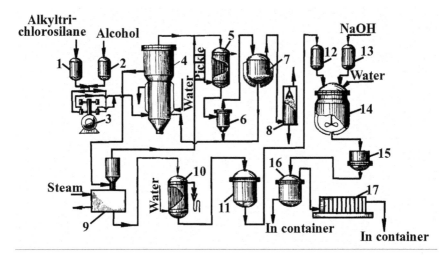

Fig. 13. Production diagram of alkyltrichlorosilane-based sodium oligoalkylsili-conates: *1, 2, 12, 13* - batch boxes; *3* - batching device; *4* - etherificator; *5, 10* - coolers; *6* - phase separator; *7* - trap; *8* - absorber; *9* - desorber; *11* - collector; *14* - reactor; *75*- nutsch filter; *16*- settling box; *17*- pressure filter

Raw stock: methyltrichlorosilane or ethyltrichlorosilane (at least 99% of the main product); sodium hydroxide (technical 42% solution); anhydrous ethyl alcohol (technical). The process can be conducted by the semicontinuous technique (Fig. 53).

To etherify methyltrichlorosilane or ethyltrichlorosilane with ethyl alcohol, etherificator *4* is continuously filled through batching device *3* with a corresponding alkyltrichlorosilane from batch box *1* and with alcohol from batch box *2*.

The etherificator is a vertical cylindrical apparatus with an expander on top; the cylindrical part has a jacket. It operates under the pressure below 0.07 MPa. The batching device is a composite apparatus consisting of a group of plunger pumps with a common drive.

The etherification is conducted at 15-35 °C. However, because the reaction releases heat, the temperature in the etherificator is regulated by sending water into the jacket of the apparatus and changing the ratio of the components. Hydrogen chloride, released in the reaction with alcohol vapours, passes through the expander of etherificator *4* and is sent into coolers *5* (the diagram shows one) cooled with salt solution. Condensed alcohol vapours are sent through phase separator *6* back into etherificator *4*, and hydrogen chloride is sent into trap *7* and then into absorber *8*, where it is absorbed with water. The formed 4% hydrochloric acid is sent to the apparatus for the neutralisation of waste waters.

Raw organotriethoxysilane is sent from the etherificator through a hydraulic gate into desorber 9, which is an igurit shelf heat exchanger (the pressure in the gas chamber is below 0.07 MPa) heated with vapour (below 0.4 MPa). The liquid phase, organotriethoxysilane, is sent into water-cooled graphite cooler *10* and into collector *11*. From the collector the product is sent into weight batch box *12*.

The saponification of organotriethoxysilane is carried out in enameled reactor *14* with an agitator and a water vapour jacket. First the reactor is filled from batch box *13* with 42% caustic soda and with a metered amount of water. After the dissolution is complete, the alkali solution receives organotriethoxysilane from batch box *12* at a temperature below 60 °C. The contents of the apparatus are agitated for 1 hour; the mixture is sampled to determine the NaOH content, solid residue and density. The prepared liquid is filtered through brass mesh in nutsch filter *15* into settling box *16*. In the settling box the product is separated from mechanical impurities (it is settled for at least 24 hours), filtered in pressure filter *17* and poured into containers.

The finished product can be poured into containers omitting the stage of filtering if the product is kept in the settling box for at least 48 hours.

The sodium oligomethyl- and oligoethylsiliconates are 30% aqueous-alcoholic solutions. They should meet the following technical requirements:

Appearance	Liquids ranging from colourless to light yellow; no sediment is tolerated
Density, g/cm^3	1.17-1.21
Alkalinity (equivalent to NaOH), % (weight)	13-17
Content, % (weight):	
of solid residue	25-35
of silicon	>4
of ethyl alcohol	
(for sodium methylsiliconate)	12-16
(for sodium ethylsiliconate)	13-18

They are used to waterproof cotton and glass cloth, construction materials and ceramics.

Preparation of sodium oligoalkylsiliconates from tank residue obtained in the production of methyl- or ethylchlorosilanes

Sodium oligoalkylsiliconates of the following common composition:

$$\text{HO}\left[\begin{array}{ccc} \overset{\displaystyle R}{\underset{\displaystyle ONa}{|}} & \overset{\displaystyle R}{\underset{\displaystyle R'}{|}} & \overset{\displaystyle R}{\underset{\displaystyle O-}{|}} \\ Si-O- & Si-O- & Si-O- \end{array}\right]_n H$$

are obtained by the hydrolytic cocondensation of the mixture of alkyltri-chlorosilanes and tank residue from the distillation of alkylchlorosilanes and by the subsequent saponification of the hydrolysate with alkali.

The production of sodium oligomethylsiliconate in this case comprises two main stages: the hydrolytic cocondensation of the methylchlorosilane mixture; the treatment of the hydrolysate with caustic soda. The reactive mixture, which consists of methyltrichlorosilane and tank residue from the distillation of methylchlorosilanes, is hydrolysed with water:

$$x\text{CH}_3\text{SiR}_3\text{R}_{3\text{-}n}\text{Cl}_n + y\text{CH}_3\text{SiCl}_3 + (x+y+1)\text{H}_2\text{O} \xrightarrow[(x+y)\text{HCl}]{}$$

$$\longrightarrow \text{HO}\left[(\text{CH}_3)\text{SiR}_{3\text{-}n}(\text{OH})_{n\text{-}1}\text{-O-Si(CH}_3)(\text{OH})\text{O}\right]_{xy} H \qquad (3.78)$$

R=CH3, H or a fragment of the molecule of a partially condensed product.

The obtained hydrolysate is treated with sodium hydroxide:

$$\text{HO}\left[(\text{CH}_3)\text{SiR}_{3\text{-}n}(\text{OH})_{n\text{-}1}\text{-O-Si(CH}_3)(\text{OH})\text{O}\right]_x H \xrightarrow[\text{xH}_2\text{O}]{\text{xNaOH}}$$

$$\longrightarrow \text{HO}\left[\begin{array}{ccc} \overset{\displaystyle CH_3}{\underset{\displaystyle ONa}{|}} & \overset{\displaystyle CH_3}{\underset{\displaystyle R}{|}} & \overset{\displaystyle CH_3}{\underset{\displaystyle O-}{|}} \\ Si-O- & Si-O- & Si-O- \end{array}\right]_x H \qquad (3.79)$$

In the production of sodium oligomethylsiliconate agitator 3 (Fig. 54) is loaded with methyltrichlorosilane and tank residue from the distillation of methylchlorosilanes. The mixture is agitated at room temperature for 0.5 hours and poured into weight batch box 4. Then hydrolyser 5 is filled with a necessary amount of water and with 10% sulfuric acid from batch box 7. After that, backflow condenser 6 is filled with water, the agitator is switched on and the reactive mixture is gradually loaded from weight batch box 4 at such speed that the temperature in the apparatus does not exceed 45-50 °C.

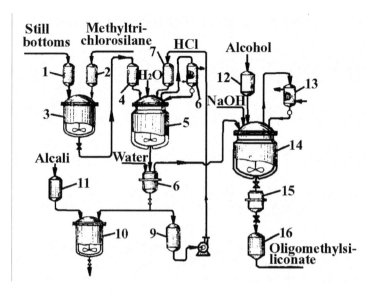

Fig. 14. Production diagram of sodium oligomethylsiliconate: *1, 2, 4, 7, 11, 12* - batch boxes; *3* - agitator; *5* - hydrolyser; *6, 13* - coolers; *8, 15* - nutsch filters; *9* - collector; *10* - neutraliser; *14* - reactor; *16* - settling box

After the whole mixture has been introducted, the jacket is filled with vapour and agitated for 1 hour at 50 °C. Then the vapour supply is stopped and water is fed to cool the reactive mixture down to 18-20 °C. The cooled mixture is poured from the hydrolyser into nutsch filter *8*.

The ratio of tank residue and methyltrichlorosilane can vary from 1:1 to 3:1. Sodium oligomethylsiliconate can also be prepared only from tank residue enriched with methyltrichlorosilane.

The hydrolysate is filtered from sulfuric acid in nutsch filter *8* with glass cloth. It is collected in collector *9* and is pumped as needed into batch box *7*. The solid hydrolysed mixture remaining in the filter is repeatedly flushed with water until the flush waters give a neutral reaction. The acidic flush waters are poured into neutraliser *10* pre-filled with a 40% alkali solution; the amount of the solution is calculated by the acid content in the flush waters.

The solid product of hydrolysis is loaded out of the nutsch filter, weighed and sent into reactor *14* through a hatch; the reactor also receives solid sodium hydroxide. After that, backflow condenser *13* is filled with water, the agitator is switched on and the reactive mixture is gradually filled with ethyl alcohol (or hydrosite) from batch box *12* at such speed that the temperature in the reactor does not exceed 80 °C. After the whole mixture has been introduced, the jacket is filled with vapour and agitated for 1-2 hours at 75-80 °C. The mixture is cooled to 20 °C and poured into nutsch filter *15*. The filtrate is collected in settling box *16*. There the finished product is settled, poured into containers and labeled.

Sodium oligomethylsiliconate is a transparent colourless liquid (d_4^{20} =1.10-1.25), has an acid reaction and is mixed with water and alcohols virtually in every ratio. The liquid does not release harmful vapours or gases.

Sodium oligoethylsiliconate can also be obtained by the technique described. Sodium oligoalkylsiliconates in the form of 5% aqueous solutions are used to waterproof construction and other materials.

3.4. Preparation of oligoorganosilazanes

Oligomer silicone compounds with the Si—N bond in the chain have recently been used more and more often. In particular, of great practical interest is the simplest representative of this class of compounds, *hexamethyldisilazane*, which is used as one of the components for certain medications, in gas chromatography and in the synthesis of organic substances.

Cyclic organosilazanes [RR'SiNH]$_n$ where R=R'=CH$_3$, C$_2$H$_5$ or R=CH$_3$, C$_2$H$_5$, R'=H, CH=CH$_2$, and $n = 3 \div 4$, and *organosilsesquiazanes* [RsiNH$_{1.5}$] $_n$, where R=C$_6$H$_5$, C$_7$H$_{15}$, C$_8$H$_{17}$, C$_9$H$_{19}$, and $n = 3 \div 5$, are of interest as waterproofing agents for various materials and as parent components in the production of polymers.

3.4.1. Preparation of hexamethyldisilazane

Hexamethyldisilazane is prepared by the ammonolysis of trimethylchlorosilane in hexamethyldisiloxane medium:

$$2(CH_3)_3SiCl + 3NH_3 \xrightarrow{2NH_4Cl} (CH_3)_3Si-NH-Si(CH_3)_3 \qquad (3.80)$$

Ammonia chloride, which precipitates in the form of fine sediment, is transformed into water-soluble NaCl and NH$_4$OH with a 15% alkali (NaOH) solution. The substances are separated from the solution when the product of the synthesis is flushed.

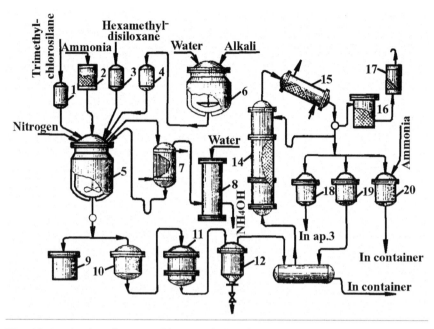

Fig. 15. Production diagram of hexamethyldisilazane: *1, 3, 4* - batch boxes; *2* - alkali-filled tower; *5, 6* - reactors; *7* - cooler; *8* - absorption tower; *9* - settling box; *10, 12* - collectors; *11* - druck filter; *13* - tank; *14* - rectification tower; *15* - refluxer; *16* - calcium chloride tower; *17* - fire-resistant apparatus; *18 - 20* - receptacles

Raw stock: trimethylchlorosilane (at least 99.6% of the main substance); hexamethyldisiloxane, the product of the hydrolytic condensation of trimethylchlorosilane (the main substance content is at least 90%, the allowable moisture content is not more than 0.07%); ammonia (at least 99.9% of the main products, not more than 0.1% of moisture); technical sodium hydroxide (at least 94% of the main product).

In the production of hexamethyldisilazane (Fig. 55) dry enameled reactor *5* cooled with water fed into the jacket is loaded with hexamethyldisiloxane and trimethylchlorosilane. The agitator is switched on and the mixture is agitated for 30 minutes. Then the reactor is filled at agitation through a bubbler with gaseous ammonia, which has been dehydrated through tower *2* filled with solid alkali. The ammonia should be fed at such speed that the temperature during the ammonolysis does not exceed 20 °C. The vapours of unreacted ammonia and the products carried with it enter backflow condenser *7* cooled with salt solution (-15 ° C); the condensate is sent back into reactor *5*, and the gaseous ammonia is sent into water-flushed absorption tower *8*.

After the ammonia has been supplied, reactor *5* is filled with nitrogen and the reactive mixture is sent through a backflow condenser into the absorption tower for 0.5-1 hours (to completely separate unreacted ammonia from the reactive mixture). Then the mixture is sampled through a hatch to determine if the ammonolysis is complete. The sample is filtered from ammonia chloride and ammonia is passed throught the filtrate. Ammonolysis is considered complete, if after the passing of ammonia there is no sediment of ammonia chloride. If there is some sediment, the ammonolysis is continued for 1.5-2 hours; then the mixture is sampled again. The ammonolysis is conducted until the process is totally completed.

Ammonia chloride is destroyed in reactor *5* immediately after the ammonolysis is finished. For this purpose, a 15% solution of NaOH is prepared in reactor *6*. The agitator is switched on and the contents of reactor *6* are agitated until sodium hydroxide dissolves completely. The alkali solution prepared in this way is sent through batch box *4* into reactor *5*, and the contents are mixed for 5-10 minutes. The reactive mixture is held for about an hour and sampled to determine the ammonia chloride destruction shown by the absence of NH4Cl in the aqueous layer. The lower layer, the aqueous NaCl solution, is poured into settling box *9*, the top layer, the hexamethyldisiloxane solution of hexamethyldisilazane, is poured through a rundown box into collector *10* and then in druck filter *11*, which operates below 0.07 MPa. The filtrate is collected into collector *12*, from where it is sent into tank *13* for rectification. The tank is heated with vapour (up to 1 MPa).

The vapours of the products rise up rectification tower *14* filled with Raschig rings, enter refluxer *15* cooled with salt solution (-15 °C), condensed and sent back into the tower. The uncondensed vapours are sent into the atmosphere through calcium chloride tower *16* and fire-resistant apparatus *17*. The tower is heated and the reflux is completely returned, until the temperature on top remains constant for 1 hour. After that, the fractions are separated.

Fraction I, hexamethyldisiloxane (the boiling point is 99.5 °C) is separated into receptacle *18*, when the temperature in the higher part of the tower does not exceed 119 °C (the temperature in the tank does not exceed 130 °C). The distilled hexamethyldisiloxane is sent again into batch box *3* to be used in ammonolysis. Fraction II (intermediate) is separated into receptacle *19*, when the temperature in the higher part of the tower does not exceed 122 °C (the temperature in the tank does not exceed 130 °C). The intermediate fraction from receptacle *19* is sent again into tank *13* for rectification.

Fraction III, hexamethyldisilazane (the boiling point is 126 °C) is collected into receptacle *20*, when the temperature in the higher part of the tower does not exceed 126 °C (the temperature in the tank does not exceed 150 °C). The tank residue is cooled in tank *13* down to 30 °C and poured into containers.

The finished product, hexamethyldisilazane, is analysed and sampled to determine its density, refraction index and nitrogen content; if it meets technical requirements, the product is poured into containers.

If hexamethyldisilazane opalesces, it is blown with gaseous ammonia for 10-30 minutes at the room temperature in receptacle *20* with subsequent filtering in a druck filter (not shown in the diagram).

Pure hexamethyldisilazane is a colourless transparent liquid (the boiling point is 126 °C) dissolves well in inert organic solvents and is easily hydrolysed with water. It is toxic and has narcotic properties. Its vapours cause the inflammation of mucous membranes and the central nervous system; they change the composition of blood.

Technical hexamethyldisilazane should meet the following requirements:

Appearance	Colourless transparent liquid
Density at 20 °C, g/cm^3	0.7730-0.7780
n_D^{20}	1.4080-1.4100
Nitrogen content, %	8-9

Hexamethyldisilazane is widely used in the pharmaceutical and chemical industries. It is also widely used in gas chromatography to modify the surfaces of supports, as well as a temporary protective group in various organic syntheses.

3.4.2. Preparation of dimethylcyclosilazanes

The synthesis of dimethylcyclosilazanes is based on the ammonolysis of dimethyldichlorosilane with ammonia in benzene:

$$n(CH_3)_2SiCl_2 \xrightarrow[2nNH_4Cl]{3nNH_3} [(CH_3)_2SiNH]_n \qquad (3.81)$$

The sediment of ammonia chloride is destroyed by treating the products of the reaction with alkali solution.

The ammonolysis of organochlorosilanes can be conducted by the continuous technique (Fig. 56). To avoid explosive concentration of the mixture of ammonia with air, the whole system is blown with nitrogen before the start. Then, to obtain gaseous ammonia, liquid ammonia is choked until the excess pressure is 0.07 MPa, heated in evaporator *6* and passed through drying tower *7* filled with alkali flakes. The dehydrated ammonia enters the lower part of bubble tower *5*. After 20 minutes of blowing the whole system the supply of ammonia is not stopped; the apparatus is filled with a solution of dimethyldichlorosilane in toluene (or petrol), which has been prepared in agitator *3*, from batch box *4*. The temperature in tower *5* should not exceed 40 °C, which is maintained by regulating the speed at which the parent reactants are supplied. The unreacted ammonia and the solvent vapours it carries enter cooler *8*. The solvent vapours condense and are collected in collector *9*, the ammonia enters trap *10*, from where the obtained ammonia water is poured into collector *11* and then sent by nitrogen flow (0.07 MPa) into the section for the purification of waste waters.

The mixture of the solvent, dimethylcyclosilazanes and ammonia chloride, obtained in tower *5*, is mixed with a 2-3% alkali solution enters the lower part of flush tower *13* from batch box *12* (an additional amount of alkali is necessary to wash ammonia chloride from the products of ammonolysis).

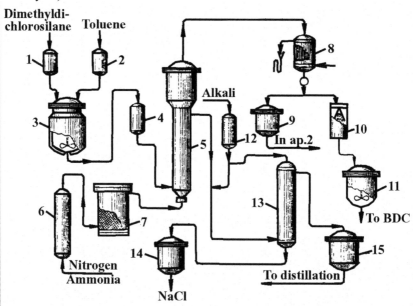

Fig. 16. Production diagram of dimethylcyclosilazanes by the continuous technique: *1, 2, 4, 12* - batch boxes; *3* - agitator; *5* - bubble tower; *6* - evaporator; *7* - drying tower; *8* - cooler; *9, 11, 14, 15* - collectors; *10* - water trap; *13* - flushing tower.

From the flushing tower, the top layer, the toluene (or petrol) solution of dimethylcyclosilazanes. is collected into collector *15*, and the lower layer, the solution of sodium chloride, is collected into collector *14*. From collector *15* the dimethylcyclosilazane solution is sent to distil the solvent and then to vacuum distillation to extract pure dimethylcyclosilazanes.

Table 19. Properties of waterproofing silicone oligomers and conditions for their application

Brand	Formula	Characteristics	Conditions for use	Recommended materials for waterproofing
GKZh-94	$\left[\begin{array}{c} C_2H_5 \\ \| \\ -Si-O- \\ \| \\ H \end{array}\right]_n$	1.3-1.42% of active hydrogen, the viscosity is 45-200 mm^2/s at 20^{0}C, pH of the aqueous extract is at least 6.	120-150^{0}C and 15-45 min, in the form of 0.5-10% aqueous emulsions or solutions in petrol, toluene, dichloroethane, carbon tetrachloride	Paper, cardboard, glass, staple and cotton cloth, metals
GKZh-94M	$\left[\begin{array}{c} CH_3 \\ \| \\ -Si-O- \\ \| \\ H \end{array}\right]_n$	1.5-1.8% of active hydrogen, the viscosity is 5-100 mm^2/s at 20^{0}C, pH of the aqueous extract is at least 6.	The same	The same
GKZh-10	$\left[\begin{array}{c} C_2H_5 \\ \| \\ -Si-O- \\ \| \\ ONa \end{array}\right]_n$	The solid residue is 30±5%, the density is 1.17-1.21 g/cm^3, pH is 13-14	120^{0}C and 1 hour (or 18-20^{0}C and 24 hours), in the form of 5% aqueous solution	Concrete, brick, cement, asbestos cement, plaster and othe construction materials

Table 19. (cont.)

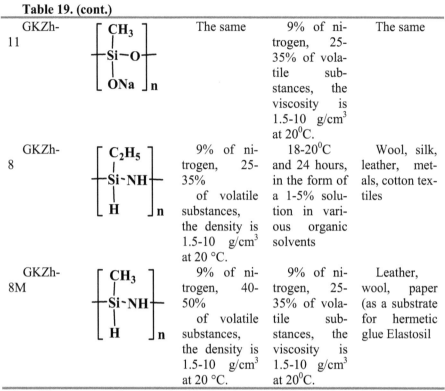

GKZh-11	$\left[\begin{array}{c} CH_3 \\ \mid \\ -Si-O- \\ \mid \\ ONa \end{array}\right]_n$	The same	9% of nitrogen, 25-35% of volatile substances, the viscosity is 1.5-10 g/cm^3 at 20^{0}C.	The same
GKZh-8	$\left[\begin{array}{c} C_2H_5 \\ \mid \\ -Si-NH- \\ \mid \\ H \end{array}\right]_n$	9% of nitrogen, 25-35% of volatile substances, the density is 1.5-10 g/cm^3 at 20 °C.	18-20^{0}C and 24 hours, in the form of a 1-5% solution in various organic solvents	Wool, silk, leather, metals, cotton textiles
GKZh-8M	$\left[\begin{array}{c} CH_3 \\ \mid \\ -Si-NH- \\ \mid \\ H \end{array}\right]_n$	9% of nitrogen, 40-50% of volatile substances, the density is 1.5-10 g/cm^3 at 20 °C.	9% of nitrogen, 25-35% of volatile substances, the viscosity is 1.5-10 g/cm^3 at 20^{0}C.	Leather, wool, paper (as a substrate for hermetic glue Elastosil

Note. The waterproofing capacity (the "pocket sample") lasts for at least 3 hours.

In the continuous process of organochlorosilane ammonolysis one should strictly observe the required speed at which the reactants are supplied, the temperature in the apparatus and the concentration of organochlorosilane in the solvent so that ammonia chloride does not clog the bubble tower. Similarly, the ammonolyses of methyldichlorosilane, ethyldichlorosilane, diethyldichlorosilane and other organochlorosilanes can yield various cyclic organosilazanes. The properties of some silicone oligomers used in waterproofing are given in Table 19.

4. Chemistry and technology of silicone polymers

4.1. Preparation of linear polyorganosiloxanes

Polyorganosiloxanes (organosiloxane elastomers) with linear molecular chains have the following formula:

$$HO \left[\begin{array}{c} R \\ | \\ Si-O \\ | \\ R \end{array} \right]_n \left[\begin{array}{c} R \\ | \\ Si-O \\ | \\ R' \end{array} \right]_m H$$

1) R and R' - methyl; 2) R - methyl, and R' - vinyl; 3) R - methyl, a R' - phenyl, etc.

They are prepared by the hydrolytic condensation or cocondensation of difunctional organochlorosilanes and subsequent polymerisation of cyclic diorganosiloxanes.

During the hydrolytic condensation diorganosiloxanes have a great tendency to form cycles. For example, dimethyldichlorosilane is hydrolysed with water (in the absence of solvents) by the mechanism of polycondensation, forming a mixture of linear and cyclic dimethylsiloxanes.

$$n(CH_3)_2SiCl_2 \xrightarrow[2nHCl]{2nH_2O} n(CH_3)_2Si(OH)_2 \xrightarrow[(n-1)H_2O]{}$$

$$\xrightarrow{} HO \left[\begin{array}{c} CH_3 \\ | \\ Si-O \\ | \\ CH_3 \end{array} \right]_n H + [(CH_3)_2SiO]_n \qquad (4.1)$$

The formation of cyclic compounds increases with the size of the organic radicals bound with silicon. For example, in the process of hydrolytic condensation methylphenyldichlorosilane and diphenyldichlorosilane form mostly cyclic products. The conditions of the hydrolysis of diorganodichlorosilanes, especially pH of the medium, are essential for cyclisation. As pH increases, i.e. the acidity of the medium decreases, the process of ring formation can be reduced but cannot be completely avoided.

That is why the most important reaction to obtain polydiorganosiloxanes with linear chains is the polymerisation of the rings that form during the hydrolytic condensation of diorganodichlorosilanes. To open the molecules of organocyclosiloxanes and obtain linear polydiorganosiloxanes, one uses the reaction of catalytic polymerisation with cationic (H_2SO_4, H_3BO_3, H_3PO_4, HOOC—COOH, KF, BF_3 и др.) and anionic (NaOH, KOH, R_3SiONa, $R_4N[OSiR_2]_nOH$, R_4NOH) initiators.

During the cationic polymerisation, e.g. with sulfuric acid, the process is the following: at the initial stage of initiation, when organocyclosiloxanes interact with sulfuric acid, the acid proton attacks the oxygen atom of the siloxane cycle. As a result of the redistribution of the electron density, the ≡Si-O bond breaks, opening the cycle and forming an active centre at the end of the chain:

$$
\begin{array}{c}
R_2Si \overset{O}{\diagdown} SiR_2 \\
| \quad\quad | \\
O \quad\quad O \\
| \quad\quad | \\
R_2Si \diagdown_O SiR_2
\end{array}
\xrightarrow{H_2SO_4}
\left[
\begin{array}{c}
R_2Si \overset{O}{\diagdown} SiR_2 \\
| \quad\quad | \\
O \quad O\text{----}^+H \\
| \quad\quad | \\
R_2Si \diagdown_O SiR_2\text{--}\bar{O}SO_2OH
\end{array}
\right]
\longrightarrow
$$

$$
\longrightarrow HO-\underset{\underset{R}{|}}{\overset{\overset{R}{|}}{Si}}-O-\underset{\underset{R}{|}}{\overset{\overset{R}{|}}{Si}}-O-\underset{\underset{R}{|}}{\overset{\overset{R}{|}}{Si}}-O-\underset{\underset{R}{|}}{\overset{\overset{R}{|}}{Si}}{}^+\bar{O}SO_2OH \qquad (4.2)
$$

The active centre continues polymerisation (growth of the chain), opening the following molecules of the cycles:

$$
HO\left[\underset{\underset{R}{|}}{\overset{\overset{R}{|}}{Si}}-O\right]_3 \underset{\underset{R}{|}}{\overset{\overset{R}{|}}{Si}}{}^+\bar{O}SO_2OH \; + \;
\begin{array}{c}
R_2Si \overset{O}{\diagdown} SiR_2 \\
| \quad\quad | \\
O \quad\quad O \\
| \quad\quad | \\
R_2Si \diagdown_O SiR_2
\end{array}
\longrightarrow
$$

$$
\longrightarrow HO\left[\underset{\underset{R}{|}}{\overset{\overset{R}{|}}{Si}}-O\right]_7 \underset{\underset{R}{|}}{\overset{\overset{R}{|}}{Si}}{}^+\bar{O}SO_2OH \qquad (4.3)
$$

The cycles transform into a linear polymer chain until a balance is established. The chain breaks due to charge transfer (when a macrocation interacts with molecules of sulfuric acid) or to the macrocation capturing anions present in the system:

$$HO + \underset{\underset{R}{|}}{\overset{\overset{R}{|}}{Si}} - O + \underset{\underset{R}{|}}{\overset{\overset{R}{|}}{Si^+}} \Big]_n \xrightarrow[\substack{H^+ \\ \\ HOSO_2O^-}]{H_2SO_4} \quad HO + \underset{\underset{R}{|}}{\overset{\overset{R}{|}}{Si}} - O + \underset{\underset{R}{|}}{\overset{\overset{R}{|}}{Si^+}} \bar{O}SO_2OH \Big]_n$$

If organocyclosiloxanes are polymerised with anionic initiators, such as α-hydroxy-ω-tetramethylammoniaoxydimethylsiloxane, the anion interacts with the silicon atom. The nucleophylic reactant forms a coordinational bond with the cycle, the Si—O bond is weakened and the cycle is opened:

$$\underset{R_2Si}{\overset{O}{\diagup}} \underset{SiR_2}{\diagdown} \quad + (CH_3)_4\overset{+}{N}\bar{O} + \underset{\underset{CH_3}{|}}{\overset{\overset{CH_3}{|}}{Si}} - O + H \Big]_n \longrightarrow$$

$$\longrightarrow \left\{ \underset{R_2Si}{\overset{O}{\diagup}} \underset{SiR_2}{\diagdown} - \bar{O} \cdots \underset{N^+(CH_3)_4}{\underset{\overset{\cdots}{|}}{}} \underset{\underset{CH_3}{|}}{\overset{\overset{CH_3}{|}}{Si}} - O + H \Big]_n \right\} \longrightarrow$$

$$\longrightarrow H + O - \underset{\underset{CH_3}{|}}{\overset{\overset{CH_3}{|}}{Si}} \Big]_n \Big[O - \underset{\underset{R}{|}}{\overset{\overset{R}{|}}{Si}} \Big]_4 \bar{O}\overset{+}{N}(CH_3)_4 \qquad (4.4)$$

The active centre interacts with the next cyclic molecule, opening it:

$$H + O - \underset{\underset{CH_3}{|}}{\overset{\overset{CH_3}{|}}{Si}} \Big]_n \Big[O - \underset{\underset{R}{|}}{\overset{\overset{R}{|}}{Si}} \Big]_4 \bar{O}\overset{+}{N}(CH_3)_4 + \underset{(CH_3)_2Si}{\overset{O}{\diagup}} \underset{Si(CH_3)_2}{\diagdown} \longrightarrow$$

$$\longrightarrow H + O - \underset{\underset{CH_3}{|}}{\overset{\overset{CH_3}{|}}{Si}} \Big]_n \Big[O - \underset{\underset{R}{|}}{\overset{\overset{R}{|}}{Si}} \Big]_4 \bar{O}\overset{+}{N}(CH_3)_4 \quad \text{etc. (4.5)}$$

In this case the cycles are also transformed into a linear polymer chain because the active centre interacts with the next cyclic molecules; it also continues until a balance is established.

In both cases of catalytic polymerisation, the chain breaks in the absence of activity, i.e. when the end groups lose the ability to bond with cyclic molecules. This can be due to the detachment of the end groups, either by the water saponification of sulfate groups (for cationic polymerisation) or by the thermolysis of tetramethylammonia groups (for anionic polymerisation).

Our industry manufactures several brands of organosiloxane elastomers: polydimethylsiloxane (SKT), polyvinylmethyldimethylsiloxanes (SKTV and SKTV-1, which differ by vinylmethylsiloxy elements), polydimethyldiethylsiloxane (SKTE), polydimethylmethylphenylsiloxane (SKTPh), low-molecular polydimethylsiloxanes (SKTN, SKTN-A and others, which differ by their molecular weight), polymethylphenylsiloxane (SKTMPh), etc.

4.1.1. Preparation of polydimethylsiloxane elastomer

The production of polydimethylsiloxane elastomer comprises three main stages: the hydrolytic condensation of dimethyldichlorosilane; the depolymerisation of the products of hydrolytic condensation; the polymerisaton of dimethylcyclosiloxanes.

The hydrolytic condensation of dimethyldichlorosilane forms a mixture of linear and cyclic oligodimethylsiloxanes according to the scheme given below.

To transform linear oligodimethylsiloxanes into cyclic oligodimethylsiloxanes, the mixture is subjected to depolymerisation. The process takes place in the presence of a catalyst (potassium hydroxide):

$$
HO \left[\begin{array}{c} CH_3 \\ | \\ Si-O \\ | \\ CH_3 \end{array} \right]_n \begin{array}{c} CH_3 \\ | \\ Si-OH \\ | \\ CH_3 \end{array} \xrightarrow[H_2O]{KOH} HO \left[\begin{array}{c} CH_3 \\ | \\ Si-O \\ | \\ CH_3 \end{array} \right]_n \begin{array}{c} CH_3 \\ | \\ Si-O^- \ K^+ \\ | \\ CH_3 \end{array} \qquad (4.6)
$$

$$
HO \left[\begin{array}{c} CH_3 \\ | \\ Si-O \\ | \\ CH_3 \end{array} \right]_{n-4} \begin{array}{c} CH_3 \\ | \\ Si-O \\ | \\ CH_3 \end{array} \begin{array}{c} CH_3 \\ | \\ Si-O \\ | \\ CH_3 \end{array} \begin{array}{c} CH_3 \\ | \\ Si-O \\ | \\ CH_3 \end{array} \begin{array}{c} CH_3 \\ | \\ Si-O \\ | \\ CH_3 \end{array} \begin{array}{c} CH_3 \\ | \\ Si-O^- K^+ \\ | \\ CH_3 \end{array} \longrightarrow
$$

$$\longrightarrow \left\{ \mathrm{HO}\!-\!\left[\begin{array}{c}\mathrm{CH_3}\\|\\\mathrm{Si}\!-\!\mathrm{O}\\|\\\mathrm{CH_3}\end{array}\right]_{n-4}\!\!\begin{array}{c}\mathrm{CH_3}\\|\\\mathrm{Si}\!-\!\mathrm{O}\!\!\!-\!\!\!\mathrm{Si(CH_3)_2}\\|\quad\ ^{+}\mathrm{K}\ \ \mathrm{O} \quad\ \mathrm{O}\\\mathrm{CH_3}\\ (CH_3)_2Si \quad Si(CH_3)_2 \\ \mathrm{O}\diagdown\ \ \diagup\mathrm{O}\\ \mathrm{Si}\\(\mathrm{CH_3})_2\end{array}\right\} \longrightarrow$$

$$\longrightarrow \begin{array}{c}(\mathrm{CH_3})_2\mathrm{Si}\!\diagup\!{}^{\mathrm{O}}\!\diagdown\!\mathrm{Si(CH_3)_2}\\|\qquad\qquad\ |\\ \mathrm{O}\qquad\qquad\ \mathrm{O}\\|\qquad\qquad\ |\\(\mathrm{CH_3})_2\mathrm{Si}\!\diagdown\!{}_{\mathrm{O}}\!\diagup\!\mathrm{Si(CH_3)_2}\end{array} + \mathrm{HO}\!-\!\left[\begin{array}{c}\mathrm{CH_3}\\|\\\mathrm{Si}\!-\!\mathrm{O}\\|\\\mathrm{CH_3}\end{array}\right]_{n-4}\!\!\begin{array}{c}\mathrm{CH_3}\\|\\\mathrm{Si}\!-\!\mathrm{O^-\ K^+}\\|\\\mathrm{CH_3}\end{array} \quad \text{etc.} \quad (4.6)$$

and stops when a balance is established in the system. The depolymerisation of linear oligodimethylsiloxanes can be summed up by the following equation:

$$\mathrm{HO}\!-\!\left[\begin{array}{c}\mathrm{CH_3}\\|\\\mathrm{Si}\!-\!\mathrm{O}\\|\\\mathrm{CH_3}\end{array}\right]_{n-1}\!\!\begin{array}{c}\mathrm{CH_3}\\|\\\mathrm{Si}\!-\!\mathrm{OH}\\|\\\mathrm{CH_3}\end{array} \underset{\mathrm{H_2O}}{\overset{\mathrm{KOH}}{\rightleftharpoons}} 0.25n[(\mathrm{CH_3})_2\mathrm{SiO}]_4 \qquad (4.7)$$

The depolymerisation forms a mixture of dimethylcyclosiloxanes (60-85% of tetramer, 13-20% of pentamer, 0.5% of hexamer and up to 5% of trimer), which is then sent to polymerisation.

The polymerisation of dimethylcyclosiloxanes is conducted in the presence of a catalyst, e.g. based on aluminum sulfate and sulfuric acid. The catalyst is used in the form of a paste of mixed dimethylcyclosiloxanes, anhydrous aluminum sulfate and sulfuric acid. The reaction proceeds thus:

$$m[(\mathrm{CH_3})_2\mathrm{SiO}]_n \xrightarrow{\mathrm{H_2SO_4}} \mathrm{HOSO_2O}\text{-}[\mathrm{Si(CH_3)_2O}]_{mn-1}\text{-}\mathrm{Si(CH_3)_2OH} \quad (4.8)$$

After the elastomer is washed with water, the macromolecule obtains end hydroxyl groups:

$$\mathrm{HOSO_2O}\text{-}[\mathrm{Si(CH_3)_2O}]_n\text{-}\mathrm{H} \xrightarrow[\mathrm{H_2SO_4}]{\mathrm{H_2O}} \mathrm{HO}\text{-}[\mathrm{Si(CH_3)_2O}]_n\text{-}\mathrm{H} \quad (4.9)$$

The production of polydimethylsiloxane elastomer relies heavily on the purity of the parent dimethyldichlorosilane, since impurities of other monomers noticeably impair the properties of the elastomer and rubbers based on it. Methyltrichlorosilane has the strongest negative effect: when dimethyldichlorosilane with impurities of this monomer is subjected to hydrolytic condensation, linear oligodimethylsiloxane develops branched methylsilsesquioxane elements $[Si(CH_3)O_{1,5}]$, which greatly reduces the elasticity and mechanical characteristics of rubbers.

For instance, rubbers based on polydimethylsiloxane obtained from methyltrichlorosilane with 55% chlorine content, have a tensile strength of 6 MPa, whereas rubbers based on the same elastomer, but obtained from methyltrichlorosilane with 55.7% chlorine content (which shows the presence of methyltrichlorosilane impurities), have a tensile strength of 2 MPa. Thus, if technical dimethyldichlorosilane contains a considerable amount of methyltrichlorosilane, it is subjected to additional rectification. Since the boiling point of methyltrichlorosilane differs from the boiling point of dimethyldichlorosilane only by 4 °C, the mixture should be rectified in towers with a large number of theoretical plates.

Our industry manufactures dimethyldichlorosilane of 99.98% purity and 54.85-55% chlorine content. This dimethyldichlorosilane can be used as raw stock in the production of polydimethylsiloxane elastomer without any additional rectification.

Dimethyldichlorosilane used in the production of polydimethylsiloxane elastomer should meet the following technical requirements:

Appearance   Transparent colourless liquid
Content, %:
of methyltrichlorosilane   ≤0.1
of chlorine   54.8-55

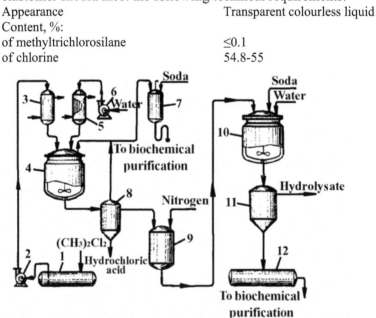

Fig. 57. Diagram of the hydrolytic condensation of dimethyldichlorosilane by the periodic technique:*1, 9, 12* - containers; *2, 6* - pumps; *3, 5* - coolers; *4* - hydrolyser; *7* - hydraulic gate; *8* - Florentine flask; *10* - neutraliser; *11* - settling box

Dimethyldichlorosilane of this quality is sent to hydrolytic condensation (Fig. 57). From container *1* dimethyldichlorosilane is sent with pump *2* into hydrolyser *4*, which is filled with filtered water cooled in cooler *5*. The hydrolyser is an enameled apparatus with an agitator and a jacket filled with salt solution. The hydrolytic condensation is conducted at 20 °C or lower and in the excess of water with 1.5:1 to 2.1 water: dimethyldichlorosilane ratio.

The hydrolytic condensation of dimethyldichlorosilane is endothermic: 31 KJ is absorbed per 1 mole of dimethyldichlorosilane; however, the dissolution of hydrogen chloride (formed by the hydrolysis of dimethyldichlorosilane) liberates 73.4 KJ per 1 mole of HCl in water. Consequently, the process has a positive total thermal effect and the heat should be diverted. If the heat is not diverted in time, the hydrolysate will contain more high-molecular linear products. This increases the viscosity and interferes with subsequent neutralisation by forming stable emulsions of hydrolysate and soda solution. Thus, apart from pre-cooling the components for hydrolytic condensation, the jacket of the hydrolyser is filled with salt solution to divert most of the heat.

If the water used for hydrolytic condensation is not enough to dissolve most of the released hydrogen chloride, the total thermal effect decreases; if the water: dimethyldichlorosilane volume ratio is from 0.35:1 to 0.5:1, it becomes zero. In this case more than 50% of the formed hydrogen chloride have to be withdrawn in gaseous form, which complicates the technological process. Besides, the hydrolysate has excessive acidity.

The properties and composition of the hydrolysate are greatly affected by the hydrodynamic mode of the process. Thus, increasing the number of agitator rotations from 300 to 1000 per minute raises dimethylcyclosiloxane content from 28 to 43% and at the same time visibly reduces the acidity and viscosity of the hydrolysate.

The hydrolysate, which is a mixture of linear and cyclic dimethylsiloxanes and hydrochloric acid, is continuously diverted from hydrolyser *4* into Florentine flask *8* for separation. The process is based on the difference of the densities of the hydrolysate (0.96 g/cm^3) and hydrochloric acid (1.12 g/cm^3) and insignificant solubility of the hydrolysate in hydrochloric acid. The hydrolysate in apparatus *8* remains in the top layer, whereas hydrochloric acid goes to the bottom. When the mode is established, hydrochloric acid is poured into tanks, and the hydrolysate with up to 0.4% of hydrochloric acid is sent into collectors *8* (the diagram shows one) and pressurised by nitrogen flow into apparatus *10* for neutralisation. Collectors 9 operate at regular intervals: while one accepts the hydrolysate, the other issues the hydrolysate for neutralisation. Hydrogen chloride from the hydrolyser and the Florentine flask is neutralised in hydraulic gate 7 with a 5% solution of powdered soda ash.

The mixture of soda and hydrolysate is agitated for 2 hours in neutraliser 10; the apparatus is filled with water to dissolve the salts formed in the neutralisation; the agitation is continued for approximately 1 hour. The amount of supplied water is 10-20 times larger than that of soda. After neutralisation the mixture is poured into settling boxes 11 (the diagram shows one) to split. The top layer (the hydrolysate with acid or alkali content not exceeding 0.01%) is sent to the following stage of elastomer production, i.e. depolymerisation. The bottom layer (the aqueous salt solution) is settled in container 12 and sent to biochemical purification.

The hydrolytic condensation of dimethyldichlorosilane forms a mixture of oligodimethylsiloxanes with similar amounts of linear and cyclic products.

The hydrolytic condensation of dimethyldichlorosilane can also be achieved by the continuous technique in the flow circuit (pump - heat exchanger - hydrolyser). The diagram is shown in Fig. 36.

There are other techniques to carry out the continuous hydrolytic condensation of dimethyldichlorosilane. Another diagram is given in Fig. 58. In this case the process takes place in a tubular reactor with a propeller agitator in the presence of a solvent and with a 1:(1.45÷1.55): (0.85÷0.75) ratio of dimethyldichlorosilane, water and solvent. Reactor 1 is continuously filled with dimethyldichlorosilane and the hydrolysing blend (water and solvent); the components are sent to the absorbing side of propeller agitator 2.

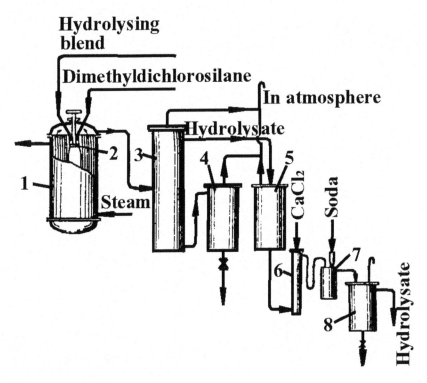

Fig.58. Production diagram of the hydrolytic condensation of dimethyldichlorosilane by the continuous technique: *1* - reactor; *2* - agitator; *3* - separation tower; *4, 5, 8* - collectors; *6* - drying tower; *7* - neutraliser

The reactive mixture is withdrawn from the top of the reactor into separation tower *3*. The hydrolysate from the top of the tower is sent into collector *5*. From there it is sent through drying tower *6* filled with calcium chloride into neutraliser *7*, which is loaded with a necessary amount of dry soda ash (continually or periodically, depending on the efficiency of the apparatus). The neutralised hydrolysate is sent into collector *8*. Hydrochloric acid, which has been settled in the bottom part of tower 3, enters collector *4*.

Using this technique of the hydrolytic condensation of dimethyldichlorosilane, one can increase the yield of dimethylcyclosiloxanes by 8-10%, but it is impossible to eliminate the by-formation of linear oligodimethylsiloxanes completely. Thus, the product of the hydrolytic condensation of dimethyldichlorosilane is sent to depolymerisation in the presence of potassium hydroxide at reduced pressure. Linear oligomethylsiloxanes are completely transformed into dimethylcyclosiloxanes.

The diagram of the depolymerisation of the hydrolysate is given in Fig. 59. The depolymerisation is conducted at 140-160°C and a residual pressure of 13.3-40 GPa at agitation in apparatus *4* with a jacket and an agitator.

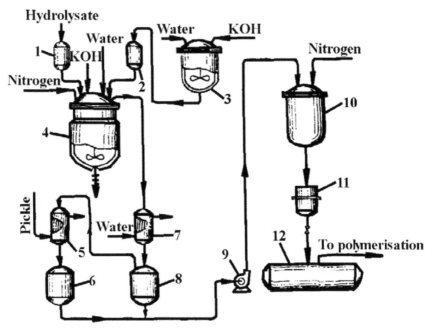

Fig. 59. Diagram of the depolymerisation of the hydrolysate: *1, 2* - batch boxes; *3* - apparatus for preparing the catalytic solution; *4* - depolymeriser; *5, 7* - condensers; *6, 8* - collectors; *9* - pump; *10* - dehydrator; *11* - nutsch filter; *12* - container

Before the reaction the depolymeriser is loaded with the hydrolysate to 20-30% of the volume and with solid potassium hydroxide (2% of the quantity of the hydrolysate) through a hatch. The agitator is switched on, a vacuum is created in the system and the jacket of the apparatus is filled with vapour. After the necessary temperature has been reached and depolymerisation has started (it is indicated by dimethylcyclosiloxane vapours in the top part of the apparatus), the hydrolysate is continuously supplied from batch box *1* and the catalyst (a 50% solution of potassium hydroxide), from batch box *2*. The catalyst is prepared in apparatus *3*.

Depolymerisate vapours (a mixture of dimethylcyclosiloxanes) enter water-cooled condenser *7*; they partially condense and are collected in collector *8*. The uncondensed vapours are sent into condenser *5* with salt solution to condense completely. Because of the components of the depolymerisate is an easily crystallised cyclic trimer, the condensation may form crystals and clog the tubes of the condenser. If the tubes are completely clogged, a reserve apparatus is put into operation, and the switched-off apparatus is defrozen with vapour. The liquid depolymerisate is poured into collector *6*. Depolymerisation continues for 7-12 days; after that a reserve depolymeriser is put into operation.

If the depolymerisate does not meet the technical requirements (mostly concerning moisture), it is sent to drying from collectors *6* and *8* by pump *9* into apparatus *10*. The drying is carried out in vacuum; to intensify the process, the depolymerisate is bubbled with anhydrous nitrogen.

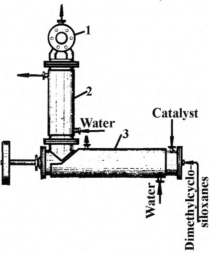

Fig. 60. Screw polymeriser: *1* - the upper part of the apparatus; *2* - the middle part of the apparatus; *3* - the lower part of the apparatus

After filtering in filter *11* the dried depolymerisate is sent into container *12* and to the next stage, polymerisation.

The product of depolymerisation is a mixture of cyclic dimethylsiloxanes with tetramer (mostly $[(CH_3)_2SiO]_4$ (the boiling point is 175 °C) with an impurity of trimer $[(CH_3)_2SiO]_3$ (the boiling point is 134 °C) and pentamer $[(CH_3)_2SiO]_5$ (the boiling point is 210 °C). It is used as raw stock in the production of polydimethylsiloxane elastomers SKT, SKTV, SKTN, etc.

The dimethylcyclosiloxane blend is polymerised with the catalyst, a mixture of dimethylcyclosiloxanes, anhydrous aluminum sulfate and sulfuric acid in the 10:5:(0.2÷0.15) weight ratio. The catalyst is prepared in a special agitator.

The continuous polymerisation of dimethylcyclosiloxanes is carried out in a screw apparatus (Fig. 60), which consists of three parts. Lower, horizontal part *3* has a blade agitator and a jacket used to maintain a temperature of 80-100 °C in the reaction zone. Middle, vertical part *2* is hollow and also has a jacket to be heated with hot water. Top, horizontal part *1* has a blade screw and two jackets: one is filled with hot water (in the direction of the polymer flow), the other is filled with cold water (to cool the polymer before unloading). Horizontal part *3* is continuously filled with a blend of dimethylcyclosiloxanes and catalyst (6-7.5% of the quantity of the original dimethylcyclosiloxanes).

The polymer is unloaded into containers through a top choke and left for 20-40 hours to "ripen". The "ripening" process is considered over, when a necessary molecular weight is achieved, as determined by the viscosity. Then the polymer is flushed in roll mills with water heated to 50 °C to eliminate the remaining catalyst and dried in vapour-heated mills at 80 °C.

Elastomer SKT should meet the following technical requirements:

Solubility in benzene	Complete
Volatile substances content at 150 °C, %, not more than	6
Reaction of the aqueous extract	Neutral
Molecular weight	$(4.0\text{-}6.5) \cdot 10^5$

4.1.2. Preparation of polyvinylmethyldimethylsiloxane elastomers

Polyvinylmethyldimethylsiloxane elastomers (SKTV) are prepared by the hydrolytic cocondensation of dimethyldichlorosilane and vinylmethyldichlorosilane with subsequent copolymerisation of the mixture of the hydrolysate and the depolymerisate obtained in the production of SKT elastomer. The process comprises two main stages: the hydrolytic cocondensation of dimethyldichlorosilane and vinylmethyldichlorosilane; the copolymerisation of vinylmethyldimethylcyclosiloxanes and dimethylcyclosiloxanes.

The hydrolytic cocondensation of dimethyldichlorosilane and vinylmethyldichlorosilane occurs in the depolymerisate medium, forming a mixture of vinylmethyldimethylcyclosiloxanes

$$m(CH_3)_2SiCl_2 + n(CH_3)(CH_2=CH)SiCl_2 \xrightarrow[4((m+n)HCl]{2(m+n)H_2O}$$

$$\longrightarrow [(CH_3)_2SiO]_m[(CH_3)(CH_2=CH)SiO]_n \qquad (4.10)$$

m+n=3÷5; m=1÷5; n=5÷1.

with impurities of linear oligomers of the following composition:

$$HO - \begin{bmatrix} CH_3 \\ | \\ Si - O \\ | \\ CH_3 \end{bmatrix}_{m-1} \begin{bmatrix} CH_3 \\ | \\ Si - O \\ | \\ CH=CH_2 \end{bmatrix}_{n} \begin{matrix} CH_3 \\ | \\ Si - OH \\ | \\ CH_3 \end{matrix}$$

The copolymerisation of vinylmethyldimethylcyclosiloxanes and dimethylcyclosiloxanes is conducted in the presence of the same catalyst as in the production of SKT according to the scheme:

$$x\,[(CH_3)_2SiO]_m[(CH_3)(CH_2=CH)SiO]_n + [(CH_3)_2SiO]_n \xrightarrow{\quad H_2SO_4 \quad}$$

$$\longrightarrow HOSO_2O - \left\{ \begin{bmatrix} CH_3 \\ | \\ Si - O \\ | \\ CH_3 \end{bmatrix}_{m-1} \begin{bmatrix} CH_3 \\ | \\ Si - O \\ | \\ CH=CH_2 \end{bmatrix}_{n} \right\}_x \begin{bmatrix} CH_3 \\ | \\ Si - O \\ | \\ CH_3 \end{bmatrix}_{ny} \begin{matrix} CH_3 \\ | \\ Si - OH \\ | \\ CH_3 \end{matrix} \qquad (4.11)$$

After flushing the catalyst with water the sulfate groups are completely saponified with water. As a result, polyvinylmethyldimethylsiloxane is formed:

$$HO - \left\{ \begin{bmatrix} CH_3 \\ | \\ Si - O \\ | \\ CH_3 \end{bmatrix}_{m-1} \begin{bmatrix} CH_3 \\ | \\ Si - O \\ | \\ CH=CH_2 \end{bmatrix}_{n} \right\}_x \begin{bmatrix} CH_3 \\ | \\ Si - O \\ | \\ CH_3 \end{bmatrix}_{ny} \begin{matrix} CH_3 \\ | \\ Si - OH \\ | \\ CH_3 \end{matrix}$$

The basic diagram of hydrolytic cocondensation and preparation of the reactive mixture for copolymerisation in the production of SKTV is given in Fig. 61.

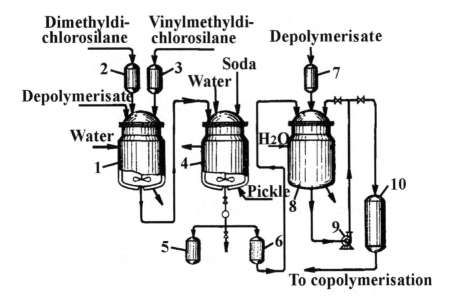

Fig. 61. Diagram of hydrolytic cocondensation and preparation of the reactive mixture for copolymerisation in the production of SKTV: *1, 8* - agitators; *2, 3, 7* - batch boxes; *4*- hydrolyser; *5, 6, 10* - collectors; *9* - pump

Enameled agitator *7* with vacuum is loaded with dimethyldichlorosilane (the boiling point is 70-70.3 °C, 54.8-55% of chlorine) and vinylmethyldichlorosilane (the boiling point is 92.5-93 °C, 50.1-50.5% of chlorine) in a certain weight ratio; it also receives the depolymerisate (the viscosity does not exceed 4 MPa, d_4^{20} = 0.95-0.96) from the synthesis of SKT. The mixture is agitated for 30 minutes at constant temperature and fed into hydrolyser *4*, which has been loaded with a required amount of distilled water. While the reactive mixture is being fed, the temperature of 30 °C is maintained in the hydrolyser by sending salt solution into the jacket. After the mixture has been fed, the agitation continues for 10-15 minutes; then the agitator is switched off and the mixture is held for 10-20 minutes. As a result, the mixture splits.

The bottom layer (22-24% hydrochloric acid) is poured through a rundown box into collector *5*, the top layer, the product of hydrolytic cocondensation, is neutralised and washed directly in the hydrolyser. The product of hydrolytic cocondensation is sampled for acidity and the hydrolyser is loaded through a hatch with a necessary amount of soda ash for neutralisation. The mixture is agitated for 2 hours; to dissolve the salts formed in neutralisation, water is added in the amount 10-20 times larger than the amount of soda. The mixture is agitated for 1 more hour; the bottom layer is separated and sent to biochemical purification. The product is washed again until the flush waters give a neutral reaction. After that the neutral product of hydrolytic cocondensation is dried in the hydrolyser (in vacuum, at a temperature not exceeding 80 °C). The dried product is loaded off into collector *6* and used as a component for copolymerisation.

The product of hydrolytic cocondensation should meet the following technical requirements:

Content, %:
of vinylmethylsiloxy elements	24±2
of moisture	≤0.01
Reaction of the aqueous extract	Neutral

The product from collector *6* is loaded into agitator *8*. The apparatus also receives the depolymerisate obtained in the production of SKT. The working mixture is prepared. To obtain SKTV, one takes 0.42 kg of the cocondensation product per 100 kg of the depolymerisate; to prepare SKTV-1, 2.1 kg of the cocondensation product per 100 kg of the depolymerisate. The mixture is apparatus *8* is agitated with pump *9* (by circulation) for 3 hours and sampled. If the content of vinylmethylsiloxy elements is appropriate, the working mixture is sent from apparatus *8* with pump *9* into collector *10* and from there to copolymerisation.

The copolymerisation of the mixture to obtain SKTV and SKTV-1 occurs similarly to the production of SKT in a screw polymeriser (see Fig. 60) and in the same conditions. The "ripening", flushing and drying of the polymer are also similar to the processes in the SKT production.

Technical SKTV and SKTV-1 should meet the same requirements as SKT. Besides, they should have a strictly determined amount of vinylmethylsiloxy elements: 0.09±0.02% for SKTV, 0.5±0.05% for SKTV-1.

The technology to prepare polydimethyldiethylsiloxane elastomer SKTE, polydimethylmethylphenylsiloxane SKTPh, poly(vinylmethyl)(methylphenyl)dimethylsiloxane SKTPhV and others is similar to that of SKTV.

All silicone elastomers have high thermal stability. When heated, they do not release toxic products and are used as main ingredients in the production of rubber compounds and heat-resistant rubbers.

4.1.3. Preparation of rubber compounds based on organosiloxane elastomers

The development of silicone elastomer technology has enabled synthetic rubber plants to obtain very convenient and practical rubber compounds.

Silicone elastomer-based rubber compounds are prepared in conventional apparatuses (closed agitators, roll mills, etc.) and consist of the following ingredients: elastomer, active fillers, vulcanising agent, stabiliser, pigment additives.

Raw stock: silicone elastomers (SKT, SKTV, SKTV-1, SKTE, SKTPh, SKTPhV, etc.) which meet technical requirements; Aerosil (pure SiO_2); dimethyldichlorosilane-modified or nonmodified zinc oxide; titanium dioxide TS; carbon white U-333 (rubber booster); iron oxide (Redoxide), a heat resistant additive which increases the stability of rubbers at high temperatures; lamp or acetylene black (a filler for conductive rubbers); dihydroxydiphenylsilane, methylphenyldimethoxysilane or branched oligoorganosiloxanes (stabilisers used in the processing of rubber compounds); chlorinated benzoyl peroxide and kumyl peroxide (vulcanising agents); pigments (Sudan-IV or other fat-soluble colourants) used to give rubbers a required colour.

The composition is prepared in an agitator and sent to mixing sheeting mills. The elastomer of a given brand is weighed and loaded into the agitator. The apparatus is filled with portions of loose ingredients (Aerosil, black, dihydroxydiphenylsilane or branched oligoorganodiphenylsiloxanes, titanium white) at certain intervals. Depending on the brand of the rubber compound dihydroxydiphenylsilane can be replaced with methylphenyldimethoxysilane (stabiliser SM-2). During the agitation the apparatus is blown with nitrogen.

Dihydroxydiphenylsilane compounds are held at 180-185 °C and agitated for 30 minutes in a vacuum (to suck the vapours of volatile substances). To warm the agitator, the jacket of the apparatus is filled with vapour. After agitation the hot mixture is loaded off to the roll mills. along with the vulcanising agent (chlorinated benzoyl peroxide or kumyl peroxide) and pigments (zinc and titanium white). The mixture is mixed with these ingredients in the mills at a temperature below 50 °C. After cooling and rolling the mixture is sent at least twice through a refiner with a thin opening. Then the mixture is sent to an extruder to strain and eliminate impurities. The straining is also repeated at least twice.

The rubber compounds prepared in this way are analysed and packed into polyethylene bags. Then they are loaded into metal cylinders or rubberised bags and sent to the consumer. Silicone rubber compounds can be stored for 3 to 6 months depending on the type of fillers and stabilisers.

Silicone rubber compounds can be molded, extruded and calendered. The vulcanisation of rubber compounds occurs in two stages: 1) in a press or steam boiler at high pressure and 120-150 °C; 2) by thermostatic control at atmospheric pressure and 200-250 °C.

The main properties of some rubber compounds and vulcanised rubbers, as well as their applications, are given in Table 20. As seen from the table, silicone elastomer-based rubbers are designed for prolonged use in a wide range of temperatures: from -50 to +250 °C, some from -70 to + 350 °C (for a short period of time). These rubbers are efficient in air, ozone and in an electric field; rubbers based on IRP-1339 and IRP-1401 compounds are also efficient in case of limited air supply. They function well in high humidity and under the influence of oxidants, hot water, vapour and low pressure. They are stable in weak-acid and weak-alkali media and are non-toxic.

Table 20. The main properties and applications of silicone rubber compounds and rubbers based on them

Brand	Vulcani-sation condi-tions	Properties of 2 mm thick vul-canisates			Tempera-ture range of applica-tion	Applications
		Hardness (IRHD), standard units	Ulti-mate tensile strength, at least	Rela-tive elonga-tion at rup-ture, %, at least		
IRP-1338	150^{0}C and 20 minutes (stage I), 200^{0}C and 6 hours (stage II)	55-70	6.5	300	From -50 to +250 ^{0}C; +300^{0}C (for a short pe-riod of time)	To produce molded and extruded goods (in-serts, caps, tubes, cords, profiles, etc.) in immobile joints (the compressive strain does not exceed 20%) in air, ozone and an electric field in all cli-matic condi-tions

Table 20. (cont.)

IRP-1339	The same	60-70	5.0	200	From -50 to +250 ^{0}C	To produce the elements of pipes with special strength requirements; molded and unmolded goods to be used in air in all climatic conditions
IRP-1340 IRP-1341 IRP-1344	" " "	60-70 65-75 50-60	3.5 6.5 6,0	140 200 200	The same " "	To produce molded and extruded goods (inserts, caps, tubes, cords, profiles) in immobile joints (the compressive strain does not exceed 30%) in air, ozone and an electric field in all climatic conditions
IRP-1354	The same	50-65	5.0	250	From -70 to +250 ^{0}C; +300^{0}C (50 hours)	To produce inserts, caps, tubes, cords, and other elements in immobile joints (the compressive strain does not exceed 20%) in air, ozone and in sunlight

Table 20. (cont.)

IRP-1400	"	65-75	5.0	200	From -50 to +250 ^{0}C	To produce the elements of pipes and dyed molded and un-molded goods to be used in air in all climatic conditions
IRP-1401	"	70-80	6.5	200	From -50 to +250 ^{0}C	To produce molded and unmolded high-hardness parts used in air, ozone and an electric field
ISKh-344	150^{0}C and 10 minutes (stage I), 200^{0}C and 6 hours (stage II)	At least 45	3.5	120	From -60 to +200, ^{0}C in air, to 100^{0}C in water	To produce heat- and cold-resistant seals and inserts used in air and in hot water; are also used in medicine
K-8	120^{0}C and 10 minutes (stage I), 200^{0}C and 6 hours (stage II)	45	6.0	250	From -50 to +250 ^{0}C	To produce heat-resistant insulating pipes by continuous vul-canisation

Silicon rubber products are widely used in aircraft, automobile, ship-building and electrical industries, as well as in electronics, farmaceutics and medicine (mitral valves, etc.). Rubber compounds based on SKT, SKTV, SKTV-1 and SKTE can be used without hardening; they are ex-truded and molded. They can function in any climatic conditions. The main properties of rubbers obtained by extruding and molding are given in Table 21.

Apart from high-molecular polyorganosiloxane elastomers, such rubbers are also based on low-molecular silicone elastomers with molecular weights from 25000 to 75000, especially polydimethylsiloxane elastomers SKTN.

Table 21. Main properties of silicone rubbers and conditions for their application

Brand	Ultimate tensile strength, MPa, at least	Relative elongation at rupture, %, at least	Brittle point, °C, not more than	Hardness (IRHD)	Conditions for use
14b-2	2.2	170	-62	-	From -60 to +250 °C in air, ozone and an electric field (the deformation is less than 10%)
14b-6 14b-15	2.5 2.5	200 200	-65 -60	-	The same, but in the -60- +200 °C range
5b-129	2.5	200	-65	_	The same, but in oil at 150 °C for 100 hours
IRP-1265 IRP-1266	2.5 2.5	250 100	-65 -65	35-55 35-55	From -60 to +250 °C in air, ozone and an electric field
IRP-1267	2.5	140	-70	40-60	The same, but in the -70- +250 °C range

4.1.4. Preparation of low-molecular polydimethylsiloxane elastomers

Low-molecular polydimethylsiloxane elastomers can be obtained by the polymerisation of a dimethylcyclosiloxane mixture in the presence of alkali:

$$m[(CH_3)_2SiO]_n + KOH \longrightarrow HO-[(CH_3)_2SiO]_{mn}-K \qquad (4.12)$$

$mn = 340 \div 1100$

The molecular weight of the elastomer depends on the amount of the catalyst; since the molecular weight is inversely proportionate to the concentration of the catalyst, the molecular weight of the polymer can be reduced by increasing the concentration, and vice versa.

The preparation of low-molecular polydimethylsiloxane elastomers often requires a considerable quantity of alkali, which calls for additional neutralisation. That is why in some cases it is advisable to use special substances (so-called chain growth regulators), which can take part in the reaction of chain transition. The part of regulators can be played by oligomethylsiloxanes, which do not have active end groups. Below we give the consumption of some regulators (g, per 1 tonne of polydimethylsiloxane):

Octamethyltrisiloxane	550-850	Tetradecamethylhexa-siloxane	500-800
Decamethyltetrasiloxane	250-500	Oligomethylsiloxane	800-
Dodecamethylpentasiloxane	400-700	PMS-50	1500

When SKTN elastomers are produced, dimethylcyclosiloxanes are polymerised with 2 N potassium hydroxide solution (0.005% of the amount of the polymerised substance); the molecular weight of the polymer is regulated by water supply (diluting the alkali). The raw stock for polymerisation can be the depolymerisate obtained at the depolymerisation stage of the product of hydrolytic condensation of dimethyldichlorosilane in the production of polydimethylsiloxane elastomer SKT.

Raw stock: depolymerisate, a mixture of dimethylcyclosiloxanes (60-85% of tetramer, up to 5% of trimer, 13-20% of pentamer, the allowable content of trifunctional impurities and moisture does not exceed 0.01%); potassium hydroxide (at least 95% of KOH); Aerosil or carbon white U-333. The production of this elastomer relies on the hermeticity of the equipment and service lines. The pressure is raised with nitrogen (0.3 MPa) and held for 1 hour. Then the pressure is reduced to atmospheric by releasing nitrogen through an air vent. Only then can polymerisation take place.

The depolymerisate is loaded out of batch box *1* (Fig. 62) into polymer-iser *2* with an agitator and jacket where a heat carrier circulates. At agita-tion the polymeriser receives a required amount of potassium hydroxide solution (catalyst). The reactive mixture is agitated at 140-180 °C until a viscosity balance is reached. After the balance has been achieved, the po-lymeriser is loaded through a hermetic box with a calculated amount of water (depending on the SKTN brand), and the mixture is agitated for 30 minutes. Then the contents of the polymeriser are cooled down to 90-100 °C by sending water into the jacket of the apparatus; the product self-flows into apparatus 3, which has been filled with a stabiliser (a mixture of SKTN with Aerosil or carbon white).

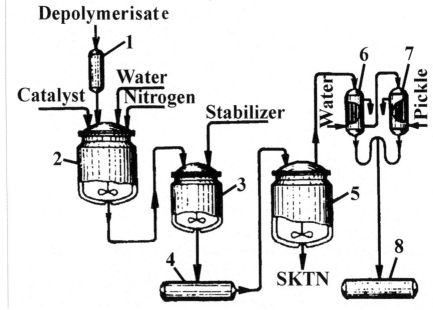

Fig. 62. Diagram of dimethylcyclosiloxane polymerisation in the production of SKTN elastomers: 1 - batch box; *2* - polymeriser; *3* - stabilisation apparatus; *4, 8* - containers; *5* - apparatus for distilling volatile substances; *6, 7* - condensers

Elastomer can be stabilised by agitating for 1-1.5 hours; after that, the polymer is tested for viscosity. If it corresponds to a given value, the polymer is sent into container *4* and on into apparatus *5*. In apparatus *5* with a jacket and agitator, at 130-165 °C and a residual pressure not ex-ceeding 90 GPa, created by a steam jet ejector unit, volatile substances and moisture, which condense in condensers *6* and *7* are distilled. The conden-sate is collected in container *8*; from there volatile dimethylcyclosiloxanes are sent as they accumulate by nitrogen flow (0.07 MPa) to repeated polymerisation.

The distillation is stopped as soon as the required viscosity and minimally allowable volatile content is reached. After the distillation, the jacket of reactor 5 is filled with water, the polymer is cooled to 60 °C and poured by nitrogen flow into special containers.

At present there are several brands of SKTN: A, B, C, D and E, which differ in their viscosity and molecular weight. Similarly to SKTN, one can obtain low-molecular polydimethylsiloxane elastomers which also contain methylphenylsiloxy elements, SKTNPh, SKTNPh-50, etc.

The most important properties and applications of low-molecular elastomers are given in Table 22.

Table 22. Important physicochemical properties of low-molecular silicone elastomers and their applications

Brand	VS-1 viscosity, (5.4 mm nozzle), s	Volatile content, % (weight), not more than:	Vulcanisation ability, h, not more than	Applications
SKTN-A*	90-150	2.5	6.0	To produce monolithic filling and enveloping compounds, sealants, foamed sealants, impregnating compositions and other materials with good dielectric characteristics operating in the -60 - 250-300^{0}C temperature range.
	151-240	2.5	6.0	
SKTN-B*	241-600	2.5	6.0	
	601-1080	7.0	6.0	
SKTN-C*	800-1600	7.0	6.0	
SKTN-D*				
SKTN-E*				
SKTN Ph	90-600	2.5-6	-	To produce monolithic filling and enveloping compounds with good dielectric characteristics operating in the -70 - 250^{0}C temperature range in high humidity.
SKTN Ph-50	-	10.0	-	To produce benzene- and oil-resistant sealants and monolithic filling and enveloping compounds with good dielectric characteristics operating in the -60 - 250-300^{0}C temperature range.

*These elastomers can be used to make dental models in dentistry.

Low-molecular silicone elastomers are viscous liquids which do not contain solvents and solidify at room temperature. The specific properties of SKTN allow one to use them as insulation against heat, moisture and electricity in various miniature and large units of machines, mechanisms and devices, as well as for thermal, electrical and vibration sealing of various devices. The physiological inertness of elastomers accounts for their wide applications in medicine.

Low-molecular silicone elastomers can be used as a base for various rubber compositions, including sealants and compounds.

4.1.5. Preparation of sealants and compounds

Sealants (sealing materials) and compounds (filling materials) are thick-flowing pastes which solidify in the cold. They are prepared by mixing low-molecular silicone elastomers with mineral ingredients. Sealants and compounds are supplemented with special catalysts which help them to solidify in the cold. Before using sealants or compounds, the surface of the product is sometimes covered with a sub-layer to improve adhesion.

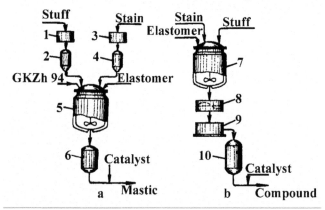

Fig. 63. Production diagram of sealants (a) and compounds (b): *1, 3* - sowers; *2, 4* - bins; *5, 7* - agitators; *6, 10* - collectors; *8* - paint mill; *9* - pressure filter

Raw stock: a) sealants are produced from elastomers SKTN, SKTN-A, SKTN-B, SKTN-C, or SKTNPh-50, titanium white TS (or zinc white M-1 or M-2), iron oxide (Redoxide), silicone liquid GKZh-94; b) compounds are produced from elastomers SKTN, SKTN-A or SKTN-B, arbon white U-333, zinc or titanium white, pigment.

The production process of sealants (Fig. 63, *a*) includes the sieving of the filler (white) through a metal mesh (No. 016 or No. 018) in closed sower *1*; and of the pigment, in sower *3* (No. 016 or No. 018). The sieved ingredients enter bins *2* and *4* and from there are sent into apparatus *5*. The apparatus is loaded with elastomer, the agitator is switched on and the mixture is agitated for 5-10 minutes. The filler is fed from bin *2* portionwise, within 1-1.5 hours. After the filler has been loaded, the mixture is agitated for 3.5 h. The temperature is regularly checked; if the mixture heats to 25 °C, the jacket of the agitator is filled with water. During the agitation, the direction of rotation is changed every 15-20 minutes. If sealants contain silicone liquid GKZh-94 and pigment, the liquid is sent into agitator *5* after elastomer and agitated for 1-2 hours. After part of the filler has been introduced from bin *2*, the stain is added from bin *4*; then, the rest of the filler. After all ingredients have been loaded, the mixture is agitated for 1 hour at cooling. The prepared paste is loaded off into collector *6* and sent to the consumer. Before use the paste is mixed with a catalyst and used in this form as a sealant. Catalysts can be metalorganic compounds, such as organic salts of dialkyl tin in tetraethoxysilane or alkylacetoxysilane.

Compounds are prepared in agitator *7* (Fig. 63, b). The apparatus, which has been heated to 50-70 °C, is filled with portions of elastomer; after agitating and heating to 100 °C it receives more filler and pigment, if necessary (also portionwise). The reactive mixture is agitated and heated to 120 °C and held at this temperature until a homogeneous product is obtained. The heating is switched off, the past is loaded off into metal containers through a lower choke and sent directly to pressure filter *9* or through paint mill *8* for better homogenisation. The prepared paste is loaded off from the pressure filter into collector *10* and sent to the consumer.

Similarly to sealants, the compound paste is mixed before use with a metalorganic or silicone catalyst; as a result, even at room temperature the paste switches from a thick-flowing to a rubberlike state and solidifies. The compound paste can be prepared with a catalyst, i.e. as a one-component system. In this case the equipment should be particularly airtight, especially the equipment used to load the compound into containers.

The most important properties and applications of silicone sealants and compounds are given in Table 23.

Silicone sealants and compounds, like rubbers, have good dielectric properties, as can be illustrated by Viksint Y-1-18:

Specific electric resistance at 20±2 °C:

volume, Ohm cm	$1 \cdot 10^{13}$
surface, Ohm	$1 \cdot 10^{13}$
Electric strength at 20±2 °C, KV/mm, at least	15.0
Dielectric loss tangent at 10^6 Hz, at least	0.02
Dielectric permeability at 10^6 Hz, not more than	3.0

As shown in Table 23, silicone sealants are used to seal various constructions and devices that operates in a wide temperature range (from -60 to 250-300 °C). Seals made from them are resistant to vibration, alternate and shock loads and atmosphere.

Table 23. Main properties and applications of silicone sealants and compounds

Brand	Density, g/cm^3	Hardness (IRHD), standard units	Ultimate tensile strength, MPa, at least	Relative elongation at rupture, %, at least	Pot life, hours	Applications
Sealant Viksint U-1-18	2.19-2.21	50-60	2.0	160	0.5-6	For surface sealing of metal joints and equipment operating in air at -60 -+300^{0}C under vibration, shock and alternate loads in various climatic conditions
Sealant Viksint U-2-28	2.20	40-50	1.8	200	3-8	For surface and intraseam sealing of riveted and welded junctions, constructions and devices which operate in the -60 -+300^{0}C temperature range for surface sealing and in the -60 - +250^{0}C temperature range for intraseam sealing

Table 24. (cont.)

Sealant Viksint U-4-21	1.35	40-50	1.5	100	0.5-5	For surface sealing of constructions and devices which operate for a long time in air at -60 -+300^{0}C
Sealant VGPh-1	1.75	At least 35	1.3	130	0.5-2	For surface sealing of metal junctions which operate in fuel T-5, T-6 and naphthyl, at -60 -+250^{0}C (VGPh-1-VGPh-2 for intraseam sealing)
Sealant VGPh-1	1.75	35-45	1.3	130	3-7	
Sealant VGO-1 and VGO-2	-	20-40	1.5	200-600	10-30 minutes	For sealing radioelectronic equipment and screw joints; for repairing articles sealed with other silicone sealants; the temperature range is from -60 to+250^{0}C
Compound Viksint K-18	1.10	50-60	1.7	80	0.5-0.6	For surface sealing of electronic and radio devices which operate in air and increased humidity at -60 -+250^{0}C
Compounds KL*	-	-	0.5-1.5	100-180	At least 10-15 min	For sealing electronic devices to protect them from humidity and air; the temperature range is from -60 to+300^{0}C.

At present industry widely manufactures glue sealant Elastosil based on low-molecular silicone elastomers. Besides elastomers, the compositions of these sealants include fillers, plasticisers, solidifiers and adhesive components; in some cases pigments are used to give them a colour.

Table 24. Main properties of Elastosil glue sealants

Brand	Appearance	Viscosity at 20 °C (before solidification), Pa/s	Time of surface film formation, min, not more than	Ultimate tensile strength, MPa	Relative elongation at rupture, %	Adhesion to metals, N/cm	Specific volume electric resistance at 20 C, Ohm cm	Dielectric permeability at 10^6 Hz and 20°C	Dielectric loss tangent at 10^6 Hz and 20 C
1101	White paste	500-800	150	1.6-2.5	140-200	20-35	$1 \cdot 10^{-4}$	4.2	0.02
1102*	Transparent colourless liquid	8-12	120	0.2-0.3	150-200	-	$1 \cdot 10^{-4}$	2.8	0.01
1106	Grey paste	800-1000	180	1.6-2.5	250-300	20-35	$1 \cdot 10^{-4}$	4.2	0.02
1110	Thick-flowing orange paste	30-50	360	0.8-1.2	100-140	10	$1 \cdot 10^{-4}$	3.0	0.005
1408**	Black paste	-	60	0.8-1.0	100-120	10-15	-	-	-
2103*	Transparent colourless liquid	2-8	120	0.1-0.2	80-100	-	$1 \cdot 10^{-4}$	2.8	0.01
2104	Green paste	400-800	90	1-1.2	100-120	10-15	$1 \cdot 10^{-4}$	3.4	0.01

*Transparency of 50 μm-thick layer of Elastosil-1102 and Elastosil-2103 in UV rays is correspondingly 75 and 98%.
*Two-component material

In the process of vulcanisation in air Elastosils form rubberlike materials and have good adhesion to steel, copper, aluminum, wood, ceramics, concrete, polymethylmetacrylate, glass and other materials. Consequently, they do not require the use of any special sublayer. Optimal physicochemical properties of Elastosils are achieved after 5-7 days of solidification at 60-75% humidity in air. The main properties of glue sealants Elastosil are given in Table 24.

Changing the molecular weight of the elastomer, the number of fillers and plastifiers, we can obtain liquid (Elastosil-1102 and 2103), thick-flowing (Elastosil-1101 and 1110) or thixotropic (Elastosil-1106 and 2104) compositions. After vulcanisation, glue sealants based on low-molecular polydimethylsiloxane elastomers (Elastosil-1101, 1102, 1106, 1110 and 1408) can operate in the -60 - +200 °C range. If they are based on polymethylphenylsiloxane elastomers (Elastosil-2103 and 2104), these materials can be used at a temperature down to -90 °C. All Elastosils have high moisture and heat resistance and, as shown in Table 24, good dielectric characteristics, etc.; Elastosil-1408 has semiconductor properties (ρ_s = $1 \cdot 10^3$ Ohm), because the filler for it is acetylene black. These properties allow glue sealants Elastosil to be used in the electrotechnical industry, mechanical engineering, instrument-making, micro- and radioelectronics, construction, etc.

4.2. Preparation of branched, ladder and cyclolinear polyorganosiloxanes

Branched (I) and ladder (II) polyorganosiloxanes:

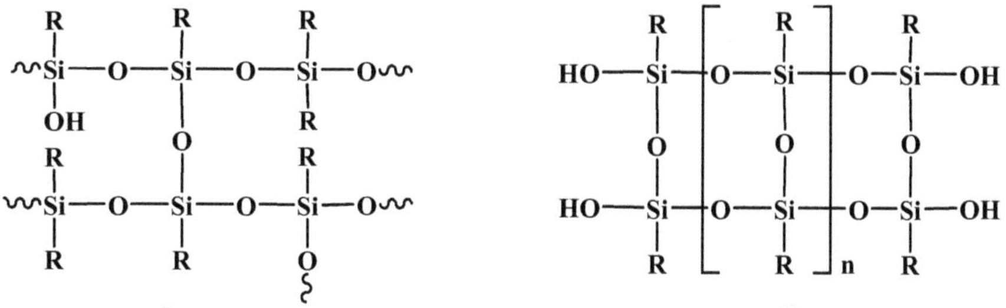

are obtained by the hydrolytic polycondensation of trifunctional silicone compounds (the synthesis of ladder polymers) or a mixture of di- and trifunctional compounds (the synthesis of branched polymers).

The properties of branched and ladder polyorganosiloxanes are determined by two factors: the functionality of parent monomers, determined by the ratio of the number of nonfunctional groups or organic radicals to one silicon atom (R:Si) and the use of functional groups (halogens, alkoxyl, acyloxyl, hydroxyl, etc.) in the synthesis. When the R:Si ratio reduces from 2:1 to 1:1, the polymers gradually become less flowing, meltable and soluble, depending on cross-linking efficiency. If R:Si = 1:1, i.e. when the raw stock is only trifunctional monomers (methyltrichlorosilane, phenyltrichlorosilane or a mixture of methyl- and phenyltrichlorosilanes), rigid polymers are formed. It means that their solutions in organic solvents (varnishes) form a three-dimensional rigid structure when they solidify.

If the functional groups are used completely, the products are mostly cross-linked, nonmeltable and insoluble; however, with the same R:Si ratio special techniques for the treatment of original organochlorosilanes can create ladder structures and help to prepare flexible high-melting or nonmentable, but soluble, products.

The introduction of difunctional elements (dimethylsiloxy, diethylsiloxy or methylphenylsiloxy elements) into the main chain of the polymer forms relatively elastic polymers, mainly cyclolinear:

Thus, as the R:Si ratio increases and the number of cross links decreases, polymers can range from glasslike to rubbery. Most cyclolinear and branched polymers are obtained when the R:Si ratio is from 1:1 to 1.6:1.

The nature of organic R groups surrounding silicon atoms is also essential for the properties of polyorganosiloxanes. The increasing of the length of alkyl radicals makes a polymer softer, increases its solubility in organic solvents and its waterproofing ability; however, it reduces its resistance to thermal oxidative destruction and heating. Phenyl radicals increase the thermal stability of a polymer. Particularly widespread are polyorganosiloxanes with methal and phenyl groups surrounding the main chain.

Alongside with such positive properties as high heat and cold resistance, good water-proofing ability and dielectric characteristics, polyorganosiloxanes have rather poor physicomechanical characteristics. To improve these properties, polyorganosiloxanes are often modified with various organic polymers (polyesters, epoxy and glyptal resins, etc.).

At present industry manufactures a wide range of branched, cyclolinear and ladder polyorganosiloxanes. which differ by the type of organic groups of radicals at the silicon atom. The production of such polyorganosiloxanes is based on hydrolytic condensation or cocondensation of organochlorosilanes or alkoxyorganosilanes and subsequent polycondensation of the products.

The main classes of branched, cyclolinear and ladder polyorganosiloxanes are:

1. polymethylsiloxanes;
2. polyphenylsiloxanes;
3. polydimethylphenylsiloxanes and polymethylphenylsiloxanes;
4. polymethylphenylsiloxanes with active hydrogen atoms and vinyl groups at the silicon atom;
5. polyphenyldiethylsiloxanes;
6. polyalkylsiloxanes with alkyl radicals C_4 and more at the silicon atom.

4.2.1. Preparation of polymethylsiloxanes and polymethylsiloxane varnishes

Contemporary polymethylsiloxanes differ in the structure of the chains formed, viscosity and gelatinisation time. Depending on the degree of cross-linkings, these characteristics are naturally different.

Gelatinisation time is the period within which a polymer transfers from a meltable and fluid state (stage A) into a nonfluid (stage B) and non-meltable (stage C) states. Gelatinisation time is usually determined with the help of a polymerisation stove at a definite temperature (180, 200 or 250 °C).

The properties of polymethylsiloxanes are also greatly determined by the conditions of the reaction and the type of the solvent used. Thus, the hydrolytic condensation of methyltrichlorosilane (the main raw stock in the production of polymethylsiloxanes) with iced water or water vapour in a medium of nonpolar solvents forms a nonmeltable and nonsoluble amorphous substance. If the condensation is carried out by pouring methyltrichlorosilane into the emulsion of water and butyl alcohol gradually (to avoid gelation), at intensive agitation and reduced temperature (about 0 °C), it forms a viscous substance, which is soluble in organic solvents and briefly heated to 150 °C loses its meltability and solubility.

This difference in the product properties can be explained by the fact that the introduction of methyltrichlorosilane into the aqueous-alcoholic emulsion in the system $CH_3SiCl_3+H_2O+C_4H_9OH$ causes three competing reactions at the same time:

$$nCH_3SiCl_3 \xrightarrow[3nHCl]{3nH_2O} n[CH_3Si(OH)_3] \xrightarrow{1.5\ H_2O} [CH_3SiO_{1.5}]_n \qquad (4.13)$$

$$CH_3SiCl_3 \xrightarrow[3HCl]{3C_4H_9OH} CH_3Si(OC_4H_9)_3 \qquad (4.14)$$

$$nCH_3Si(OC_4H_9)_3 \xrightarrow[3nC_4H_9OH]{3nH_2O} n[CH_3Si(OH)_3] \xrightarrow{1.5\ H_2O} [CH_3SiO_{1.5}]_n \qquad (4.15)$$

These reactions greatly differ in speed: the speeds of the first two are commensurable; however, the third one occurs much slower. As a result, the hydrolytic condensation of methyltrichlorosilane in butil alcohol can be presented by the following overall scheme:

$$(n+m)CH_3SiCl_3 + 2nH_2O + mC_4H_9OH \xrightarrow{3(n+m)HCl} (n+m) \begin{bmatrix} CH_3 \\ | \\ HO-Si-OH \\ | \\ OC_4H_9 \end{bmatrix} \longrightarrow$$

$$\xrightarrow{(n+m-1)H_2O} HO \begin{bmatrix} CH_3 \\ | \\ Si-O \\ | \\ OH \end{bmatrix}_n \begin{bmatrix} CH_3 \\ | \\ Si-O \\ | \\ OC_4H_9 \end{bmatrix}_m H \qquad (4.16)$$

We can see that the butoxyl groups block some places where hydroxyl groups can be formed, thus hampering crosswise chain growth, i.e. prevent complete cross-linking of the polymer and gelation. forming branched polymers.

If the hydrolytic condensation of methyltrichlorosilane is conducted in a mixture of nonpolar solvents (toluene, benzene) and the polar solvents that do not react with methyltrichlorosilane (acetone), it forms *polymethyl-silsesquioxane*, mostly of the ladder structure:

$$\begin{bmatrix} CH_3 & CH_3 \\ | & O & | \\ HO-Si & \diagup\diagdown & Si-O-H \\ | & & | \\ O & & O \\ | & & | \\ HO-Si & \diagdown_O\diagup & Si-O-H \\ | & & | \\ CH_3 & CH_3 \end{bmatrix}_n$$

It should be kept in mind, however, that such polymers have a defective structure; since the polymers are a mixture of polymer homologues, the mixture also contains macromolecules, which consist of branched links, i.e. the so-called "link-varied polymer" is formed[1].

As we have already said, the reaction of the hydrolytic condensation of methyltrichlorosilane is largely determined by the type of the solvent used. In the presence of nonpolar solvents in water the process occurs at high speed; the formed polymer precipitates. In case of polar solvents, which dissolve the monomer, polymer and water, the condensation takes place in a homogeneous medium, because methyltrichlorosilane is well-soluble in alcohols and ethers. It rules out the possibility of precipitation; therefore, we obtain a soluble branched or ladder polymer.

In practice, the hydrolytic condensation of methyltrichlorosilane is carried out with a mixture of a polar organic solvent (acetone, alcohol) with a nonpolar one (toluene, benzene).

Polymethylsiloxane varnishes are toluene-butanol solutions of the products of the hydrolytic condensation of methyltrichlorosilane, partially etherificated with butyl alcohol, whereas polymethylsilsesquioxane varnishes are toluene-acetone solutions of the products of the hydrolytic condensation of methyltrichlorosilane.

Preparation of polydimethylsiloxane varnishes

The production of polymethylsiloxane varnishes comprises three main stages: (Fig. 64): the partial etherification of methyltrichlorosilane with butyl alcohol; the hydrolytic condensation of partially etherified methyltrichlorosilane; the distillation of the solvent and preparation of the varnish.

Methyltrichlorosilane is etherified in enameled hydrolyser 5, which is loaded with a necessary amount of methyltrichlorosilane (at least 99% of the main fraction) and toluene (half of the methyltrichlorosilane volume). The mixture is agitated for 30 minutes and checked for chlorine content (it should be 46.9±0.5%). Then the hydrolyser is filled with butyl alcohol at such speed that the temperature in the apparatus does not rise above 30 °C (for partial etherification one should take slightly more than 1 mole of C_4H_9OH per 1 mole of CH_3SiCl_3).

If there is an excess of methyltrichlorosilane, at the stage of hydrolytic condensation the product can enter a nonmeltable and nonsoluble state or form a structure with a greater number of cross links than can be tolerated; if there is an excess of butyl alcohol, the polymer, on the contrary, will have few cross links and therefore a small molecular weight. The stage of

---

[1] More detailed information on link-varied polymers can be found in *Link-varied polymers* (Raznozvennost polymerov) by Korshak VV (1977) Nauka Moscow [in Russian]

partial etherification largely determines the viscosity and gelatinisation time of the polymer.

The hydrogen chloride released in the reaction can be withdrawn into a water-flushed hydroejector (not shown in the diagram) and in the form of weak hydrochloric acid sent to biochemical purification.

After butyl alcohol has been introduced, the reactive mixture is heated to 60±5 °C within 2-3 hours and held at this temperature for 3-5 hours. The standing is finished when the chlorine content in silicon chloroethers (a by-product of the partial etherification of methyltrichlorosilane) is 15-42% depending on the type of the varnish. The mixture is cooled to 20-30 °C by sending water into the jacket of the hydrolyser. The cooled chloroethers are sent through a siphon into weight batch box 6.

Hydrolytic condensation is also conducted in hydrolyser 5. It is filled with a necessary amount of water and additional quantities of toluene and butyl alcohol, to make the conditions for chloroether hydrolysis milder.

The mixture is agitated for 10-15 minutes; then, chloroethers are sent from batch box 6 through a siphon at such speed that the temperature in the hydrolyser does not exceed 20±3 °C. After all the chloroethers have been introduced, the reactive mixture is kept at the temperature of hydro-lytic condensation for 6-8 hours. Most of the hydrogen chloride released during hydrolytic condensation dissolves in water; the excess is withdrawn into a water-flushed hydroejector (not shown in the diagram) and in the form of weak hydrochloric acid sent to biochemical purification.

After the standing the agitator is switched off and the organic layer is thoroughly separated from the aqueous layer. The aqueoud layer is settled and poured through a run-down box into neutraliser 7, from where it is sent after more settling through a hydraulic gat to biochemical purification. The toluene-butanol solution of silanol is flushed with water until the flush waters give a neutral reaction (the number of flushes depends on the acidity of the flush waters).

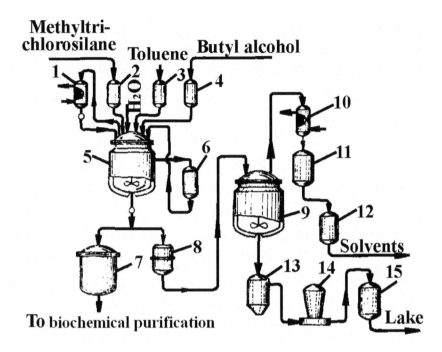

Fig. 64. Production diagram of polymethylsiloxane varnishes: *1, 10* - coolers; *2 - 4, 6* - batch boxes; *5* - hydrolyser; *7* - neutraliser; *8* - nutsch filter; *9* - distillation tank; *11* - receptacle; *12, 15* - collectors; *13* - settling box; *14* - ultracentrifuge

The flushed silanol is sampled to determine the solid residue, which should be 15-24%, and the gelatinisation time at 200±2 °C (the standard value is 0.5-60 minutes). After that, silanol is sent with vacuum through nutsch filter *8* into tank *9*, which is an enameled apparatus with an agitator and a water vapour jacket, to distil the solvents. Before the distillation the product of hydrolytic condensation is held at 20-25 °C to eliminate traces of moisture. The distillation is conducted at 45-50 °C and a residual pressure of 105-80 GPa until the solid residue is 40-55%. Then the jacket of the tank is filled with water and the varnish is cooled to 20-25 °C. During the distillation of the solvents, the product of hydrolytic condensation is further condensed to form a higher-molecular polymer.

The distilled solvents (the mixture of toluene and butanol) are sent into cooler *10* to condense and are collected in receptacle *11*, from where they are poured into collector *12*. The finished varnish is analysed to determine solid residue (40-55%), drying time (not more than 40 minutes at 125±5 °C) and solidification time (not more than 60 minutes); after that it is sent from the tank into settling box *13*. In the settling box the finished varnish is separated from mechanical impurities at ambient temperature, purified in ultracentrifuge *14* and directed into collector *15*. In the process of centrifuging the varnish is sampled every 30 minutes to monitor the appearance.

Polymethylsiloxane varnishes are toluene-butanol solutions of polymers with colour ranging from yellow to brown. Their viscosity at 20 °C (according to the VS-4 viscosimeter) should not exceed 60 s. The gelatinisation time may vary (depending on the method of hydrolytic condensation) in a rather wide range, from 0.5 to 60 minutes; the solid (nonvolatile) residue content is from 40 to 55%. The varnish layer should dry at 125±5 °C for not more than 40 minutes.

Polymethylsiloxane varnishes are mainly used in plastic composites to manufacture heat- and arc-resistant parts, which can operate in high humidity, as well as to paint the units and parts of electric machines and transformers.

Polymethylsiloxanes can also be obtained by the hydrolytic copolycondensation of methylchlorosilanes of various functionality. Thus, by the hydrolytic copolycondensation of methyltrichlorosilane and dimethyldichlorosilane one can obtain thermosetting polymethylsiloxane, which quickly solidifies at a relatively low temperature (approximately 150 °C) without any catalysts. It is used as a binding agent in pressure compositions, as well as as a 5% petrol solution to treat the surface of metal pressure molds to facilitate their splitting when molding polymer materials.

The hydrolytic copolycondensation of trimethylchlorosilane and trimethylchlorosilane is used to obtain polymethylsiloxane, the xylene solution of which can be used as an additive to eliminate flotation (splitting) of pigments in enamel films.

Production of polymethylsilsequioxane varnish by continuous technique

Polymethylsilsesquioxane is prepared by the hydrolytic condensation of methyltrichlorosilane by the continuous technique in a mixture of solvents (acetone and toluene) according to this scheme:

$$(n+m)CH_3SiCl_3 \xrightarrow[3(m+n)HCl]{(2n+1,5m)H_2O} [CH_3SiO_{1.5}]_m [CH_3SiO(OH)]_n \qquad (4.17)$$

m=4÷8, n=1÷2

i.e. the reaction forms ladder polymethylsilsesquioxane with a defective structure, because it also contains branched elements with hydroxyl groups at the silicon atom.

The production of polymethylsilsesquioxane varnish (Fig.65) comprises the following main stages: the preparation of raw stock and reactive mixture; the hydrolytic condensation of methyltrichlorosilane and flushing of the hydrolysate; the distillation of the solvents; the finishing and completion of the varnish; the regeneration of acetone.

Apparatus *3*, which has an agitator, is filled with methyltrichlorosilane and toluene. The mixture is agitated for 30 minutes, analysed to determine the chloroion content and poured by self-flow into weight batch box *5*.

The hydrolytic condensation of methyltrichlorosilane is carried out in hydrolyser 6, which is a shell-and-tube heat exchanger cooled with salt solution (-15 °C). Before introducing it into the hydrolyser, the reactive mixture is mixed (in its bottom part) with acetone; this mixture then enters the capillaries. At the same time the bottom part of the hydrolyser is filled with water. The reaction takes place in the tubes of the apparatus. The product of hydrolytic condensation is cooled and through the top of the hydrolyser is sent into tower 7, which is a Florentine flask, to split into the aqueous and organic layers.

The bottom layer, the acidic aqueous-acetone solution, is sent through a hydraulic gate to be neutralised in apparatus *11*; the top layer, the toluene solution of the product of hydrolytic condensation, is sent into flusher *12*, which is also a Florentine flask. At the same time the flusher is filled with a 10-20% solution of sodium chloride, heated to 50-60 °C. The processed solution of sodium chloride is sent through a hydraulic gate into neutraliser *11*, and the toluene solution of the product of hydrolytic condensation is sent into flusher-receptacle *13*, where the second flushing takes place.

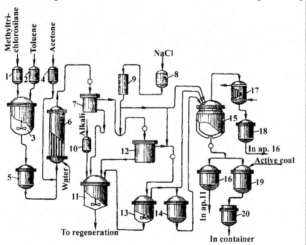

Fig. 65. Production diagram of polymethylsilsesquioxane varnish by the continuous technique: *1, 2, 4, 5, 8, 10* - batch boxes; *3* - agitator; *6* - hydrolyser; *7, 12* - Florentine flasks; *9* - heat exchanger; *11* - neutraliser; *13* - flusher-receptacle; *14, 16, 18* - collectors, *15* - vacuum distillation tank; *17* - cooler; *19* - balancing tank; *20* - druck filter

For this purpose apparatus *13* is filled with a solution of sodium chloride at agitation. The contents of the apparatus are agitated for 5-10 minutes and settled. The bottom layer, the sodium chloride solution, is poured into neutraliser *11*, and the top layer, the toluene solution of the product of hydrolytic condensation, is loaded into collector *14* or sent directly into vacuum distillation tank *15*.

The toluene solution of the product of hydrolytic condensation with pH of the nonaqueous solution not lower than 5 and the nonvolatile content of 40±3% is sent for completion into collector *14*; the toluene solution of the product of hydrolytic condensation with pH of the nonaqueous solution lower than 5 and the nonvolatile content of less than 37% is sent into tank *15* for additional flushing and partial distillation of the solvent. The additional flushing in the tank is carried out with the solution of sodium chloride (like the second flushing); then the contents of the tank are heated to 40-60 °C and held for 2-10 hours. The bottom layer, the settling products, is poured off and collected in collector *16*. Then a vacuum is created in the tank (the residual pressure is 0.05-0.07 MPa) and the solvent is partially distilled (until the nonvolatile content is 40+3%). The distilled solvent vapours are condensed in cooler *17* and collected in collector *18*. If the moisture content is above 1.5%, the solvent is sent into collector *16* for subsequent additional drying. The toluene solution of polymethylsilsesquioxane is poured from receptacle *14* or tank *15* into balancing tank *19*, where several batches of varnish are mixed and balanced. The finished varnish is analysed to determine the nonvolatile content, pH of the nonaqueous solution, viscosity and gelatinisation time.

If pH is lower than 5, the varnish is treated with kil or active coal. Balancing tank *19* is loaded with kil or coal (5-12% of the varnish amount); the mixture is agitated for 1-2 hours. If pH is 5÷7, the varnish is sent by nitrogen flow (0.3 MPa) at agitation through druck filter *20* into containers.

The acid aqueous layers and flush waters from apparatuses *7, 12, 13, 16* are sent into apparatus *11* and neutralised with 20-42% alkali. The contents of the neutraliser are agitated for 30 minutes and sampled for acidity. If pH is 6.5-8.5, the neutral aqueous-acetone solution is sent with a batching pump to regenerate acetone.

Polymethylsilsesquioxane varnish is a solution of polymethylsilsesquioxane in a mixture of solvents (acetone and toluene).

Technical polymethylsilsesquioxane varnish should meet the following requirements:

Appearance	Homogeneous solution ranging in colour from colourless to yellow
Nonvolatile content, %	40±3
Viscosity at 20 °C, mm^2/s, not less than	1.5
pH of the nonaqueous extract	5-7

Gelatinisation time, minutes, not less than   1

Polymethylsilsesquioxane varnish can be used a base for heat-resistant and flame-proof pressure materials.

4.2.2. Preparation of polyphenylsiloxanes and polyphenylsiloxane varnishes

Polyphenylsiloxanes are produced from alkoxyphenylsilanes, triacetoxy-phenylsilane, pure phenyltrichlorosilane or undistilled phenylchlorosilanes. Depending on the functionality of the parent monomer and the preparation technique, polyphenylsiloxanes can be branched or ladder. Often, to improve the properties of polyphenylsiloxanes, they are modified with various organic substances (by mechanical mixing or chemically).

Preparation of polyphenylsiloxanes from alkoxyphenylsilanes

Alkoxyphenylsilanes, which are used as raw stock in the production of polyphenylsiloxanes and polyphenylsiloxane varnishes, can be produced by the Grignard technique and etherification of phenyltrichlorosilane with alcohol.

If phenylethoxysilanes are produced by the Grignard technique, the production of polyphenylsiloxanes and polyphenylsiloxane varnishes comprises three main stages: the synthesis of phenylethoxysilanes; the hydrolytic condensation of phenylethoxysilanes; the distillation of the solvent and the preparation of varnish.

The synthesis of phenylethoxysilanes is carried out according to the reaction:

$$Si(OC_2H_5)_4 + C_6H_5CI + Mg \longrightarrow C_6H_5Si(OC_2H_5)_3 + C_2H_5OMgCI \quad (4.18)$$

However, the phenylation of tetraethoxysilane is a complicated process, which forms not only phenyltriethoxysilane, but also diphenyldiethoxysilane and triphenylethoxysilane. The products of phenylation also include unreacted tetraethoxysilane and diphenyl, which is formed due to the by-interaction of chlorobenzene with magnesium.

The hydrolytic condensation of phenylethoxysilanes is carried out in butyl alcohol; that is why at this stage we observe a partial re-etherification of phenylethoxysilanes with butyl alcohol and partial condensation of the re-etherification products. Generally, the process can be described by the following scheme:

$$nC_6H_5Si(OC_2H_5)_3 + mC_4H_9OH + nH_2O \xrightarrow[nC_2H_5OH]{}$$

$$\longrightarrow [C_6H_5Si(OC_2H_5)_n(OC_4H_9)_m(OH)_{3-(m+n)}] \xrightarrow{\;H_2O\;}$$

$$(4.19)$$

The hydrolytic condensation is conducted in an acid medium (in the presence of hydrochloric acid); therefore, it is accompanied by the destruction of ethoxy magnesium chloride formed:

$$C_2H_5OMgCI + HCI \longrightarrow C_2H_5OH + MgCI_2 \qquad (4.20)$$

The distillation of the solvent entails further polycondensation:

$$\xrightarrow[\;(n-1)C_4H_9OH\;]{nC_2H_5OH}$$

$$(4.21)$$

The diagram of the hydrolytic condensation of phenylethoxysilanes is given in Fig. 66.

To prepare the reactive mixture, agitator *1* is filled with a necessary amount of chlorobenzene and tetraethoxysilane (the content of the main substance is 97%). Then ethyl bromide is added to activate the synthesis. The reactive mixture is agitated for 30-60 minutes, sampled (to determine the chlorobenzene and tetraethoxysilane ratio) and pumped into weight batch box *5*. After that, reactor *7* is loaded through a hatch with magnesium chipping and with part of the reactive mixture from weight batch box

5 at agitation. The contents of the apparatus are heated to 120 °C; after that, the heating is switched off and more of the mixture is sent from weight batch box *5* to stimulate the spontaneous reaction. Due to exothermicity the temperature in the apparatus may spontaneously rise up to 150 - 180 °C. If after 10 minutes the temperature does not rise to the required level, more of the reactive mixture should be introduced.

After that, the mixture from batch box 5 is gradually sent into the reactor; the temperature is maintained at 140-160 °C. After the whole mixture has been introduced, the mixture is agitated in the reactor at the same temperature for 7-8 hours. The agitator is stopped and the reactive mixture is sampled to determine the phenylethoxysilane content.

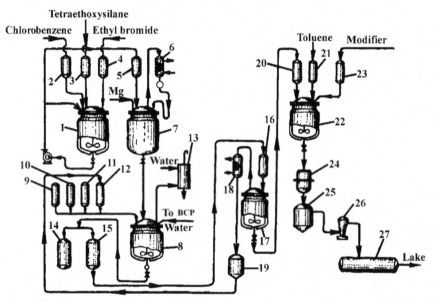

Fig. 66. Production diagram of polyphenylsiloxane varnishes: *1* - agitator; *2 - 5, 9 - 12, 16, 20, 21, 23* - batch boxes; *6, 18* - coolers; *7* - reactor; *8* - hydrolyser; *13* – faolite tower; *14, 15* - collectors; *17* - distillation tank; *19, 25* - settling boxes; *22* - apparatus for varnish preparation; *24* - nutsch filter; *26* - ultracentrifuge; *27* – container

If the phenylethoxysilane content is less than 65% (equivalent to phenyltriethoxysilane), the mixture is held at agitation for a few more hours, until the content of phenylethoxysilanes is as required.

After a positive analysis, the reactive mixture (a mixture of phenylethoxysilanes) is cooled down to 60-70 °C and sent into hydrolyser *8*. The hydrolyser has already been filled with toluene and butyl alcohol from batch boxes *9* and *10*. Phenylethoxysilanes should be loaded at agitation and cooling of the reactive mixture. At 30 °C 25-30% hydrochloric acid is sent slowly and gradually from weight batch box *11*. A temperature of 30-40 °C is maintained in the hydrolyser. After hydrochloric acid has been loaded, the mixture stands for 3 hours. The released hydrogen chloride is absorbed with water in faolite tower *13*, and the reactive mixture is diluted with water until the acidity of the aqueous layer is at least 40 g of HCl/l.

The aqueous layer after 0.5-1 hours of settling in hydrolyser *8* is poured into collector *14* and then sent to obtain hydrosite, which is used as a solvent in the production of water-proofing silicone liquids. The mixture which remains in the hydrolyser is washed with water several times, until it is totally neutral. After the flushing the silanol solution is sent into intermediate collector *15* and from there through batch box *16* into tank *17*, where the solvent is distilled to be collected in settling box *19*. The time of distillation is monitored by the amount of distilled solvent. The solvent (a mixture of toluene, butyl alcohol and chlorobenzene with an impurity of water) is separated from water in settling box *19*. After 12-24 hours of settling, the bottom (aqueous) layer is sent to biochemical purification, and the mixture of solvents is loaded into batch box *12* to be used at the stage of hydrolytic condensation.

After the distillation of the solvent, tank residue, which is a mixture of polyphenylsiloxanes, is cooled in apparatus *17* down to 50-60 °C and tested for varnish concentration. Then the product is sent out of the tank with vacuum through batch box *20* into apparatus *22*. The apparatus is also filled from batch box *21* with a calculated amount of toluene (or xylene) to prepare a 50% solution of varnish. The mixture in apparatus *22* is agitated, heated to 60-70 °C and held at this temperature for 3 hours.

The prepared polyphenylsiloxane varnish is filtered through nutsch filter *24* and sent into settling box *25*. Sometimes the varnish is subjected to prior modification. For this purpose, apparatus *22* is filled with a necessary amount of glyptal polymer or polybutylmethacrylate. In this case the mixture is agitated for 2 hours and sent through the nutsch filter into settling box *25*. The varnish is held there for 3 days and centrifuged. The finished varnish is collected in containers *27*.

Polyphenylsiloxane varnishes are light-brown to brown liquids without visible mechanical impurities. The volatile content is 45-85% (depending on the brand); the acidity number is 3-15 mg KOH/g.

These varnishes can be used to produce heat-resistant enamels, which are prepared directly before use by mixing varnishes with aluminum power (usually 5 weight parts of powder per 100 weight parts of varnish). This enamel can be used for parts which operate at 450-500 °C (automobile radiators, calorifers, ventilation units, etc.). Polyphenylsiloxane can be used after diluting it with toluene as a water-proofing agent to impregnate fishing nets and other materials.

The most important properties of some polyphenylsiloxanes and polyphenylsiloxane varnishes are given in Table 25.

Below let us use poly{[4-[2-(hydroxyphenyl)propyl-2]-phenyl]phenyl}siloxane as an example and illustrate the production of polyphenylsiloxanes from trialkoxyphenylsilane (in our case tributoxyphenylsilane) synthesised by the second technique, the etherification of phenyltrichlorosilane. The production comprises the following main stages: the synthesis of tributoxyphenylsilane and its partial hydrolytic condensation; the re-etherification of polybutoxyphenylsiloxane with diphenylolpropane.

Table 25. Most important properties of polyphenylsiloxane varnishes KO-85, KO-815 and polymer F-9

Characteristics	KO-85	KO-815	F-9
Solid residue, % (weight):	15-17	≥30	55
VS-1 viscosity, (2.5 mm nozzle), at 20°C, s, at least	20	12	32±6
Acidity number, mg KOH/g, not more than	3	10	15
Weight increase of capron fiber after treatment with polymer F-9, % (weight), at least	-	-	12.5

Tributoxyphenylsilane is synthesised by the etherification of phenyltrichlorosilane with butyl alcohol:

$$C_6H_5SiCl_3 + 3C_4H_9OH \xrightarrow[3HCl]{} C_6H_5Si(OC_4H_9)_3 \qquad (4.22)$$

The partial hydrolytic condensation of the obtained tributoxyphenylsilane forms polybutoxyphenylsiloxane:

$$nC_6H_5Si(OC_4H_9)_3 \xrightarrow[2nC_4H_9OH]{(n+1)H_2O} HO\left[\begin{array}{c} C_6H_5 \\ | \\ Si-O \\ | \\ OC_4H_9 \end{array}\right]_n H \qquad (4.23)$$

The re-etherification of polybutoxyphenylsiloxane with diphenylolpropane occurs thus:

$$
HO\!\!\left[\begin{array}{c} C_6H_5 \\ | \\ Si\!-\!O \\ | \\ OC_4H_9 \end{array}\right]_n\!\!\!H \;+\; n\;HO\!-\!\!\left\langle\!\!\!\bigcirc\!\!\!\right\rangle\!\!-\!\!\underset{CH_3}{\overset{CH_3}{\underset{|}{\overset{|}{C}}}}\!\!-\!\!\left\langle\!\!\!\bigcirc\!\!\!\right\rangle\!\!-\!OH \xrightarrow[\;nC_4H_9OH\;]{}
$$

$$
\xrightarrow{\hspace{2cm}} HO\!\!\left[\begin{array}{c} C_6H_5 \\ | \\ Si\!-\!O\!-\!\! \\ | \\ O\!-\!\!\left\langle\bigcirc\right\rangle\!\!-\!\underset{CH_3}{\overset{CH_3}{\underset{|}{\overset{|}{C}}}}\!\!-\!\!\left\langle\bigcirc\right\rangle\!\!-\!OH \end{array}\right]_n\!\!\!H \qquad (4.24)
$$

The production diagram of poly{[4-[2-(hydroxyphenyl)propyl-2]-phenyl]phenyl}siloxane is given in Fig. 67.

Tributoxyphenylsilane is produced when phenyltrichlorosilane interacts with butyl alcohol in the 1:3.3 mole ratio; the synthesis is carried out in enameled apparatus 7. Phenyltrichlorosilane (the 196-202 °C fraction, the chlorine content is 49-50.5%) self-flows from batch box 2 into pre-dried apparatus 1. The agitator is switched on, and anhydrous butyl alcohol is sent from batch box 3 at such speed that the temperature in the reactor does not exceed 30-35 °C. After the whole alcohol has been introduced, the reactive mixture is heated to 90-100 °C (gradually, for 7-8 hours). At this temperature and agitation the mixture is held for 7-10 hours and sampled for the chlorine content, which should not exceed 1.5%. The obtained product is cooled to 80-85 °C by sending water into the jacket of the reactor. The hydrogen chloride released in the reaction is withdrawn into a water-flushed hydroejector and in the form of weak hydrochloric acid sent to biochemical purification.

Partial hydrolytic condensation can be carried out in the same apparatus 1. It is filled with a necessary amount of water within 0.5-1 hours, when the temperature of the reaction medium is 80-85 °C. The temperature in the reactor is increased to 95-100 °C, the reactive mixture is held at agitation for 7-8 hours and cooled down to 20-25 °C. The cooled polybutoxyphenylsiloxane mixed with butyl alcohol, which has been formed in the process of hydrolytic condensation, is sent with vacuum through nutsch filter 5 into distillation tank 6. In the tank polybutoxyphenylsiloxane is "clarified" at 50-60 °C, poured into batch box 7 and sent with vacuum into reactor 8. Re-etherification is carried out in stainless steel apparatus 8 with a water vapour jacket and an anchor agitator. After loading polybutoxyphenylsiloxane, the apparatus is filled with a necessary amount of diphenylolpropane (the melting point is 154-156 °C) through a hatch. Then the temperature is gradually (at the speed of 10-15 degrees per hour) increased to 135 °C; butyl alcohol is intensively distilled. Alcohol vapours are sent into cooler *9* to condense. The condensate is collected in receptacle *10* and from there poured into collector *11*. After drying, butyl alcohol can be re-used in the synthesis.

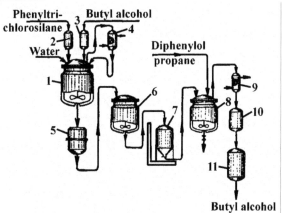

Fig. 67. Production diagram of poly{[4-[2-(hydroxyphenyl)propyl-2]-phenyl]phenyl}siloxane: *1* - etherificator; *2, 3, 7* - batch boxers; *4, 9* - coolers; *5* - nutsch filter; *6* - distillation tank; *8* - reactor; *10* - receptacle; *11* - collector

When the temperature in reactor *8* is 135 °C and the distillation of butyl alcohol stops, the temperature is raised further (to 145-160 °C) at a residual pressure of 330-200 GPa. In these conditions re-etherification is accompanied by further condensation of the polymer. The apparatus is periodically sampled to check the degree of polymer condensation by the gelatinisation time at 250 °C before stage C. If the time does not exceed 6 minutes, condensation is considered finished (the given polymerisation degree is attained). The supply of vapour into the jacket is immediately stopped and the hot polymer is loaded off.

Poly{[4-[2-(hydroxyphenyl)propyl-2]-phenyl]phenyl}siloxane is a solid dark brown mass (the volatile content is not more than 7-8%, the polymerisation time at 250 °C not more than 6 minutes). The polymer dissolves well in a mixture of toluene and tetraethoxysilane. It is used as a binding agent for plastic laminates and other composites.

Preparation of polyphenylsilsesquioxane from triacetoxyphenylsilane

The synthesis of ladder polyphenylsilsesquioxane:

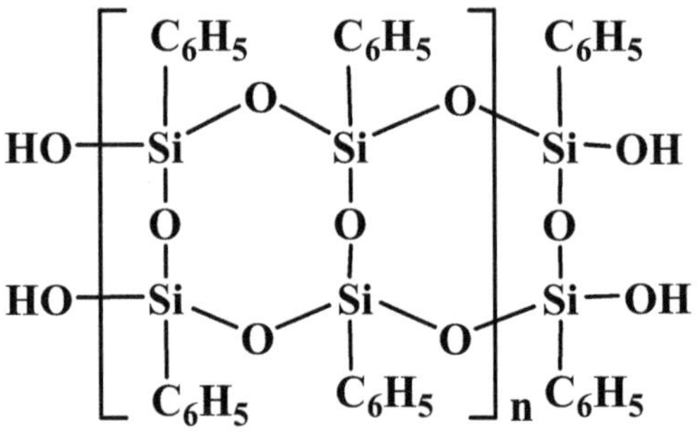

$$
\begin{bmatrix}
\text{C}_6\text{H}_5 & \text{C}_6\text{H}_5 \\
| & | \\
\text{HO}-\text{Si} & \text{Si} \\
\end{bmatrix}
$$

comprises three main stages:

1. The acetylation of phenyltrichlorosilane with potassium acetate:

$$\text{C}_6\text{H}_5\text{SiCl}_3 + 3\text{CH}_3\text{COOK} \xrightarrow[\text{3KCI}]{\text{C}_6\text{H}_5\text{CH}_3} \text{C}_6\text{H}_5\text{Si(OCOCH}_3)_3 \qquad (4.25)$$

2. The alcoholysis of the obtained triacetoxyphenylsilane with methyl alcohol:

It should be noted that alcoholysis is sometimes complicated and its direction depends on the reaction medium. Thus, if pH > 2, the reaction forms mostly trimethoxyphenylsilane, i.e. it is an exchange reaction rather than alcoholysis:

$$\text{C}_6\text{H}_5\text{Si(OCOCH}_3)_3 + 3\text{CH}_3\text{OH} \longrightarrow \text{C}_6\text{H}_5\text{Si(OH)}_3 + 3\text{CH}_3\text{COOCH}_3 \qquad (4.26)$$

If the medium is more acid, i.e. pH is 1-2, the reaction is an alcoholysis and forms mostly trihydroxyphenylsilane. However, in this case the reaction can be incomplete and apart from trihydroxyphenylsilane form a product of incomplete alcoholysis, acetoxydihydroxyphenylsilane $\text{C}_6\text{H}_5\text{Si(OH)}_2\text{OCOCH}_3$. Thus, in the preparation of polyphenylsilsesquioxane at the alcoholysis stage it is important to maintain the pH of 1 to 2.

3. The distillation of toluene, which is accompanied by the condensation of trihydroxyphenylsilane, thus forming ladder polyphenylsilsesquioxane. Since the second stage of alcoholysis forms not only trihydroxyphenylsilane, but also acetoxydihydroxyphenylsilane, it is natural that the condensation forms ladder polyphenylsilsesquioxane with a defective structure (a link-varied polymer with a certain number of linear and branched elements in the chain).

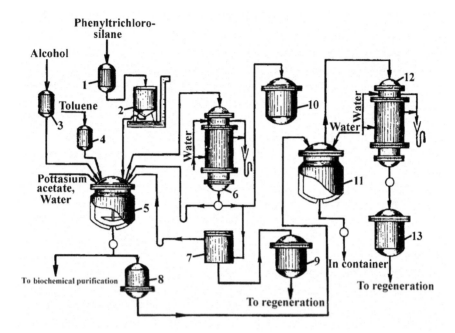

Fig. 68. Production diagram of polyphenylsilsesquioxane from triacetoxyphenylsilane: *1 - 4* - batch boxes; *5* - reactor; *6, 12* - coolers; *7* - Florentine flask; *8* - nutsch filter; *9, 10, 13* - collectors; *11* - distillation tank

The production process of polyphenylsilsesquioxane (Fig. 68) comprises the following main stages: the drying of potassium acetate; the acetylation of phenyltrichlorosilane; the alcoholysis of triacetoxyphenylsilane and flushing of silanol; the distillation of toluene and condensation of the products of alcoholysis.

Crystal potassium acetate used in the synthesis of triacetoxyphenylsilane contains crystalline hydrate water. Since acetylation should be carried out in an anhydrous medium, potassium acetate is subjected to azeotropic drying by the technique described on page 177.

The azeotropic drying of potassium acetate is carried out with toluene, fed into reactor *5* from batch box *4*. The excess of toluene is sent through Florentine flask *7* into collector *9* and from there to regeneration.

After the anhydrous potassium acetate has been dried and its exact weight has been measured, acetylation reactor *5* receives at agitation from batch box *1* through a siphon under the layer of toluene potassium acetate solution a required amount of phenyltrichlorosilane. The temperature (30-50 °C) is regulated by changing the speed of phenyltrichlorosilane supply and intensity of cooling reactor *5* (by sending water into the jacket). After the introduction of phenyltrichlorosilane the mixture in apparatus *5* heated to 45-55 °C and held at this temperature for 2 hours. The jacket of the apparatus is filled with vapour (0.3 MPa) to heat the mixture.

Because pH of the medium is so essential for the alcoholysis of triacetoxyphenylsilane, at this stage the acidity of the reactive mixture is monitored. pH should be about 1÷2. If pH >2, reactor 5 is filled with an additional quantity of phenyltrichlorosilane from batch box 2, reducing the temperature to 30-40 °C prior to that. After phenyltrichlorosilane has been added, the mixture in the apparatus is held at 45-55 °C at least for 1 more hour and then checked for acidity and appearance. If the analysis is positive (pH 1÷2), the mixture is sampled and sent to the laboratory, where triacetoxyphenylsilane is subjected to alcoholysis with methyl alcohol to determine the onset temperature for the distillation of methylacetate (58-68 °C in the reactive mixture). If at this temperature methylacetate does not distil, reactor 5 is filled from batch box 2 with an additional amount of phenyltrichlorosilane. The sample is chosen in such a way that the solid and liquid phases are approximately 1:1.

The second stage, the alcoholysis of triacetoxyphenylsilane with methyl alcohol, is also conducted in reactor 5. If the analysis is positive ((pH is 1-2) and the temperature has reached the level of methylacetate distillation, the reactor receives methyl alcohol from batch box 3. The quantity of alcohol is determined depending on the weight of anhydrous potassium acetate. The temperature during alcoholysis is increased from 35 to 55 ° and regulated by sending cold water into the jacket of the reactor and changing the speed at which alcohol is introduced. After the whole of alcohol has been introduced, it is necessary to test pH of the medium (pH 1÷2), temperature when the distillation of methylacetate begins (58-68 °C in the reactive mixture) and appearance. If the results are positive, the temperature in reactor 5 is raised to 75 °C, and the methylacetate released during alcoholysis is distilled. The jacket of the apparatus is filled with vapour (0.3 MPa) to heat the mixture. During the first 30 minutes cooler 6 operates in the "self-serving mode"; then it is switched into the direct operation mode. The vapours of distilled methylacetate enter water cooler 6. There they condense and flow into receptacle 10. The end of distillation is determined by the quantity of distilled methylacetate and the temperature not exceeding 75 °C. After distillation vapour supply is stopped, the jacket of reactor 5 is filled with water and the mixture is cooled to 40-70 °C. To dissolve the potassium chloride formed, the reactor is filled with a calculated amount of water. The reactive mixture is agitated for 15 minutes and held at 40-70 °C. The lower layer (water with dissolved potassium chloride) is sent to biochemical purification; the organic layer (toluene solution of silanol) is washed with water until its pH is 5.5÷5.7.

A vacuum is created with a vacuum pump (not shown in the diagram) in distillation tank *11*. With its help, the washed toluene solution of silanol is sent into tank *11* through filter *8* with a metal mesh. Before distilling toluene, the mixture in the tank is washed with water and analysed (pH 6.5÷7) and "clarified", i.e. released from traces of moisture by holding for 4 hours at 70-95 °C.

The settled water is poured off, and the "clarified" reactive mixture is distilled to obtain toluene at a temperature below 100 °C and a residual pressure of 800±80 GPa. The end of distillation is determined by complete stop in toluene extraction, as monitored through a run-down box and analytically. The distilled toluene enters water cooler *12*, which, like cooler *6*, is a vertical cylindrical apparatus of the vessel in a vessel type. Toluene vapours condense there and are collected in collector *13*. The distillation of toluene is accompanied by the main process of silanol condensation.

The finished product, hot polyphenylsilsesquioxane, is poured through the lower choke of tank *11* into clean dry containers. Before pouring, the product is sampled to check if it meets the technical requirements.

Polyphenylsilsesquioxane is a solid product with colour ranging from light yellow to brown. It should meet the following technical requirements:
Content, %:

of volatile substances	≤3
of silicon (equivalent to SiO_2)	43-47
Gelatinisation time at 200±2 °C, hours	1-7
pH of the aqueous extract	6-7
Solubility	
in ethyl alcohol	Complete; tolerable opalescence
in mixtures of ethyl alcohol with acetoneor ethylacetone	The same
Softening point, °C, not more than	100

Polyphenylsilsesquioxane is used as a binding agent for various textolites.

Preparation of polyphenylsilsesquioxane by the hydrolytic condensation of phenyltrichlorosilane

This production method uses phenyltrichlorosilane as raw stock; hydrolytic condensation is carried out in toluene. The production of polyphenylsilsesquioxane and polyphenylsilsesquioxane varnish comprises two main stages: the hydrolytic condensation of phenyltrichlorosilane and the polycondensation of the obtained product; the distillation of the solvent and the preparation of varnish.

The hydrolytic condensation of phenyltrichlorosilane and the polycondensation of the obtained product occur according to the following scheme:

$$nC_6H_5SiCl_3 \xrightarrow[3nHCl]{3nH_2O} n[C_6H_5Si(OH)_3] \xrightarrow[1.5nH_2O]{} [C_6H_5SiO_{1.5}]_n \qquad (4.29)$$

The basic diagram of the semicontinuous production of polyphenyl-silsesquioxane and polyphenylsilsesquioxane varnish is given in Fig. 69. Hydrolytic condensation is conducted in hydrolyser *2* with an anchor or gate agitator. The mixture of phenyltrichlorosilane with toluene is continuously introduced from weight batch box *1* and interacts with water. The volume ratio of the components should remain the same and equal 1:(3±0.2). It is advisable to carry out hydrolytic condensation at 50-70 °C. The released hydrogen chloride partially dissolves in water, and partially is withdrawn through faolite piping into a water-flushed hydroejector and in the form of weak hydrochloric acid sent into the acid drainage (to biochemical purification). The finished product at the stage of hydrolytic condensation is the silanol solution with 15-20% of the polymer.

The product is continuously sent from apparatus *2* into the middle part of separator *3* to be separated from water. Silanol from the top part of the separator is continuously sent to be flushed with water in apparatus *4*. Acid waters from the separator and flusher are withdrawn through a siphon into collector *12*. Between the separator and the flusher there is a hydroejector, which is filled with water for washing silanol, heated to 50-70 °C. Silanol from the top part of the separator is continuously sent into the middle part of separator settling box *5*. The separated water is periodically poured into receptacle *11*. By sending vapour (0.3 MPa) into the jacket or coil of the apparatus, a temperature of 70-90 °C is maintained in separator settling box *5*. Here silanol is completely separated from water and partially "clarified". The contents of the separator settling box are periodically sampled to determine the acidity of silanol. pH of the aqueous extract of silanol should be from 5 to 7.

From the top part of separator settling box *5* silanol continuously enters one of two condensator tanks *6* (the diagram shows one), switched as they fill. These enameled apparatuses with gate agitators and water vapour jackets are used for the distillation of the solvent and partial condensation of the product. If it is necessary to obtain modified varnishes, it is possible to introduce various organic additives into apparatus *6* from batch box *7* at this stage. As one of condenser tanks *6* accumulates, silanol is "clarified" before the distillation of the solvent, i.e. separated from traces of moisture by settling at 80-90 °C.

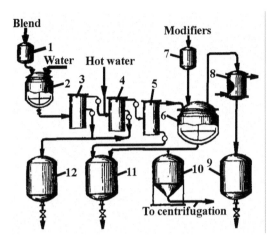

Fig. 69. Production diagram of the semicontinuous production of polyphenyl-silsesquioxane obtained by the hydrolytic condensation of phenyltrichlorosilane: *1, 7* - batch boxes; *2* - hydrolyser; *3* - separator; *4* - flusher; *5* - separator settling box; *6* - distillation condenser tank; *8* - cooler; *9, 11, 12* - collectors; *10* - settling box

The separated water is poured off into collector *11*; after "clarifying" si-lanol is analysed to determine dry residue and acidity. The distillation of the solvent is conducted at 90-115 °C and the residual pressure of 265 GPa until the solid residue content is at least 30-35%. In the process of distilla-tion silanol condenses further.

When a certain polymer content (30-35%) is observed, the distillation is stopped and the jacket of the condenser is filled with water. Varnish is cooled down to 30-40 °C and self-flows into settling box *10*. After settling and separating mechanical impurities, varnish is purified in the ultracentri-fuge and sent into the collector.

In the production of polyphenylsilsesquioxane varnish by this technique raw stock can be not only phenyltrichlorosilane, but also undistilled raw phenyl with the following characteristics:

n_D^{20}	1.165-1.332
Composition, %:	
phenyltrichlorosilane	$\geq$70-75
chlorobenzene	$\leq$2
tank residue	$\leq$20
phenyldichlorosilane	the rest

Polyphenylsilsesquioxane varnish is a transparent liquid ranging in col-our from light yellow to light brown (the viscosity is 10-20 s according to the VS-4 viscosimeter, the acidity number should not exceed 10 mg of KOH/g).

Polyphenylsilsesquioxane varnishes are used as binding agents for heat-resistant, protective and decorative enamels. To increase low adhesion, which is characteristic of silicone varnishes, enamels are often supplemented with various organic modifiers, pigments and fillers used in the varnish-and-paint industry.

There are two types of enamels, industrial and household. Industrial enamels are produced with toluene or xylene as solvents; household enamels, with petrol or butylacetate. Silicone enamels are attractive, tough, atmosphere-resistant, durable and exceptionally hard.

4.2.3. Preparation of polymethylphenylsiloxanes, polydimethylphenylsiloxanes and varnishes based on them

Polydimethylphenylsiloxanes are prepared by the hydrolytic cocondensation of di- and trifunctional organochlorosilanes and subsequent polycondensation of the obtained products.

When di- and trifunctional organochlorosilanes cocondense in an acid medium, there are favourable conditions for the interaction (according to the scheme of intermolecular condensation) of structure I cyclic compounds, formed as a result of intermolecular dehydratation of alkyl- and aryltrihydroxysilanes, with linear products of the hydrolytic condensation of diorganodichlorosilanes. As a result, as the products polycondense further, we find structure II cyclolinear polymers:

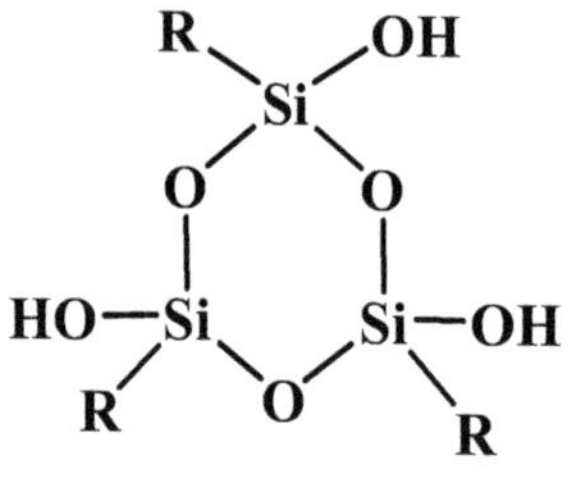

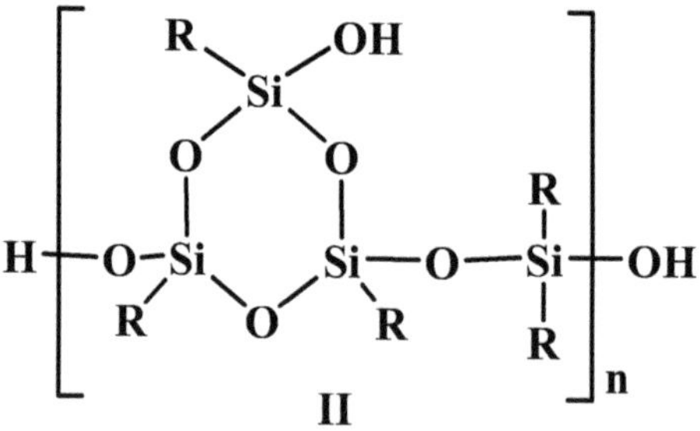

Structure II polymers are relatively elastic; when polymethylphenylsiloxanes are obtained by the hydrolytic cocondensation only of trifunctional monomers (e.g., methyl- and phenyltrichlorosilanes), there are polymers with low elasticity. Polydimethyl- and polymethylphenylsiloxanes can be modified with organic polymers (polyester, epoxy) or silicone substances, e.g. methyl(phenylaminomethyl)diethoxysilane. The modification of polydimethyl- and polymethylphenylsiloxanes improves some properties of these polymers and varnishes based on them; in particular, it considerably increases adhesion and mechanical durability of varnish films.

The production of polydimethyl- and polymethylphenylsiloxanes and varnishes based on them comprises two main stages: hydrolytic cocondensation; solvent distillation and filtering of varnish.

Raw stock: organochlorosilanes: methyltrichlorosilane, dimethyldichlorosilane, methylphenyldichlorosilane or phenyltrichlorosilane with the main substance content of at least 99.0-99.8%; toluene (the boiling point is 109.5-111 °C, $d_4^{20} = 0.865\pm0.002$); butyl alcohol (the boiling point is 115-118 °C); polyester, the product of the polycondensation of ethylene glycol with phthalic and maleic anhydride:

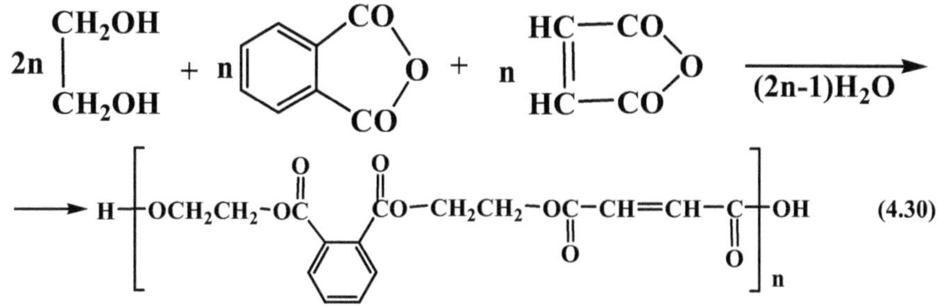

$$ (4.30) $$

modified with castor oil (the viscosity according to VS-4 at 20 °C should not exceed 20 s); epoxy polymer, the product of the interaction of epichlorohydrin with diphenylolpropane:

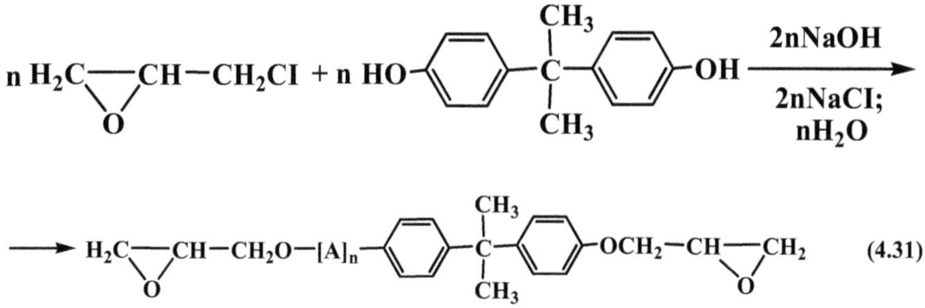

$$ (4.31) $$

where:

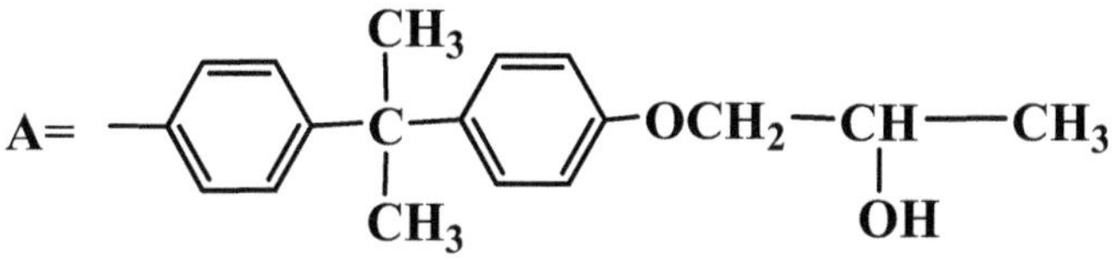

contains not more than 21% of epoxy groups, or methyl(phenylaminomethyl)diethoxysilane (the boiling point is 115-120 °C at 16 GPa).

At present polydimethylphenylsiloxane and polymethylphenylsiloxane varnishes are produced both by the periodic and by the continuous technique.

The periodic technique has significant drawbacks:

1. low equipment capacity (1 tonne of varnishes per year requires about 23 m^3 of industrial space);
2. low labour productivity (approximately 4 tonnes of varnish per year per one worker);
3. the multi-staged process, which requires mutliple loading and reloading of semiproducts and finished varnish from one apparatus into another. This often disrupts the technological process, fouls up the products and pollutes the air in the worksite;
4. the difficulty in automating the process;
5. high cost of the obtained polyorganosiloxanes and varnishes based on them

However, in some cases (for small-scale production) it is economically advisable to use the periodic technique.

The semicontinuous and continuous technique for the production of polydimethylphenylsiloxane and polymethylphenylsiloxane varnishes have no such drawbacks.

Some brands of polydimethylphenylsiloxane varnishes can be produced from raw phenyl rather than phenyltrichlorosilane, a nonrectified mixture of phenylchlorosilanes (the phenyltrichlorosilane content is at least 77-80%, the chlorine content is 43-47%), using the semicontinuous technique.

In the semicontinuous production process of polydimethylphenylsiloxane varnish (Fig. 70) agitator 5 is filled with raw phenyl and dimethyldichlorosilane from the batch boxes. The mixture is agitated for 0.5-1 hours and sampled for chloroions. Then, the apparatus is filled with a necessary amount of toluene and the reactive mixture is agitated for 0.5-1 hour.

Continuous hydrolytic cocondensation is conducted in hydrolyser 8, which has a water vapour jacket, an anchor agitator and siphons for introducting the reactive mixture. From agitator 5 the reactive mixture is sent by nitrogen flow (the pressure does not exceed 0.07 MPa) into weight batch box 6 and from there continuously fed into the hydrolyser by two siphons. At agitation the hydrolyser is also continuously filled with water, pre-heated to 50-90 °C in heater 7 by sending vapour (0.3 MPa) into the space between the pipes. It is necessary to observe a constant ratio of the reactive mixture and water introduced, which should be (1:3)±0.2 (by volume).

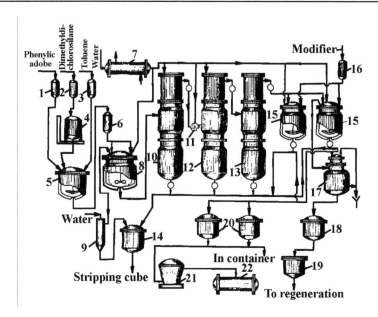

Fig. 710. Semicontinuous production diagram of polydimethyl(phenyl)siloxane varnish: *1 -4, 6, 16* - batch boxes; *5* - agitator; *7* - heater; *8* - hydrolyser; *9, 11*- hydroejectors; *10, 13* - separators; *12* - flusher; *14, 18* - collectors; *15* - polycondensation apparatuses; *17* - cooler; *19, 20* - settling boxes; *21* - ultracentrifuge; *22* - container

The temperature of hydrolytic cocondensation (50-70 °C) is regulated by changing the temperature of water and the speed at which the reactive mixture is introduced. The released hydrogen chloride partially dissolves in excess water, and partially is withdrawn through faolite piping into water-flushed hydroejector *9* and in the form of weak hydrochloric acid sent into collector *14*.

The mixture of the product of hydrolytic cocondensation (silanol) and water from the bottom part of the hydrolyser continuously enters the middle part of separator *10*, where silanol and water separate. The aqueous layer from the separator is analysed (to determine its acidity). Silanol from the top part of the separator is continuously sent into the middle part of flusher *12*. There is hydroejector *11* in this direction, which is filled with water from heater *7* for washing silanol. The quantity of water sent to flushing can vary depending on pH of silanol.

The separator and flusher are steel enameled cylindrical apparatuses with flat lids, spherical bottoms and siphons, through which water is continuously withdrawn into collector *14*. Additional flushing of silanol is carried out in apparatuses *10* and *12* by sending water into diffusers mounted in the top section of the apparatuses. Silanol from the top part of flusher *12* continuously enters the middle part of separator settling box *13*, which is a cylindrical apparatus with a flat lid, spherical bottom and water vapour jacket. By sending vapour (0.3 MPa) into the jacket of the apparatus, a temperature of 70-90 °C is maintained in separator settling box *13*. Silanol is completely separated from water, partially clarified and from the top part of the separator settling box continuously enters one of the two switching (as they fill) apparatuses *15*. The water separated from silanol is poured out from the lower part of separator settling box *13* into collector *14*. The contents of separator settling box *13* are periodically sampled to determine acidity. pH of the aqueous extract of silanol should be from 5 to 7 according to the universal indicator.

Further polycondensation of the the product of hydrolytic cocondensation (silanol) is carried out as toluene is distilled in apparatuses *15*, which are steel enameled apparatuses with anchor agitators and water vapour jackets. As one of apparatuses 15 is filled, silanol is "clarified" before the distillation of toluene, i.e. separated from traces of moisture, by settling at 70-90 °C. The remaining water is also poured into receptacle *14*.

The liquid from collector 14 is loaded through a siphon, as it accumulates, into a distillation tank (not shown in the diagram), where it is flushed, "clarified" and tempered.

When pH of silanol is lower than *5*, there is a possibility of additional flushing in apparatuses *15* with water sent from heater *7*.

After "clarifying" silanol is sampled to determine the solid residue content, acidity and density. If necessary, before distilling the solvent, apparatus *15* is filled with organic or silicone modifier from batch box *16*. After the modifier has been loaded, the contents of the apparatus are agitated for about 0.5 hours and toluene is distilled at 90-115 °C and a residual pressure of 600-670 GPa until the solid residue content is as required.

The distilled toluene vapours enter water cooler 17, a steel enameled vertical heat exchanger of the "vessel in a vessel" type with the heat exchange surface of 4 m^2. There toluene condenses and is collected in collector *18*, from where it self-flows into settling box *19*. The settled toluene is sent to regeneration and re-used in manufacture.

Whether the toluene distillation and product polycondensation is finished, is determined by viscosity or gelatinisation time, and by solid residue. The jacket of apparatus *15* is filled with water to cool the varnish to 40 °C; the cooled varnish self-flows into settling boxes *20*. After settling, the varnish is sent to ultracentrifuge *21*, and the waste accumulated in the bottom part of the settling boxes is periodically loaded off into containers and used to paint factory equipment.

The final purification of the varnish from mechanical impurities and gel takes place in the ultracentrifuge, from where the purified varnish is pumped into container *22*.

For some brands of polydimethylphenylsiloxane varnishes (KO-921, KO-922, KO-923) the polycondensation stage is replaced with polymerisation. In this case the polymerisation of the product of hydrolytic cocondensation is carried out in a solution of potassium hydroxide.

The Si—O bond in cyclosiloxane fragments is broken under the influence of the nucleophylic agent. First, the nucleophylic agent (the hydroxyl group) is coordinationally bonded with the silicon atom in the chain:

$$(4.32)$$

The complex is instable; the siloxane bond breaks, forming an active centre:

The chain grows until a balance is achieved, finally forming a branched polymer.

The polymerisation is carried out in a reactor with an agitator and a water vapour jacket in the presence of a catalyst (10-12% alcohol solution of potassium hydroxide) at 20+5 °C. The polymerisation is continued until the product attains a certain viscosity; after that, the reactor is loaded with a required amount of dimethyldichlorosilane to break the chain, with an addition of toluene to dilute the varnish. The polymer is treated with dimethyldichlorosilane for 3-5 hours at agitation; the end of the stage is monitored by the universal indicator. The varnish is filtered to eliminate mechanical impurities and potassium chloride and sent to repeated toluene distillation, which is continued until the varnish attains the necessary viscosity and polymer content.

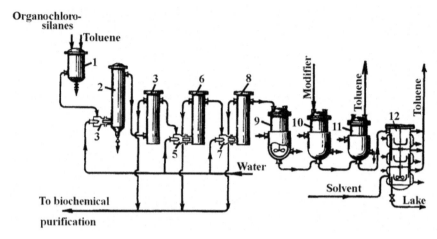

Fig. 71. Production diagram of polydimethylphenylsiloxane and polymethylphen-ylsiloxane varnishes by the continuous technique: *1* - weight batch box; *2* - tower; *3, 5, 7* - hydroejectors; *4, 6, 8* - Florentine flasks; *9* - container; *10* - agitator; *11* - distillation tank; *12* - condensation apparatus

Polydimethylphenylsiloxane and polymethylphenylsiloxane varnishes, which are in ever greater demand, are also produced by the continuous technique (Fig. 71).

A toluene solution of organochlorosilane mixture is sent from weight batch box *7* into jet mixer (hydroejector) *3* with a certain amount of water. The consumption of the components is monitored by rotameters. The reaction of hydrolytic cocondensation takes place in the mixing chamber of hydroejector *3*. To complete the hydrolytic cocondensation, the reactive mixture is sent into tower *2*, from where the mixture is poured into Florentine flask *4*. There the products of hydrolytic cocondensation and hydrochloric acid split.

The acid is sent to biochemical purification, and the hydrolysate is subjected to two-stage flushing with water in hydroejectors *5* and *7*. The hydrolysate is flushed until pH is 5÷6 and separated from flush waters in Florentine flasks *6* and *8* and in container *9*.

From container *9* the flushed hydrolysate is sent to partially distil the solvent into tank *11*; to produce modified polymethylphenylsiloxane varnishes, it is first mixed with polyester or epoxy polymer in apparatus *10*, and then sent into the tank. From the tank, the hydrolysate is sent to condensation into three-sectioned apparatus *12*. In the first section the solvent is additionally distilled and the product of the hydrolytic cocondensation is partially condensed; in the second section it is further condensed at 125-

180 °C (depending on the varnish brand); in the third section the polymer is dissolved to prepare a varnish of required concentration.

Tower apparatus 12 is separated with cross dividers into three sections. Each section has an anchor agitator, mounted on the common shaft, and a vapour jacket. Liquid travels from one section into another through inner piping. This construction allows for continuous condensation, excluding the danger of sudden gel formation. All the main apparatuses in the continuous production of varnishes are cascaded; thus, the main product can self-flow throughout the whole technological process. Comparing the continuous process of varnish production with the periodic process, we can note the following advantages of the former:

1. equipment capacity increases more than tenfold; the production of varnishes by the continuous technique raises the efficiency above 100 kg/h, which corresponds to approximately 2 m^3 of industrial space per 1 tonne of varnish produced in a year.
2. labour productivity increases tenfold: one technological thread can be serviced by one man per shift; thus, the production of varnish per one worker is more than 40 tonnes a year.
3. cost coefficients are greatly reduced due to smaller losses of raw stock and semiproducts;
4. the continuous process is easy to control and therefore can be fully automated.

All these advantages of the continuous process help to decrease the manufacturing costs of varnish.

Polydimethylphenylsiloxane and polymethylphenylsiloxane varnishes are transparent liquids with colour ranging from light yellow to light brown. They can be dissolved in toluene, benzene, xylene and other non-polar organic solvents; but are not soluble in water and alcohols.

The properties of these varnishes first of all depend on the parent monomers. As we have mentioned above, the production of varnishes can be based on both difunctional monomers (e.g. dimethyl- or methyl-phenyldichlorosilane) and trifunctional monomers (methyltrichlorosilane, phenyltrichlorosilane, etc). As the content of difunctional monomer in a mixture of organochlorosilanes increases, the varnish film becomes more elastic; however, its hardness decreases and the time of transition into the nonmeltable and nonsoluble state is reduced. Increasing the trifunctional monomer content (phenyltrichlorosilane) in the mixture increases thermal stability and gloss of the film); at the same time, the time of transition into the nonmeltable and nonsoluble state is greatly increased in comparison with varnishes containing methylsisequioxane elements.

Polydimethylphenylsiloxane and polymethylphenylsiloxane varnishes are widely used in the electrotechnical industry. Some of them (e.g. varnishes based on methyl- and phenyltrichlorosilanes) can be used in certain conditions as binding agents for pressure materials; however, these varnishes enter the nonmeltable and nonsoluble state only after prolonged heating and are impractical for the production of plastic laminates.

The most important properties of some insulating polydimethylphenylsiloxane and polymethylphenylsiloxane varnishes are given in Table 26.

Table 26. . Important properties of some insulating polydimethylphenylsiloxane and polymethylphenylsiloxane varnishes

Characteristics	Nonmodified varnishes					
	KO-926	K-43	KO-08	KO-921	KO-922	KO-923
Nonvolatile content, %:	50±2	-	30-35	50±2	50±2	50-55
VS-4 viscosity at 20°C, s	14-22	-	16 (VS-1, (2.5 mm nozzle)	17-27	17-27	2-3* (10% solution)
Drying time of varnish film on copper at 200 °C, minutes, at least	30 (at 110^{0}C)	-	60 (at 100^{0}C)	15 (at 20^{0}C)	60 (at 20^{0}C)	30
Thermal elasticity of varnish film on copper at 200 °C, hours, at least	-	140	-	75	200	100 (at 220^{0}C)
Electric strength of varnish film, KV/mm, at least at 20±2 ^{0}Cat 200±2 ^{0}C	- -	- 50	80-120 30-50	70 35 (at 180^{0}C)	70 35 (at 180^{0}C)	50 25
After 24 hours at 20^{0}C and 95-98% relative humidity	-	-	60-80	35	35	25

Table 26. (cont.)

Specific volume electric resistance of varnish film, Ohm·cm, at least at 20±2 °C, at 200±2 °C, After 24 hours at 20°C and 95-98% relative humidity	- -	- -	$4\cdot10^{14}$ - $1\cdot10^{14}$	$1\cdot10^{14}$ $1\cdot10^{12}$ $1\cdot10^{13}$	$1\cdot10^{14}$ $1\cdot10^{12}$ $1\cdot10^{13}$	$4\cdot10^{13}$ $1\cdot10^{11}$ $1\cdot10^{11}$

*In relative units.

Characteristics	Modified varnishes						
	KO-075	KO-915	KO-916	KO-917	KO-918	KO-919	KO-945
Nonvolatile content, %	34-40	60	60	65±5	65	65±5	70±1
VS-4 viscosity at 20°C	12 (VS-1, (2.5 mm nozzle)	30-70	45-65	100-180	75-125	70-170	50-125
Drying time of varnish film on copper at 200 °C, minutes, at least	-	90	15	90	90 (for aluminum)	60	24
Electric strength of varnish film, KV/mm, at least at 20±2 °C, at 200±2 °C, After 24 hours at 20°C and 95-98% relative humidity	- - -	70 28 50	70 30 25	50 25 25	50 25 30	70 35 35	70 35 (at 180°C) 40

Table 26. (cont.)

Specific volume electric resistance of varnish film, Ohm·cm, at least							
at 20±2 °C,	$1 \cdot 10^{13}$	$1 \cdot 10^{14}$	$1 \cdot 10^{13}$	$1 \cdot 10^{14}$	$1 \cdot 10^{14}$	$1 \cdot 10^{14}$	$1 \cdot 10^{14}$
at 200±2 °C,	$1 \cdot 10^{11}$	-	$1 \cdot 10^{11}$	$1 \cdot 10^{12}$	$1 \cdot 10^{12}$	$1 \cdot 10^{12}$	$1 \cdot 10^{12}$
After 24 hours at 20°C and 95-98% relative humidity	$1 \cdot 10^{11}$	$1 \cdot 10^{13}$	$1 \cdot 10^{12}$	$1 \cdot 10^{13}$	$1 \cdot 10^{13}$	$1 \cdot 10^{13}$	$1 \cdot 10^{13}$

Polymethyl(phenyl)silsesquioxane   and   polymethyl(phenyl)-silsesquioxane varnish can also be obtained from a mixture of triacetoxy-methylsilanes and triacetoxyphenylsilanes.

4.2.4. Production of polymethyl(phenyl)silsesquioxane and polymethyl(phenyl)silsesquioxane varnish from triacetoxymethyl- and triacetoxyphenylsilanes

The synthesis of laddeer polymethyl(phenyl)silsesquioxane

comprises three main stages:

1. The acetylation of methyl- and phenyltrichlorosilanes with potassium acetate:

$$\text{Toluene}$$
$$CH_3SiCl_3 + C_6H_5SiCl_3 + 6CH_3COOK \xrightarrow{\hspace{2cm}}$$
$$\xrightarrow{\hspace{1cm}} CH_3Si(OCOCH_3)_3 + C_6H_5Si(OCOCH_3)_3 + 6KCl \hspace{2cm} (4.33)$$

2. The alcoholysis of silicon acetoxyderivatives with methyl alcohol to form methyl- and phenyltrihydroxysilanes:

$$CH_3Si(OCOCH_3)_3 + C_6H_5Si(OCOCH_3)_3 + 6CH_3OH \xrightarrow{\hspace{2cm}}$$
$$\xrightarrow{\hspace{1cm}} CH_3Si(OH)_3 + C_6H_5Si(OH)_3 + 6CH_3OCOCH_3 \hspace{2cm} (4.34)$$

The direction of the reaction of alcoholysis depends on the medium. An acid medium forms trihydroxysilanes, whereas a neutral medium is favourable for an exchange reaction, which forms organotrimethoxysilanes:

$$CH_3Si(OCOCH_3)_3 + C_6H_5Si(OCOCH_3)_3 + 6CH_3OH \longrightarrow$$

$$\longrightarrow CH_3Si(OH)_3 + C_6H_5Si(OH)_3 + 6CH_3COOH \qquad (4.35)$$

or products of incomplete alcoholysis like:

$$CH_3Si(OH)_2OCOCH_3, \qquad CH_3Si(OH)(OCOCH_3)_2$$

Thus, in the preparation of polymethyl(phenyl)silsesquioxane at the alcoholysis stage it is important to maintain the pH of the medium approximately at 2.

3. The condensation of trihydroxymethyl- and trihydroxyphenylsilanes takes place during the distillation of toluene, in which the synthesis occurs. As a result, we obtain ladder polymethyl(phenyl)silsesquioxane. Since there is a certain amount of the products of incomplete alcoholysis formed at the stage of alcoholysis, the process forms a ladder copolymer, i.e. a copolymer with a certain number of branched and linear elements in the chain.

Polymethyl(phenyl)silsesquioxane production comprises the following stages: the drying of potassium acetate in toluene; the preparation of the mixture of methyl- and phenyltrichlorosilanes and its acetylation; alcoholysis; the flushing and filtration of silanol mixture; the distillation of toluene and condensation of alcoholysis products; the preparation of varnish, its settling and centrifuging.

Raw stock: methyltrichlorosilane (MTCS)[at least 95% (vol.) of the 65-67 °C fraction, 69.8-71.15% of chlorine]; phenyltrichlorosilane (PTCS) [at least 95% (vol.) of the 196-202 °C fraction, 49-50.5% of chlorine]; toluene [at least 98% (vol.) of the 109.5-111 °C fraction, $d_4^{20} = 0.865\pm0.003$]; poisonous methanol [at least 99% (vol.) of the 64.7-67 °C fraction, $d^{20} = 0.79 \div 0.798$]; ethyl alcohol [at least 99.6% (vol.)].

Crystal potassium acetate used in the synthesis of acetoxysilanes, as mentioned above (see page 177), contains crystalline hydrate water. Since acetylation should be carried out in an anhydrous medium, potassium acetate is subjected to azeotropic drying by the technique described on page 177.

The azeotropic drying of potassium acetate is carried out directly in synthesis reactor 1 (Fig. 72), which is loaded through a hatch with a necessary amount of potassium acetate and with toluene from batch box 5. The drying is conducted at 100-110 °C. After drying and determining the exact weight of anhydrous potassium acetate by the weight difference of the loaded product and distilled water, the process of acetylation is started. A mixture of methyl- and phenyltrichlorosilanes is prepared in weight batch box *4*. At 30-35 °C the organochlorosilane mixture is sent into reactor *1* from weight batch box *4* at such speed that the temperature in the reactor does not exceed 35 °C. After the whole mixture has been introduced, the reactive mixture is kept at 45-55 °C for 2-2.5 hours. After that, the pH of the medium is checked.

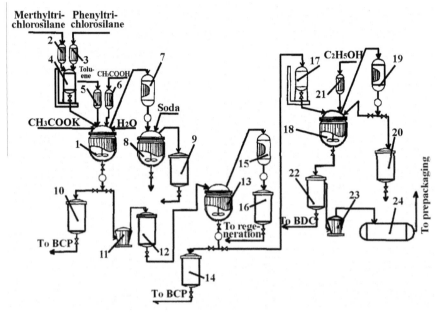

Fig. 11. Production diagram of polymethyl(phenyl)silsesquioxane from triacetoxyderivatives of methyl- and phenylsilanes: *1* - synthesis reactor; *2, 3, 5, 6, 21* - batch boxes; *4, 17* - weight batch boxes; *7, 15, 19* - coolers; *8* - neutraliser; *9, 12, 14, 16, 20* - collectors; *10, 22* - settling boxes; *11, 23* - ultracentrifuges; *13* - distillation tank; *18* - condensation apparatus; *24* - container

If the medium is neutral, the reactive mixture in reactor I is cooled by sending water into the jacket of the apparauts to 30-35 °C and the reactor receives an additional amount of organochlorosilane mixture.

When the medium becomes acid, at 45-48 °C reactor *1* receives a calculated amount of poisonous methanol from batch box *6* at such speed that the temperature in the reactor does not exceed 50°C. Then, at 67-75 °C the released methylacetate is sent from reactor *1* through cooler *7* into neutraliser *8* with an agitator.

In case if at 67-75 °C methylacetate is not distilled, reactor *1* receives from weight batch box *4* an additional quantity of organochlorosilanes to create an acid medium in the reactive mixture; after that the reactive mixture is heated to 67 °C and methylacetate is distilled.

In neutraliser 8 methylacetate is neutralised by mixing it with dry soda loaded through a hatch. After neutralisation methylacetate is settled in apparatus 8, flows through a siphon out of the top part of the neutraliser into collector 9, is poured into containers and sent to the consumer.

After distilling methylacetate, reactor 1 is loaded with a required amount of water for complete dissolution of KCl and the reactive mixture is agitated for 5-10 minutes. The agitator is stopped and the reactive mixture is settled for 1.5-2 hours. The lower acid aqueous layer is poured through a run-down box into settling box *10* and sent to biochemical purification; the organic layer (silane mixture) is washed with water until it gives a neutral reaction. The neutral solution of silanols self-flows from reactor *1* through a run-down box to centrifuging into the lower part of ultracentrifuge *11*, where mechanical impurities are deposited. The clarified silanol from the ultracentrifuge enters collector *12* and from there is sent to the distillation of the solvent into distillation tank *13*. Silanol in tank *13* is heated with vapour sent into the jacket of the tank and settled to separate the entrained moisture. The settled water is poured through the lower drain of the tank into collector *14* and is sent to biochemical purification. The distillation of toluene from tank *13* is carried out at 75-85 °C and a residual pressure of 800-500 GPa, until the condensate no longer enters the run-down box below cooler *15*. Toluene is distilled at the speed of 50-60 l/hour. After partial distillation of toluene the resin content should be 45-50%. The distilled toluene enters collector 16 and is sent to regeneratiion.

The mixture of organohydroxysilanes (silanols), distilled to a 45-50% concentration, is vacuum-pressurised from tank *13* into weight batch box *17* and sent to condense in condenser *18*, which has an agitator and a water vapour jacket. The condensation of the silanol mixture is carried out in vacuum under 285-340 GPa at 90-95 °C, until toluene no longer enters the run-down box on the drain line from cooler *19*. The toluene enters collector 20 and is sent to regeneration.

To obtain resin, the hot product of condensation is poured through the lower drain of condenser *18*; to obtain varnish, the resin in condenser *18* is dissolved with ethyl alcohol, which self-flows into the apparatus from batch box *21*. While the resin is dissolved, cooler *19* operates in the inverse mode. The obtained varnish is loaded by vacuum into settling box 22, where it is settled at ambient temperature for a long time (24-48 hours) to separate mechanical impurities. There is also a possibility for additional centrifuging in ultracentrifuge 23 for complete elimination of mechanical impurities, as well as "clarification" of the varnish. The finished polymethyl(phenyl)silsesquioxane varnish is sent from the centrifuge into container *24* and packaged.

Polymethyl(phenyl)silsesquioxane resin is a solid mass ranging in colour from light brown to brown; polymethyl(phenyl)silsesquioxane varnish is a 50-60% solution of the resin in ethyl alcohol.

They should meet the following technical requirements:

	Varnish	Resin
Appearance	Transparent light brown liquid. Opalescence is tolerated.	Solid melt, from light brown to brown.
Content, %:		
of solid residue .(resin content)	50-60	-
of SiO_2	25.5-33	51-55
of volatile substances	-	≤ 3
Gelatinisation time before stage B, at 200 °C, minutes	60-180	Not more than 180
pH of the aqueous extract	5-7	5-7
Compatibility with bakelite varnish	Complete	Complete
Softening point, °C	-	60-85

Polymethyl(phenyl)silsesquioxane resin and varnish based on it are used to manufacture pressure compositions and special parts.

4.2.5. Preparation of polymethylphenylsiloxanes with active hydrogen atoms and vinyl groups at the silicon atom

Polymethylphenylsiloxanes with active hydrogen atoms and vinyl groups at the silicon atom are industrially manufactured without solvents and are of great practical interest, since they do not release volatile substances when processed into goods. The process occurs in the presence of a catalyst (H_2PtCl_6 or dimethylformamide) according to the mechanism of hydride attachment.

$$\equiv Si-H + CH_2=CH\text{-}Si\equiv \longrightarrow \equiv Si-CH_2\text{-}CH_2\text{-}Si\equiv \qquad (4.36)$$

That is why articles made from these polymethylphenylsiloxanes do not shrink.

At present industry produces two types of these polymers:

1. the two-component polymers PFVG-1 and PFVG-2. PFVG-1 consists of silanol C and silanol D, whereas PFVG-2 consists of products 136-13 and 139-14 (or PFGS), Silanol C and product 136-13 are polyorganosi-

loxanes with vinyl groups at the silicon atom; silanol D, products 139-14 and PFGS are polyorganosiloxanes with active hydrogen atoms at the silicon atoms.

2. one-component polymers MFVG-3, MFVG-5 and MFVG-35, which contain both active hydrogen atoms and vinyl groups at the silicon atoms. Because the main chains of the polymers also contain difunctional elements (dimethylsiloxy-, methylphenylsiloxy-, vinylmethylsiloxy-), these polymers are cyclolinear and therefore elastic.

Preparation of polymethylphenylsiloxanes with active hydrogen atoms and vinyl groups at the silicon atom comprises the following stages: the hydrolytic cocondensation of organochlorosilanes; the flushing and filtering (during the distillation the products of hydrolytic cocondensation polycondense). In the production of two-component polymers after the distillation of the solvent there is an additional stage when the components are mixed.

Raw stock: dimethyldichlorosilane (at least 99% of the main fraction); methylphenyldichlorosilane (at least 98% of the main fraction); diphenyldichlorosilane (at least 98% of the main fraction); methyldichlorosilane (at least 98% of the main fraction); vinylmethyldichlorosilane (at least 99.6% of the main fraction); phenyldichlorosilane (at least 99% of the main fraction); phenyltrichlorosilane (at least 99% of the main fraction), toluene (coal or oil).

The preparation of polymethylphenylsiloxanes with active hydrogen atoms and vinyl groups at the silicon atom (Fig, 73) begins with the preparation of the reactive mixtures in enameled mixer *8* with an agitator and a water vapour jacket. Different polymers require different organochlorosilanes in different mole ratios.

It is necessary to batch the mole ratios of organochlorosilanes exactly; therefore, they are in turns sent from batch boxes *2, 3, 4, 5* and *6* into batch box *7* and after weighing self-flow into mixer *8*. The mixer is also filled with a calculated amount of toluene from batch box *1*. After that the reactive mixture is agitated with cooling for appoximately 30 minutes, after which the chloroion content is determined. The prepared mixture of organochlorosilanes and toluene is sent by nitrogen flow ($\leq$0.07 MPa) into weight batch box *9*.

The process of hydrolytic cocondensation is carried out in hydrolyser *10*, which is (just as vacuum distillation tank *13*) an enameled apparatus with an agitator and a water vapour jacket.

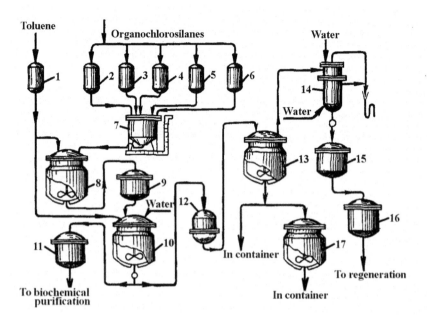

Fig. 12. Production diagram of polymethyl(phenyl)siloxanes with active hydrogen atoms and vinyl groups at the silicon atom: *1 - 7, 9* - batch boxes; *8, 17* - mixers; *10* - hydrolyser; *11, 15* - collectors; *12* - nutsch filter; *13* - vacuum distillation tank; *14* - cooler; *16* - settling box

If cocondensation is conducted to obtain two-component polymers and one-component polymer MFVG, the hydrolyser is cooled with salt solution (-15 °C); to obtain one-component polymers MFVG-5 and MFVG-35, the jacket of the hydrolyser is filled with vapour (0.3 MPa). The hydrolyser is loaded with a calculated amount of water and with a required amount of toluene from batch box 1; the reactive mixture is sent from weight batch box 9 at such speed that in the production of silanol C, products 136-13, PFGS and MFVG-3 the temperature in the apparatus is 10-12 °C; in the production of silanol D and product 139-14 the temperature is maintained at 2-5 °C, and in the production of products MFVG-5 and MFVG-35, not lower than 40 °C. After the whole of reactive mixture has been introduced, it is agitated for 0.5-1.5 hours at 10-12 °C (for the production of two-component polymers and MFVG-3) or at 40-45 °C (for the production of MFVG-5 and MFVG-35). The agitator is stopped and the reactive mixture is held for 1-1.5 hours. After settling the lower aqueous layer is analysed for acidity, which should be 110-300 g of HCl per 1 l (depending on the polymer brand), poured through a run-down box into collector *11* and sent to biochemical purification. The top layer, the toluene solution of the product of hydrolytic cocondensation, is agitated and flushed with water several times until it gives a neutral reaction. After the final flushing it is very thoroughly settled and the flush water is poured off so that it does not remain in the solution.

If the reactive mixture splits inadequately or slowly, while the product of cocondensation is flushed, the hydrolyser receives an additional amount of toluene from batch box *1* at each stage of flushing.

Most of the hydrogen chloride released during hydrolytic cocondensation dissolves in the excess of water sent to the cocondensation; the unabsorbed hydrogen chloride is withdrawn from the hydrolyser into a faolite water-flushed hydroejector (not shown in the diagram) and in the form of weak hydrochloric acid sent to biochemical purification.

The neutral product in the hydrolyser is sampled to determine the solid residue content (10-15%), filtered through portable filter *12* with a metal mesh and poured into vacuum distillation tank *13* to distil toluene. The distillation is carried out in two stages: the first takes place under atmospheric pressure to 50-55% concentration of the product; the second, at a residual pressure of 860-870 GPa and 80-100 °C until the solid residue content is not less than 97%. Toluene vapours are sent into water cooler *14* to condense. Toluene is poured into collector *15*. Raw toluene with impurities of volatile products and water self-flows from receptacle into settling box *16*, where it is separated from them. Then it is sent to regeneration and re-used in manufacture.

Table 27. Technical requirements to polymethylphenylsiloxanes with active hydrogen atoms and vinyl groups at the silicon atom

Char-acteris-tics	Two-component polymers					One-component poly-mers		
	Sil anol-D	Si-lanol-D	13 9-14	1 36-13	PF GS	MF VG-3	MF VG-35	MF VG-5
Ap-pearance	Viscous resinous products with colour ranging from light yellow to yellow with opalescence							
Solid residue, %	At least 97					At least 97		
VS-1 viscos-ity, (5.4 mm noz-zle), min	3-20	18-60	-	-	-	2-3	1-3	3-10

Table 27. (cont.)

Ac-tive hydrogen content, H_{act}, %	-	0.55 -0.6	0.08-0.1	-	0.0 7-0.09	0.13 -0.18	0.12 -0.16	0.13 -0.17
Bromine value, g Br_2/ 100 g	35 -41	-	-	1 7-24	-	21-29	19-26	20-27
So-lidification degree, % (weight)	-	-	-	-	-	80	80	80

After the distillation of toluene, the product of polycondensation is cooled in tank *13* to 25-30 °C, analysed to check if it meets the technical requirements and loaded into containers; the obtained two-component polymers are send (in equimolar quantities) into mixer *17*, where they are agitated at 45-50 °C for 8 hours.

Two-component PFVG-1 is obtained by mixing silanol C and silanol D, whereas PFVG-2 is prepared by mixing products 136-13 and 139-14 (or PFGS),

Polymethylphenylsiloxanes with active hydrogen atoms and vinyl groups at the silicon atom should meet the technical requirements given in Table 27.

Polymethylphenylsiloxanes with active hydrogen atoms and vinyl groups at the silicon atom are used to manufacture glass fibres (by the contact molding technique), glue compositions and filling compounds.

4.2.6. Preparation of polyphenyldiethylsiloxanes and polyphenyldiethylsiloxane varnishes

Polyphenyldiethylsiloxanes are obtained by the hydrolytic cocondensation of phenyl- and ethylethoxysilanes synthesised by the Grignard technique. They are high-molecular substances mostly of the following structure:

$$\left[\begin{array}{c} \underset{\underset{C_2H_5}{|}}{H}{-}\underset{C_2H_5}{\overset{C_2H_5}{\underset{|}{\overset{|}{Si}}}}\cdots \end{array}\right]_n$$

By dissolving polyphenyldiethylsiloxanes in toluene, petrol or a mixture of petrol and turpentine, one can obtain polyphenyldiethylsiloxane varnishes, which differ not only by the type of the solvent, but also by the ratio of phenylsiloxy and ethylsiloxy elements in the polymer.

The production of polyphenyldiethylsiloxane varnishes comprises three main stages: the synthesis of phenyl- and ethylethoxysilanes; the hydrolytic cocondensation of phenyl- and ethylethoxysilanes; the distillation of the solvent and the preparation of varnish.

As mentioned above, phenylethoxysilanes are synthesised by the Grignard technique with the ratio $Mg:(C_2H_5O)_4Si$ of 1:1. First chlorobenzene interacts with metallic magnesium to form phenyl magnesium chloride: Then, tetraethoxysilane is phenylated:

$$C_6H_5MgCI + Si(OC_2H_5)_4 \longrightarrow C_6H_5Si(OC_2H_5)_3 + Mg(OC_2H_5)CI \quad (4.37)$$

However, the reaction between phenyl magnesium chloride and tetra-ethoxysilane is a complicated process, which forms not only phenyltriethoxysilane, but also diphenyldiethoxysilane and triphenylethoxysilane.

$$C_6H_5Si(OC_2H_5)_3 \xrightarrow[Mg(OC_2H_5)CI]{C_6H_5MgCI} (C_6H_5)_2Si(OC_2H_5)_2 \xrightarrow[Mg(OC_2H_5)CI]{C_6H_5MgCI} (C_6H_5)_3SiOC_2H_5 \quad (4.38)$$

Apart from phenylethoxysilanes, it can also form diphenyl:

$$2C_6H_5CI + Mg \xrightarrow{\quad} (C_6H_5)_2 \quad (4.39)$$
$$MgCI_2$$

Ethylethoxysilanes are also synthesised by the Grignard technique, but with the ratio of magnesium and ethylethoxysilane of 1.7:1 to 2:1. First, ethylchloride interacts with magnesium. Then, the reaction of ethyl magnesium chloride and tetraethoxysilane yields a mixture of ethyltriethoxysilane, diethyldiethoxysilane and triethylethoxysilane.

$$Si(OC_2H_5)_4 \xrightarrow[Mg(OC_2H_5)CI]{C_2H_5MgCI} C_2H_5Si(OC_2H_5)_3 \xrightarrow[Mg(OC_2H_5)CI]{C_2H_5MgCI}$$

$$\longrightarrow (C_2H_5)_2Si(OC_2H_5)_2 \xrightarrow[Mg(OC_2H_5)Cl]{C_2H_5MgCl} (C_6H_5)_3SiOC_2H_5 \qquad (4.40)$$

The hydrolytic cocondensation of the obtained phenyl- and ethylethoxysilanes is carried out in an acid medium. This also forms hydroxyphenylethylsiloxane (silanol):

$$n(C_6H_5)_xSi(OC_2H_5)_{4-x} + m(C_2H_5)_xSi(OC_2H_5)_{4-x} \longrightarrow$$

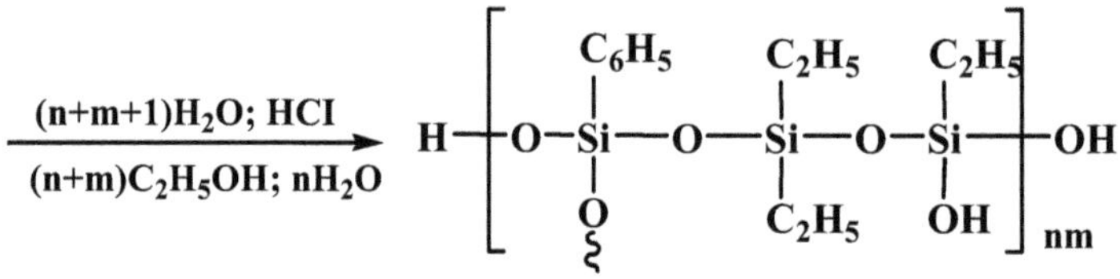

$$\xrightarrow[(n+m)C_2H_5OH;\, nH_2O]{(n+m+1)H_2O;\, HCl} \quad (4.41)$$

whereas ethoxy magnesium chloride under the influence of hydrochloric acid disintegrates. The silanol formed is subjected to polycondensation at 180-200 °C to form high-molecular polyphenyldiethylsiloxane. It is accompanied by intermolecular condensation (due to hydroxyl groups).

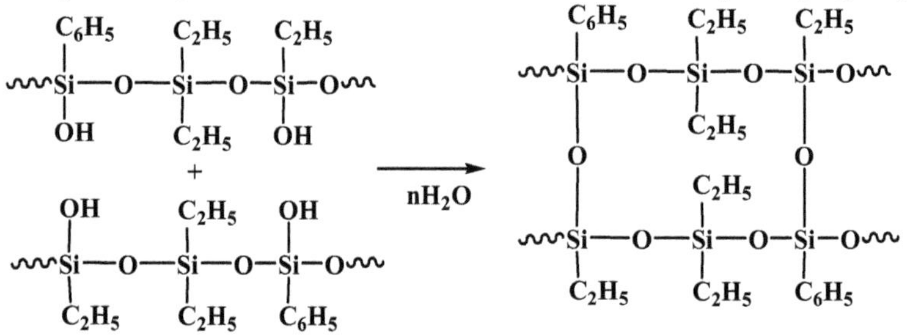

$$(4.42)$$

and oxidative condensation, which breaks off ethyl groups and forms a high-molecular branched polymer with fragments of cyclic structure.

The production diagram of polyphenyldiethylsiloxane varnish is given in Fig. 74. The reactive mixtures for the synthesis of phenyl- and ethylethoxysilanes are prepared correspondingly in mixers *29* and *12*. Mixer *29* receives from weight batch boxes *19, 20, 21* and *22* correspondingly chlorobenzene, tetraethoxysilane, ethyl bromide and diethyl ether. Ethyl bromide activates the Grignard reaction (it is taken in the amount of 3% of the reactive mixture); diethyl ether catalyses it (its consumption is 0.5% of the quantity of the reactive mixture) and is used in case the reaction occurs not intensively enough. The mixture is agitated for 0.5-1 hours and sampled for the content of the components. The finished mixture is pumped into weight batch box *23*. Mixer *12* receives from weight batch boxes *1, 2, 3* correspondingly toluene, ethylchloride, and tetraethoxysilane. The mixture is cooled with salt solution, agitated for 0.5-1 hours and sampled to determine the content of the components. The mixture is pumped into weight batch box *4*.

Phenylethoxysilanes are synthesised in reactor *30*, which is loaded with magnesium chipping and receives a small amount of reactive mixture through weight batch box *23*. The reactor is heated to 130 °C by sending vapour into the jacket; then the vapour supply is stopped and batch box *23* sends some more mixture to activate the reaction. Due to exothermicity the temperature in the apparatus may spontaneously rise up to 150 -180 °C, which shows the beginning of the reaction. If the reaction does not begin, an additional amount of reactive mixture is introduced. After that, the temperature is reduced to 140 °C and the mixture is gradually sent from weight batch box *23* at such speed that the temperature in the apparatus is maintained at 140-160 °C.

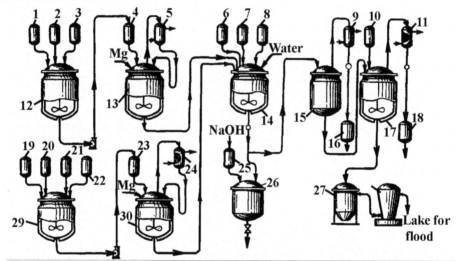

Fig. 13. Production diagram of polyphenyldiethylsiloxane varnish: *1-4, 6-8, 10, 19 - 23, 25* - batch boxes; *5, 9, 11, 24* - coolers; *12, 29* - mixers; *13, 30* - reactors; *14*-hydrolyser; *15* - distillation tank; *16, 18* - receiving tanks; *17*- condensation apparatus; *26* - neutraliser; *27*- settling box; *28* - ultracentrifuge

After the reactive mixture has been introduced, it is kept in the reactor at 150-160 °C for 6-8 hours and sampled to determine the phenylethoxysilane content. The content of phenylethoxysilanes (equivalent to phenyltriethoxysilane) should not be less than 65%. After the synthesis the reactive mixture is cooled to 40-60 °C by sending water into the jacket of the apparatus. The cooled mixture of phenylethoxysilanes, the so-called phenyl paste, is sent into hydrolyser *14* to hydrolytic cocondensation.

Ethylethoxysilanes are synthesised in reactor *13*. It is loaded with magnesium (in the form of chipping) and the apparatus is heated to 40-50 °C by sending vapour into the jacket. Some reactive mixture is added from weight batch box *4* to activate the reaction. Due to exothermicity the temperature in the apparatus may spontaneously rise up to 90 -110 °C, which shows the beginning of the reaction. The reactive mixture is cooled to 45-50 °C and at this temperature the reactor gradually receives the rest of the mixture at such speed that the temperature in the apparatus does not exceed 50 °C.

After the whole mixture has been introduced, the content of the unreacted ethylchloride is determined, which should not exceed 4%. If the ethylchloride content is more than 4%, the reactive mixture is held at 45-50 °C for 1.5-2 hours and analysed again. If the analysis is positive, the mixture in the reactor is heated within 3 hours to 100 °C and held at this temperature for 1.5-2 hours. The formed mixture of ethylethoxysilanes *(ethyl paste)* is cooled down to 60 °C and sent into hydrolyser *14* to hydrolytic cocondensation.

First, hydrolyser 14 receives toluene and butyl alcohol from batch boxes 6 and 7; then, from apparatuses 30 and 13, phenyl and ethyl paste. The mixture is agitated in the hydrolyser for 30 minutes. The temperature during the introduction of phenyl- and ethylethoxysilanes and agitation should not rise above 40 °C. Then, at 40 °C 25-30% hydrochloric acid is gradually sent from weight batch box *8*. The reactive mixture in the apparatus is heated to 60-70 °C; at this temperature they hydrolyser receives water. After adding a necessary amount of water, the mixture in the hydrolyser is settled for 0.5-1 hour and the acidity of the aqueous layer is determined. The acidity should not be less than 20 g of HCl/l; if it is lower, some acid is added from batch box *8*.

To obtain polyphenyldiethylsiloxane varnishes, it is possible to fill the hydrolyser with a mixture of diethyldichlorosilane and ethyltrichlorosilane instead of ethyl paste. In this case there is no need to add hydrochloric acid to the reactive mixture, since the hydrolytic cocondensation of phenyl paste with ethylchlorosilanes forms hydrogen chloride, which dissolves in water.

When the reactive mixture is settled, the mixture splits. The lower, water acidic is poured through a run-down box into neutraliser *26*, which has been loaded with a 40% solution of NaOH from batch box *25*.

The neutralised water toluene solution, which contains ethyl and butyl alcohol (hydrosite), can be used as a solvent in the production of silicone water-proofing liquids, polyhydrideorganosiloxanes and sodium organosiliconates.

The top layer, the solution of the product of hydrolytic cocondensation in toluene, is flushed with warm water (50-60 °C) several times, until pH is 6÷7. The neutral silanol solution enters distillation tank *15*, where it is heated with vapour to 60-80 °C and additionally settled for 2 hours to separate the remaining water. After that the temperature in the tank is raised to 130 °C and the solvent is distilled. Its vapours condense in cooler 9 and are collected in receptacle 16 (the solvent is a mixture of butyl alcohol and alcohol, which can be re-used at the stage of hydrolytic cocondensation instead of toluene). The solvent is distilled until the concentration of silanol is 85±5%. The product is cooled to 50-60 °C and sent into apparatus 17 to condense.

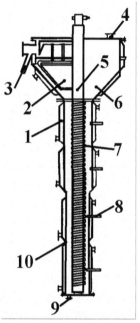

Fig. 14. Reactor of the continuous organomagnesium synthesis of organoethoxysilanes: *1* - cooling jacket; *2* - rake element; *3* - choke for loading off the products of the reaction; *4* - choke for loading magnesium; *5* - agitator shaft; *6* - separator; *7* - spiral (hollow pipe); *8* - thermocouple pocket; *9* - choke for loading the reactive mixture; *10* - casing of the reactor

After supplying silanol, the heating of apparatus 17 is switched on and the solvent is distilled, sending air through the bubble at the speed of 3 m^3/h at 150 °C. The solvent vapours are condensed in cooler *11* and collected in receptacle 18. The distillation of the solvent occurs simultaneously with the condensation of silanol. During the distillation the temperature in apparatus *17* is increased to 170-200 °C, and the speed of air supply is raised to 30 m^3/h. The mixture is periodically sampled to determine the viscosity and polymerisation time of the polymer. After the distillation air supply is stopped, the reactor receives a necessary amount of the solvent from batch box *10* and the mixture is cooled to 60 °C. The solution (varnish) is mixed and the polymer content is determined, which should be 50-70%. From apparatus *17* the varnish enters settling box *27*, where it is separated from mechanical impurities, and then on to ultracentrifuge *28*. The purified and "clarified" varnish is sent to be poured into containers and labelled.

In recent years there has been a great interest to the continuous processes of organomagnesium synthesis of silicone monomers. This is caused by the increasing consumption scale of organomagnesium compounds and the synthesis of silicone monomers used in the production of polyphenyldiethylsiloxanes.

The continuous organomagnesium synthesis of organoethoxysilanes has been designed by Russian scientists B.A.Klokov, M.V.Sobolevsky et al. They propose a vertical apparatus with an original agitator (Fig.75), where the agitating element is cooled spiral *7*, which is mounted on the shaft of agitator *5*. This helped to increase the productivity of the apparatus (to 100 kg/h of magnesium), to reduce the load on the agitator mechanism (by 30%) and ensure stable and long-term operation (at least a month). The apparatus is periodically loaded through choke 4 from a bunker with pelleted (1-20 mm) magnesium (with an addition of 1 to 5% of larger (20-30 mm) particles, since in 700 mm diameter industrial reactors spiral agitation is efficient only in the presence of larger magnesium particles in the apparatus). The reactive mixture consisting of ethylchloride or chlorobenzene and tetraethoxysilane solution is continuously pumped with a batching pump through choke *9* from the container. The products of the synthesis are continuously sent through choke *3* to the hydrolysis unit for subsequent preparation of polyphenyldiethylsiloxanes. The heat of the reaction is withdrawn with water sent into the jackets of reactor *1*. In the process ethylchloride (or chlorobenzene) reacts fully.

Polyphenyldiethylsiloxane varnishes are solutions of phenyldiethylsiloxane polymer in petrol, toluene or a mixture of petrol and turpentine. They are liquids ranging in colour from yellow to brown. The varnishes are heat-resistant; they are used to produce insulating materials for temperatures up to 180 °C and high humidity. Heat-resistant insulation of polyphenyldiethylsiloxane varnishes is widely used in electric engines, in rotors of turbogenerators, etc.

Table 28. Main properties of electroinsulating polyphenyldiethylsiloxane varnishes

Characteristics	EF-1	EF-3BSU	EF-5B	EF-5T
VS-1 viscosity at 20°C, s	15-70	15-70	15-70	15-70
Nonvolatile content, %, at least	40	40	40	40
Drying time of varnish film on copper at 220 °C, hours, not more than	1	2	2	2
Thermal elasticity of varnish film on copper at 200 °C, hours, not more than	-	40	20	20
Electric strength of varnish film after baking* at 200⁰C, KV/mm, at least				
at 20±2 ⁰C,	65	75	75	75
at 180±2 ⁰C,	25 (at 200⁰C)	35	30	30
Specific volume electric resistance of varnish film after baking*, Ohm·cm, at least				
at 20±2 ⁰C,	$1 \cdot 10^{14}$	$1 \cdot 10^{14}$	$1 \cdot 10^{14}$	$1 \cdot 10^{14}$
at 200±2 ⁰C,	$1 \cdot 10^{12}$	$1 \cdot 10^{12}$	$1 \cdot 10^{12}$	$1 \cdot 10^{12}$
After 24 hours at 20⁰C and 95-98% relative humidity	$1 \cdot 10^{12}$	$1 \cdot 10^{12}$	$1 \cdot 10^{12}$	$1 \cdot 10^{12}$

*1 hour for EF-1 varnish, 6 hours for varnishes EF-3BSU, EF-5B and EF-5T.

The main properties of some electroinsulating polyphenyldiethylsiloxane varnishes are given in Table 28.

4.2.7. Preparation of polyalkylsiloxanes with higher alkyl radicals at the silicon atom and varnishes based on them

Additions of polyalkylsiloxanes with higher alkyl radicals (C_4 and higher) efficiently eliminate pore and bubble formation in the impregnation of windings of electric machines with polyorganosiloxane varnishes. Specifically, good additives for polyorganosiloxane varnishes are polyhexyl-, polyheptyl- and polynonylsiloxane varnishes.

These polyalkylsiloxanes are prepared by the interaction of alkyltrichlorosilanes with alcohols and subsequent hydrolytic polycondensation of the obtained products.

For example, polynonylsiloxane is synthesised by the hydrolytic condensation of partially etherified nonyltrichlorosilane and subsequent polycondensation of the obtained product. The production process comprises two main stages: the partial etherification of nonyltrichlorosilane and the hydrolytic condensation of the product; the distillation of the solvent and preparation of the varnish.

The partial etherification of nonyltrichlorosilane with butyl alcohol is carried out at 75-80 °C and the mole ratio of the parent substances of 1:1.5.

$$2C_9H_{19}SiCI_3 + 3C_4H_9OH \xrightarrow[3HCI]{} C_9H_{19}Si(OC_4H_9)CI_2 + C_9H_{19}Si(OC_4H_9)_2CI \qquad (4.43)$$

The partially etherified products seem to partially condense at the synthesis temperature to form bis[(butoxy)nonylchloro]disiloxane (I):

$$
\begin{array}{cc}
CI & CI \\
| & | \\
C_9H_{19}-Si-CI \;+\; C_4H_9O-Si-C_9H_{19} \xrightarrow{\;\;C_4H_9CI\;\;} \\
| & | \\
OC_4H_9 & OC_4H_9
\end{array}
$$

$$
\xrightarrow{}
\begin{array}{cc}
CI & CI \\
| & | \\
C_9H_{19}-Si-O-Si-C_9H_{19} \\
| & | \\
OC_4H_9 & OC_4H_9
\end{array}
\qquad (4.44)
$$

The hydrolytic condensation of partially etherified product I occurs thus:

$$
\begin{array}{cc}
CI & CI \\
| & | \\
C_9H_{19}-Si-O-Si-C_9H_{19} \xrightarrow[8nHCI;\; 6nC_4H_9OH]{8nH_2O} \\
| & | \\
OC_4H_9 & OC_4H_9
\end{array}
$$

$$
\xrightarrow{}
\left[
\begin{array}{cccc}
C_9H_{19} & C_9H_{19} & C_9H_{19} & C_9H_{19} \\
| & | & | & | \\
HO-Si-O-Si-O-Si-O-Si-OC_4H_9 \\
| & | & | & | \\
O & O & O & O \\
| & | & | & | \\
C_4H_9O-Si-O-Si-O-Si-O-Si-OH \\
| & | & | & | \\
C_9H_{19} & C_9H_{19} & C_9H_{19} & C_9H_{19}
\end{array}
\right]
\qquad (4.45)
$$

Then product II is subjected to thermal polycondensation and transforms into polynonylsiloxane, mostly with ladder structure.

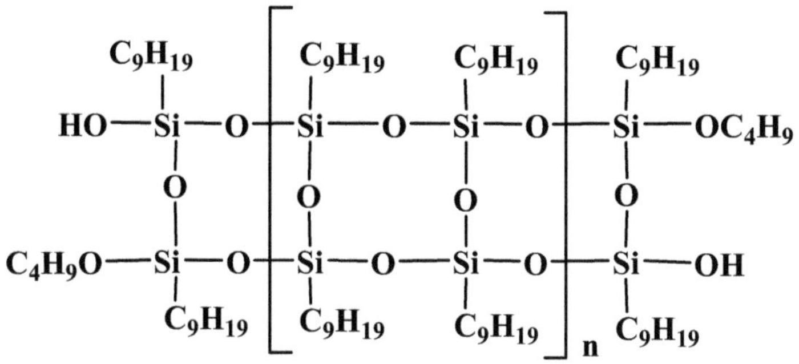

$$
\begin{array}{c}
\underset{\substack{| \\ HO-Si-O}}{C_9H_{19}}
\left[
\begin{array}{ccc}
C_9H_{19} & C_9H_{19} \\
| & | \\
Si-O-Si-O \\
| & | \\
O & O
\end{array}
\right]
\begin{array}{c}
C_9H_{19} \\
| \\
Si-OC_4H_9 \\
| \\
O
\end{array}
$$

The production diagram of polynonylsiloxane varnish is given in Fig. 76. The main apparatus is spray tower 3 consisting of five rings (sections) less than 150 mm in diameter and two rings 250-300 mm in diameter with run-down boxes. The rings are lined with fluoroplast. The tower can be heated by sending hot water or vapour into the ring jackets. Both the etherification of nonyltrichlorosilane anfd the hydrolytic condensation of the etherified product take place in the tower.

To carry out the etherification, a mixture of nonyltrichlorosilane and toluene (the weight ratio is 1:1.8) is prepared in mixer 9. Toluene is sent into the mixer from batch box 7; part of it is poured into tower 3 to fill 2/3 of its height.

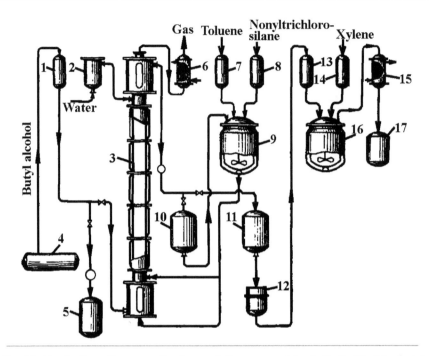

Fig. 15. Production diagram of polynonylsiloxanevarnish: *1, 7, 8, 13, 14* - batch boxes; *2* - boiler mixer; *3* - section tower; *4* - container; *5, 10, 11* - collectors; *6, 15* - coolers; *9* - mixer; *12* - nutsch filter; *16* - reactor; *17* - receptacle

The mixer is loaded with nonyltrichlorosilane from batch box *8*, the agitator is switched on and the mixture is agitated for 15-20 minutes. From container *4* butyl alcohol is sent into batch box *1*. The ring jackets of tower *3* are filled with vapour; when the temperatue in the tower rises to 75-80 °C, the toluene solution of nonyltrichlorosilane is sent from mixer *9* and butyl alcohol is sent from batch box *1* through fluoroplast nozzles in the bottom part of the tower. The solution of nonyltrichlorosilane and butyl alcohol is sent in the mole ratio of 1:1.5. The product of partial etherification is continuously withdrawn from the top of the tower through a run-down box into collector *10*.

During etherification, the reactive mixture is tested to determine the content of butoxyl groups. If the contents of these groups is less than 40%, the speed at which butyl alcohol is supplied is increased to 2.4-3.3 l/h, so that the mole ratio of nonyltrichlorosilane and butyl alcohol corresponds to 1:2.

Hydrolytic condensation is carried out in the same tower *3*. The etherified product is sent from collector *10* by nitrogen flow (0.07 MPa) into mixer *9*. Before hydrolysis the tower is filled to 2/3 of its heights with the etherified product from the mixer; the ring jackets are filled with vapour so that the temperature in the tower during the hydrolytic condensation is maintained at 75-80 °C. After the temperature has been achieved, the etherified product is continuously fed into the lower part of the first ring, and hot water (from boiler mixer 2) into the top part of the tower. The product of hydrolytic condensation is continuously withdrawn from the top part of the tower through a run-down box into collector 11; the released water is withdrawn from the bottom of the tower through a run-down box into collector 5.

The weight ratio of the water and etherified product sent into the tower should be 1:10. During the hydrolytic condensation the mixture is sampled to determine the content of chlorine, butoxyl and hydroxyl groups and the flush water is tested for acidity. The finished product should have not more than 2% of chlorine, 15% of butoxyl groups and 3% of hydroxyl groups. If the chlorine content is above the norm, the product is subjected to repeated hydrolytic condensation.

If the tests are positive, the product from collector 11 is filtered in nutsch filter 12 and sent into batch box 13. The distillation of the solvent and polycondensation of the product of hydrolytic condensation is carried out in reactor 16, a vertical cylindrical apparatus with a jacket, agitator and direct cooler 15. The reactor receives from batch box 13 toluene solution of the product of hydrolytic condensation; the agitator is switched on; the mixture is heated to 120-130 °C and the mixture of toluene and butyl alcohol is distilled into receptacle 17. After the distillation, the temperature in the reactor is increased within 1.5-2 hours to 250 °C; at 250-260 °C and continuous agitation polycondensation is conducted. The distilled butyl alcohol, water and partially toluene are condensed in cooler 15 and collected in receptacle 17. After 3-4 hours of polycondensation the contents of the reactor are cooled to 60 °C and analysed.

The finished product is polynonylsiloxane with not more than 1% of hydroxyl groups and at least 2.5% of butoxyl groups; there should be no chlorine. In case it contains chlorine or more than 1% of hydroxyl groups, polycondensation should be continued for 1.5-2 hours at 250-260 °C. The total amount of the distilled solvent and volatile substances during the distillation and polycondensation is 60-70%.

The varnish is prepared in the same reactor, 16: at agitation xylene is introduced from batch box 14 to obtain a 50% solution of the varnish; the solution is mixed for 1.5-2 hours at the temperature of about 100 °C to obtain a homogeneous mixture.

Polynonylsiloxane varnish is a homogeneous transparent liquid with colour ranging from light yellow to brown.

The varnish should meet the following technical requirements:

Content, %:

of butoxyl groups	2.5
of hydroxyl groups	≤ 1.0
of chlorine	Absent
of solid residue	50 ± 1
VS-4 viscosity, s	10-12
Mechanical impurities	Absent

Polynonylsiloxane is used as a modifying agent in silicone varnishes.

Similarly to polynonylsiloxane varnish, one can obtain polyhexyl-, polyheptyl- and other polyalkylsiloxane varnishes with higher alkyl radicals at the silicon atom.

4.3. Preparation of polyorganosilazanes and polyorganosilazane varnishes

Nowadays polyorganosilazanes are used more and more often. Until recently, due to the small hydrolytic stability (in acid and neutral media) of compounds with a Si—N bond, the possibility of their practical application seemed doubtful. However, the increased ability to hydrolyse and the chemical activity of the Si—N bond in polyorganosilazanes predetermined their industrial applications. It was found that polyorganosilazanes are hydrolysed when kept in air even at room temperature, with 80-85% of silazane bonds replaced by siloxane bonds. Probably, due to the high gas permeability of siloxane film the ammonia released during the hydrolysis is withdrawn out of the system without disrupting the film, even in case of considerable thickness (1-2 mm).

The hydrolysis forms *polyorganosilazoxanes*, i.e. polymers which contain both Si—N and Si—O bonds, are stable ahd have good physicomechanical characteristics and adhesion to various materials.

At present, owing to the fundamental research in the field by Russian scientists (Prof. D.Y.Zhinkin et al, Academician K.A.Andrianov), our industry produces polyorganosilazanes industrially, because they have found a wide application as impregnators for glass cloth, binding agents for plastics, modifying agents for cable rubbers as well as solidifying agents for epoxy and epoxysilicone polymers.

Polyorganosilazanes are obtained by the ammonolysis or coammonolysis of various organochlorosilanes and subsequent polycondensation of the products of ammonolysis.

4.3.1. Preparation of polymethyldimethylsilazanes and polymethyldimethylsilazane varnishes

Polymethyldimethylsilazanes are synthesised by the coammonolysis of a mixture of methyltrichlorosilane and dimethyldichlorosilane in the 3:1 mole ratio with gaseous ammonia in toluene solution:

$$3nCH_3SiCl_3 + n(CH_3)_2SiCl_2 + 16,5nNH_3 \longrightarrow$$

$$\longrightarrow \{[CH_3Si(NH)_{1.5}\text{-}[(CH_3)_2SiNH]\}_n + 11nNH_4Cl \qquad (4.46)$$

After coammonolysis, to obtain varnish of required concentration (50-80%, depending on the brand), toluene should be distilled from the product. The distillation is accompanied by the polycondensation of the coammonolysis product and the formation of polymethyldimethylsilazane of the following composition:

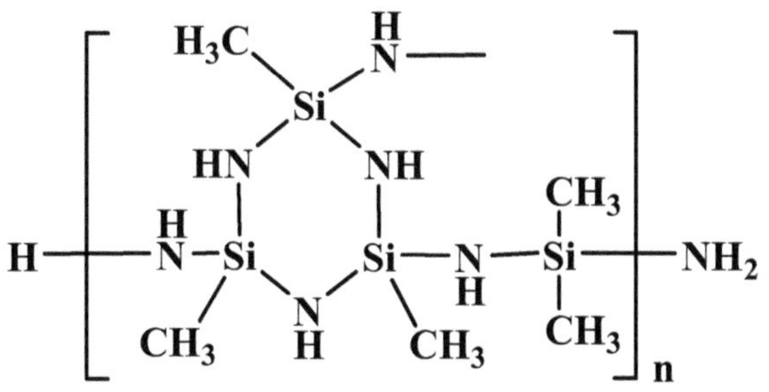

The production process (Fig.77) comprises two main stages: the coammonolysis of methyltrichlorosilane and dimethyldichlorosilane; the preparation of varnish of a required concentration.

Apparatus *1* is filled with methyltrichlorosilane, dimethyldichlorosilane and toluene in the required ratio. The mixture is agitated for 30 minutes and sent by nitrogen flow (0.07 MPa) into reactor *5*, an enameled apparatus with an agitator, inverse condenser *6*, a water vapour jacket and a bubbler to feed ammonia. To avoid explosive concentration of the mixture of ammonia with air, the whole system is blown with nitrogen before the start of synthesis. The inverse condenser is filled with salt solution, the reactor is filled with reactive mixture from the agitator, the jacket of the apparatus is filled with water.

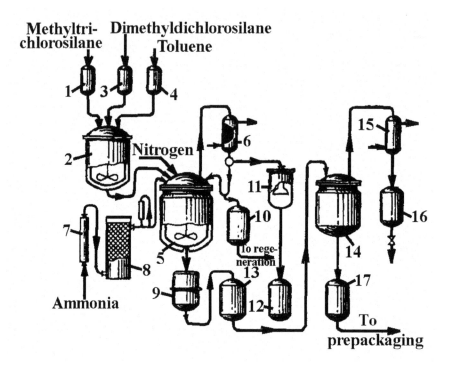

Fig. 16. Production diagram of polymethyldimethylsilazane varnishes: *1* - agitator; *2-4* – batch boxes; *5* - reactor; *6, 15* - condensers; *7* - evaporator; *8* - drying tower; *9* - nutsch filter; *10, 12, 13, 17* - collectors; *11* - trap; *14* - distillation tank; *16* – receptacle

The agitator is switched on and gaseous ammonia is sent through the bubbler at such speed that the temperature in the reactor does not rise above 25 °C. To produce gaseous ammonia, liquid ammonia is choked until the pressure is 0.07 MPa, and sent into evaporator *7* heated with water. Gaseous ammonia passes through drying tower *8* filled with solid alkali and on through the rotameter into the reactor for coammonolysis. After 15-20 hours the supply of ammonia is stopped and the contents of the reactor are sampled.

The sample is put in a special airproof metallic sampler, filtered from ammonia chloride; gaseous ammonia is sent through the filtrate. The reaction is considered complete, if after the passing of ammonia there is no sediment of ammonia chloride. In case there is sediment, coammonolysis is continued and the sample is taken again after 1-2 hours.

This ends the process of coammonolysis. The unreacted ammonia with an impurity of toluene enters backflow condenser *6*, where toluene vapours condense and are collected in collector *10*. Toluene can be re-used in manufacture. Ammonia from condenser 6 enters water trap *11*, from where the ammonia water is poured into collector *12* and sent by nitrogen flow (0.07 MPa) into the section for the purification of waste waters.

After coammonolysis the product in filtered in nutsch filter 9 filled with coarse calico, filter paper and glass cloth. At a pressure below 0.2 MPa the solution is filtered from the sediment of ammonia chloride. The filtered solution enters collector *13* and then distillation tank *14*.

The sediment of ammonia chloride can be separated not only by filtering, but also when the reactive mixture is treated with aqueous alkali in reactor *5* at agitation. In this case ammonia chloride interacts with the solution of sodium hydroxide; the solution of sodium chloride can be easily separated by settling.

This technique for separating the sediment increases the yield of the polymer, increases its molecular weight and reduces the active hydrogen content. It must be explained by the fact that alkali treatment of the coammonolysis product hydrolyses mostly amino groups, which are situated at the ends of the polymer chain, whereas silazane bonds remain virtually untouched. Besides, this technique for separating the sediment naturally eliminates the necessity to filter the product of coammonolysis.

The product of coammonolysis is distilled to separate toluene in enameled tank *14* with a water vapour jacket, condenser *15* and receptacle *16*. The distillation is carried out at 110-115 °C until the solid residue content is 50-80% (depending on the brand of varnish). The finished product, polymethyldimethylsilazane varnish, is loaded at 20-30 °C out of the tank into collector *17*.

Polymethyldimethylsilazane forms a glossy film on various substrates. When held in air, the silazane bonds in the film are replaced with siloxane bonds due to humidity, which increases the stability of the product.

Technical *polymethyldimethylsilazane varnish* should meet the following requirements:

Appearance	Liquid (from colourless to dark yellow)*
Solid residue, %	50-80
Appearance of the varnish film	Transparent, colourless, glossy, no streaks.
The drying time of the film at 20±2 °C, hours, not more than	5
Hardness at 200±5 °C, relative units, notless than	0.6
Heat resistance at 200±5 °C, hours, not less than	10

*A white crystalline sediment is tolerated.

Below we give the dielectric characteristics of polymethyldimethylsilazane film dried for 24 hours at room temperature (the measurements were taken at 20 and 200 °C):

	At 20 °C	At 200 °C
Dielectric loss tangent at 10^6 Hz	0.0031	0.0018
Dielectric permeability at 10^3 Hz	3.9	4.1
The same, after the film has been subjected to 95-98% relative humidity	3.7	-
Electric strength, KV/mm	136.0	114.0
The same, after the film has been subjected to 95-98% relative humidity	76.0	-
Specific volume electric resistance, Ohm cm	$1.8 \cdot 10^{14}$	$1.9 \cdot 10^{14}$
The same, after the film has been subjected to 95-98% relative humidity	$2.5 \cdot 10^{12}$	-

Polymethyldimethylsilazane varnish is used to impregnate glass cloths and glass fibres (to waterproof them and give dielectric properties), as a solidifying agent for epoxy and epoxysilicone polymer compositions, as well as waterproof transparent protective coating for hardened silicate glass.

Similarly, the coammonolysis of a mixture of dimethyldichlorosilane and vinyltrichlorosilane can yield *polyvinylmethyldimethylsiloxane varnish*, which has a relatively high adhesion to metals, glass and rubber based on silicone elastomers; this varnish is also used in glue compositions. The coammonolysis of a mixture of dimethyldichlorosilane and phenyltrichlorosilane produces *polydimethylphenylsilazane varnish*.

4.3.2. Preparation of polydimethylphenylsilazane and polydimethylphenylsilazaboroxane varnishes

Polydimethylphenylsilazane varnishes (types A and B) are obtained by the coammonolysis of dimethyldichlorosilane and phenyltrichlorosilane (in the equimolar ratio) with gaseous ammonia in toluene medium:

$$n(CH_3)_2SiCl_2 + nC_6H_5SiCl_3 + 7{,}5nNH_3 \xrightarrow[5nNH_4Cl]{}$$

$$\xrightarrow{\hspace{2cm}} [C_6H_5Si(NH)_{1.5}-(CH_3)_2SiNH-]_n \qquad (4.47)$$

The formed ammonia chloride is disintegrated with sodium hydroxide solution. After coammonolysis, if the required concentration of the products (87-97%) has been achieved, the toluene is distilled from the reactive mixture. This is accompanied by the polycondensation of the coammonolysis product and mostly forms a cyclolinear product.

Polydimethylphenylsilazaboroxane varnish is obtained by the interaction of synthesised polydimethylphenylsilazane with boric acid:

$$[C_6H_5Si(NH)_{1.5}-(CH_3)_2SiNH-]_n + H_3BO_3 \xrightarrow[3NH_3]{}$$

$$\longrightarrow [C_6H_5Si(NH)_{1.5}-(CH_3)_2SiNH-]_{n-3} [-C_6H_5Si(NH)_{1.5}-(CH_3)_2SiO-]_3B \qquad (4.48)$$

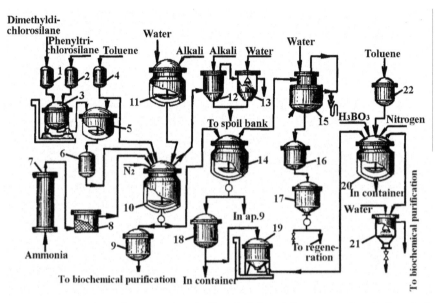

Fig. 17. Production diagram of polydimethylphenylsilazane and polydimethylphenylsilazaboroxane varnishes: *1 - 4, 6, 19, 22* - batch boxes; *5* - agitator; *7* - evaporator; *8* - tower; *9, 16, 18* - collectors; *10, 11* - reactors; *12, 13, 21* - traps; *14, 20* - distillation tanks; *15* - cooler; *17-* settling box

In this case condensation is accompanied by secondary reactions, for example:

$$2NH_3 + 4H_3BO_3 \longrightarrow (NH_4)_2B_4O_7 + 5H_2O \qquad (4.49)$$

This reaction forms small amounts of sodium tetraborate and water, which partially hydrolyses polydimethylphenylsilazane:

$$[C_6H_5Si(NH)_{1.5}-(CH_3)_2SiNH-]_n \xrightarrow{mH_2O}$$

$$\longrightarrow [C_6H_5Si(NH)_{1.5}-(CH_3)_2SiNH-]_{n-1} [-C_6H_5Si(NH)_{1.5}-(CH_3)_2SiO-]_m \qquad (4.50)$$

The production of polydimethylphenylsilazane and polydimethylphenylsilazaboroxane varnishes (Fig.78) comprises the following main stages: the coammonolysis of dimethyldichlorosilane and phenyltrichlorosilane and the treatment of the reactive mixture with alkali solution; the distillation of the solvent and the preparation of polydimethylphenylsilazane varnishes of the required concentration; the reaction of polydimethylphenylsilazane with boric acid.

The reactive mixture is prepared in enameled agitator 5 with an anchor agitator. One should strictly observe the mole ratios of dimethyldichlorosilane and phenyltrichlorosilane; therefore, they are dispensed through weight batch box 3. Dimethyldichlorosilane and phenyltrichlorosilane are alternately poured out of the batch boxes into batch box 3, where they are weighed and poured into the agitator in necessary amounts. The organochlorosilane mixture is agitated for 30 minutes; after that it is checked for chlorine content, which should be 51.5+1% or for organochlorosilane composition (chromatographically). If the analysis is positive, the agitator receives a necessary amount of toluene from batch box 4 by self-flow; the reactive mixture is agitated for 30 minutes. The ready mixture is sent by nitrogen flow (<0.07 MPa) into weight batch box 6.

Coammonolysis is carried out in enameled reactor 10 with a water vapour jacket, anchor agitator and bubblers (for introducing ammonia and sodium hydroxide solution and for nitrogen blowing). To avoid the formation of an explosive mixture of ammonia with air, the entire system is blown with nitrogen before coammonolysis. Thoroughly dried reactor 10 receives from batch box 4 self-flowing toluene and a mixture of organochlorosilanes and toluene from weight batch box 6; after that, the free volume of the reactor is blown with nitrogen for 5 more minutes. The agitator is switched on; the jacket of the reactor is filled with cooled water and the reactive mixture is cooled down to 15-28 °C. At this temperature ammonia is sent through the bubbler at such speed that the temperature in the reactor does not exceed 29-30 °C. Before that, liquid ammonia is choked until the excess pressure is 0.07 MPa and sent to coammonolysis through vapour-heated evaporator 7 (0.3 MPa) and drying tower 8 filled with solid sodium hydroxide.

After a calculated amount of ammonia has been introduced, the reactor is blown with nitrogen for 0.5-1.5 hours at a speed of 1.5-3 m³/h to eliminate the unreacted ammonia and organochlorosilanes; the reactive mixture is sampled. The sample is analysed to determine the completeness of coammonolysis by the technique described with the production of polymethyldimethylsilazanes. The offset of coammonolysis is accompanied by a sharp drop in temperature in the reactor (below 20 °C). The unreacted ammonia and organochlorosilane vapours are carried away by nitrogen and consequtively captured in traps 12 and 13 with sodium hydroxide solution and water.

After coammonolysis, to destroy the ammonia chloride formed in the process, reactor 10 is loaded at agitation from apparatus 11 by vacuum or

nitrogen flow (the pressure is below 0.07 MPa) with prepared 17±3% alkali solution (sodium hydroxide). The mixture in reactor *10* is agitated for
15-30 minutes, the agitator is stopped and the reactive mixture is held for
2-3 hours. The destruction of ammonia chloride is monitored visually, until
NH$_4$Cl crystals vanish in the sample from the lower layer of the reactive
mixture.

The lower layer (aqueous solution of sodium chloride) is poured through
a run-down box into collector *9*; the upper layer (a 10% toluene solution of
dimethylphenylsilazane) is poured into distillation tank *14*.

The distillation of toluene and polycondensation of dimethylphenylsilazane is conducted in tank *14*, which, like tank *20*, is a steel enameled apparatus with a water vapour jacket and an anchor agitator. Before distilling
toluene, the dimethylphenylsilazane mixture in the distillation tank is
"clarified", i.e. released from traces of moisture by holding for 1-3 hours at
50-70 °C. The jacket of the apparatus is filled with vapour (0.3 MPa) to
heat the tank. The settled water from the lower part of the tank is poured
through a run-down box into collector *9*; the "clarified" dimethylphenylsilazane solution is analysed to determine solid residue content. After that,
the distillation of toluene is started.

Toluene is distilled at 70-120 °C (liquid) and a residual pressure of
145±70 GPa until the solid residue content is 87-97% (depending on the
varnish type). Toluene vapours from the distillation tank are sent into water cooler *15* to condense. Toluene is poured into collector *16*; from there
raw toluene enters settling box *17*, where it settles, is separated from water
and sent to regeneration. Regenerated toluene can be re-used in manufacture. After the distillation of toluene the heating of the tank is stopped and
the jacket is filled with water. The obtained polydimethylphenylsilazane
varnish is cooled down to 60 °C and filtered through cotton and metal
mesh into collector *18*.

The finished varnish is analysed to check if it meets the technical requirements; if the results are positive, it is poured into containers or
(polydimethylphenylsilazane varnish of A type) loaded with vacuum
through batch box *19* into distillation tank *20*. There the varnish is heated
to 75-80 °C and at this temperature and at agitation the tank is filled
through a hatch with a calculated amount of boric acid (3% of the weight
of 100% polydimethylphenylsilazane). The mixture in the tank is heated
to 150±3 °C and agitated at this temperature for 1.5-2 hours (until the
sample stops foaming). After the reaction is over, the mixture in the tank is
cooled to 100 °C; the released ammonia is blown away with nitrogen (0.07
MPa). It enters trap *21*; the tank is loaded from batch box *2* with toluene to
obtain a 57-63% solution of polydimethylphenylsilazaboroxane in toluene.
The finished varnish is cooled (at agitation) to ambient temperature, ana-

lysed to check if it meets the technical requirements and poured into containers.

Polydimethylphenylsilazane varnishes (A and B types) and polydimethylphenylsilazaboroxane varnish should meet the following technical requirements:

	Polydimethylphenylsilazane varnishes		Polydimethylphenylsilazaboroxane varnish
	type A	type B	
Appearance	Resinous products ranging in colour from light yellow to dark yellow*		Cloudy liquid ranging in colour from light yellow to brown**
Solid residue, %	87-92	97	57-63
Content, %:			
of silicon	23.5-27	24	15-20
of nitrogen	13-17	14	8-12
Solubility in toluene		Complete	

*No mechanical impurities **Sediment is tolerated.

Polydimethylphenylsilazane varnish of A type is used as a binding agent for glass fibres; B type is used as a modifying agent to improve electroinsulating properties of rubbers used in the cable industry. Polydimethylphenylsilazaboroxane varnish is used as a solidifying agent and stabilier for polyorganosiloxanes and materials based on them.

4.4. Preparation of polyelementorganosiloxanes and polyelementorganosiloxane varnishes

First polyelementorganosiloxanes were obtained in Russia in 1947 (by K.A.Andrianov). It was found that the introduction of other elements (e.g. aluminum, titanium, boron, iron) into the siloxane chain considerably affects the polymer characteristics. At present, mainly owing to the research carried out by Academician K.A.Andrianov and Prof. A.A.Zhdanov et al, industry manufactures polyalumorgano-, polyironorgano- and polytitaniumorganosiloxanes. These polymers are used as binding agents in the production of heat-resistant plastics, plastic laminates and other materials; as solidifying agents for organic and silicone polymers; as modifying agents for various polymers. They also present considerable interest as film-forming agents used for precision molding of metals, as well as thermostabilisers for polymer coatings. Further research into the synthesis and properties of polyelementorganosiloxanes will undoubtedly expand the industrial production of these polymers.

4.4.1. Preparation of polyalumophenylsiloxane and polyalumophenylsiloxane varnish

Polyalumophenylsiloxane is synthesised by the hydrolytic condensation of phenyltrichlorosilane:

$$nC_6H_5SiCl_3 \xrightarrow[3nHCl]{3nH_2O} n[C_6H_5Si(OH)_3] \xrightarrow[1.5H_2O]{} (C_6H_5SiO_{1.5})_n \quad (4.51)$$

the interaction of the obtained product with alkali to obtain sodium dihydroxyphenylsilanolate

$$(C_6H_5SiO_{1.5})_n + nNaOH + 0,5nH_2O \longrightarrow nC_6H_5Si(OH)_2ONa \quad (4.52)$$

and subsequent exchange decomposition of sodium dihydroxyphenylsilanolate with alumopotassium alum:

$$3nC_6H_5Si(OH)_2ONa + nKAl(SO_4)_2 \longrightarrow$$

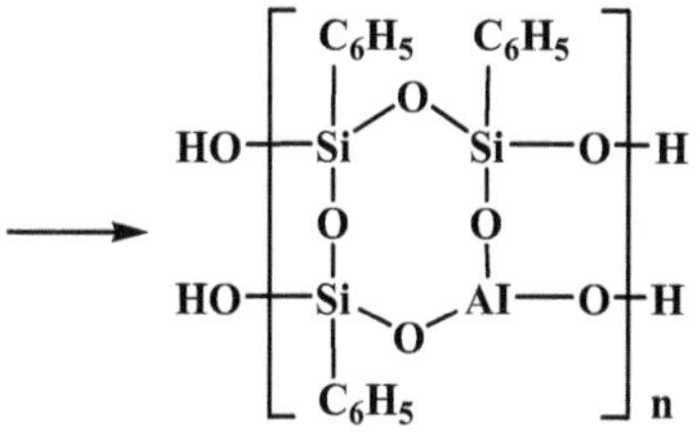

$$+ 1,5nNa_2SO_4 + 0,5K_2SO_4 + (n-2)H_2O \quad (4.53)$$

The polyalumophenylsiloxane obtained is a polydisperse mixture of polymer homologues consisting of the abovementioned ladder macromolecules and branched macromolecules of the following common formula:

$$\{[C_6H_5Si(OH)_{2-x}O_{0.5x}(O)]_3Al\}_n$$

x=0÷1, n=1÷4

The production of polyalumophenylsiloxane and polyalumophenylsiloxane varnish (Fig.79) comprises three main stages: the hydrolytic condensation of phenyltrichlorosilane; the treatment of the products with alkali and the preparation of polyalumophenylsiloxane varnish.

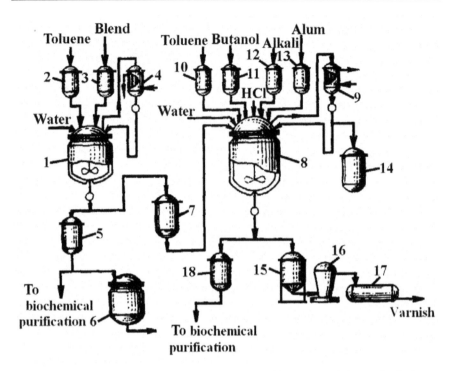

Fig. 18. Production diagram of polyalumophenylsiloxane varnish: *1* - hydrolyser; *2, 3, 7, 10 - 13* - batch boxes; *4,9* - coolers; *5, 14, 18* - collectors; *6* – distillation tank; *8* - reactor; *15*- settling box; *16* - ultracentrifuge; *17*- container

The hydrolytic condensation of toluene solution of phenyltrichlorosilane is conducted in hydrolyser *1*, which is an enameled apparatus with an agitator and a water vapour jacket. The hydrolyser is filled with water and toluene from batch box *2*. The jacket of the apparatus and cooler *4* are also filled with water; the agitator is switched on and a prepared mixture of phenyltrichlorosilane and toluene is introduced from weight batch box *3* through a bubbler under the liquid layer at such speed that the temperature in the hydrolyser does not rise above 40 °C. The hydrolysis temperature is regulated by changing the speed of reactive mixture supply and sending water into the jacket of the hydrolyser. After all the reactive mixture has been introduced, the contents of the hydrolyser are agitated for 30 minutes. Then the product of hydrolytic condensation, the toluene solution of silanol, is settled to separate water for 1.5-2 hours. The lower acid aqueous layer is poured through a run-down box into collector *5*; the product of hydrolytic condensation is flushed in the hydrolyser with warm water (about 40 °C) until it gives a neutral reaction. Flush waters are also poured into receptacle *5*. The liquid in collector *5* is settled and then sent by vacuum

into distillation tank *6* to be re-used in manufacture. The obtained neutral
solution of the product of hydrolytic condensation is heated with vapour
sent into the jacket of the hydrolyser; at 60-80 °C it is settled to separate
moisture for 2-3 hours. The clarified product is analysed (the solid residue
content should be 20-26%) and through weight batch box *7* sent into reactor *8*.

The reactor is an enameled apparatus with an agitator and a water vapour jacket. The production of sodium dihydroxyphenylsilanolate is carried out in butanol and toluene or ethanol and toluene medium at 35-50 °C.
The consumption of other components is calculated by the amount of the
loaded condensation product. After loading the product of condensation,
the reactor is filled with toluene and butanol (or ethanol and toluene) from
batch boxes *10* and *11*. The ratio of the solvents should be 1:1.4 to obtain
10% silanol solution. The calculation takes into account toluene contained
in the product of hydrolytic condensation. The loaded mixture is agitated
in the reactor for 30 minutes; after that it receives 20% alkali solution from
batch box *12* at agitation. The reaction forms sodium dihydroxydiphenylsilanolate and water.

The sodium dihydroxydiphenylsilanolate is heated to 73-75 °C by sending vapour into the jacket of the apparatus; at this temperature a prepared
8-10% solution of alumopotassium alum from batch box *13*. After the
alum has been introduced, the reactive mixture is agitated at 73-75 °C for
approximately 2 more hours; cooler *9* works in the inverse mode. The mixture is cooled to 40 °C and held for 1.5-2 hours. The water layer is poured
through a run-down box into collector *18* and sent to biochemical purification. The obtained polyalumophenylsiloxane is sampled for free alkali,
which is neutralised with 2% hydrochloric acid.

The amount of 2% hydrochloric acid (*X*, g) required to neutralise free
alkali in the varnish is calculated by the formula

$$X = \frac{36.46AB}{40 \cdot 2}$$

where 36.46 is the molecular weight of hydrochloric acid; *A* is the content of free alkali in varnish, %; *B* is the amount of varnish, g; 40 is the
molecular weight of sodium hydroxide; 2 is the concentration of hydrochloric acid, %.

The reactor is filled at agitation with calculated portions of hydrochloric
acid and the reaction of the medium is tested. It should be neutral (pH is
5.5-7). The neutral solution of varnish is washed with water at 60 °C from
SO_4^{-2} ions until the analysis with barium chloride is neutral.

If during the water flushing the organic layer is difficult to separate from
the aqueous layer, one should add butanol or ethanol. When SO_4^{-2} ions are

flushed, it should be taken into account that they may be contained in the water loaded into apparatus *8*. Therefore, the end of the flushing should be determined by the clouding of the aqueous extract of varnish in the presence of $BaCl_2$ (compared to the check sample).

The solvent is distilled from polyalumophenylsiloxane in the same apparatus, *8*. Before the distillation the product is clarified at 45-50 °C. The settled water is poured into collector *18*; the clarified product (after switching inverse cooler *9* into the direct mode) is distilled to separate the toluene and butanol or ethanol mixture at a residual pressure of 800±65 GPa. The distillation temperature gradually rises to 90 °C. The distillation is considered finished when the resin concentration in the varnish is 40-65%. The vapours of the distilled solvent enter water cooler *9*. There they condense and flow into receptacle *14*.

After distillation the varnish is cooled by sending water into the jacket of the reactor. The finished varnish is sampled to determine the solid residue and alkali content and gelatinisation time. If the characteristics meet the technical requirements, the varnish is sent into settling box *15*. After settling the varnish is centrifuged to separate mechanical impurities completely and sent into container *17*.

Polyalumophenylsiloxane varnish is a solution of polyalumophenylsiloxane in a mixture of toluene and butanol. Its flammability, boiling point and inflammation point, as well as explosive limits of varnish vapours mixed with air, mostly depend on the properties of the solvents used.

The varnish should meet the following technical requirements:

Solid residue, %	40-65
Gelatinisation time at 200±2 °C, minutes, not more than	5
Si:Al ratio in the polymer	from 3:1 to 6:1

Polyalumophenylsiloxane varnish is used as a binding component in the production of heat-resistant plastics and fiberglass and as a hardener for organic and silicone polymers.

Similarly to polyalumophenylsiloxane varnish, one can also obtain other polyalumoalkyl(aryl)siloxane varnishes.

4.4.2. Preparation of polyironphenylsiloxane and polyironphenylsiloxane varnish

Polyironphenylsiloxane, as well as polyalumophenylsiloxane, is a polydisperse mixture of macro chains of ladder type I and branched type II structure.

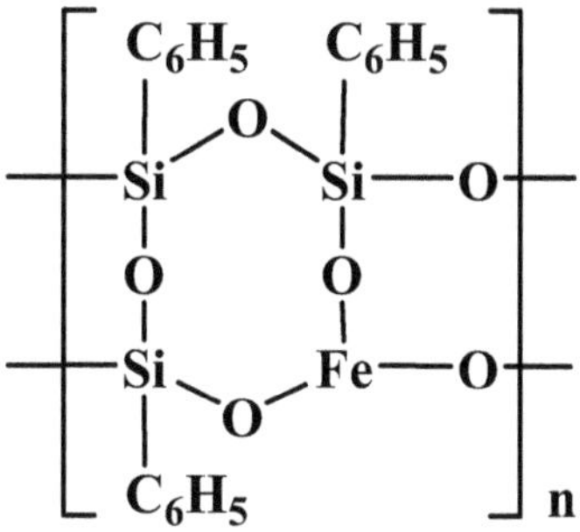

$$\left[\begin{array}{c} \overset{C_6H_5}{\underset{\mid}{}} \quad \overset{C_6H_5}{\underset{\mid}{}} \\ -Si \overset{O}{\diagdown} Si-O- \\ \mid \qquad \mid \\ O \qquad O \\ \mid \qquad \mid \\ -Si \diagdown_{O} Fe-O- \\ \mid \\ C_6H_5 \end{array} \right]_n \qquad \mathbf{I}$$

$$\{[C_6H_5Si(OH)_{2-x}O_{0.5x}(O)]_3Fe\}_n$$

$x=0 \div 1, \; n=1 \div 4$

Polyironphenylsiloxane is obtained in three stages. The first two stages
are similar to those of polyalumophenylsiloxane. The third stage is the exchange destruction of ethanol-toluene solution of anhydrous ferric iron
chloride and sodium dihydroxyphenylsilanolate:

$$3nC_6H_5Si(OH)_2ONa + nFeCI_3 \longrightarrow$$

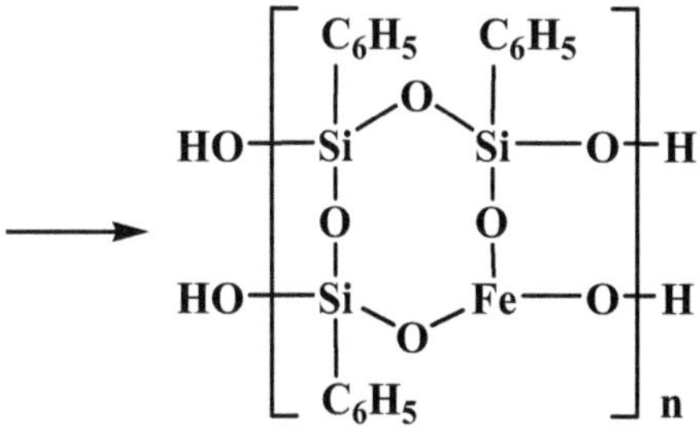

$$\longrightarrow \left[\begin{array}{c} \overset{C_6H_5}{\underset{\mid}{}} \quad \overset{C_6H_5}{\underset{\mid}{}} \\ HO-Si \overset{O}{\diagdown} Si-O-H \\ \mid \qquad \mid \\ O \qquad O \\ \mid \qquad \mid \\ HO-Si \diagdown_{O} Fe-O-H \\ \mid \\ C_6H_5 \end{array} \right]_n \quad + \; 3NaCI + (n-2)H_2O \qquad (4.54)$$

The technological diagram of the production of polyironphenylsiloxane
varnish is similar to the diagram given in Fig. 79. The difference in the
production of polyironphenylsiloxane is that during the exchange destruction sodium dihydroxyphenylsilanolate, which is synthesised in apparatus
8, receives from batch box 13 at 30-40 °C for 1-2 hours a pre-prepared
ethanol-toluene solution of anhydrous ferric iron chloride. After the salt
has been introduced, the reactive mixture is agitated at 30-40 °C for 2
hours and settled for 1-2 hours. The obtained solution of polyironphenylsiloxane is sampled to determine free alkali content, which is neutralised
with 20% ethanol-toluene solution of anhydrous ferric iron.

The necessary amount of $FeCl_3$ solution is calculated by the amount of
free alkali in the polyironphenylsiloxane solution. The amount of $FeCl_3$ solution (X, g) is calculated according to the formula:

$$X = \frac{162.2AB}{3 \cdot 40C}$$

where 162.2 is the molecular weight of $FeCl_3$; A is the content of free alkali in varnish; B is the amount of varnish, g; 40 is the molecular weight of NaOH; C is the concentration of $FeCl_3$ solution, %.

The reactive mixture is agitated in apparatus *8* for 30 minutes at 20-40 °C and settled for 1-2 hours. The separated neutral polymer solution is decanted out of apparatus *8* into a distillation tank (not shown in the diagram); the sediment of sodium chloride in apparatus *8* is dissolved with water; the flush waters are sent to biochemical purification.

The distillation of the solvent (the mixture of toluene and butanol) is carried out in a tank at a temperature not exceeding 50 °C and a residual pressure of 460±70 GPa. The distillation is considered complete when the nonvolatile content is 30-50%. The finished varnish is cooled in the distillation tank to 40 °C and analysed for the content of solid residue, alkali, silicon and iron.

Polyironphenylsiloxane varnish is a solution of polyironphenylsiloxane in a mixture of toluene and ethanol. The varnish should meet the following technical requirements:

Appearance	Motile dark brown liquid
Solid residue, %	30-50
Density of 40% solution, g/cm^3	1.040-1.050
Content in solid residue, %:	
of silicon	14-17
of iron	10-12

Polyironphenylsiloxane varnish is used as a catalyst for the chlorination of aromatic hydrocarbons in the nucleus and as a thermostabiliser for antiadhesion silicone coatings.

Of particular interest is the substance containing silicon and titanium atoms in the chain, which is a star-shaped polymer, oligotetrakis(methylphenylsiloxanohydroxy)titanium.

4.4.3. Production of oligotetrakis(methylphenylsiloxanohydroxy)titanium

Oligotetrakis(methylphenylsiloxanohydroxy)titanium (TMPht) is obtained by two-stage synthesis. The first stage is the reaction of methylphenylcyclotrisiloxane with methanol NaOH solution to synthesise oligo-α-sodiumoxy-ω-hydroxymethylphenylsiloxanes:

$$[CH_3(C_6H_5)SiO]_3 + NaOH \xrightarrow{CH_3OH} NaO-\underset{\underset{C_6H_5}{|}}{\overset{\overset{CH_3}{|}}{Si}}-O-R \quad \quad (4.55)$$

$$n$$

The parent methylphenylcyclotrisiloxane is obtained by the hydrolytic condensation of methylphenyldichlorosilane:

At the second stage the reaction of oligo-α-sodiumoxy-ω-hydroxymethylphenylsiloxanes with $TiCl_4$ in benzene yields TMPhT:

$$4NaO\left[\overset{\overset{CH_3}{|}}{\underset{\underset{C_6H_5}{|}}{Si}}-O\right]_3 R + TiCl_4 \xrightarrow{4NaCl} Ti\left[\left(O-\overset{\overset{CH_3}{|}}{\underset{\underset{C_6H_5}{|}}{Si}}\right)_x OR\right]_4 \quad (4.56)$$

where $R = -H, -CH_3; x = 2 \div 15$ (mostly 2-5).

Raw stock: methylphenylcyclotrisiloxane with the main substance content of at least 90% (weight); tetrachloridetitanium (at least 99% of $TiCl_4$); poisonous methanol (not more than 0.05% of water); sodium hydroxide (solid, 100%, analysis pure), with the main substance content of at least 98%); benzene ($d_4^{20} = 0.875 \div 0.880$ g/cm^3, moisture content is not more than 0.05%).

Periodic TMPhT production (Fig.80) comprises the following main stages: the production of sodium hydroxide solution in methyl alcohol; the preparation of oligo-α-sodiumoxy-ω-hydroxymethylphenylsiloxanes; the preparation of titanium tetrachloride solution in benzene; the synthesis of TMPhT; the filtration of TMPhT from sodium chloride; the distillation of the solvent and extraction of TMPhT; the filtration of TMPhT, the product.

Apparatus 1, thoroughly washed with benzene-methanol mixture and dried, is loaded with a necessary amount of methyl alcohol from batch box 2; the agitator is switched on and the apparatus is loaded with sodium hydroxide through a hatch. The agitation is continued until NaOH is completely dissolved. Then synthesis reactor 7 is loaded at agitation from batch box 3 with methylphenylcyclotrisiloxane; inverse cooler 5 is switched on and pre-prepared NaOH solution in methyl alcohol is supplied from agitator 1. The temperature during the introduction of the solution should not exceed 40 °C. After loading the reactive mixture is heated to 40-45 °C and held at this temperature for 4-5 hours. After that the reactive mixture is cooled to 10-20 °C; benzene self-flows from batch box 4 and the agitation is continued for 1 more hour.

To synthesise TMPhT, a benzene solution of $TiCl_4$ is prepared in thoroughly washed and dried agitator 6. For this purpose agitator 6 is loaded with benzene from batch box 4 and with a calculated amount of TiCl₄. For better mixing, TiCl₄ is loaded portionwise.

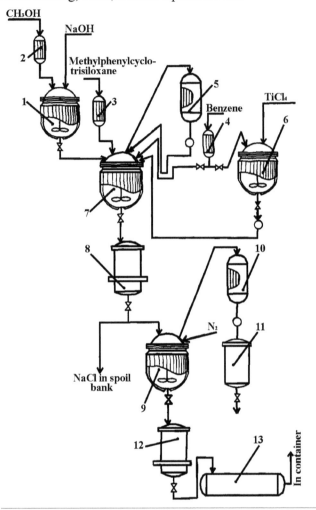

Fig.   19.   Production   diagram   of   oligotetrakis(methylphenyl-siloxanohydroxy)titanium: *1, 6* - agitators; *2-4* – batch boxes; *5, 10* - coolers; *7*-synthesis reactor; *8, 12* – vacuum filters; *9* – distillation apparatus; *11* – solvent collector; *13* – finished product collector

The process is conducted at agitation for 1 hour. The solution is prepared directly before introducing the mixture into the reaction. The synthesis of TMPhT is carried out in synthesis reactor *7*. The reactor with a pre-prepared mixture of oligo-α-sodiumoxy-ω-hydroxymethylphenylsiloxanes is filled at agitation from agitator *6* with benzene solution of TiCl₄ until the reaction of the reactive mixture is neutral. The temperature conditions for the synthesis of TMPhT (10-25 °C) are maintained by regulating the speed at which TiCI₄ is introduced and by sending water into the jacket of synthesis apparatus 7. The quantity of the TiCI₄ solution introduced is monitored with the help of a run-down box located below agitator 6. After the entire benzene TiCI₄ solution has been introduced and the reaction has become neutral, the mixture is agitated for 30 more minutes and periodically sampled to determine pH of the medium.

If the medium remains neutral, the reactive mixture is held for 1 more hour and then checked for pH again. The agitator is switched off and the obtained mixture of TMPhT solution with sodium chloride is sent by nitrogen flow to filtrate in vacuum filter 8. Filtering is conducted in apparatus 9 at a residual pressure of 66.6-93.3 GPa. Filtering paper with NaCl sediment is hauled away to refuse disposal.

After that the solvents are distilled in apparatus 9. In the system of solvent condensation (cooler 10 and receptacle 11) a vacuum is created (66.6-93.3 GPa). Cooler 10 is filled with cooled water. The solvents (benzene and methanol mixture) are distilled at agitation and a temperature in apparatus 9 not exceeding 45 °C. The solvent vapours are condensed in cooler 10 and collected in receptacle 11. The process of distillation is monitored through a run-down box on the solvent intake line. After the solvents are no longer released, the vacuum is released by nitrogen in cooler 10 and the product is sampled to check its appearance.

If the product is transparent, the distillation is finished; if it is milky white, the distillation is continued until transparent liquid is formed; then the product is sampled to determine the volatile content and VS-1 viscosity. The volatile content should not exceed 10%; the VS-1 viscosity should be 20-90 s at 20 °C.

To purify TMPhT from mechanical impurities, the product is subjected to additional filtration in vacuum filter 12. The ready product is loaded out of vacuum filter 12 into collector 13.

Oligotetrakis(methylphenylsiloxanohydroxy)titanium (TMPhT) is a light yellowish liquid. It is a mixture of oligomer homologues which contain 2-15 methylphenylsiloxy elements and can be easily dissolved in organic solvents.

Technical TMPhT should meet the following requirements:

Appearance	Transparent liquid; yellow, no mechanical impurities
Content, %:	
of titanium	1.5-3.0
of silicon	18.0-24.0
of volatile substances	≤10

Conditional VS-1 viscosity at 20 °C, s   20-90

TMPhT is used as a binding agent in various heat-protection compositions.

4.4.4. Preparation of polytitaniumpentenylsiloxane

Polytitaniumpentenylsiloxane is synthesised in two stages. The first stage
includes the hydrolysis of pentenyltrichlorosilane and partial condensation
of the hydrolysis product:

$$C_5H_9SiCl_3 + 3H_2O \xrightarrow[3HCl]{} C_5H_9Si(OH)_3 \qquad (4.57)$$

$$nC_5H_9Si(OH)_3 \xrightarrow[nH_2O]{} [C_5H_9Si(OH)O-]_n \qquad (4.58)$$

The second stage is the heterofunctional polycondensation of polypentenylhydroxysiloxane formed at the first stage and tetrabutoxytitanium:

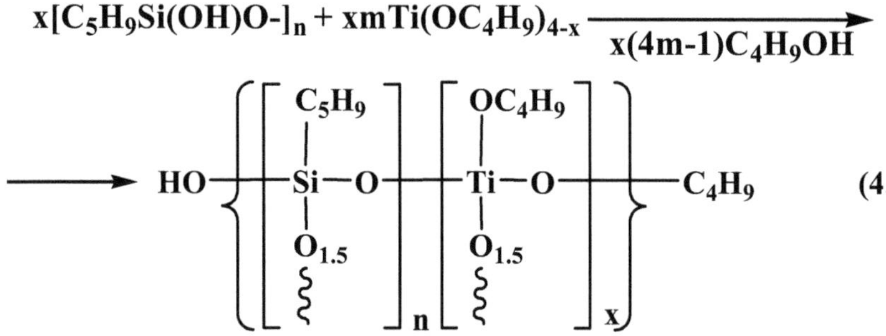

$$x[C_5H_9Si(OH)O-]_n + xmTi(OC_4H_9)_{4-x} \xrightarrow[x(4m-1)C_4H_9OH]{}$$

$$\longrightarrow HO \left\{ \left[\begin{matrix} C_5H_9 \\ | \\ Si-O \\ | \\ O_{1.5} \\ \end{matrix} \right]_n \left[\begin{matrix} OC_4H_9 \\ | \\ Ti-O \\ | \\ O_{1.5} \\ \end{matrix} \right] \right\}_x C_4H_9 \qquad (4.59)$$

n:m=from 4:1 to 5:1

Raw stock: pentenyltrichlorosilane (the boiling point is 158-168 °C, the
chlorine content is 51.8-52.8%); tetrabutoxytitanium (the boiling point is
179-186 °C at 13.3 GPa, the titanium content is 13.8-17%); toluene (the
boiling point is 109.5-111 °C , $d_4^{20} = 0.865\pm0.002$); acetone (the boiling
point is 56.2 °C, $d_4^{20} = 0.7908$).

The production process of polytitaniumpentenylsiloxane (Fig.81) comprises two main stages: the hydrolytic polycondensation of pentenyltrichlorosilane; the heterofunctional polycondensation of the polypentenylhydroxysiloxane formed and tetrabutoxytitanium.

Hydrolytic polycondensation is carried out in enameled hydrolyser 5
with a water vapour jacket and an agitator. The apparatus is loaded with a
calculated amount of water, toluene and acetone. The agitator is switched
on, and pre-prepared mixture of pentenyltrichlorosilane and toluene is sent
from batch box 3 at such speed that the temperature in the hydrolyser does
not exceed 35 °C. After the entire solution of pentenyltrichlorosilane has

been introduced, the mixture is agitated at 20-30 °C for 1 more hour. Then
the mixture is settled (the agitator is switched off) and split. The lower acid
aqueous layer is separated and poured through a run-down box into collector 6; the top layer (solution of the hydrolysis product) is flushed warm
water (about 40 °C) until it gives a neutral reaction (pH is 6-7). The neutral
product is sent through a run-down box into nutsch filter 7.

The filtered solution is sent into vacuum distillation tank 9. There a residual pressure of 15-30 GPa is created; at a temperature not exceeding 50
°C the solvent (a mixture of toluene and acetone) is distilled. The distillation is carried out until the polymer concentration is 85-90% and monitored by solid residue. The distillation of the solvent is accompanied by the
polycondensation of the hydrolysis product and forms polypentenylhydroxysiloxane.

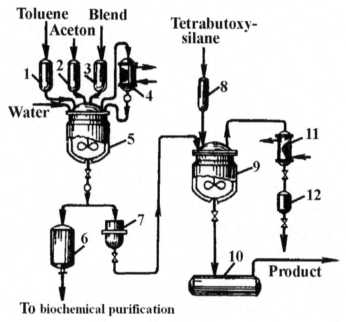

Fig. 20. Production diagram of polytitaniumpentenylsiloxane: *1 - 3, 8* – batch
boxes; *4, 11* - coolers; 5 - hydrolyser; *6,12* - collectors; 7 - nutsch filter; *9* – vacuum distillation tank; *10* - container

At the end of the distillation the content of hydroxyl groups in polypentenylhydroxysiloxane is determined; it should not be less than 7%. The distilled mixture of toluene with acetone is condensed in cooler *11* and flows
into collector *12*.

The heterofunctional polycondensation of polypentenylhydroxysiloxane and tetrabutoxytitanium is carried out in the same distillation tank *9*, which also contains the obtained polypentenylhydroxysilane. It is loaded at agitation with a necessary amount of tetrabutoxytitanium.

The quantity of tetrabutoxytitanium is calculated by the amount of polypentenylhydroxysiloxane and its number of hydroxyl groups. For example, we have 150 kg of polypentenylhydroxysiloxane with 7.5% of hydroxyl groups; i.e. the number of hydroxyl groups in polypentenylhydroxysiloxane is $0.075 \cdot 150 = 11.25$ kg. It follows that:

$$X = \frac{11.25 x 73.12}{17.01} = 48.36$$

where X is the quantity of tetrabutoxytitanium, kg; 73.12 is the molecular weight of the butoxyl group; 17.01 is the molecular weight of the hydroxyl group.

After the whole of tetrabutoxytitanium has been introduced, a residual pressure of 860-890 GPa is created in tank *9*; heterofunctional polycondensation is carried out at constant agitation, gradually increasing the temperature to 140-150 °C. The process is monitored by the relative viscosity of the 10% toluene solution. When the relative viscosity is 1.25-1.35, the process is stopped and the finished product, polytitaniumpentenylsiloxane, is loaded at 100 °C into container *10*.

Polytitaniumpentenylsiloxane is a product which can be dissolved in common organic solvents and ranges in colour from light brown to brown. It should meet the following technical requirements:

Relative viscosity of 10% toluene solution   1.25-1.35
Si:Ti ratio in the polymer   from 4:1 to 5:1

Polytitaniumpentenylsiloxane is used as a modifier for various organic and silicone polymers.

Similarly to polytitaniumpentenylsiloxane, one can also obtain other polytitaniumorganosiloxanes.

4.4.5. Safety measures, fire prevention and health measures in silicone production

Safety engineering, fire prevention and health measures are vital in the chemical industry, particularly in the production of silicone compounds.

At plants producing silicone and other elementorganic compounds and at other chemical facilities safety measures and fire prevention are an integral part of the production process. Apart from engineering instructions, it is as necessary to observe safety and fire prevention instructions, as well as

technological regulations. Besides a comprehensive description of the
properties of raw materials, semiproducts and the finished product, the
characteristics of the main and secondary chemical processes, the descrip-
tions of the equipment, etc., regulations should also include special chap-
ters describing safe methods for conducting the process, techniques to test
the quality of raw materials, most important technological parameters connected with health, safety and fire prevention measures, as well as faulty
operations with possible reasons and remedies.

The production of many silicone compounds makes use of substances
which have a harmful effect on human body; technological processes are
often carried out at high temperatures and pressures. Therefore, safety
measures are especially important.

Major hazards that can lead to accidents include the following:

1. poisoning with harmful substances;
2. chemical burns (with acids, liquid ammonia, alkali, chlorosilanes and
   other aggressive chemical substances) and thermal burns (boiling solutions, hot water, vapour, inflamed gases, incandescent contact mass);
3. frostbites (with liquid ammonia, solid carbon dioxide and other refrigerants);
4. mechanical injuries (cuts, scrapes, bruises, dislocations, bone fractures)
   caused by the violation of safety rules when servicing moving mechanisms, machines, elevators and other equipment;
5. electric shock (when servicing electric equipment or coming in contact
   with naked cables and wires);
6. the danger of being run over by rail, automotive or other kind of transport on the premises;
7. contamination of water bodies with waste waters.

Industrial poisoning. The production of silicone products uses substances harmful for human health. These are inorganic substances (ammonia, chlorine, sodium and potassium hydroxides, sulfuric and hydrochloric
acids, hydrogen chloride) and organic compounds of various types, such as
hydrocarbons (methane, benzene and its homologues), chlorine derivatives
(methyl- and ethylchloride, chlorobenzene), alcohols (methyl, ethyl, n-
butyl, hydrosite), acetone, pyridine, etc. The information about their toxicity, explosion hazard, effect on human body, as well as maximum allowable concentrations of gases and vapours in the air at workplace can be
found in special references.(Ryabov 1970). A comprehensive description
of silicone substances is given in Table 29.

Table 28. Toxicity of silicone compounds and main substances used in their production

Expl.lim. is the explosive limits, MAC is the maximum allowable concentration in air at workplace.

Substances and their characteristics	Effect on human body	Defensive measures
Contact mass is a solid grey (fresh) or black (discharge) substance The presence of carbon contributes to high flammability in air MAC =4 mg/m^3 (for fresh mass)	Prolonged inhalation of the dust irritates the respiratory tract; chronical exposure causes silicosis	Breathing mask
Silicon tetrachloride is a colourless or yellowish liquid. It is easily hydrolysed, releasing hydrogen chloride. MAC=1 mg/m^3.	It irritates mucous membranes of eyes and upper airways. Skin contact causes burns.	Special gas mask, rubber gloves, goggles
Methylchlorosilanes (methyltrichlorosilane, dimethyldichlorosilane, methyldichlorosilane, trimethylchlorosilane or a mixture of them) are colourless or yellowish liquids. They are easily hydrolysed, releasing hydrogen chloride. Expl.lim. =1.2-91% (vol.). MAC=1 mg/m^3.	They irritate mucous membranes of eyes and upper airways. Affect the nervous system. Skin contact causes burns.	The same
Ethylchlorosilanes (ethyltrichlorosilane, diethyldichlorosilane, triethylchlorosilane, etc. or a mixture of them) are colourless or yellowish liquids. They are easily hydrolysed, releasing hydrogen chloride. MAC=1 mg/m^3.	The same	The same
Phenylchlorosilanes (phenyltrichlorosilane, phenyldichlorosilane, diphenylchlorosilane, etc.) are transparent, yellowish motile liquids. They are easily hydrolysed, releasing hydrogen chloride. Expl.lim. =0.8-77.5% (vol.). MAC=1 mg/m^3.	"	"
Hydroxyorganosiloxanes are motile liquids or solid crystal products. They are easily condensed, releasing water.	They cause fatty degeneration of liver, burn haemorrhages in lungs and pneumonia. No effect on skin has been noted.	Special gas mask or breathing mask, rubber gloves

Table 29. (cont.)

Aminoorganoalkoxysilanes [methyl(phenylaminomethyl)didiethylaminomethyltriethoxysilane, γ-aminopropyltriethoxysilane, etc.] are motile yellowish or dark maroon liquids. In the presence of water they gradually hydrolyse. Under normal conditions do not form explosive mixtures due to small volatility. MAC=1 mg/m^3.	They have a narcotic and irritating effect; cause skin irritation. Prolonged exposure may cause chronic poisoning of internal organs.	Special gas mask, ethylethoxysilane, rubber gloves.
Hexamethyldisilazane is a nonmotile liquid; can be hydrolysed with water, releasing ammonia. Volatile, flammable. MAC=5 mg/m^3.	Narcotic. Its vapours cause the inflammation of mucous membranes and the central nervous system; they change the composition of blood.	Special gas mask
Silicone and elementosilicone varnishes (solvents are toluene, acetone, butyl alcohol, etc.).	The effect on human body is determined by the solvent content.	The same

To prevent poisonings and professional or chronic diseases, it is necessary to conduct all stages of the technological process excluding direct contact of workers with toxic substances. Therefore, special attention should be given to wide implementation of automatic and distance control of production, complete airtightness of equipment and piping, mechanisation of transport and loading of harmful substances (e.g. the use of pneumatic transport to load contact mass for organochlorosilane synthesis excludes all direct human contact with harmful dust); absorption of released harmful gases and dust in special devices (traps, towers, scrubbers, etc.).

Equally important are supplementary safety measures like ventilation devices, personal protective equipment, overalls, etc. The efficiency of ventilation devices depends on the air circulating factor. The necessary air circulating factor is established in each case depending on the amount of harmful substances in air and their toxicity. In the production of silicone monomers and polymers the factor should not fall below 3; for especially harmful workshops, the factor should be as high as 20.

If in some production divisions due to construction characteristics of the equipment or other reasons it is impossible to exclude direct human contact with harmful substances, the decisive part is played by supplementary safety measures.

In case of skin contact with organochlorosilanes or other poisonous substances one should wipe the affected area clean and wash it under a strong stream of water.

Chemical and thermal burns. The causes of chemical burns include the effect of concentrated acids, alkali, liquid ammonia, chlorosilanes and other aggressive substances. Thermal burns are caused with boiling solutions, hot water, vapour, inflamed gases, incandescent contact mass. Preventive and protective measures mostly include strict observance of all the established technological regulations and equipment maintenance order.

The use of necessary protection devices is obligatory. Since eyes are particularly vulnerable to burns, one should use protective goggles. All workers in contact with acids and alkali should wear proper overalls, rubber boots, gloves, etc.

In case of acid or alkali burn it is necessary to wash the affected area immediately under a strong jet of water and then with a 2-3% soda solution (for acid or chlorosilane burns) or 2-3% acetic acid (for alkali burns). It should be kept in mind that the degree of the burn can be considerably reduced if the affected areas are treated timely and correctly. If caustic drops are found in eyes, they should also be washed.

Mechanical injuries (cuts, scrapes, bruises, dislocations, broken bones, etc.) are as a rule caused by the violation of safety rules when servicing moving mechanisms, machines, elevators or faulty equipment; they can also be caused by breakdown of equipment or piping.

To prevent accidents, all pressure apparatuses (reactors, tanks) should be subjected to state technical inspection

for examination and testing. They can be commissioned only after obtaining the necessary permission. It is also necessary to see to the proper functioning of manometers and safety valves in pressure apparatuses.

Bruises are signified by swellings, haematomas and pain. The bruised should be given first before-doctor aid, such as applying a cold wash and pressure bandage to the bruised area (to reduce inner haemorrage).

In case of dislocations, it is vital to find the most comfortable and immobile position for the dislocated limb. In case of small dislocations after giving the patient first aid he should be taken to the doctor, since dislocations are easier to set within the first few hours after the injury. If a bone is broken, the victim needs to be immobilised. If no such position is to be found when the victim feels the least pain, the injured limb should be splinted. If the fracture is open, sterile bandage should be applied.

In case of cuts and other wounds the skin around the wound should first of all be covered with iodine; sterile bandage should be applied. The

wound may not be bathed in any other solution because an infection can
enter it. First aid is given with the help of a first-aid kit at the workplace.

Electric shock can be general or local, i.e. electric current passing
through the body either shocks the entire system or burns some body parts.
General electric shock is the most dangerous of the two.

All personnel should be aware of the rules and know how to give first
aid to a victim of electric shock. In this case electicity should be immediately switched off; if it is not possible, the wires should be cut with scis-
sors of a knife with insulated handles. The victim should be given artificial
respiration at once.

Fire prevention

The probability of fires and explosions at plants producing various silicone monomers and polymers is significant, since many kinds of raw materials, intermediate and finished products are flammable and explosive.
Therefore, silicone production belongs to fire risk category A because it
makes use of flammable raw materials (the flash point is lower than +28
°C).

Flammable and explosive substances. Many substances can burn in
the dry state in the presence of oxygen in air. However, only those are considered flammable that require special precaution measures.

We list below the most flammable and explosive substances among
those used in silicone production.

One of the most flammable and explosive gases is hydrogen, which is
released in the production of hydrideorganochlorosilanes. It interacts with
oxygen in air to form detonating gas with very wide explosive limits [4.15-
75% (vol.)]. Explosions of detonating gas are rare in manufacture, since
hydrogen is much lighter than air and quickly escapes; however, they are
quite possible when hydrogen is formed in a closed space (e.g. in an autoclave during the production of methylphenyldichlorosilane). The explosion may be caused by a spark produced by a chance steel object hitting
steel parts or walls. Ammonia and methane, which forms in the direct synthesis of methylchlorosilanes, are also exposive.

Flammable liquids include organic solvents widely used in silicone production, such as benzene, toluene, xylene (a mixture of isomers), chlorobenzene, methyl, ethyl and butyl alcohols.

Fire risks are presented by low-boiling silicone monomers (e.g. methyldichlorosilane) and raw stock for preparing various silicone monomers
(e.g. trichlorosilane), because these liquids have very low flash and self-
inflammation points and wide explosive limits:

	Flash point °C	Self-inflammation point, °C	Explosive limits, % (vol.)
Methyldichlorosilane	Below -70	175	0.2-91
Trichlorosilane	Below -50	175	1.2-90.5

Pure methyldichlorosilane does not inflame by shock; however, it immediately inflames by contact with minium, lead dioxide, copper and silver oxides. Pure trichlorosilane does not self-inflame in air (excluding the possibility of spark formation by electrostatic charge); neither does it self-inflame by shock. However, since technical trichlorosilane almost always contains dichlorosilane SiH_2Cl_2 (the boiling point is 8.3 °C), capable of self-inflaming by shock, trichlorosilane can also inflame by shock. Thus, if technical trichlorosilane contains more than 0.2% of dichlorosilane, one should avoid shocks and pushes when it contacts air.

Another flammable solid substance is discharge contact mass, which can also self-inflame.

Fire and explosion prevention measures. First and foremost, the prevention of fires and explosions requires strict observance of technological regulations and safety and fire prevention rules. Explosive gases and vapours must not form mixtures in air. The equipment should be airproof, the ventilation should be efficient enough to eliminate flammable gases and vapours quickly. When transporting and loading flammable liquids, it is preferable to use piping. Flammable products should be handled with care. They should not be spilled or scattered; if it does happen, they should be immediately washed away with water or collected.

It is necessary to prevent chemical reactions from becoming too active; e.g. the hydrolysis or ammonolysis of organochlorosilanes or the Grignard reaction for the synthesis of alkyl(aryl)alkoxysilanes.

Fire extinction measures. Most silicone monomers and oligomers of various technological mixtures and products containing organic solvents are highly flammable liquids; high-molecular silicone and elementosilicone compounds belong to combustible substances or hard-combustible products; incombustible products are only mixtures containing inorganic substances (water, mineral salts, metal or silicon oxides).

Inflamed silicone products should be extinguished with the most common fire extinguishing means. In most cases it can be water. Many organochlorosilanes, which are easily hydrolysed with water, should be extinguished by sprayed water and air-mechanical foam. Because of the high toxicity of organochlorosilanes, the products of their hydrolysis and combustion (chlorine, fine Aerosil and hydrogen chloride), fire extinguishing should be carried out using all necessary means for individual protection.

When hydrogen-containing chlorosilanes are extinguished with water and foam, it should be remembered that their hydrolysis forms hydrogen; therefore, there should be some measures to prevent explosions, e.g. withdrawal of hydrogen with the helf of aspirators or forced draft.

Because the hydrolysis of organochlorosilanesis is acommpanied by the formation of highly flammable silicone products, water is not an effective fire extinguisher in this case. Nor can it be used for many organic solvents widely used in the production of silicon products, or products containing an organic solvent (technological mixtures, varnishes, lubricants, compounds, etc.). In these cases it is recommended to use air mechanical foam. Foam is more advisable for organochlorosilanes than water, since it reduces the amount of released toxic substances and flame buildup (the ratio of flame height at the initial moment of extinguishing and the maximum height at burning).

Air mechanical foam is the most universal fire extinguisher in silicone production. The introduction of 20-30% (weight) of urea into the solution of foaming agent improves its performance and neutralising characteristics.

A classification of silicone products has been suggested. According to this classification, all silicone products can be divided into four groups.

1. Polymer and oligomer silicone products belonging to the category of combustible and hard-combustible substances (e.g. different brands of PMS, SKTN, compounds). They can be extinguished by all standard means: compact and sprayed water; air mechanical foam; powder compositions; inert gases; carbon dioxide and refrigerants like tetrafluorodibromethane or trifluorobromethane.

2. Solutions of polymer and oligomer silicone products in organic solvents (e.g. varnishes, lubricants, technological mixtures). The fire risk of these products is determined by the inflammability of the solvent, which generally belongs to highly flammable liquids. Solutions of silicone products in water-insoluble organic substances, e.g. varnishes with solvents (benzene, toluene) should be extinguished with air mechanical foam. In some cases powder compositions can be used. Water, as a rule, has no effect. Silicone products dissolved in water-soluble or partially soluble substances (technological mixtures based on lower alcohols, acetone, GKZh-11) should be extinguished with water foam and powder compositions. Water foam substances are effective when diluted with water below a certain concentration. The most efficient foam-forming substances are "Universalny" and "Foretol".

3. Monomer organosilanes and oligomer organosiloxanes belonging to highly flammable liquids. According to the nature of their interaction

with water, they can be divided into two subgroups: highly combustible substances that are not hydrolysed with water (e.g. tetraethylsilane, hexamethyldiethoxysilane) and that are hydrolysed with water, forming highly combustible liquids (e.g., organosilanes, hexamethyldisilazane). They should be extinguished with air mechanical foam containing such foam forming agents as "Universalny" and "Foretol".

4. Hydrogen-containing silicone compounds present the greatest fire risks before and after hydrolysis (e.g., trichlorosilane, methyldichlorosilane, dimethylchlorosilane), because apart from hydrogen chloride their hydrolysis forms hydrogen. Therefore, rooms where these products are either stored or used should be provided with facilities preventing the formation of explosive concentration of hydrogen both in daily work and during or after the extinguishing of fires.

Hydrideorganosilanes should be extinguished with air mechanical foam and in some cases with sprayed water (the intensity of water supply from the top should be limited).

A rather universal substance for volume extinction of fire is carbon dioxide, which is kept in liquid form in vessels under excess pressure. Carbon dioxide devices and fire extinguishers are used to extinguish highly flammable liquids. If air contains 12-15% (vol.) of CO_2, the fire stops. 1 kg of liquid CO_2 produces 500 l of gas. Carbon dioxide is sent out of vessels as gas and "snow". Snow CO_2 has a strong cooling effect and is usually supplied directly onto the burning surface.

Industrial health measures

Workshops and other rooms at factories should be equipped with ventilation, central heating, light, water and sanitation according to health regulations in industry. Workers and engineers should periodically undergo medical examination according to the factory schedule.

Biochemical purification of waste waters in silicone production

At present much attention is given to environmental protection.

Waste waters in the production of silicone monomers and polymers is an important economic issue. It is connected with environmental protection from air contamination and pollution of rivers, seas and other water bodies. Especially important in silicone production is the purification of acid waste waters from organic impurities (toluene, benzene, methyl alcohol, etc.).

Major silicone factories, which flush waste waters into waste disposal works, produce silicone monomers, liquids, varnishes, ethylsilicates and

other silicone and elementosilicone polymers. An effective technique for
purifying industrial waste waters is their biochemical purification.

The process of biochemical purification of industrial waste waters consists of biochemical destruction of organic impurities under the influence
of biocenosis, the complex of all microorganisms in waste disposal works.
The major part is played by bacteria which are capable of accumulating
into zoogleas, i.e. active silt in the form of dark brown flakes. These flake-
like aggregates form mostly by the interaction of fibrillar polymers produced and adsorbed by bacterial cells. The mobility of the flakes is connected with the metabolic processes in the cells, which change the surface
of the flakes.

Active silt is an agglomeration of bacteria, including ray fungi, sapro-
legnia, yeast, as well as protozoa and some complex organisms (rotifera,
worms, larvae, etc.). The qualitative and quantitative ratio of different mi-
croorganism in active silt varies depending on the composition of waste
waters and environment. The composition of the organisms characterises
the function of waste disposal works. Efficient active silt contains 8-10
types of microorganisms; one of them may slightly predominate. Active
silt is most stable in neutral and low-alkalinity media; that is why waste
waters from silicone factories with hydrogen chloride are first neutralised
with pre-prepared lime milk.

$$CaO \xrightarrow{\ H_2O\ } Ca(OH)_2 \xrightarrow{\ 2HCl\ } CaCl_2 + 2H_2O \quad (4.60)$$

After that, neutralised waters are subjected to biochemical purification.

The mechanism of biochemical purification can be conventionally divided into three stages: 1) the movement of organic material in the liquid
to the surface of the microbe cell; 2) the diffusion of organic material
through semipermeable membranes with the help of carrier molecules, or
special coferments; 3) the metabolism of diffused products. The third stage
in the microbe cell consists of two simultaneous and interconnected processes: the oxidation of organic substances and the synthesis of cytoplasm,
i.e. bacterial cell.

1. The oxidation of organic substances (the process of cell breathing) occurs thus:

$$C_xH_yO_z + (x+0.25y-0.5z)O_2 \longrightarrow xCO_2 + 0.5yH_2O + \Delta E \quad (4.61)$$

2. The synthesis of the bacterial cell (the process of feeding) occurs thus:

$$nC_xH_yO_z + nNH_3 + n(x+0.25y-0.5z-5)O_2 \longrightarrow$$

$$\longrightarrow n(C_5H_7NO_2) + n(x-5)CO_2 + 0.5n(y-4)H_2O - \Delta E \quad (4.62)$$

with subsequent oxidation of cell material (the cell breathing process):

$$n(C_5H_7NO_2) + 5nO_2 \longrightarrow 5nCO_2 + 2nH_2O + nNH_3 + \Delta E \qquad (4.63)$$

In this case x, y and z are positive numbers or zero; ΔE is the amount of released or absorbed energy; the ratio between the coefficients changes with the chemical composition of bacterial cytoplasm.

Thus, as a result of microorganisms feeding on organic substances, biomass increases and organic substances are broken down into H_2O and CO_2.

The synthesis of the bacterial cell uses not only organic substances, but also biogenic elements, mostly nitrogen and phosphorus, and insignificant amounts of potassium, magnesium, calcium, sulfur and iron. These substances can be usually found in waste waters. Nitrogen and phosphorus are sent into waste disposal plants in the form of aqueous solutions of ammonia sulfate and superphosphate (without mechanical impurities) to stimulate the growth of bacteria and oxidation of carbon-containing substances. If there is a lack of nitrogen or phosphorus, silt deposits and grows slowly; if there is a lack of dissolved oxygen and biogenic additives, the speed and completeness of biochemical degradation decrease.

To preserve a certain composition of the silt (contributing to quick oxidation), it is regenerated. The undissolved organic substances sorbed by silt are oxidated, and the number of viable microorganisms in active silt increases. Regeneration is carried out in the presence of oxygen in air and biogenic additives to support the biochemical processes in the body of the bacterial cell.

The technological process of the biochemical purification of waste waters in silicone production comprises the following stages: the neutralisation of acid waste waters; the averaging and mechanical purification of waste waters; the biochemical purification of waste waters and the dehydration of the sediment (Fig. 82).

Quicklime (in the form of lumps) is periodically loaded with slusher *1* into apparatuses *2* for slaking; the apparatuses are filled with technical water through vapour-heated shell-and-tube heat exchanger *3*. Slaking forms lime milk and solid residue (cinder). The cinder is sent out of apparatuses *2* with facilities for forced unloading into transporter *4*; lime milk flows down a chute into screw trap *5* and reservoirs *6*. Small pieces of cinder are sent from the trap with screws to the transporter. The cinder in the form of dirt is transported into bunker *7* and periodically removed. Lime milk in reservoirs *6* is constantly mixed so that the suspension does not settle. The weight ratio of $Ca(OH)_2$ is determined by the density of the sample of lime milk from the reservoirs; it should be 10 - 18%.

Acid waste waters from silicone production enter neutraliser *9*, a concrete container with two agitators, lined with acid-resistant tile from epoxy
resin. Here the waste waters are neutralised with lime milk, which is automatically sent from reservoirs *6* with pumps *8*. The pumps automatically
switch off when pH in the neutraliser is 8-10 and switch on when pH is below 5. Neutral waste waters continuously self-flow through a collector into
receiving reservoir *10*; as they accumulate, they are automatically sent
with pumps *11* for averaging and mechanical purification.

When the sediment accumulates, it is sent out of neutraliser *9* with
pump *12* into dirt settling box *13*; from there the aqueous layer is poured
into reservoir *10*. The settling box is periodically cleaned and the cinder
dirt is removed.

The neutralised waste waters are sent with pumps *11* through a pressure
pipeline into receiving chamber *14* (flow damper) and then into the distribution pan of averager *15*. The averager is a ten-section open concrete
monolithic reservoir with a bubble system. The homogeneity of waste waters distribution through sections *1-10* is determined visually. To accelerate
the averaging process, waste waters are introduced through shield gates in
the A row into sections *1-5* and in the B row into sections *6-10*. The averaging is achieved by constant bubbling of the waste waters with air entering through the distribution system of the pipeline.

The averaged waste waters with mechanical impurities (in the suspension and colloid states) are poured into the diagonal pan of averager *15* and
are sent into distribution pans of parallel primary settling boxes *16*. The
settling boxes are open concrete monolithic cylindrical reservoirs with a
conical bottom. Waste waters enter the lower part of the settling boxes, the
conical shield, which plays the role of a bumper. Exiting the central pipes
of the settling boxes, the waste waters slow down, change the direction and
slowly move up. Part of the impurities settles on conical bottoms, and suspended impurities (gellike) rise to the surface of the clarified waste waters.

The sediment from the conical bottoms of the settling boxes is continuously withdrawn into reservoir *17*. It is partially removed as dirt and partially sent with pumps *18* and *26* to installation *27* to dehydrate. At the
stage of biochemical purification the clarified waste waters from settling
boxes *16* self-flow into the top channel of the clarified water of aerotank
10 . The aerotank is a six-section open concrete precast reservoir with
bubble and aeration systems. All sections of the aerotank function parallel
to each other. Each consists of two corridors: the oxidation corridor, where
the process of biochemical purification takes place, and the regeneration
corridor, which serves to prepare silt for purification in the oxidation corridor.

Solutions of ammonia suphate and superphosphate (biogenic additives) are prepared in tank *20*, an open concrete apparatus with a bubble system.

The quantity of superphosphate and ammonia sulfate is calculated by the phoshorus (8-10 mg/l) and nitrogen (12-40 mg/l) requirement of the bacteria. It depends on the contamination degree of the waste waters and their quantity by the formula:

$$P = \frac{DV \cdot 100}{tn}$$

where P is the weight of superphosphate or ammonia sulfate; D is the portion of phosphorus or nitrogen, kg/m^3; V is the daily consumption of waste waters, m^3; t is the period between alternating temperings, h; n is the amount of phosphorus or nitrogen in dry biogen, %.

The hatch grate of the first section of tank *20* is loaded with superphosphate; the spraying device is filled with water. Undissolved superphosphate and the solution flow into the first section. It is filled with water (to half of the space) and air (through a bubbler) to agitate the solution and dissolve the rest of the superphosphate. Then the solution is settled. The clarified solution is pumped into the second section of tank *20*. The hatch grate of the second section is loaded with ammonia sulfate; the spraying device is filled with water.

The concentrated solution is poured into the tank of the second section, which is also filled with water (up to the top container). The solution of ammonia sulfate and superphosphate prepared thus

is continuously sent with pumps *21* into aerotank *19*. A break of no longer than 30 minutes in the supply of biogenic additives is tolerated.

Waste waters from the top channel of the clarified water from the aerotank are poured into the oxidation corridor, which is filled with prepared silt out of the regeneration corridor.

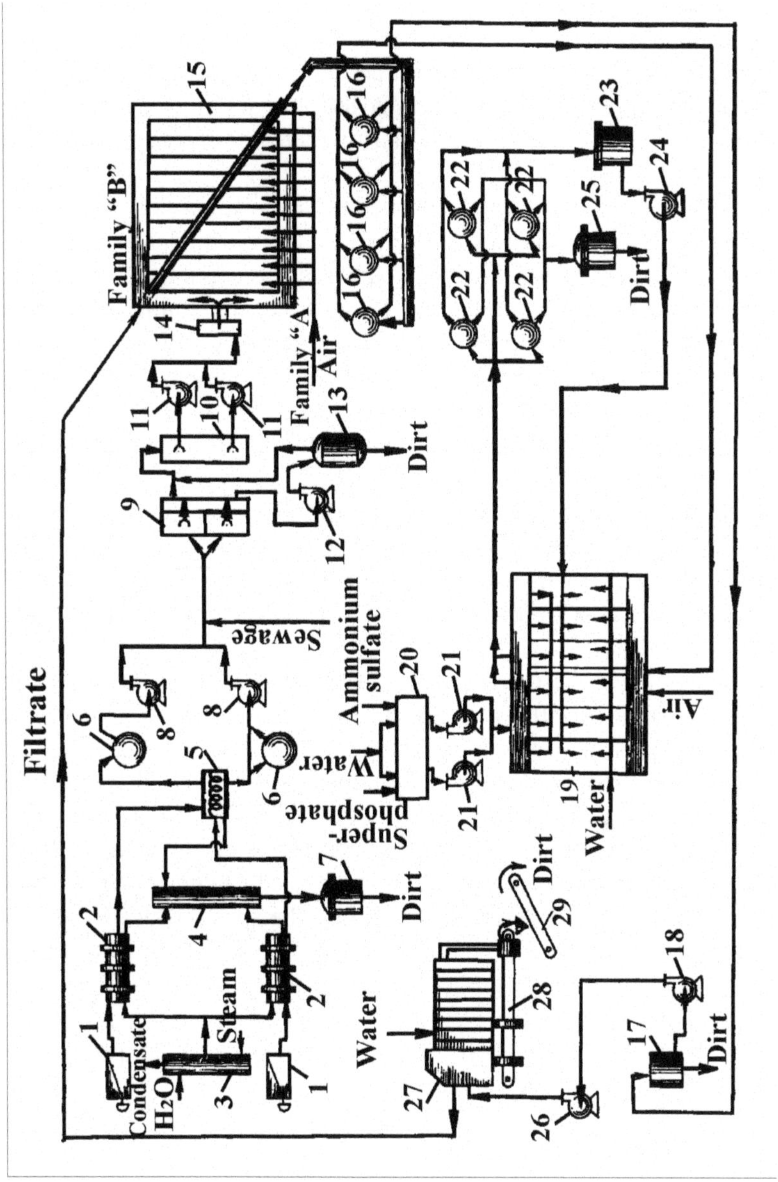

Fig. 21. Diagram of the biochemical purification of waste waters: 1 – slushers; 2 – slaking apparatuses; 3 – heat exchanger; 4, 28, 29 - transporters; 5 – screw trap; 6, 10, 17, 23 – reservoirs; 7 - bin; 8, 11, 12, 18, 21, 24, 26 – pumps; 9 - neutraliser; 13, 16, 22, 25 - settling boxes; 14 – damping chamber; 15 – averager; 19 – aero-tank; 20 – biogenic tank; 27 – pressure filter

Silt adsorbs impurities from waste waters; the life activity of aerobic microorganisms destroyes fine suspensions, colloid and dissolved organic substances. At the same time the reproduction of microorganisms and increase of their sorbing surface increases the amount of active silt. The aerotank is filled with water through filtering plates or other aerators in air piping.

The oxidation corridor releases purified waste waters with active silt. They self-flow (through the lower channel of the purified waste waters) into secondary settling boxes *22*, where active silt is separated. According to their structure and principle, these settling boxes are similar to settling boxes *16*. Active silt is constantly withdrawn into reservoir *23*, from where it is pumped with pump *24* to regeneration in the aerotank in the amount of up to 50% of the waste waters sent to purification. The regeneration corridor is filled with air (through aerators) and solution of biogenic additives (through a system of pipelines).

The regeneration time is determined by the time in which silt passes from the beginning of the regeneration corridor to the beginning of the oxidation corridor. Purified waste waters self-flow out of settling boxes *22* through pans into dirt settling box *25*. Dirt is removed from there as it accumulates.

The installation for the dehydration of sediment functions periodically and can recycle up to 40-50% of sediment from primary settling boxes *16*. The installation consists of "step-up" pump *26*, automatic pressure filter *27* and transporters *28* and *29*. Pressure filter *27* is used for poor-filtering fluid suspensions with the liquid phase concentration up to 500 g/l.

The filter operates in the "filtering – pressing – drying – unloading" mode. The sediment from the averaging stage is sent with pumps *18* and *26* out of reservoir *17* to pressure filter *27* into the space between the diaphragm and filtering cloth. The sediment remains on the cloth, whereas the filtrate is pressurised through the cloth, enters the diversion collector and self-flows into averager *15*.

After biochemical purification, waste waters in silicone production should meet the following requirements:

pH	6.5-8.5
Content, mg/l, not more than:	
of iron	0.5
of copper	Absent
of silicon	7
of aromatic hydrocarbons	Absent
of alcohols	35
of chlorides	3000
of suspended substances	200
of sulfates	80

of solid residue	6000
Dissolved oxygen, mg/l, at least	4
Colouring	Absent

Waste waters purified from organic impurities by this technique are absolutely harmless.

5. Chemistry and technology of other elementorganic compounds

Due to the growing practical importance of elementorganic compounds, their chemistry and technology are developing at unprecedented speed. Elementorganic substances are used in various fields of technology and economy. E.g., the simplest organoaluminum compounds, aluminumtrial-kyls, are used as a component of complex catalysts in the production of valuable isotactic polyolefines. Some organophosphorus and organotin substances have proved efficient against agricultural pests. Tetraethyl- and tetramethyllead are still used as antiknock additives in fuels, etc. This enumeration of the applications of elementorganic compounds, by no means complete, accounts for the recent development of their industrial production.

5.1. Preparation of organoboron compounds

Some of the most useful organoboron compounds are ethers of boric acid, *trialkyl(aryl)borates*. There are several techniques for producing these ethers:

Interaction of borax with alcohols:

$$Na_2B_4O_7 + 6ROH \xrightarrow[3H_2O]{} 2B(OR)_3 + 2NaBO_2 \qquad (5.1)$$

The formed sodium metaborate can be transformed back into borax with carbon dioxide:

$$4NaBO_2 + CO_2 \longrightarrow Na_2B_4O_7 + Na_2CO_3 \qquad (5.2)$$

Interaction of boric anhydride with alcohols:

$$0.5B_2O_3 + 3ROH \xrightarrow[1.5H_2O]{} B(OR)_3 \qquad (5.3)$$

Interaction of boric acid with alcohols:

$$H_3BO_3 + 3ROH \xrightarrow[3H_2O]{} B(OR)_3 \qquad (5.4)$$

when the released water is continuously withdrawn out of the reaction field in the form of azeotropic mixture with alcohol (taken in excess).

Of practical interest are the last two techniques for producing trialkylborates. It should be noted however, that the industrial reaction of boric anhydride with alcohols is fraught with difficulties. E.g., when powderlike boric anhydride is introduced into alcohol, clumps should be broken lest they kill the reaction. If boric anhydride is used in the form of pieces, the reaction should be conducted under an increased pressure and alcohol should be dehydrated beforehand. Thus, the most technological method is to obtain trialkyl(aryl)borates by the interaction of boric acid with alcohols.

Preparation of trimethylborate

The synthesis of trimethylborate is based on the reaction of methyl alcohol with boric acid:

$$CH_3OH + H_3BO_3 \xrightarrow[3H_2O]{} B(OCH_3)_3 \qquad (5.5)$$

Trimethylborate forms an azeotropic mixture with the excess of methyl alcohol, which boils approximately at 55 °C. To separate the mixture and extract pure trimethylborate, there are several techniques: to flush methyl alcohol with sulfuric acid; to separate methyl alcohol with the help of calcium chloride, lithium chloride or other alcohol-soluble salts; to distil methyl alcohol (in the form of azeotropic mixture with trimethylborate, which boils under a lower temperature than both components) with subsequent extraction of trimethylborate with mineral oil, etc.

A rather simple and convenient method to separate methyl alcohol and trimethylborate is the extraction technique. Since trimethylborate is highly reactive, its extraction requires well-purified oils, e.g. salve base. Salve base is a highly convenient agent, since it practically does not dissolve methyl alcohol; in the presence of trimethylborate the solubility of methyl alcohol increases in proportion to the trimethylborate content.

Raw stock: boric acid (colourless crystals, the melting point is 169 °C); methyl alcohol (the boiling point is 64.5 °C, d_4^{20} =0.7868); base salve.

The production of trimethylborate comprises two main stages (Fig. 83): the synthesis and extraction of trimethylborate; the distillation and rectification of trimethylborate.

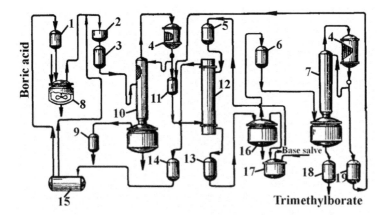

Fig. 83. Production diagram of trimethylborate: *1, 3, 5* – batch boxes; *2* - filter: *4* -coolers; *6, 9, 11, 13, 14,18, 19* - collectors; *7* - rectification tower; *8* – apparatus for preparing the solution; *10* – synthesis tower; *12* – extraction tower; *15, 17*-containers; *16* – distillation tank

Before the synthesis, the tank of tower *10* is loaded with methyl alcohol and the tank is heated to the boiling point of alcohol (≈ 65 °C). The middle part of the tower is filled with 19-20% solution of boric acid from batch box *3*. The solution is prepared in apparatus *8* and filtered in filter *2*.

Tower *10* operates at atmospheric pressure; the top part is filled with nichrome wire packing. At first, the tower operates returning the reflux until the temperature on top stabilises at 55 °C. The azeotropic mixture containing approximately 75% of trimethylborate is separated into collector *11* from the top of the tower at a speed that corresponds to that at which the solution of boric acid is fed. The tower tank is filled with 70% methyl alcohol; the alcohol is reloaded into collector *9* and sent to the distillation of methyl alcohol returned to the synthesis.

The extraction of trimethylborate is carried out with base salve in vertical tower *12* (Fig. 84), which is separated into alternating zones of mixing and settling. The settling zones are filled with Raschig rings; the mixing zones have blade agitators (the role of extractors can also be played by tilting and horizontal apparatuses).

The base salve from tower *12* is sent from the top, from batch box *5*; the azeotropic mixture of trimethylborate and methyl alcohol is sent from below, from collector *11*.

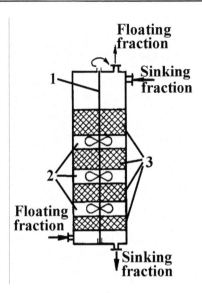

Fig. 84. Vertical extractor: *1* - shaft; *2* – mixing sections with agitators; *3* – settling sections with packing

Accordingly, the solution of trimethylborate in base salve is separated from the lower part of the tower into collector *13*; methyl alcohol is separated from the top of the tower into collector *14*. The base salve solution contains 36-40% of trimethylborate and 5-6% of methyl alcohol. From collector *14* alcohol with 15-25% of trimethylborate is sent into container *15* and then into apparatus *8* to prepare the solution of boric acid. The degree of trimethylborate extraction under these conditions is 90-94%.

Out of collector *13* the base salve solution of trimethylborate is sent into tank *16*, where at 200 °C trimethylborate is distilled. The distilled fraction, which contains 88-90% of trimethylborate, is collected into collector *6* and sent into the tank of rectification tower *7*; the base salve from tank *16* is sent through container *17* back into batch box *5*. During rectification all methyl alcohol is separated in the form of azeotropic mixture with trimethylborate and collected in collector *19*; trimethylborate remains in the tower tank. The azeotropic mixture is sent through collector *11* for repeated extraction into tower *12*; the ready product, 98.5-99.5% trimethylborate, is sent from the tank of tower *7* into collector *18*.

Trimethylborate is a colourless transparent liquid which boils at 68.7°C. It is used as gaseous welding flux and as a parent component for some boron derivatives, such as sodium and potassium boranes, trimethoxyboroxol, etc. E.g., trimethoxyboroxol can be rather easily formed when trimethylborate interacts with boron anhydride.

$$B(OCH_3)_3 + B_2O_3 \longrightarrow CH_3OB \underset{O-BOCH_3}{\overset{O-BOCH_3}{\diagdown}} O \tag{5.6}$$

Trimethoxyboroxol is a colourless viscous liquid which solidifies at about 10 °C. Of particular interest is the use of trimethoxyboroxol for extinguishing burning metals (sodium, lithium, potassium, magnesium, zirconium, titanium). When these metals burn, the temperature rises considerably.

Table 30. Physicochemical properties of trialkylborates

Substance	Boiling point, °C	d_4^{20}	n_D^{20}
$B(OCH_3)_3$	68,7	0.9200	1.3543
$B(OC_2H_5)_3$	117-118	0.8590 (at 26°C)	1.3723 (at 28 – 29.5 °C)
$B(OC_3H_7-n)_3$	177	0.8560	1.3933
$B(OC_4H_9-n)_3$	228-229	0.8740	1.4078
$B(OC_4H_9-$secondary$)_3$	195,4-195,8	0.8264 (at 25 °C)	1.3960
$B(OC_4H_9-$tertiary$)_3$	88-89 (at 70 GPa)	0.8530	1.3879
$B(OC_5H_{11})_3$	274-276	0.8522 (at 27 °C)	1.4183
$B(OC_6H_{13})_3$	310-311	0.8470 (at 28 °C)	1.4248
$B(OC_7H_{15})_3$	185-186 (at 2.66 GPa)	0.8398	1.4280
$B(OC_8H_{17})_3$	200 (at 2.66 GPa)	0.8430	1.4359

and the fire is difficult to extinguish by conventional means. The use of water, chlorinated hydrocarbons and carbon dioxide in these cases is impossible, because at the flame temperature they interact with metal, forming highly flammable or toxic gaseous products. Trimethoxyboroxol is efficient because when sprayed at the flames it burns and releases boron oxide, which covers the metal with a glassy film, thus closing the air supply and killing the fire.

Similarly to trimethylborate, the reaction of a corresponding alcohol with boric acid or boric anhydride yields other trialkylborates: triethyl-, tri-n-propyl-, tri-n-butyl-, tri-*sec*-butyl-, tri-*tret*-butyl-, tri-n-pentyl-, triisopentyl-, tri-n-hexylborates, etc. Some physicochemical properties of important trialkylborates are given in Table 30.

5.2. Preparation of organoaluminum compounds

Recently, organoaluminum compounds have been increasingly used in industry and technology.

Organoaluminum compounds are widely used in the production of polyolefines and stereoregular elastomers (as components of catalyst complex), as raw stock in the production of higher alcohols and carboxylic acids, as additives for reactive fuels, etc.

Among organoaluminum compounds, trialkyl derivatives of aluminum are of the most practical interest. These compounds can be obtained by several techniques.

1. The effect of alkylhalogenides on alumomagnesium alloy:

$$6RHal + Al_2Mg_3 \longrightarrow 2AlR_3 + 3MgHal_2 \qquad (5.7)$$

2. Organomagnesium synthesis with Grignard reagents:

$$AlCl_3 + 3RMgHal \longrightarrow AlR_3 + 3MgHalCl \quad (5.8)$$

The reaction should be conducted in hydrocarbon rather than ether, since in ether it forms etherates of trialkyl(aryl)aluminum.

3. With the help of organomercuric compounds:

$$2Al + 3RMgHal \longrightarrow 2AlR_3 + 3Hg \qquad (5.9)$$

4. With the help of organolithium compounds:

$$AlHal_3 + 3RLi \longrightarrow AlR_3 + 3LiHal \qquad (5.10)$$

5. Joining aluminum hydride and olefines:

$$AlH_3 + 3CH_2{=}CHR \longrightarrow Al(CH_2CH_2R)_3 \quad (5.11)$$

6. The interaction of aluminum with alkylhalogenides and dehalogenation of the products:

$$2Al + 3RCl \longrightarrow AlR_2Cl \cdot AlRCl_2 \xrightarrow{3Na} AlR_3 + 3NaCl + Al \qquad (5.12)$$

$$2(AlR_2Cl \cdot AlRCl_2) + 3Na \longrightarrow 3AlR_2Cl + 3NaCl + Al \qquad (5.13)$$

7. Direct synthesis from aluminum, hydrogen and olefines:

$$Al + 1{,}5H_2 + 3CH_2{=}CHR \longrightarrow Al(CH_2CH_2R)_3 \qquad (5.14)$$

Of the techniques mentioned above, direct synthesis seems the most promising. It is a convenient and economical technique for the industrial production of trialkyl derivatives of aluminum, because the raw stock is not so scarce. Besides, the reaction releases small amounts of by-products.

5.2.1. Preparation of triethylaluminum

The direct synthesis of triethylaluminum

$$AI + 1,5H_2 + 3CH_2=CH_2 \longrightarrow AI(C_2H_5)_3 \qquad (5.15)$$

can be carried out in one or two stages. The two-stage synthesis of triethylaluminum is carried out thus: aluminum reacts with hydrogen and triethylaluminum, forming diethylaluminum hydride:

$$3AI + 1,5H_2 + 2AI(C_2H_5)_3 \longrightarrow 3AI(C_2H_5)_2H \qquad (5.16)$$

which can be extracted if necessary. At the second stage diethylaluminum hydride interacts with ethylene, forming triethylaluminum:

$$3AI(C_2H_5)_2H + 3CH_2=CH_2 \longrightarrow 3AI(C_2H_5)_3 \qquad (5.17)$$

The separation of the process into two stages is explained by fact that the reaction takes place at a temperature above 100 °C and triethylaluminum reacts with ethylene, forming higher aluminumtrialkyls and higher olefines:

$$AI(C_2H_5)_3 \xrightarrow{mCH_2=CH_2} AI \begin{cases} (CH_2CH_2)_x\text{-}C_2H_5 \\ (CH_2CH_2)_y\text{-}C_2H_5 \\ (CH_2CH_2)_z\text{-}C_2H_5 \end{cases} \qquad (5.18)$$

$x+y+z=m$

$$AI \begin{cases} (CH_2CH_2)_x\text{-}C_2H_5 \\ (CH_2CH_2)_y\text{-}C_2H_5 \\ (CH_2CH_2)_z\text{-}C_2H_5 \end{cases} \xrightarrow[AI(C_2H_5)_3]{3CH_2=CH_2} 3CH_2=CH\text{-}(CH_2CH_2)_{m-1}\text{-}C_2H_5 \qquad (5.19)$$

the hydrogenation of triethylaluminum and higher olefines:

$$AI(C_2H_5)_3 \xrightarrow[C_2H_6]{H_2} AI(C_2H_5)_2H; \quad RCH_2=CH_2 \xrightarrow{H_2} RCH_2CH_3 \qquad (5.20)$$

Thus, the conditions for the formation of by-products are close to the conditions of the main process, which impairs one-stage synthesis. However, the one-stage production of triethylaluminum is more convenient due to the simplicity of the technological process. The one-stage production of triethylaluminum nevertheless requires the speed of the main reaction to be much higher than the speed of secondary reactions.

The one-stage synthesis should be conducted at 135 °C, 5 MPa, the 0.71:1 ratio of aluminum and triethylaluminum and the 1:1 ratio of ethylene and hydrogen. First, aluminum power is activated to eliminate the oxide film which hampers the contact of aluminum with hydrogen and triethylaluminum.

The activation can be carried out by the chemical technique (with various reagents), the physical technique (with ultrasound; with inert gas used to disperse liquid aluminum) or the mechanical technique (fine dispersion in a cavity, ball or vibration mill).

The chemical activation of aluminum. The reactor is loaded with a suspension of aluminum in petrol and activating additive (triethylaluminum or triethylaluminum mixed with aluminum chloride). The mixture is agitated and heated in hydrogen to 160-200 °C and held at this temperature for 10 hours. After the activation the reactor is cooled, the excess hydrogen is withdrawn and the synthesis is started.

The activation of aluminum with ultrasound or dispersion of liquid aluminum. The suspension of powder aluminum in petrol or n-geptane without oxygen is subjected to ultrasound; the tough oxide film on the surface of aluminum is removed and aluminum becomes reactive. The second activation technique is the dispersion of liquid aluminum with argon or purified nitrogen flow into a finely dispersed state. It should be noted, however, that the most reactive aluminum powder for direct synthesis is the powder alloyed with transition metals (titanium, zirconium, niobium, tantalum) with the size of particles from 10 to 125 μm.

Activation of aluminum in a cavity, ball or vibration mill. This activation technique should be used in nitrogen and in 5% solution of triethylaluminum in n-heptane, because aluminum suspension is easily transported through pipes and activated aluminum is protected from oxidation (with oxygen in air) during transportation and storage. Besides, wet grinding is less explosive than dry. Aluminum powder should be activated in the mill for 20-30 hours. The aluminum ground in a vibration mill is the most active.

Direct synthesis. In the production of triethylaluminum by the one-stage technique (Fig.85) autoclave *6* with a jacket and shielded electric drive agitator is loaded with a suspension of aluminum powder (ground and activated in vibration mill *3*) in n-heptane from batch box *5* and with a necessary amount of triethylaluminum from collector *10*. The contents of the autoclave are heated to 135 °C; the autoclave recieves a 1:1 mixture of ethylene and hydrogen.

After that, the reactive mixture is held for 10 hours at 5 MPa. After the synthesis the mixture is cooled, the excess gases are withdrawn and the contents of the autoclave are sampled. In case there is diethylaluminum hydride in the products, they are subjected to additional ethylation by passing ethylene through the reactive mixture at 75 °C and 0.5-1 MPa for 1.5 hours. The mixture from the autoclave is sent into collector *7* and centrifuge *8*. The product is collected in collector *10*. The mixture of aluminum-trialkyls obtained in this way contains up to 90% of triethylaluminum.

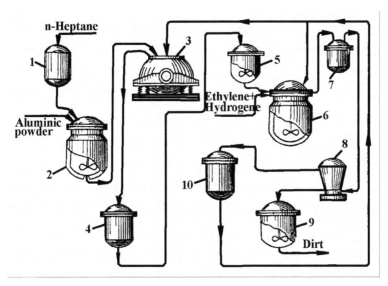

Fig. 85. Production diagram of triethylaluminum by the one-stage technique: *1,5*-batch boxes; *2* - agitator; *3* – vibration mill; *4, 10* - collectors; *6* - autoclave; *7,9*-receptacles; *8* – centrifuge

Triethylaluminum is a colourless motile transparent liquid (the boiling point is 128-130 °C at 66.5 GPa), which dissolves in hydrocarbons and is sensitive to humidity and oxygen. Triethylaluminum has a dimeric structure:

$$(C_2H_5)_2Al \underset{CH_2-CH_3}{\overset{CH_2-CH_3}{<\quad>}} Al(C_2H_5)_2$$

This molecule is stable even in the gaseous stage and dissociates only above 100 °C. In dilute solutions aluminumtrialkyls dissociate with time; e.g.,triethylaluminum dissolved in benzene dissociates into monomer within 6-8 hours.

5.2.2. Preparation of triisobutylaluminum

Like the synthesis of triethylaluminum, the direct synthesis of triisobuty-laluminum can be carried out in two stages, releasing diisobutylaluminum hydride:

$$Al + 1,5H_2 + 2Al(C_4H_9\text{-izo})_3 \longrightarrow 3Al(C_4H_9\text{-izo})_2H \qquad (5.21)$$

$$Al(C_4H_9\text{-izo})_2H + CH_2{=}C(CH_3)_2 \longrightarrow Al(C_4H_9\text{-izo})_3 \qquad (5.22)$$

as well as in one stage:

$$Al + 1,5H_2 + 3CH_2{=}C(CH_3)_2 \longrightarrow Al(C_4H_9\text{-izo})_3 \qquad (5.23)$$

since in these conditions isobutene virtually does not interact with tri-isobutylaluminum.

The synthesis is carried out at 5-6 MPa and the temperature not exceeding 150 °C; the reaction speed is much higher than in the synthesis of triethylaluminum. The processes can be carried out in tower apparatuses and in a cascade of reactors (Fig. 86).

A suspension of toluene solution of triisobutylaluminum and activated aluminum powder is made in apparatus 4. The ready suspension is sent by nitrogen flow into collectors 5. For a continuous dosage of the suspension, there are two collectors: the suspension enters one and is continuously pumped out of the other with pump 6 through collector 8 for synthesis.

The synthesis is carried out in a cascade of three consequtive reactors 9 of the same type. The parent substances are continuously sent through collector 8 into first reactor 9; from there, the products of the reaction flow into the second and the third reactors. The given level in the first two reactors is supported by pouring the products through the piping; for the third, a pressure regulator is used with a gate located on the outlet pipe. Strict observance of the level is necessary to keep the calculated synthesis time.

The outlet gases (unreacted hydrogen, isobutene and nitrogen) from the first reactor are released through cooler 10, which is cooled with bobbin oil. From the cooler the gases enter separator 11 and are released into the atmosphere after the hydraulic gate and bumper (not shown in the diagram) or sent into the recovery system. After the second reactor the pressure is reduced to the atmospheric level and the solution of triisobutylaluminum is sent into collector 12. The solution in the collector is pumped into settling box 13 or for centrifuging.

This technique gives a high yield of triisobutylaluminum; the degree of aluminum conversion can reach up to 90% in case of a cascade of four reactors equal in volume.

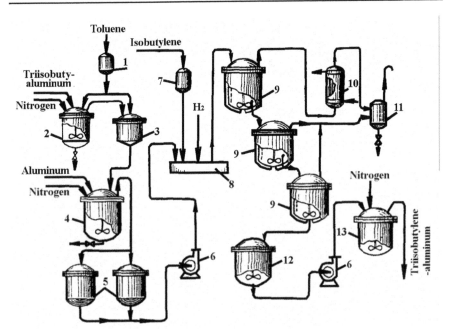

Fig. 86. Diagram of continuous production of triisobutylaluminum: *1, 2, 3, 7* – batch boxes; *4* – apparatus for preparing suspension; *5, 12* - collectors; *6* - pumps; *8*- collector; *9*- reactors; *10*- cooler; *11* – separator; *13*- settling box

Triisobutylaluminum, which is used as a component of a catalyst for the production of olefines, should contain not more than 0.01% finely dispersed solid particles. The product obtained by direct synthesis contains up to 5% of solid impurities, mostly 0.1-1 µm aluminum particles. If this product is purified by centrifuging, even subsequent settling (for 24 hours or more) cannot reduce the concentration of solid particles to the required level; therefore, it is advisable to replace centrifuging with kieselguhr or perlite filtering.

Triisobutylaluminum is a colourless transparent liquid (the boiling point is 138 °C at 6.6 GPa), which dissolves well in hydrocarbons. When heated up to 140-160 °C in vacuum (the residual pressure is ≈ 35 GPa), triisobutylaluminum disintegrates into isobutene and diisobutylaluminum hydride. Unlike lower trialkyl derivatives of aluminum with the direct carbon chain (e.g. triethylaluminum), aluminumtrialkyls with a branched chain (triisobutylaluminum is one of them) are monomeric.

Trialkylderivatives of aluminum (particularly triethyl-, tripropyl- and triisobutylaluminum) are widely used due to their exceptional catalytic properties.

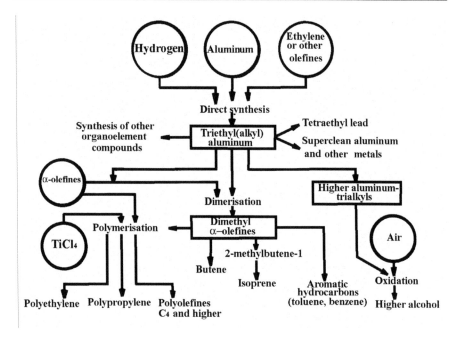

Fig. 87. Most important syntheses based on aluminumtrialkyls

They are used as a component in the Ziegler-Natta catalytic system (AlR₃•TiCl₄, etc.) in the polymerisation and copolymerisation of olefines. Besides, the high reactivity of these compounds makes them practical for many interesting syntheses (Fig. 87).

Equally important are higher aluminumtrialkyls as raw stock in the production of primary higher alcohols. Higher aluminumtrialkyls are compounds with alkyl groups C_6 or higher at the aluminum atom.

5.2.3. Preparation of higher aluminumtrialkyls

At present higher aluminumtrialkyls are obtained by the polymerisation of olefines with lower aluminumtrialkyls, when Al—C groups are attached at the C=C bond.

E.g., higher aluminumtrialkyls are synthesised by the polymerisation of ethylene with triethylaluminum. At 90-120 °C and 4.5-12 MPa ethylene is stagewise attached to triethylaluminum, forming a complex mixture of higher aluminumtrialkyls:

$$AI(C_2H_5)_3 \xrightarrow{mCH_2=CH_2} AI \begin{cases} (CH_2CH_2)_x\text{-}C_2H_5 \\ (CH_2CH_2)_y\text{-}C_2H_5 \\ (CH_2CH_2)_z\text{-}C_2H_5 \end{cases} \qquad (5.24)$$

$n=(x+y+z)=15\div20$

Insignificant impurities of heavy metals (e.g. nickel) in a reactive medium or equipment material can cause a secondary displacement reaction which forms higher olefines:

$$AI \begin{cases} CH_2CH_2R \\ R' \\ R'' \end{cases} + CH_2=CH_2 \rightleftharpoons AI \begin{cases} C_2H_5 \\ R' \\ R'' \end{cases} + RCH_2=CH_2 \qquad (5.25)$$

The higher olefines are found in final products, which greatly hampers the purification of higher alcohols. Besides, in some conditions higher olefines can interact with triethylaluminum; that is why the equipment for the synthesis of higher aluminumtrialkyls should be manufactured from O steel or copper. Raw stock, especially ethylene, should be dried with great care, because even the slightest traces of moisture lead to the formation of paraffines.

Raw stock: ethylene [0.8-1.5% (vol.) of ethane, not more than 0.5% (vol.) of acetylene, not more than 0.5% (vol.) og methane]; triethylaluminum [9.4-10.3% (weight) of active aluminum, 4.1-4.5% (weight) of hydroxy compound, the calculated amount of triethylaluminum is 39.6-43.5% (weight), the rest is diethylaluminum hydride]. The diagram of a pilot unit for the continuous production of higher aluminumtrialkyls is given in Fig. 88.

Out of batch box *2* triethylaluminum (in the form of 10-12% isooctane or petrol solution) is pumped with batching pump *4* into the top part of absorber *3*. The middle of the absorber is filled under the pressure of 3-4 MPa with ethylene, which has been dried and purified in the system of subsequent towers with active aluminum oxide and active coal. In the absorber the triethylaluminum solution is presaturated with ethylene to ensure exact operation of the batching equipment. The triethylaluminum:ethylene mole ratio can be from 1:9 to 1:25 depending on the desired distribution of alkyl groups in aluminumtrialkyls.

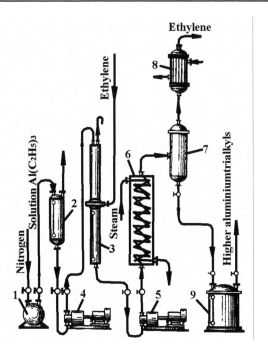

Fig. 88. Diagram of a pilot unit for the continuous production of higher aluminumtrialkyls: *1* - container; *2* – batch box; *3* - absorber; *4, 5* - pumps; *6* - polymeriser; *7* - separator; *8* - cooler; *9* – collector

The triethylaluminum solution, saturated with ethylene, is pumped with pump *5* out of the absorber and sent into polymeriser *6*. The optimal conditions for polymerisation are: 105-110 °C, 10-12 MPa, the reaction time is 6-8 hours. It should be kept in mind that when the temperature rises to 125 °C and pressure to 12.5 MPa, the considerable acceleration and release of much heat (~92.2 KJ/mole) can cause an explosion. The mixture from the polymeriser is sent into separator *7*, where liquid products are separated from the unreacted ethylene. The ethylene is withdrawn through backflow condenser *8*, where the solvent vapours condense; the solution of higher aluminumtrialkyls enters collector *9*. In this case the conversion degree of ethylene is 83-85 %; the yield of higher aluminumtrialkyls is 75-77 %.

Higher aluminumtrialkyls are used in the production of primary higher fatty alcohols. This presupposes the oxidation of aluminumtrialkyls and the hydrolysis of the aluminum alcoholates formed.

Of practical interest among organoaluminum compounds are also alkylaluminumhalogenides, especially ethyl-, propyl- and isobutylaluminumchlorides.

5.2.4. Preparation of ethylaluminumchlorides

Ethylaluminumchlorides can be obtained by the so-called sesquichloride technique, the reaction between aluminum and ethylchloride. This forms an equimolar mixture of ethylaluminumchlorides, *triethylaluminum sesquichloride*.

$$2AI + 3C_2H_5CI \longrightarrow (C_2H_5)2AICI + C_2H_5AICI_2 \qquad (5.26)$$

To separate pure diethylaluminumchloride from the mixture of ethylaluminumchlorides, it is subjected to "symmetrisation" in the presence of sodium:

$$2C_2H_5AICI_2 + 3Na \longrightarrow (C_2H_5)_2AICI + 3NaCI + AI \qquad (5.27)$$

it can be accompanied by a secondary interaction of ethylaluminumchloride and sodium chloride which forms a complex compound, $Na[Al(C_2H_5)Cl_3]$. The process of the secondary reaction largely depends on the ratio of metal sodium and triethylaluminum sesquichloride: when the ratio is 0.125:1, the secondary reaction is the least intensive.

The production process of pure diethylaluminumchloride (Fig. 89) comprises the following main stages: the activation of aluminum; the synthesis of triethylaluminum sesquichloride and its "symmetrisation"; the settling of diethylaluminumchloride.

Reactor *3* with a shielded electric drive agitator, which has been dried with nitrogen, is loaded with a necessary amount of aluminum powder, petrol and 50-60% petrol solution of triethylaluminum sesquichloride. The agitator is switched on, the mixture is heated to 50-60 °C; aluminum is activated by adding ethyl bromide gradually and at constant temperature. The heated reactor is filled with ethylchloride at such speed that the given temperature is maintained. After ethylchloride has been supplied, the mixture is held at reaction temperature for 1-2 hours to complete the synthesis. The obtained product is cooled and loaded off into collector *7*.

The "symmetrisation" of the equimolar mixture of ethylaluminumchloride or triethylaluminum sesquichloride is carried out with the help of metallic sodium. Sodium can be mixed with sesquichloride according to the scheme "sodium into sesquichloride" or vice versa, "sesquichloride into sodium". Better results are achieved when the reactants are mixed according to the "sesquichloride into sodium" scheme with 0.125:1 ratio of sodium to sesquichloride.

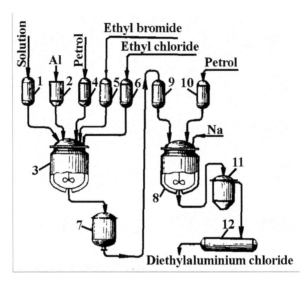

Fig. 89. Production diagram of diethylaluminumchloride: *1,4-6, 9, 10* – batch boxes; *2* - bin; *3,8*- reactors; *7* - collector; *11* - settling box; *12* – container

The "symmetrisation" of triethylaluminum sesquichloride takes place in reactor *8*. The suspension of sodium in petrol is prepared there; sesquichloride enters the reactor from collector *7* through batch box *9* at 125-135 °C and agitation. Then the mixture is agitated at the synthesis temperature for 2-3 hours to complete the reaction; then it is cooled to ambient temperature. The obtained diethylaluminumchloride solution is settled for 2-3 hours to separate dirt; the purified product enters container *12*.

The remaining diethylaluminumchloride is extracted by washing the dirt with petrol; after that, the dirt is burned or extinguished. When the sesquichloride: sodium ratio = (0.125÷1.140): 1, dirt contains mostly sodium chloride. It easily suspends in hydrocarbons and can be transported through piping.

This technique for preparing diethylaluminumchloride presents few problems. However, aluminum powder should be activated, otherwise the formation of triethylaluminum sesquichloride starts much later but proceeds autocatalytically, releasing a lot of heat, which greatly increases the pressure in the reactor and even causes explosions. The formation stage of diethylaluminumchloride is not so active; however, the process should be strictly controlled (because the reaction releases heat) and the ion should be supplied from the components gradually.

Considerable difficulties are presented when solid impurities (dirt) are separated from the diethylaluminumchloride solution. Impurities as a rule consist of finely dispersed particles of sodium chloride and highly flammable aluminum, which settle from the reactive mixture very slowly. The settling can be promoted by granulation. After granulation the dirt consisting of aluminum and the NaCl · Al(C$_2$H$_5$)Cl$_2$ complex is settled and transported through piping. The dirt is slaked with methyl alcohol.

$$AI + 3CH_3OH \longrightarrow AI(OCH_3)_3 + 1,5H_2O \qquad (5.28)$$

$$NaCI \cdot AI(C_2H_5)CI_2 + CH_3OH \longrightarrow NaCI + AI(OCH_3)CI_2 + C_2H_6 \quad (5.29)$$

This slaking technique does not require any special equipment.

Diethylaluminumchloride is a colourless transparent liquid (the boiling point is 65-66 °C at 16 GPa), which dissolves well in organic solvents. It actively reacts with hydrides of alkali metals; chlorine in this case is replaced with hydrogen, forming diethylaluminumchloride, which is an efficient reducing agent. Diethylaluminumchloride is "symmetrised" under the influence of sodium amalgam; the yield of triethylaluminum is virtually theoretical. Therefore, the production of triethylaluminum through triethylaluminum sesquichloride is also a convenient industrial technique.

Alkylaluminumchlorides can also be obtained by the reaction of aluminum chloride with corresponding trialkyl derivatives of aluminum. E.g., the interaction of triethylaluminum with aluminum chloride can yield both diethylaluminumchloride and ethylaluminumchloride:

$$2AI(C_2H_5)_3 \xrightarrow{AICI_3} 3(C_2H_5)_2AICI \qquad (5.30)$$

$$AI(C_2H_5)_3 \xrightarrow{2AICI_3} 3C_2H_5AICI_2 \qquad (5.31)$$

These reactions require anhydrous aluminum chloride and a 50-80% petrol solution of triethylaluminum. The process is carried in a cylindrical steel apparatus, which has a propeller agitator with a shielded electric drive. The synthesis takes place at a temperature not exceeding 60 °C for 2-3 hours. Then the reaction products are subjected to filtering or distillation.

Alkylaluminumchlorides are widely used as catalysts in the production of polyethylene and low-pressure polypropylene, as well as in the polymerisation of dienes and α-olefines. Organoaluminum compounds are as a rule highly reactive substances.

Some physicochemical properties of important aluminumtrialkyls and alkylaluminumhalogenides are given in Table 31.

5.2.5. Safety measures in the production of organoaluminum compounds

Taking into consideration: a) the specific properties of organoaluminum compounds, especially lower aluminumtrialkyls, and their hydride-, halogene- and alkoxy derivatives, which are highly flammable in air and explode at contact with water; b) the use of hydrogen, ethylene, isobutene, ethylene, isobutene, ethylchloride, sodium and aluminum (finely dispersed and active, which can self-inflame in air), the production of organoaluminum compounds can be considered one of the most dangerous chemical productions. Therefore, safety measures and fire prevention are especially important.

When dealing with organoaluminum compounds, it is necessary:

- to prevent any leakage or contact with other reactive substances;
- to provide for the protection of the workers from any contact with organoaluminum compounds;
- to operate in inert gas, which has been dried and purified from oxygen [oxygen content should not exceed 0.01 (vol.), moisture content should not exceed 0.1 mg/l].

Table 31. Physicochemical properties of aluminumtrialkyls and alkylaluminum-halogenides

Substance	Boiling point, °C	d_4^{20}	n_D^{20}
Al(CH$_3$)$_3$	125-126 (at 1006 GPa)	0.7520	—
Al(C$_2$H$_5$)$_3$	48-49 (at 1.3 GPa)	0.8370	1.4800 (at 6.5 °C)
Al(C$_3$H$_7$-n)$_3$	65 (at 0.13 GPa)	-	-
Al(C$_4$H$_9$-n)$_3$	149-150 (at 4 GPa)	-	-
Al(C$_4$H$_9$-iso)$_3$	86 (at 13.3 GPa)	0.7859	1.4494
Al(C$_6$H$_{13}$-n)$_3$	105 (at 10^{-3} GPa)	-	-
[CH$_3$AlCl$_2$]$_2$	94.5-95 (at 133 GPa); Melting point is 72.7 °C	-	-
[(CH$_3$)$_2$AlCl]$_2$	119-121	1.0000	-
[CH$_3$AlBr$_2$]$_2$	124-139; melting point is 79 °C	-	-
[(CH$_3$)$_2$AlBr]$_2$	54 (at 22.6 GPa)	-	-

Table 31. (cont.)

$Al(C_2H_5)_2Cl$	125-126 (at 66 GPa)	20 $1.0059(d_{20}^{20})$	-
$Al(C_2H_5)_2Br$	84.2-85 (at 4 GPa)	–	–
$Al(C_4H_9\text{-}iso)_2Cl$	152 (at 13.3 GPa) Melting point is 39.5 °C	0.9088	1.4506
$Al(C_2H_5)Cl_2$	60-62 (at 0.66 GPa); Melting point is 30-32 °C	-	-

Before operation the apparatuses and equipment should be blown with nitrogen to eliminate all traces of oxygen. At the end of each line for withdrawing blowing gases there should be special stop gates or hydraulic shutters with a 250-mm layer of oil to prevent air inflow into the system. Besides, there should be a large-diameter emergency line for quick pressure release. The hermeticity of the equipment is first and foremost achieved due to reactors with shielded electric drive agitators, immersed pumps, gate fittings and sealed flanges. Before operation, all equipment and piping should be thoroughly pressurised.

Pipes for pure aluminumtrialkyls should be short and mounted so that the product flows freely and completely (no "pockets"; tilted towards the storage container). Pipings for organoaluminum compounds should be located at a distance from those for other reactive substances, water and vapour. The production of organoaluminum compounds requires apparatuses without lower drains; the product can be sent from one apparatus into another by nitrogen flow through immersed pipes.

To prevent the inverse flow of the product, backflow gates should be used. Workshops for organoaluminum compounds should not be fitted with water and vapour piping. Heat carriers can be only inert hydrocarbons (pentane, mineral oil, etc.). Tanks and reactors should have drains. Dirt can be separated with highly productive hermeticised centrifuges.

Aluminumtrialkyls and their derivatives do not self-detonate; the danger can be significantly reduced by diluting them with certain hydrocarbons. Thus, industry mostly uses organoaluminum compounds in the form of solutions. A very important stage is mixing them with organic solvents (benzene, hexane, heptane). Chlorinated hydrocarbons are not used as solvents, because they violently react with organoaluminum compounds. Organoaluminum compounds can be mixed with a solvent in a separate container or directly in the pipeline. Mixing in a separate container is simpler and safer; however, mixing in the pipeline visibly reduces the number of containers for storing the solution of the organoaluminum compound.

Apparatuses, piping and fittings for the production of organoaluminum compounds can be made from all major steel brands. Lead and aluminum are not stable toward organoaluminum compounds. Of packing materials the most convenient are copper and Teflon; paronite is stable towards aluminumtrialkyls at low temperatures. Cotton, woolen and capron cloth cannot resist organoaluminum compounds even in inert gases; ceramic materials, on the other hand, are inert towards them.

Before filling containers for organoaluminum compounds should be thoroughly dried, purified from impurities, blown and pressurised with inert gas.Organoaluminum compounds should be stored at a temperature below 30 °C and a pressure of 0.03-0.05 MPa; the containers must not be subjected to direct sunlight or heat from other sources.

Pure organoaluminum compounds are pyrophoric (capable of self-ignition) at these temperatures: from -68 °C for triethylaluminum to -40 °C for triisobutylaluminum and for -64 °C for diethylaluminumchloride. When organoaluminum compounds are diluted, the self-inflammation point of the solutions increases to +20 °C at 50-70% concentration (diluted even more, organoaluminum compounds are not pyrophoric).

Below we give the maximum concentrations (%) of some organoaluminum compounds in benzene, at which the mixtures are not pyrophoric:

Triethylaluminum	15	Ethylaluminumchloride	30
Diethylaluminumchloride	16	Triethylaluminum sesquichloride	20

Spilled solutions of aluminumalkyls are very dangerous, because the solvent evaporates very quickly due to the heat released in the interaction of aluminumalkyls with the oxygen in air. Besides, when the temperature increases, organic compounds disintegrate releasing olefines; with the increase of temperature the speed of disintegration abruptly increases.

Because of their flammability, organoaluminum compounds require special attention to fire extinguishers in their production. In case of inflammation of large amounts of organoaluminum compounds, carbon dioxide is inefficient. Carbon tetrachloride cannot be used either, because it reacts with organoaluminum compounds (forming toxic fumes) and does not extinguish the flame. Extinguishing fire caused by organoaluminum compounds is so difficult due to the danger of repeated inflammation at contact with air.

Good fire extinguishers are dry sand, diatomaceous earth and cement; however, they should be used in large quantities to absorb the liquid and create a 100-150 mm protective layer, the air access to the burning product is stopped. An efficient fire extinguisher is porous vermiculite, a micalike material (the apparent density is 0.112 g/cm^3), which floats to the surface of organoaluminum compounds and puts out the flame. This fire can also be extinguished by sodium chloride treated with 0.05% fuchsin. Aluminumalkyl solutions with less than 10% concentration can also be extinguished with finely sprayed water or air mechanical foam.

Wastes of organoaluminum compounds are also dangerous, since they contain active aluminum, organic solvents and residues of organoaluminum compounds. The wastes are oiled and destroyed in special furnaces.

If organoaluminum compounds contact skin, they cause persistent burns. Their 40% solutions cause heavy burns (5-10% solutions do not). If an organoaluminum compound contacts skin, it should be washed away with petrol or kerosene (not water!) within the first 5-8 seconds of contact; later washing is not so effective. The affected area is wiped with alcohol and bandaged.

Apart from the local effect, organoaluminum compounds have a general effect on human body. When these substances burn or interact with moisture in air, they release smoke containing finely dispersed aluminum oxide, which when inhaled can cause a condition similar to flu in symptoms. Usually the person feels unwell for not more than 36 hours; then the condition quickly improves. When aluminumtrialkyls burn or interact with oxygen and moisture in air, they also form other harmful substances.

Below we give maximum allowable concentrations (MAC) of some disintegration products of triethyl- and triisobutylaluminum and their chlorides (in mg/m^3):

Metallic aluminum (aerosol)	2	Isobutene	100
Aluminum oxide	2	Ethyl alcohol	1000
Hydrogen chloride	5	Isobutyl alcohol	20
Ethane	300	Heptane (solvent)	300
Ethylene	300	Petrol (solvent)	300
Isobutane	300		

The greatest danger is presented by aluminum aerosol and oxides; they are followed by the alcohols and hydrocarbons. Hydrogen chloride, which has been adsorbed on aerosol particles, is very toxic. Besides, aerosol particles can have small amounts of undecomposed organoaluminum compound, which enters airways and lungs and causes burns of mucous membranes. Organoaluminum compounds are irritants.

The protection from organoaluminum fumes requires thorough ventilation of all indoor rooms. In case of accidents a breathing mask should be worn. Triisobutylaluminum fumes require common gas masks; fumes of alkylaluminumchlorides require an antiacid gas mask (because the decomposition of alkylaluminumchlorides forms hydrogen chloride). All people who work with organoaluminum compounds should undergo clinical examination for timely discovery of any damage to lungs and bronchi.

5.3. Preparation of organotitanium compounds

Among widely used organotitanium compounds are tetraalkoxytitaniums, obtained by the etherification of titanium tetrachloride with alcohols:

$$TiCI_4 + 4ROH \xrightarrow[4HCI]{} Ti(OR)_4 \qquad (5.32)$$

The production techniques of titanium tetrachloride are described below in detail.

Preparation of titanium tetrachloride

Titanium tetrachloride was prepared for the first time in 1825 by chlorine acting on titanium at high temperature. Since that time more production techniques have been devised, e.g. the chlorination of titanium carbide and the effect of hydrogen chloride on metallic titanium at a temperature above 300 °C.

The most widespread technique, however, is the chlorination of titanium dioxide with chlorine or chlorine-containing substances (carbon tetrachloride, chloroform, sulfuryl chloride, phosphorus oxychloride, silicon tetrachloride). These reactions give high yield at high temperatures (800 °C and more); the chlorination with free chlorine occurs at noticeable speed only in the presence of reducing agents (e.g., coal). If there is a lack of coal, the reaction forms carbon dioxide; if there is an excess of coal, it releases carbon oxide:

$$TiO_2 + 2CI_2 \longrightarrow \begin{array}{l} \xrightarrow{\quad C \quad} TiCI_4 + CO_2 \\[2mm] \xrightarrow{\quad 2C \quad} TiCI_4 + 2CO \end{array} \qquad (5.33)$$

Industrially titanium tetrachloride can be produced both from titanium dioxide by the technique described above and from titanium- and iron-containing ores: rutile (TiO_2), which contains about 60% of titanium and up to 10% of iron, ilmenite ($FeO \cdot TiO_2$ with 25-35% of titanium) or titaniferous magnetites (the mechanical mixture of ilmenite and magnetite or loadstone).

The most widespread minerals are ilmenite and rutile. Major deposits of rutile are found in Mexico, Australia, India and the USA; ilmenite deposits are situated in India, Australia, Indonesia, Africa, South America and the USA. Russia and Ukraine can also boast large deposits of ilmenite and titaniferous magnetite. The most convenient raw stock for the production of titanium tetrachloride is rutile with 91-99% of TiO_2. However, rutile production in Russia is limited and rather expensive; that is why our industry uses ilmenite ore.

Raw stock: ilmenite ore (about 50% of TiO_2, 40% of FeO, 5% of SiO_2, about 0.5-1% V_2O_5, 4-4.5% of other impurities); evaporated chlorine (at least 99.6% of chlorine, not more than 0.02% of moisture).

The ore is first of all subjected to reducing fusion in arc or ore thermal furnaces at a temperature about 1600 °C. The process of melting depending on temperature forms intermediate oxides Ti_2O_3 and Ti_3O_5 ($TiO_2 \cdot Ti_2O_3$), which can dissolve iron protoxide and ilmenite, as well as form solid products with titanium oxide and protoxide. The final products of reducing fusion are cast iron and titanium dross.

The production of titanium tetrachloride comprises the following main stages: the preparation of raw stock; the chlorination of raw stock; the condensation of chlorination products; the purification of technical titanium tetrachloride.

Depending on the chosen production technique, the preparation of raw stock entails either the production of bricks from titanium raw stock and coke or the grinding of these components to make up the furnace charge. Chlorination is carried out with evaporated or diluted chlorine (e.g., gaseous chlorine obtained in the production of electrolytic magnesium) in shaft electric furnaces, salt melt furnaces or in apparatuses with a fluidised layer.

Chlorination in shaft electric furnaces. A shaft furnace is a steel apparatus lined on the inside with a layer of diabase on liquid glass and a layer of special low porosity chamotte brick. The lower part of the furnace has two rows of coal electrodes (three in each row). The lower part of the furnace is filled with coal packing to 400-700 mm above the top electrodes. The melted metal chlorides flow down the heated packing. They are collected in the lower zone of the furnace and periodically unloaded.

Chlorination in shaft furnaces allows one to extract almost all of titanium (97-98%) out of the furnace charge. The extraction degree of other oxides depends on the chlorination temperature and properties of the extracted component. E.g., if silicon dioxide is in the mixture in the form of quartz, its chlorination degree is 10-20%; if silicon dioxide is part of silicate, it is clorinated by 80% and more. Aluminum oxide in the form of corundum chlorinates only slightly; alumosilicates chlorinate almost completely.

It is very important to maximise the chlorination of the furnace charge components. If nonchlorinated residue accumulates, it noticeably increases the resistance in the furnace (because the melt impregnates the unreacted residue), impairs the distribution of chlorine in the furnace and causes other serious problems.

The bricks should be held in shaft electric furnaces for 8 to 12 hours. The productivity of a shaft furnace is determined by its inner diameter and the height of the furnace charge layer. The yield of titanium tetrachloride from 1 m^2 of the furnace section is 2-2.5 tonnes per 24 hours.

Chlorination in salt melt. For chlorination in melted salts, it is especially important to use raw titanium with large quantities of the impurities (calcium, magnesium, manganese oxides) which form low-melting chlorides in the process. The medium is a melt of potassium ans sodium chlorides (the eutectic mixture of these salts melts at 660 °C).

The chlorinator is a rectangular shaft furnace lined with chamotte brick (the shafts can be uni- and multichamber). The lower part of the chlorinator contains structures for sending chlorine into gas distribution devices; the side walls of the apparatus have graphite conductive electrodes to keep the electrolyte melted during the launch or temporary stop of the chlorinator. There are tapholes to pour off the melt. The top lid of the chlorinator has chokes for loading the charge and for adding the melt, as well as a branch pipe for withdrawing the reaction gases. The furnace charge (titanium dross and oil coke) is loaded onto the melt layer.

Chlorination in melt greatly depends on the degree to which the reducing agent (coke) has been ground. The optimal size of oil coke grains is 0.1 mm: coarser grinding does not allow for sufficient extraction of titanium, finer grinding causes some coke to be carried away with reaction gases. During melt chlorination the oxygen from the oxides in the titanium mixture interacts with coke to form mostly carbon dioxide rather than carbon oxide, like in shaft furnaces. This increases the concentration of titanium tetrachloride in reaction gases and thus improves the conditions for condensation. Besides, the decrease of carbon oxide content in gases increases the safety of the process and reduces the probability of phosgene formation.

The chlorination of titanium mixture is exothermal, the temperature in the furnace spontaneously rises; therefore, the chlorinator is fashioned with special elements for heat withdrawal. They are water-cooled graphite blocks with steel pipes inside. Excess heat can also be withdrawn by spraying the melt with some of the obtained liquid titanium tetrachloride; the heat in this case is spent on heating, evaporating and overheating $TiCl_4$ to the temperature of the outlet vapour and gas mixture.

In the course of chlorination the melt is gradually enriched with chlorides of calcium, magnesium and other metals; this increases its viscosity and impairs the reaction conditions. Because of this the melt should be periodically renewed. When part of the melt is poured off, the nonclorinated residue is also withdrawn, which is an important advantage of this process in comparison with chlorination in shaft furnaces. During the unloading of the melt, the chlorination process of the titanium mixture should remain continuous.

Thus, the chlorination of titanium raw stock in salt melt helps to build very productive apparatuses and avoid labour-intensive operations of preparing and backing bricks, partially increases the conditions for further condensation of the vapour and gas mixture, considerably reduces the concentration of carbon oxide in outlet gases and helps to withdraw the nonchlorinated residue from the reactive zone continuously. Disadvantages of chlorination in salt melt include increased losses of titanium with discharge melt (because small particles of the furnace charge are carried away with reaction gases) and increased amounts of solid chlorides.

Chlorination in a fluisidised layer. The use of a fluidised layer for the chlorination of titanium raw stock offers one of the opportunities to intensify the production of titanium tetrachloride.

The apparatus for the chlorination of titanium raw stock in a fluidised layer is a cylindrical shaft lined with thick silica brick. There are several shelves with gas distribution grates located one above another for better chlorine consumption and titanium extraction. The parent mixture is loaded onto the top shelf where it is partially chlorinated with unreacted chlorine coming from below and is sent onto the next shelf through a pipe. The most intensive chlorination occurs on the lower shelf.

If raw stock contains a lot of Ca, Mg, Mn oxides, etc., the process may form chlorides of these metals, which leads to the clotting of particles and disturbs the conditions of the fluidised layer. This changes the conditions of weight and heat exchange, and the process has to be stopped. That is why the chlorination of the charge with considerable amounts of calcium and magnesium oxides should be carried out at a temperature not exceeding 600 °C, i.e. at the temperature not higher than the temperature at which the most low-melting eutectic mixture of the chlorides is formed. Naturally, the speed of the process under these conditions is noticeably reduced. When chlorinating rutile ores, which contain basically no impurities to form low-melting chlorides, the temperature can be increased to 900-1000 °C thus accelerating the process.

Titanium carbide is also an efficient kind of raw stock for chlorination in apparatuses with a fluidised layer. Powderlike titanium carbide is a high-melting noncaking compound (the melting point is 3140 °C). Another advantage of this raw stock is that it can be chlorinated without a reducing agent (the making-up of the charge mixture is no longer necessary); chlorination can be carried out at sufficient speed at 300-400 °C.

Fluidised layer apparatuses offer efficient weight and heat exchange, quick temperature adjustment throughout the layer and high speed of the process even at relatively low temperatures. Similarly to chlorination in melt, the process in a fluidised layer makes bricks unnecessary and helps to create a continuous process. The productivity of fluidised layer apparatuses depending on chlorination temperature is 5-10 tonnes of $TiCl_4$ from 1 m^2 of the apparatus section per 24 hours.

The choice of the chlorination technique and equipment for the process greatly depends on the compositon of raw stock for chlorination. For shaft furnaces and fluidised layer apparatuses, it is advisable to chlorinate titanium raw stock with relatively small amounts of oxides of calcium, magnesium, manganese and other metals which form low-melting chlorides in chlorination. On the other hand, in chlorination in salt melt these oxides do not have any significant effect on the process.

Attention should be given to the chlorination technique developed at Illinois Technological Institute (the USA), which drastically differs from the common high-temperature chlorination of titanium raw stock. The technique is based on the use of base ilmenite ores (containing less than 40% of TiO_2) and allows the process to be carried out at low temperatures. The chlorinating agent is hydrogen chloride.

Ilmenite ore is treated with sulfuric acid; the sediment of iron sulfate is filtered off. The solution is cooled to 0 °C, saturated with hydrogen chloride and mixed with solid potassium chloride. This forms a deposit of potassium chlorotitanate K_2TiCl_6., which is not soluble in water. The salt is separated (in the filter) and sent into a rotating furnace to be heated to 300-500 °C and disintegrated. The disintegration releases vapours of titanium tetrachloride, which are sent to condense; the sediment of potassium chloride is returned into the cycle. The titanium tetrachloride produced by this technique is very pure, since most impurities remain in the mother waters.

The chlorination of titanium raw stock forms a complex mixture, which apart from $TiCl_4$ contains other solid and gaseous substances, such as $SiCl_4$, $VOCI_3$, $AlCl_3$, $FeCl_3$, $FeCl_2$, $CaCl_2$, $MgCl_2$, $MnCl_2$, CCl_4, CO, CO_2, HCl, N_2, $COCl_2$, etc. The extraction of titanium tetrachloride out of this multicomponent mixture is one of the most difficult stages of the process. It is complicated by the fact that many of these compounds tend to interact, forming solid and liquid products. Besides, all chlorides in the mixture are hygroscopic and in the presence of moisture hydrolyse. That is why the condensation system can function normally only if the apparatuses and piping are airproof.

After the condensation system the gases contain mostly CO and CO_2, as well as some impurities (Cl_2, $TiCl_4$, $SiCl_4$, $AlCl_3$, HCl, $COCl_2$, $VOCl_3$, etc.). Before discharge in the atmosphere, the gases are subjected to sanitary cleaning in scrubbers sprayed with water or lime milk and containing copper for capturing chlorine. The following reactions take place in the scrubbers:

$$2AICI_3 + 3H_2O \longrightarrow AI(OH)CI_2 + AI(OH)_2CI + 3HCI \quad (5.34)$$

$$2VOCI_3 + Cu \longrightarrow CuCI_2 + 2VOCI_2 \quad (5.35)$$

$$nSiCI_4 + 2nH_2O \longrightarrow (SiO_2)_n + 4nHCI \quad (5.36)$$

$$CI_2 + 2CuCI_2 \longrightarrow Cu_2CI_2 \quad (5.37)$$

$$COCI_2 + H_2O \longrightarrow CO_2 + 2HCI \quad \text{and etc.} \quad (5.38)$$

If the content of silicon tetrachloride in discharge gases is increased, its utilisation becomes profitable. For this purpose there are absorbers mounted before the sanitary scrubbers. The absorbers are sprayed with cold titanium tetrachloride. The obtained mixture of $TiCl_4$ and $SiCl_4$ is sent to further rectification.

Technical titanium tetrachloride contains dissolved and suspended impurities. The former include gases (N_2, Cl_2, $COCl_2$), some chlorides ($AlCl_3$, $FeCl_3$, $SiCl_4$, $SnCl_4$), oxychlorides ($VOCl_3$, $TiOCl_2$, $SOCl_2$) and organic compounds (CCl_4, etc.), as well as a finely dispersed mixture of iron chlorides, aluminum, calcium and magnesium. The composition and quantity of impurities depend on the quality of raw stock and the conditions of the process. Titanium tetrachloride is purified from suspended particles by filtering or centrifuging, and from dissolved impurities, by fractional distillation.

In recent years, techniques for deeper purification of $TiCl_4$ have been devised. E.g., it is treated with various substances (fats or oils, acids, alcohols, etc.) and heated until it chars. The sediment is removed by filtering. This technique yields very pure titanium tetrachloride.

Titanium tetrachloride can also be purified from impurities by the continuous technique. The installation for continuous purification consists of several vertical pipe coolers. Liquid products of the reaction are sent into the first cooler, which is located a little higher than the rest, where the mixture is cooled at agitation to -3 - -5°C. After that, the mixture is abruptly cooled to -20 - -23.5 °C; the solution deposits crystals of Si_2Cl_6 and $VOCl_3$. The deposited crystals remain in the primary cooler, and the solution self-flows into the secondary coolers, where it is gradually cooled from -23 to -27 °C; titanium tetrachloride deposits as white sediment. It is collected in the secondary coolers and washed with water. The $TiCl_4$ thus purified is 99.92% pure.

Titanium tetrachloride is a colourless transparent liquid (the boiling point is 136 °C); it is easily decomposed with water forming hydrogen chloride and titanium dioxide. It joins the moisture in air to form white suffocating fumes, which are drops of hydrochloric acid.

Titanium tetrachloride is used as raw stock in the production of tetraalkoxy(aroxy)titaniums and of spongy titanium, which is produced by the reduction of titanium tetrachloride with magnesium. Besides, TiCl₄ is used in the production of pure titanium dioxide.

5.3.1. Preparation of tetrabutoxytitanium

Tetrabutoxytitanium is obtained by the etherification of titanium tetrachloride with butyl alcohol:

$$TiCl_4 + 4C_4H_9OH \xrightarrow[4HCI]{} Ti(OC_4H_9)_4 \qquad (5.39)$$

The released hydrogen chloride should be neutralised to avoid the reversible formation of titanium chloroethers:

$$Ti(OC_4H_9)_4 + 2HCI \rightleftharpoons (C_4H_9O)_2TiCI_2 + 2C_4H_9OH \qquad (5.40)$$

To neutralise hydrogen chloride, gaseous ammonia is introduced into the reactive mixture.

Raw stock: titanium tetrachloride (at least 99% of TiCl₄); butyl alcohol (the boiling point is 116.5-118.5 °C, $d_4{}^{20} = 0.809$-0.810, $n_4{}^{20} = 1.3993 \div 1.3995$); gaseous ammonia.

The periodic production of tetrabutoxytitanium (Fig. 90) comprises three main stages: the etherification of titanium tetrachloride with butyl alcohol; the neutralisation of hydrogen chloride with ammonia; the distillation of the excess of butyl alcohol and the filtering of the finished product.

Before the start all equipment is throroughly washed with butyl alcohol. Etherification is conducted in reactor 6 with a jacket and agitator. First, the reactor is loaded with butyl alcohol; the agitator is switched on; the jacket is filled with cooling salt solution.

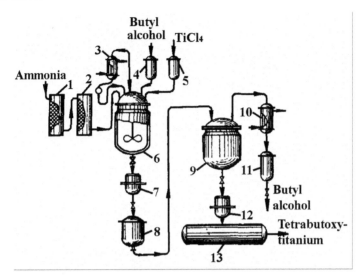

Fig. 90. Production diagram of tetrabutoxytitanium by the periodic technique: *1* – tower with soda lime; *2* - tower with sodium hydroxide; *3, 10* - coolers; *4, 5* – batch boxes; *6* - reactor; *7, 12* - nutsch filters; *8* - collector; *9* – vacuum distillation tank; *11* - receptacle; *13* – container

After that, titanium tetrachloride is slowly introduced. While it is introduced, the temperature in the reactor should not rise above 30 °C and the pressure should remain the same (when the temperature is above 30 °C and the pressure is above atmospheric, the supply of TiCl₄ should be slowed down or stopped altogether, and continued only after the temperature drops to 20 °C). After the whole of titanium tetrachloride has been fed, the reactive mixture is agitated for 1.5-2 hours and the reactor is continuously cooled with salt solution.

Usually the etherification of titanium tetrachloride is carried out with a large excess of butyl alcohol (10-12 moles of C_4H_9OH per 1 mole of $TiCl_4$); otherwise the reaction might form secondary substitution products, titanium chloroethers, i.e. the reaction of substitution will be incomplete. The interaction of $TiCl_4$ and alcohol releases hydrogen chloride; to neutralise it, the reactive mixture receives dehydrated ammonia at constant agitation. Ammonia is dehydrated in tower *1* with soda lime and tower *2* with sodium hydroxide. During neutralisation the temperature of the reactive mixture should not exceed 30 °C; the pressure should be 0.1 MPa. When the temperature and pressure increase, the supply of ammonia should be reduced; if the temperature and pressure continue to increase, the supply of ammonia should be stopped until the given parameters are established again.

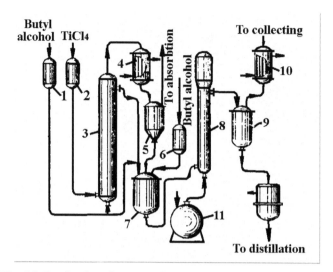

Fig. 91. Production diagram of tetrabutoxytitanium by the continuous technique: *1, 2, 6* – batch boxes; *3* – tower for partial etherification; *4, 10* - coolers; *5* - trap; *1,9*- collectors; *8* - bubble tower; *11* - container; *12* - nutsch filter

The process of neutralisation is considered complete if the temperature of the reactive mixture begins to drop when ammonia is passed through it. After that the reactive mixture is agitated for 2 more hours, passing ammonia through it. After neutralisation the mixture is poured into nutsch filter *7* and filtered at excess pressure not exceeding 0.07 MPa. Ammonia chloride is washed with butyl alcohol sent from batch box *4* through the reactor. The filtrate (the solution of tetrabutoxytitanium in butyl alcohol) is collected in intermediate collector *8* and from there sent into vacuum distillation tank *9*. At 80-90 °C and a residual pressure of 350-200 GPa in the tank the whole of butyl alcohol is distilled. It is condensed in cooler *10* and collected in receptacle *11*, from where it is returned to production. The end of the distillation is determined by the refraction index; for the finished product, tetrabutoxytitanium, it should be 1.4800-1.5050.

Tetrabutoxytitanium from the vacuum distillation tank is sent into nutsch filter *12*, where it is filtered from residual ammonia chloride and mechanical impurities. The filtered product is collected in receptacle *13*.

Tetrabutoxytitanium can also be produced by the continuous technique (Fig. 91). The process is conducted in two stages; ammonia is used as the acceptor of the released hydrogen chloride.

Tower *3* receives titanium tetrachloride from batch box *2* for partial etherification. After the tower has been filled approximately to 1/5, the supply of TiCI$_4$ is not stopped and the tower receives butyl alcohol from batch box *1* at such speed that the temperature in the tower does not exceed 70 °C.

The released hydrogen chloride passes through cooler *4* and trap *5* with butyl alcohol and is sent to absorption. The partially etherified product, dibutoxydichlorotitanium, is sent out of the tower into intermediate collector *7*, which is filled with butyl alcohol from batch box *6*. Out of collector *7* the mixture enters bubble tower *8* to complete etherification; the reaction takes place in the presence of ammonia. A temperature of 50 °C is maintained in tower *8*.

The mixture obtained in the tower is sent into collector *9*, and the excess of unreacted ammonia is withdrawn through collector *9* and cooler *10* for capturing. The products of the reaction are sent to filtering or centrifuging to separate ammonia chloride and (if necessary) to distillation in order to extract pure tetrabutoxytitanium.

Tetrabutoxytitanium is a transparent light yellow liquid (the boiling point is 179-186 °C at 13.3 GPa, the flash point is 40 °C). It can be easily dissolved in most organic solvents.

Technical tetrabutoxytitanium should meet the following requirements:

Appearance	Transparent viscous yellowish brown liquid
n_D^{20}	1.4800-1.5050
Content, %:	
of titanium	13.8-17
of chlorine	≤ 0.3
of iron	≤ 0.01

Tetrabutoxytitanium is used in the production of varnishes for the electrotechnical industry, in the production of polybutoxytitaniumoxane, as well as a solidifying agent for polymer compositions.

5.3.2. Preparation of polybutoxytitaniumoxane

Polybutoxytitaniumoxane is obtained by the partial hydrolytic condensation of tetrabutoxytitanium:

$$nTi(OC_4H_9)_4 + (n+1)H_2O \longrightarrow$$

$$\longrightarrow HO\left[\begin{array}{c} OC_4H_9 \\ | \\ Ti-O \\ | \\ OC_4H_9 \end{array}\right]_n H + 2nC_4H_9OH \qquad (5.41)$$

It can be produced in the installation used to prepare tetrabutoxytitanium (see Fig. 90). After preparing tetrabutoxytitanium, reactor *6* receives out of batch box *4* a mixture of water and butyl alcohol as necessary for hydrolytic condensation. When the mixture is added, the temperature of the reactive mixture should be 25-30 °C. The obtained mixture is agitated for 1.5-2 hours; the reactor is constantly cooled with salt solution. After filtering and flushing, the product from collector 8 is sent into vacuum distillation tank 9, where butyl alcohol is distilled at 90-100 °C and a residual pressure of 350-200 GPa. Similarly to the synthesis of tetrabutoxytitanium, the end of the distillation is determined by the refraction index, which should be 1.5150-1.5170 when the product is finished.

After alcohol has been distilled, the tank receives a necessary amount of toluene or xylene to prepare a 50-55% solution of polybutoxytitaniumoxane. After the solvent has been introduced, the mixture in the tank is agitated approximately for 15 minutes and then sent into nutsch filter 12. From there the finished product, filtered from mechanical impurities, enters container 13.

Polybutoxytitaniumoxane is a transparent viscous liquid, ranging in colour from yellow to light brown.

Polybutoxytitaniumoxane is used as a binding agent for coatings and as a special additive for paints and varnishes.

5.4. Preparation of organotin compounds

Among organotin compounds the most widely used are various organic salts of dialkyl tin, such as diethyl- and di-n-butyltin dicaprylates, di-n-butyltin dilaurate, di-n-butyltin distearate, etc. As a rule, these compounds are obtained through tin alkylhalogenides.

Tin alkylhalogenides can be prepared by several techniques, the most important of which are direct synthesis and organomagnesium synthesis.

Direct synthesis is based on the interaction of alkylhalogenide with tin oxide.

$$4SnO + 5RHal \xrightarrow{\text{Cu}} R_2SnHal_2 + RSnHal_3 + 2SnO_2 + RR \qquad (5.42)$$

This technique is of practical interest first and foremost for the synthesis of tin methylhalogenides; since no suitable conditions have yet been found for the production of higher alkylhalogenides with any significant yield.

The raw stock for direct synthesis can be both tin oxide and metallic tin.

$$3Sn + 6RHal \longrightarrow RSnHal_3 + R_2SnHal_2 + R_3SnHal \qquad (5.43)$$

The reaction occurs rather easily in the presence of a catalyst, crystalline iodine (the chemical technique) or under ionising γ-radiation (the radiation chemical technique). The use of metallic tin allows higher tin alkylhalogenides to be obtained.

5.4.1. Preparation of tin octylbromides by the chemical technique

The direct synthesis of tin octylbromides by the interaction of metallic tin and octylbromide is conducted in the presence of crystalline iodine and triethylamine:

$$3Sn + 6C_8H_{17}Br \longrightarrow C_8H_{17}SnBr_3 + (C_8H_{17})_2SnBr_2 + (C_8H_{17})_3SnBr \quad (5.44)$$

Raw stock: metallic tin (grey powder, 99.5% of the main substance); octylbromide (the boiling point is 201-204 °C); crystalline iodine (the melting point is 112.5-114 °C); triethylamine (the boiling point is 89-90 °C).

The production of tin octylbromides (Fig. 92) comprises three main stages: the synthesis of tin octylbromides and the filtering of the reactive mixture to separate unreacted tin; the distillation of excess octylbromide; the filtering of the mixture of tin octylbromides and its distillation.

Cylindrical reactor 4 with a jacket, shielded electric drive agitator and inverse cooler 3 receives a given amount of octylbromide from batch box 1. The hatch receives powderlike tin from transportable bin 2, triethylamine (3-4% of the reactive mixture) and iodine (0.3-0.4%). The hatch is closed and the agitator is switched on. The contents of the apparatus are heated to 170-180 °C and agitated at this temperature for 4-5 hours. Then the reactive mixture is cooled to 20 °C and sent into filter 5 to separate unreacted tin. The filtrate is sent by nitrogen flow into tank 7 to distil excess octylbromide. The solid residue, unreacted tin, is flushed twice with acetone in the filter. After flushing acetone is collected in collector 6, and the flushed tin is used again in the production.

The distillation of octylbromide in tank 7 is carried out at 60-80 °C and a residual pressure of 5.2-6.6 GPa. Octylbromide is collected in collector 10 and used again in the synthesis of tin octylbromides. After the distillation of octylbromide the mixture of tin octylbromides and solid residue (complex triethylamine salt, which forms sediment during the distillation) is cooled in the tank down to 20-30 °C and sent into filter 8 to separate the solid residue.

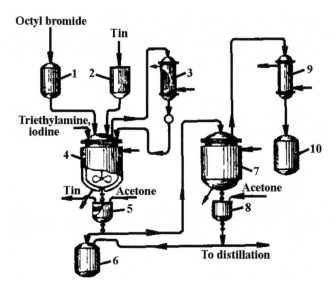

Fig. 92. Production diagram of tin octylbromides: *1* - batch box; *2* - bin; *3,9*- coolers; *4* - reactor; *5, 8* - filters; *6, 10* - collectors; *7*- distillation tank

The residue of complex salt is dissolved in acetone and collected in collector 6, the filtrate, which is a mixture of tin octylbromides, is sent to distillation (if necessary) to extract individual substances.

The mixture of tin octylbromides is a liquid ranging in colour from yellow to brown, which contains 70-80% of dioctyltindibromide, 15-25% of dioctyltintribromide and 5-10% of trioctyltinbromide. The tin content in the mixture is 22.5-23.5%; $d_4^{20} = 1.3604 \div 1.390$; the solidification point is from 0 to -6 °C.

5.4.2. Preparation of dibutyltindibromide by the radiation chemical technique

The production process of dibutyltindibromide is based on the interaction of butylbromide with metallic tin at 85-90 °C in the presence of activating additives (butyl alcohol or water) and under the influence of ionising γ-radiation:

$$2C_4H_9Br + Sn \longrightarrow (C_4H_9)_2SnBr_2 \qquad (5.45)$$

The main reaction is accompanied by some by-processes:

$$3C_4H_9Br + 2 Sn \longrightarrow (C_4H_9)_3SnBr + SnBr_2 \qquad (5.46)$$

$$4C_4H_9Br + 2 Sn \longrightarrow (C_4H_9)_3SnBr + C_4H_9SnBr_3 \qquad (5.47)$$

Raw stock: tin (powder of 10-40 μm particles with 99.5% of the main substance); butylbromide ("pure"). The radiation chemical synthesis of dibutyltindibromide is very sensitive towards impurities in parent reactants; therefore, they should be thoroughly purified. Thus, before being sent into the reactor, tin powder is washed from organic impurities with acetone, treated with diluted hydrochloric acid to remove oxide film, washed with distilled water and dried.

The reactor for radiation chemical synthesis with an agitator (about 200 rotations per minute) is located in the operation chamber. Inside the reactor there is a cavity for introducing sources of γ-radiation. The role of the radiation source is played by ^{60}Co in airproof stainless steel ampules; the activity of ^{60}Co radiation is about 3000 eq Ra. The consumption degree of the energy of radiation when the sources are placed in the cavity, or the radiation coefficient of efficiency, varies from 17 to 19%. The equipment is encased in a special box made of stainless steel and plexiglass; the box has intensive ventilation.

The reactor is draft-loaded with corresponding amounts of butylbromide, tin and butyl alcohol. Then the apparatus is closed with a lid fashioned with an agitator, moved into the operation chamber and connected to all piping. After that the jacket of the reactor is filled with a heat carrier with a temperature of 90-95 °C, the contents of the apparatus are heated at agitation to 85-90 °C, the operation chamber is closed with a protective plug and sources of γ-radiation are introduced into the chamber. After certain periods of time the mixture in the reactor is sampled with the help of a special sampler. 8-10 hours after the radiation has started, the reaction is completed and the sources of radiation are taken out of the chamber. The reactive mixture is loaded off and the apparatus is prepared for the next operation.

The production of dibutyltindibromide by this technique can also be carried out in the so-called pseudocontinious mode. In this case, after the reaction has developed and the concentration of the target product in the mixture is 60-70%, the agitator is periodically stopped for a short time so that tin settles down. Then, part of the reactive mixture is sucked off and the apparatus is loaded with an additional amount of reactants.

After the operation is over, the reactive mixture is filtered in vacuum to eliminate tin particles and poured into a receptacle. After the separation of solid particles the reactive mixture has the following composition: 65-80% of dibutyltindibromide, 1-2% of dibutyltintribromide and 4-7% of tributyltinbromide; the rest is unreacted butylbromide and butyl alcohol.

The maximum yield of dibutyltindibromide, which is 80-85%, is achieved with the help of an activator, butyl alcohol, in the quantity of 3%

of the reactive mixture; water can also be quite a good activator. The increased yield and accelerated formation of dibutyltindibromide in the presence of these hydroxyl-containing substances seem to be caused by the fact that they are good solvating agents, which weaken the carbon-halogene bond and thus increase the reactivity of alkylhalogenide, e.g. butylbromide. Incidentally, the effect of radiation also seems to amount to the weakening of the carbon-halogene bond in alkylhalogenide molecules.

The ratio of parent reactants is also essential for the speed at which the target product accumulates, as well as for the yield. The optimal Sn:C$_4$H$_9$Br ratio is 1:1.2. In the pseudocontinuous process of dibutyltindibromide production the duration of synthesis increases from 8-10 to 40-50 hours (any further increase reduces the efficiency of the process).

The radiation chemical technique of tin alkylhalogenide production remains not very promising due to economic reasons (special equipment is needed) and rigid safety requirements (additional safety measures against the effect of radiation are necessary).

Of greater practical interest for the production of tin alkylhalogenides is organomagnesium synthesis, which is based on the interaction of alkyl-magnesiumhalogenides with tin tetrachloride. This technique can also yield tin tetraalkyl derivatives. Since these substances are semiproducts in the production of organic salts of dialkyl tin, their organomagnesium synthesis is described along with the production of diethyl tin dicaprylate.

5.4.3. Preparation of diethyl tin dicaprylate

The production of diethyl tin dicaprylate comprises four main stages: the production and rectification of tetraethyltin; the production of diethyl tin dichloride; the production and drying of diethyl tin oxide; the production of diethyl tin dicaprylate.

Tetraethyltin is synthesised by Grignard technique in two stages. The interaction of metallic magnesium with ethyl bromide in diethyl ether gives the Grignard reactant, ethylmagnesiumbromide. Further interaction of ethylmagnesiumbromide with tin tetrachloride forms tetraethyltin:

$$4C_2H_5MgBr + SnCI_4 \longrightarrow Sn(C_2H_5)_4 + 2MgBr_2 + 2MgCI_2 \qquad (5.48)$$

The synthesis of diethyl tin dichloride is based on the reaction of tetraethyltin with tin tetrachloride:

$$Sn(C_2H_5)_4 + SnCI_4 \longrightarrow 2(C_2H_5)_2SnCI_2 \qquad (5.49)$$

The synthesis of diethyl tin oxide is carried out by the interaction of diethyl tin dichloride with potassium hydroxide:

$$(C_2H_5)_2SnCl_2 + 2KOH \longrightarrow (C_2H_5)_2SnO + 2KCl + H_2O \quad (5.50)$$

Diethyl tin oxide enters the reaction with caprylic acid to form diethyl tin dicaprylate:

$$(C_2H_5)_2SnO + 2CH_3(CH_2)_6COOH \xrightarrow[H_2O]{} (C_2H_5)Sn[OCO(CH_2)_6CH_3]_2 \quad (5.51)$$

Raw stock: magnesium chipping; ethyl bromide (the boiling point is not less than 35 °C, d_4^{20} = 1.420÷1.445); diethyl ether (d_4^{20} = 0.714÷0.715); benzene (the density is 0.8770-0.8791 g/cm^3); tin tetrachloride (anhydrous, the boiling point is 114-115 °C); caprylic acid (the boiling point is 239-240 °C, d_4^{20} = 0.9089); potassium hydroxide (technical product). Ethyl bromide, diethyl ether and benzene are dried before use with calcium chloride.

The production diagram of diethyl tin dicaprylate is given in Fig. 93. Before synthesis the whole system is washed, dried and pressurised with nitrogen (the pressure is 0.05 MPa). After that reactor 1 is loaded through a hatch with magnesium and several grams of crystalline iodine (the initiator of the reaction), the agitator is switched on and inverse cooler 4 is filled with water. Then the reactor is filled at room temperature with part of the reactive mixture (ethyl bromide, diethyl ether and benzene) from batch box 2 to stimulate the reaction.

After the reaction has been stimulated, as shown by the temperature increase to 40-60 °C, the remaining part of the mixture is fed at such speed that the temperature in the reactor does not exceed 70 °C. After the mixture has been supplied, the reactor is held at 65-75 °C for 1.5-2 hours; then, the jacket of the apparatus is filled with water and the contents are cooled to 15-20 °C.

Tetraethyltin is synthesised in the same reactor, 1. For this purpose the reactor, which already contains ethylmagnesiumbromide, receives a solution of tin tetrachloride in benzene (1:1 ratio) out of batch box 3 at agitation and operating inverse cooler 4 at such speed that the temperature in the apparatus does not exceed 75-80 °C. After the solution of SnCl$_4$ has been supplied, the contents of the reactor are held for 3 hours at this temperature and then cooled to 15-20 °C. The reactive mixture is loaded through the lower choke into enameled reactor 5 to decompose the unreacted ethylmagnesiumbromide with a solution of hydrochloric acid.

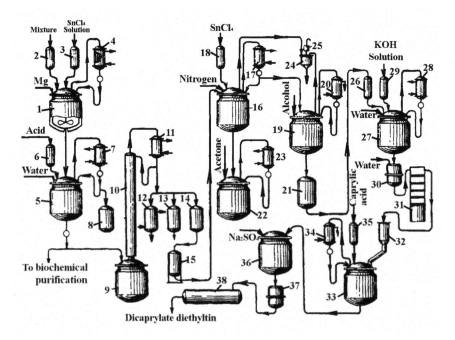

Fig. 93. Production diagram of diethyl tin dicaprylate: *1, 5, 16, 27, 33* - reactors; *2, 3, 6, 15, 18, 26, 29, 32, 35* - batch boxes; *4, 7, 11, 17, 20, 23, 28, 34* - coolers; *8, 21* – collectors; *9* - distillation tank; *10* - rectification tower; *12* - freezer; *13, 14* – receptacles; *19, 22* – apparatuses with agitators; *24* – oil gate; *25* - fire-resistant apparatus; *30, 37* - nutsch filters; *31* - shelf draft; *36* -dehydrator; *38* – container

The jacket of reactor *5* is filled with cold water, the agitator and inverse cooler *7* are switched on; after the temperature reaches 10-15 °C the apparatus is fed a necessary amount of water at such speed that the temperature does not rise above 40 °C. Then, the solution of hydrochloric acid is supplied out of batch box *6* into reactor *5*, just as gradually, so as not to exceed the 40 °C limit. The agitation is continued for 5-10 minutes; then the agitator is stopped and the mixture is settled for 15-20 minutes. The settled lower layer, the aqueous solution of magnesium salts, is sent through the lower choke to biochemical purification; the top layer, the benzene tetraethyltin solution, remains in reactor *5* for benzene distillation.

Before the distillation the inverse cooler is switched into the direct operation mode. Benzene vapours are condensed in cooler *7* and collected in receptacle *8*. The distillation is carried out below 100-120 °C in apparatus *5* (depending on the pressure of the vapour sent into the jacket). The distilled benzene is dried with calcium chloride and can be used to prepare the parent mixture for organomagnesium synthesis; the concentrated benzene solution of tetraethyltin is sent from apparatus *5* into distillation tank *9*.

The rectification of tetraethyltin is carried out in vacuum. Before rectification the whole system is pressurised with nitrogen (at atmospheric pressure); cooler *11* is filled with water, freezer *12* is filled with salt solution; after that, the tank receives the solution of tetraethyltin from reactor *5*. The vacuum pump is switched on; when the residual pressure in the system is 50-70 GPa, benzene is distilled and collected in the freezer. After the distillation is over, the jacket of the tank is filled with vapour. Until the temperature in the vapours is constant, the tower operates without refluxing. After constant temperature has been established, the first fraction is separated into receptacle *13*. This fraction (a mixture of tin ethyl bromides) is separated below 54 °C (at a residual pressure of 50-70 GPa); the target fraction, tetraethyltin, is separated above 54 °C into receptacle *14*.

Technical tetraethyltin is a liquid which boils at 174- 178 °C; its density is 1.19-1.20 g/cm^3; the allowable chlrone content is not more than 1%.

Tetraethyltin from receptacle *14* is sent into weight batch box *15*, from where it is loaded into reactor *16*. The agitator is switched on, inverse cooler is filled with water and tin tetrachloride is fed out of batch box *18* at such speed that the temperature in the reactor gradually (within 20-30 minites) rises to 190- 200 °C. After the whole of SnCl$_4$ has been loaded, the heating is switched on and the reactor is kept at 200-210 °C for 1.5 hours. After that, pure diethyl tin dichloride is distilled out of the reactive mixture. For that purpose, inverse cooler 11 is switched into the direct mode. Diethyl tin dichloride is distilled in nitrogen flow below 235 °C. The vapours condense in cooler 17, where a temperature of 95-100 °C is maintained, since diethyl tin dichloride melts at 84-85 °C. The condensate enters apparatus 19, which has been filled with ethyl alcohol. The condensate is separated when inverse cooler *20* is in operation and the agitator works at such speed that the temperature in appratus *19* does not exceed 40 °C.

After distillation the tank residue from reactor *16* is sent by nitrogen flow at 260 °C into apparatus *22*, which has been filled with acetone. The discharge is carried out at agitation; inverse cooler *23* is switched off. All devices at this stage should operate in nitrogen flow; they contact the atmosphere through oil gate *24* and fire-resistant apparatus *25*. From apparatus 19 the alcohol solution of diethyl tin dichloride enters reactor *27*. The agitator and inverse cooler *28* are switched on and a 30% solution of potassium hydroxide is sent from batch box 29 at such speed that the temperature in the reactor does not exceed 60 °C (the alkali is fed taking into account the 10% excess by the stoichiometric quantity). After potassium hydroxide has been loaded, the reactive mixture is agitated for 20-30 minutes at 60-70 °C. The obtained suspension of diethyl tin oxide is poured into nutsch filter *30*; the reactir is flushed with water, which is also poured into the filter. The product in the filter is pressed and washed until the flush waters give a neutral reaction. The washed product is loaded out of the filter onto metal drying trays in shelf dryer 31 until they reach constant weight at 60-80 °C. The dried product enters weight batch box *32* with a screw batcher.

Reactor *33* is loaded with caprylic acid from batch box 35 and heated to 65-75 °C with the agitator and backflow cooler *34* switched on. Then the reactor is filled with diethyl tin oxide out of batch box *32* at the same temperature. After it has been fed, the contents of the apparatus are agitated for 4-5 hours at the same temperature, and then cooled to 20-25 °C. The obtained technical diethyl tin dicaprylate is poured into apparatus 36 to be dried with anhydrous sodium sulfate. The drying takes 20-30 hours; after that the dried diethyl tin dicaprylate is filtered in nutsch filter *37* and collected in container *38*.

Table 32. Physicochemical properties of organic salts of dialkyl tin

Product	Formula	Boiling point (at 4-5.2 GPa), °C	n_D^{20}
Diethyl tin di-caprylate	$(C_2H_5)_2Sn(OCOC_7H_{15})_2$	176-182	–
Di-n-butyltin di-caprylate	$(n-C_4H_9)_2Sn(OCOC_7H_{15})_2$	215-220	1.4653-1.4681
Di-n-butyltin di-caprynate	$(n-C_4H_9)Sn_2(OCOC_9H_{19})_2$	Undistillable liquid	1.4675-1.4701
Di-n-butyltin di-laurate	$(n-C_4H_9)_2Sn(OCOC_{11}H_{23})_2$	- 22-24 (melting point)	-
Di-n-butyltin distearate	$(n-C_4H_9)_2Sn(OCOC_{17}H_{15})_2$	48-50 (melting point)	-
Di-n-butyltin di-maleate	$(n-C_4H_9)_2Sn$ OCOCH ‖ OCOCH "	103-105 (melting point)	-
Di-n-butyltin dioleate	$(n-C_4H_9)_2Sn(OCOC_{17}H_{33})_2$	Undistillable liquid	1.4812-1.4816

Diethyl tin dicaprylate is a high-boiling yellow liquid $(d_4^{20} = 1.14 \div 1.15$, the allowable chlorine content does not exceed 0.2%).

Dethyltin dicaprylate is used as a stabiliser in polymers and as a catalyst in the process of their production. Besides, the tetraethoxysilane solution of diethyl tin dicaprylate is used as a vulcaniser for elastomers and for the production of silicone compounds and sealants.

Similarly to diethyl tin dicaprylate, one can also obtain dicaprylate, dicaprynate and distearate of di-n-butyltin and its other organic salts.

The physicochemical properties of some organic salts of dialkyl tin are given in Table 32.

5.5. Preparation of organolead compounds

Organolead compounds have been recently widely used. They are used in agriculture (as pesticides), in medicine, as well as in various chemical processes as catalysts for the polymerisation of vinyl monomers, chlorination of hydrocarbons, etc. Lead tetraalkyl derivatives are used as additives for engine fuels due to their antiknock properties.

Lead tetraalkyl derivatives are synthesised by several techniques:

The reaction of alkylchloride or alkylbromide with lead and sodium alloy:

$$4RHal + 4PbNa \longrightarrow PbR_4 + 3Pb + 4NaHal \qquad (5.52)$$

1. The interaction of alkylmagnesiumhalogenide obtained by organomagnesium synthesis with lead chloride:

$$10RMgBr + 5PbCl_2 \longrightarrow PbR_4 + Pb_2R_6 + 2Pb + 10MgBrCl \qquad (5.53)$$

The yield of tetraalkyllead depends on the conditions of the reaction: at a temperature below 20 °C the yield of hexaalkyldilead Pb_2R_6 is increased (up to 40%); on the contrary, at a temperature above 20 °C hexaalkyldilead decomposes into $PbR_4 + Pb$, accordingly increasing the yield of tetraalkyllead.

2. The alkylation of anhydrous lead acetate with trialkylaluminum:

$$6(CH_3COO)_2Pb + 4R_3Al \longrightarrow 3PbR_4 + 3Pb + 4(CH_3COO)_3Al \qquad (5.54)$$

3. The electrolysis of the aqueous-alcoholic solution of alkylbromide or alkyliodide and sodium hydroxide (the cathode liquid). The cathode can be made from porous lead, the anode from graphite. The mechanism of the process seems to be the following:

$$4RHaI \longrightarrow 4R\cdot +2HaI_2$$

$$Pb + 4R\cdot \longrightarrow PbR_4$$

$\left.\phantom{\begin{matrix}a\\b\end{matrix}}\right\}$ **On the cathode**   (5.55)

$HaI_2 + 2NaOH \longrightarrow NaHaI + NaOHaI + H_2O$ **(On the anode)**   (5.56)

The formed tetraalkyllead is separated by the distillation with water vapour.

Practical importance is also attached to the electrolysis of complex metalorganic substances on the lead anode. E.g., Ziegler suggested an industrial technique for the production of lead tetraalkyl derivatives through readily accessible complex aluminum compounds.

5.5.1. Preparation of tetraethyllead

Tetraethyllead can be produced by chemical and electrochemical techniques.

One of the chemical techniques for the preparation of tetraethyllead is based on the interaction of lead and sodium alloy with ethylchloride.

$$4PbNa + 4C_2H_5CI \longrightarrow Pb(C_2H_5)_4 + 3Pb + 4NaCI \qquad (5.57)$$

This overall reaction, however, does not show all the stages of the process, since apart from the substances mentioned it forms hexaethyldilead, ethylene, ethane and butane, which demonstrates some secondary reactions. In particular, at the first stage of the reaction ethylchloride seems to interact with metallic sodium forming free ethyl radicals, i.e. the reaction uses the radical mechanism and ethyl radicals are responsible for further synthesis.

$$C_2H_5CI \xrightarrow[NaCI]{Na} C_2H_5\cdot \xrightarrow{Pb} [PbC_2H_5] \xrightarrow{C_2H_5\cdot} Pb(C_2H_5)_2 \qquad (5.58)$$

Diethyllead decomposes at the synthesis temperature, forming tetraethyllead and metallic lead. The main reaction is accompanied by some by-processes: E.g., we find triethyllead, which can easily turn into a dimer, hexaethyldilead.

$$Pb(C_2H_5)_2 + C_2H_5\cdot \longrightarrow Pb(C_2H_5)_3 \longrightarrow 0.5Pb_2(C_2H_5)_6 \qquad (5.59)$$

The dimer disintegrates at a temperature above 20 °C to form tetraethyllead.

$$2Pb_2(C_2H_5)_6 \longrightarrow 3Pb(C_2H_5)_4 + Pb \qquad (5.60)$$

The synthesis is based on lead and sodium alloy (9:1 ratio) or this alloy with an addition of 0.3-1.2% of potassium. The process of alloy production is based on mixing molten metals: lead, sodium and potassium (metallic potassium is obtained by the interaction of potassium chloride with metallic sodium), After melting the metals together in an electric furnace at 400-550 °C the alloy is pelleted. The finished lead and sodium alloy takes the form of grey steel discs not more than 33 mm in diameter. The content of active alkali metals should be 9.65-10.1% (including 0.3-1.2% of potassium); the density is 5.1 g/cm^3; the melting point is 367 °C. The alloy actively reacts with water and oxidises in air; therefore, it should be stored in air- and waterproof bins under a layer of oil.

Tetraethyllead can be synthesised in reactors of various construction and various volume. The process comprises two stages (Fig. 94): the synthesis of tetraethyllead; its distillation from the reactive mixture with water vapour.

The process is carried out in reactor 5, which is a steel apparatus with a water vapour jacket, an agitator, expander 4 and cooler 3. After the alloy has been loaded, the reaction is stimulated by pouring part of ethylchloride (35-50 1 per 1 tonne of the alloy) into the reactor from batch box 2 and sending hot water into the jacket. During the heating ethylchloride interacts with the alloy, which is accompanied by noticeable growth in temperature and pressure. As soon as the temperature in the reactor reaches 60 °C and the pressure is 0.15 MPa, the heating is stopped and the jacket is filled with water for cooling. Further temperature rise in the apparatus continues due to the exothermicity of the reaction; a rise of 90 °C is tolerated.

As ethylchloride enters the reaction, the temperature and pressure in the reactor increase, abruptly at first, then more slowly. After the temperature and pressure no longer rise, ethylchloride is fed from batch box 2 at such speed that the temperature in the reactor does not exceed 90 °C. If the temperature rises above 90 °C, the apparatus is filled with mineral oil from batch box 1 to dilute the reactive mixture and reduce the activity of the reaction. During the synthesis it is necessary to withdraw the released heat continuously and rather intensively. It is done by sending cooling water into the jacket of the reactor and due to the evaporation of unreacted ethylchloride.

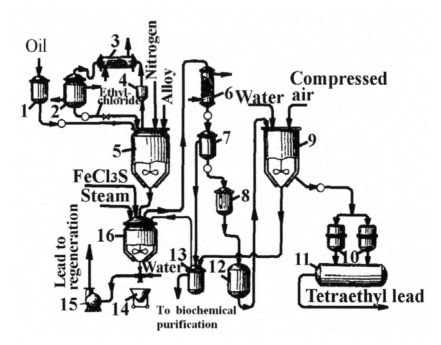

Fig. 94. Production diagram of tetraethyllead by the chemical technique: *1, 2, 8* - batch boxes; *3, 6* - coolers; *4* - expander; *5* - reactor; *7* - separator; *9* -flusher; *10* - filters; *11* - container; *12* - collector; *13* - trap; *14* – trolley bin; *15* - pump; *16* - distillation tank

Vapours of ethylchloride are sent out of the reactor into expander *4*, where solid particles deposit, and then condense in cooler *3*. The condensate is returned to the synthesis through batch box *2*. After the whole of ethylchloride has been supplied, the mixture in the reactor is agitated for 1 hour at 70-90 °C to complete the reaction. If in the absence of heat withdrawal the temperature begins to drop, it signifies the end of the reaction. If a rise in temperature is observed, it means that the reaction is not finished. In this case the process is conducted at 70-90 °C for a longer period, until heat release begins.

After the reaction is complete, the excess of ethylchloride is sent into batch box *2* (the heating is not stopped). The supply of ethylchloride into the reactor is stopped, and the pressure in the system is gradually reduced to atmospheric. The distillation of ethylchloride is considered complete if during the next 10-15 minutes the distillate stops entering batch box *2*. Then, for a more thorough elimination of ethylchloride vapours, the system is blown with nitrogen (the pressure is 0.05 MPa). After the distillation the reactive mixture is cooled down to 60 °C and sampled through a hatch when the agitator is stopped. The reaction is considered finished when the content of active alkali metal does not exceed 0.1%.

The mixture is loaded out of the reactor into tank 16 to distil tetraethyllead. The tank should already be filled with ground sulfure and ferric iron chloride. Iron chloride reduces the alkalinity of the dross and improves its consistency due to the formation of the colloid solution of iron hydroxide; ground sulfure is uniformly distributed through the dross, also improving its consistency and preventing clotting of lead particles.

The distillation of tetraethyllead from the reactive mixture is conducted by sending live steam through it. In the beginning of the distillation the steam pressure is 0.02-0.04 MPa; then it increases and can be brought to 0.1-0.2 MPa. The temperature of the steam should not exceed 130 °C. The distilled vapours of tetraethyllead and water enter cooler 6, where they condense, and flow into separator 7. There, due to the difference in the densities, the aqueous condensate is separated from tetraethyllead and flows into trap 13, from where it is sent to biochemical purification. The distilled tetraethyllead is sucked into batch box 8 and sent to water flushing and air treatment.

After the distillation of tetraethyllead, the dross from apparatus 16 (lead, sodium chloride, oil, some impurities and water) is cooled and loaded into special trolley bins 14. The solution of sodium chloride and oil from the bins are sucked after settling into special montejuses, from where through traps they are sent to waste disposal works. The dross is pumped with pump 15 into the furnace to regenerate lead.

To flush tetraethyllead, the product is sent from batch box 8 into collector 12 and from there by nitrogen flow into flusher 9. There tetraethyllead is purified from stable impurities by water flushing and oxidating with oxygen in air. The flusher has a a propeller agitator and a ring bubbler with a 25-60 m³/h speed of air passage. The flusher is filled with water and at agitation with tetraethyllead from collector 12 (in the 1.5:1 volume ratio to water). The mixture is agitated for 30 minutes; after that the bubbler is filled for 3 hours with compressed air. Tetraethyllead treated in this way is kept in the flusher for 5-10 hours; after that it is filtered and collected in container 11.

The water separated in the flusher is poured into trap 13. The tetraethyllead found there is periodically (as soon as it accumulates) sent into distillation tank 16; the water is sent to biochemical purification.

Technical tetraethyllead should meet the following requirements:

Appearance	Transparent colourless or slightly yellowish liquid without sediment
d_4^{20}, at least	1.570
Content, %:	

of tetraethyllead ≥ 92
of ethylchloride ≤ 0.6

Tetraethyllead is the main component for the production of antishock compounds.

The disadvantages of the chemical technique of tetraethyllead production are the low degree of reactant transformation (75% of the lead does not enter the reaction) and secondary formation of sodium chloride. Besides, it is very problematic to extract tetraethyllead out of the spongelike mass, the mixture of lead and sodium alloy and sodium chloride completely. The process is periodic and hence difficult to automate. Electrochemical techniques for the synthesis of tetraethyllead seem to be more promising in this aspect.

Electrolysis of complex salts. The essence of the technique is the following. The mixture of potassium tetraethylaluminate and complex salt of potassium phluoride with triethylaluminum is subjected to electrolysis; the anode is lead, the cathode is mercury. Electrolysis dissolves the lead anode, forming tetraethyllead; the cathode is covered with potassium amalgam:

$$KAl(C_2H_5)_4 + KF \cdot Al(C_2H_5)_3 + Pb + xHg \longrightarrow$$

$$\longrightarrow Pb(C_2H_5)_4 + K(Hg)_x + KF \cdot 2Al(C_2H_5)_3 \qquad (5.61)$$

Tetraethyllead is separated; the electrolyte is treated with ethylene and sodium hydride:

$$KF \cdot 2Al(C_2H_5)_3 + C_2H_4 + NaH \longrightarrow KF \cdot Al(C_2H_5)_3 + NaAl(C_2H_5)_4 \quad (5.62)$$

The obtained sodium tetraethylaluminate is treated with potassium amalgam:

$$NaAl(C_2H_5)_4 + K(Hg)_x \longrightarrow KAl(C_2H_5)_4 + Na(Hg)_x \qquad (5.63)$$

which formed on the cathode during electrolysis: Potassium tetraethylaluminate is sent to electrolysis again; the obtained sodium amalgam is decomposed with water into alkali and hydrogen, which is used in the production of sodium hydride. This makes up a closed three-stage cycle of tetraethyllead production (the production is called the *three-cycle arrangement*).

The overall reaction that takes place in the electrochemical synthesis of tetraethyllead is the following:

$$Pb + 4Na + 4C_2H_4 + 4H_2O \longrightarrow Pb(C_2H_5)_4 + 4NaOH \qquad (5.64)$$

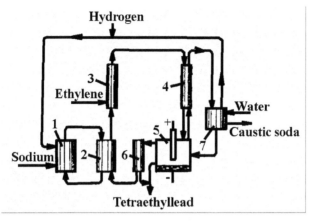

Fig. 95. Production diagram of tetraethyllead by electrolysis of complex salts: *1* - *4* – synthesis apparatuses; *5* - electrolyser; *6* - trap; *7* – apparatus for decomposing sodium amalgam

The general diagram of tetraethyllead production by the electrolysis of complex salts is given in Fig. 95. Electrolyser *5* is filled with low-temperature melt, a mixture of potassium tetraethylaluminate and the complex salt of potassium fluoride with tetraethylaluminum. The anode is a lead rod; the cathode is mercury. In the electrolyser the lead anode is dissolved, forming tetraethyllead; the cathode is covered with potassium amalgam. After the process is over, the electrolyte is separated from tetraethyllead. It enters trap *6*; the electrolyte, the complex salt $KF \cdot 2Al(C_2H_5)_3$, is treated with ethylene and sodium hydride, which has been obtained in apparatus *1* from hydrogenand metallic sodium. This forms sodium tetraethylaluminate; it is treated in apparatus *4* with potassium amalgam which has formed on the cathode during electrolysis. The thus obtained potassium tetraethylaluminate is sent from apparatus *4* to electrolysis again; sodium amalgam is sent into apparatus *7* to be decomposed with water into alkali and hydrogen. The hydrogen from apparatus *7* enters apparatus *1* for the production of sodium hydride.

The electrolysis of complex salts seems rather convenient economically; however, technologically and from the point of view of fire safety it is less advantageous than the technique based on the electrolysis of Grignard reactant.

Electrolysis of Grignard reactant. The essence of the process is the following. First, Grignard reactant (methyl- or ethylmagnesiumchlorde)is synthesised in the solvent medium, a mixture of tetrahydrofurane and dibutyl ether of diethylene glycol. The obtained alkylmagnesiumchloride is subjected to electrolysis in the apparatus where the role of the cathode is played by steel walls, and that of the anode, lead balls. Electrolysis forms magnesium chloride, metallic magnesium and tetraethyllead.

$$4RMgCl \longrightarrow 4R\cdot + 2MgCl_2 + 2Mg \left.\begin{array}{l} \\ \\ Pb + 4R\cdot \longrightarrow PbR_4 \end{array}\right\} \text{On the anode} \qquad (5.65)$$

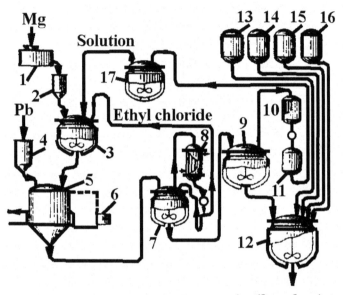

Antiknock mixture

Fig. 96. Production diagram of tetraethyllead by electrolysis of Grignard reactant: *1* - mill; *2, 4* - bins; *3* - reactor; *5* - electrolyser; *6* – electric current rectifier; *7* – apparatus for separating ethylchloride; *8, 10* - cooler; *9* - apparatus for separating tetraethyllead; *11* - collector; *12* - agitator; *13 - 16* - batch boxes; *17* – apparatus for purifying solvents

Metallic magnesium reacts with excess alkylchloride, transforming again into Grignard reactant:

$$Mg + RCl \longrightarrow RMgCl \qquad (5.66)$$

In the production of tetraethyllead by the electrolysis of Grignard reactant (Fig. 96) metallic magnesium is ground in mill *1*, sent through bin *2* into reactor *3*, which is also filled with a required amount of ethylchloride from apparatus *7* and solvent (the mixture of tetrahydrofuran and dibutyl ether of diethylene glycol) from apparatus *17*. The reactive solution containing ethylmagnesiumchloride, excess ethylchloride and a solvent, is sent from reactor *3* into electrolyser *5*, the walls of which play the role of the cathode. The anode (lead balls or pellets) are sent into the hydrolyser from bin 4. During the electrolysis the lead balls dissolve and tetraethyllead is formed (the yield can be 96%); magnesium chloride and metallic magnesium are formed on the cathode. To prevent closure of the electric circuit caused by the formation of metallic magnesium, an excess of ethylchloride is introduced into the electrolyser. Metallic magneisum reacts with excess alkylchloride, transforming again into Grignard reactant.

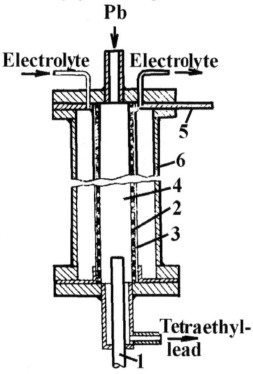

Fig. 97. Industrial electrolyser for the production of tetraethyllead: 1 – graphite anode; 2 – steel cathode; 3 - mesh; 4 – anode chamber filled with lead balls; 5 – cathode bus; 6- casing

Then the products of electrolysis are sent into apparatus *7* to be separated from ethylchloride returned into the reactor. Out of apparatus *7* the products move into apparatus *9* where tetraethyllead is purified from solvents and enters apparatus *12* to be mixed with various additives (dibromoethylene, dichloroethylene, toluene solution of the dye, oxidiser) which comprise antishock mixtures. The solvents from collector *11* are sent into apparatus *17* for purification and then back into reactor *3*.

Figure 97 shows one of the most original constructions of an electro-lyser. The peculiarity of this apparatus is the use of small lead balls or pel-lets in chamber *4* as the anode. The casing of the electrolyser is steel tube *6* up to 6 m long; there is a steel perforated cylinder in the tube, cathode *2*. The inner side of the cathode is connected with mesh 3 of nonconductive material (Teflon or ceramics); it is 3-4 mm thick. The power to the cathode is supplied with insulated bus *5*.

Graphite anode *1* is in the central part of the apparatus. All this space is filled with small lead balls. Thus, the anode-cathode spacing is determined by the thickness of insulating mesh *3*. The electrolyte continuously circu-lates through the tank. The formed tetraethyllead does not dissolve in the electrolyte; it is collected in the lower part of the anode chamber and is pe-riodically withdrawn to purification. In order to replenish the reacted lead, new portions of lead pellets are periodically introduced through the choke, just like ethylmagnesiumchloride.

This construction largely solves the problem of continuous electrolysis without opening the electrolyser;at the same time, the anode-cathode spac-ing always remains constant.

Table 33. Physicochemical properties of lead tetraalkyl derivatives

Substance	Boiling point, °C	d_4^{20}	n_D^{20}
$Pb(CH_3)_4$	110	1.9952	1.5120
$Pb(C_2H_5)_4$	82*	1.6524	1.5198
$Pb(C_3H_7\text{-}n)_4$	126*	1.4419	1.5094
$Pb(C_4H_9\text{-}n)_4$	159*	-	-
$Pb(C_5H_{11}\text{-iso})_4$	-	1.2337	1.4947
*At 17.3 GPa			

Tetraethyllead is a transparent colourless liquid; in small concentrations its vapours have a sweetish fruity smell; in large concentrations it has an unpleasant odour. Tetraethyllead boils at 200 °C (with decomposition); $d_4^{20} = 1.6524$; $n_D^{20} = 1.5198$. It does not dissolve in water but can be easily dissolved in alcohols, diethyl ether, petrol, acetone and other organic sol-vents.

Like all organic lead compounds, tetraethyllead is very toxic. It can en-ter the body through skin, lungs or digestive tract; thus, the production and use of tetraethyllead requires airproof equipment, overalls and gas masks.

Similarly to tetraethyllead, one can also obtain other tetraalkyl and tetraaryl derivatives of lead. Some physicochemical properties of impor-tant tetraalkyl derivatives of lead are given in Table 33.

Tetramethyllead has been also put forward as an antishock additive for engine fuel. Comparison tests have shown that tetramethyllead is more efficient that tetraethyllead, especially in petrols with high content of aromatic hydrocarbons. Tetraalkyl lead derivatives can also be used in the production of lead alkylhalogenides and their derivatives.

5.6. Preparation of organophosphorus compounds

Organophosphorus compounds each year gain more and more practical importance. As we know, a very important condition for increasing crop yield is the use of chemical plant defences, or so-called pesticides, to combat pests and disease. Many organophosphorus compounds can be effective pesticides. They have also proved to be very poweful insecticides and acaricides. Besides, organophosphorus compounds are good plastifyers (tricresyl-, tributyl- and triphenylphosphates) and stabilising agents (alkylaryl ethers of phosphorous or pyrocatechinphosphorous acid) for polymer materials.

The raw stock for the synthesis of practically important organophosphorus compounds is phosphorus trichloride, phosphorus chlorooxide and phosphorus thiotrichloride.

5.6.1. Preparation of phosphorus trichloride, phosphorus chlorooxide and phosphorus thiotrichloride

Phosphorus trichloride. The main raw stock for the production of phosphorus trichloride is yellow phosphorus obtained from phosphorite, quartzite and coke:

$$Ca_3(PO_4)_3 + 5C + 3SiO_2 \longrightarrow 2P + 5CO + 3CaSiO_3 \qquad (5.67)$$

Coke is a reducing agent to bind oxygen; quartzite reduces the melting point of the furnace charge. The yellow phosphorus obtained in this way (powder ranging in colour from light yellow to brownish green) should meet the following requirements: 99.7-99.9% (soluble), 0.1-0.3% of the residue which does not dissolve in carbon disulfide.

The preparation of phosphorus trichloride is based on the interaction of phosphorus with free chlorine in PCI$_3$ solution:

$$4P + 3CI_2 \xrightarrow{\text{PCl}_3} 2PCI_3 + 2P \qquad (5.68)$$

The necessary condition for chlorination is that the reaction should take place in an excess of free phosphorus; with its lack phosphorus trichloride chlorinates to form pentachloride. The formed PCI$_5$ interacts with free phosphorus and transforms back into PCI$_3$. However, since this reaction occurs very actively and releases a great amount of heat, loading phosphorus into a reactor with phosphorus pentachloride will cause an abrupt increase of temperature and pressure, and the apparatus may explode. It is especially dangerous when raw stock contains moisture, because water energetically decomposes phosphorus tri- and pentachlorides and forms phosphorous acids and hydrogen chloride. When phosphorus chlorides interact with water in a closed space, the quick growth of pressure (owing to the formation of a lot of vapours) can destroy the equipment. Excess content of nonsoluble impurities (SiO_2, H_2SiO_3, H_3PO_3) is also undesirable, because they contaminate the equipment.

Phosphorus trichloride is a colourless transparent motile liquid (the boiling point is 76 °C); it fumes in air. It mixes with diethyl ether, petrol, chloroform, carbon disulfide and dichloroethane in all ratios: It can be easily destroyed with water, acids and alcohols. Phosphorus trichloride vapours hydrolyse even in humid air.

Phosphorus trichloride is very poisonous; it causes serious skin burns and can cause death when enters the human body (even in small quantities). The maximum allowable concentration of phosphorus trichloride vapours in industry is 0.5 mg/m^3.

There are two types in which phosphorus trichloride is manufactured, A and B. Technical characteristics for these products are given below:

Characteristics	Type A	Type B
Main substance content, %, not less than	97.5	97.5
Density at 20 °C, g/cm^3	1.575-1.585	1.575-1.585
Allowable impurities content, %, not more than		
phosphorus chlorooxide	2.5	5
free phosphorus	Absent	

Phosphorus trichloride can be used to produce not only organophosphorus compounds, but also the chlorooxide of phosphorus pentachlorde, as well as methylchloride and other chloroorganic products, as a chlorinating agent.

Phosphorus chlorooxide. POCl$_3$ is obtained by the oxidation of phosphorus trichloride:

$$2PCI_3 + O_2 \longrightarrow 2POCI_3 \qquad (5.69)$$

in a cylindrical reactor with an agitator. It is continuously fed with phosphorus trichloride and oxygen, as well as free phosphorus as a catalyst (since PC1$_3$ contains some PCI$_5$, free phosphorus is an agent that contributes to the transformation of phosphorus pentachloride into trichloride). Oxygen is fed into the lower part of the reactor through a bubbler. A temperature of 40-60 °C is maintained in the apparatus. The released heat can be withdrawn by flushing the walls of the apparatus with cold water. When the temperature falls, the oxidation process is stopped. The duration of the process is from 40 to 60 hours. The reactor has a protective membrane, because the oxidation can happen very quickly and release a lot of heat.

Phosphorus chlorooxide is a colourless transparent liquid (the boiling point is 105.8 °C). It can be easily dissolved in organic solvents and hydrolysed with water. Phosphorus chlorooxide is raw stock for the production of tricresyl-, tributyl- and triphenylphosphates, as well as other organophosphorus compounds.

Phosphorus thiotrichloride $PSCl_3$ is produced in industry by direct interaction of sulfure and phosphorus trichloride at a temperature about 180 °C in the presence of $AlCl_3$ as a catalyst.

Phosphorus thiotrichloride is a colourless transparent liquid (the boiling point is 125 °C). It easily dissolves in benzene, carbon tetrachloride and chloroform; it is slowly hydrolysed with water.

Phosphorus thiotrichloride is raw stock for the production of thiophos, mercaptophos and some other organophosphorus compounds.

5.6.2. Preparation of tricresylphosphate

The basis for the production of tricresylphosphate is the etherification of phosphorus chlorooxide with cresol in the presence of a catalyst, anhydrous magnesium chloride:

$$POCl_3 + 3CH_3C_6H_4OH \xrightarrow[3HCl]{MgCl_2} PO(OC_6H_4CH_3)_3 \qquad (5.70)$$

The reaction takes place very easily; however, it is always accompanied with by-processes such as the incomplete etherification of phosphorus chlorooxide:

$$POCl_3 + CH_3C_6H_4OH \xrightarrow{HCl} PO(OC_6H_4CH_3)Cl_2 \qquad (5.71)$$

$$POCl_3 + CH_3C_6H_4OH \xrightarrow[2HCl]{} PO(OC_6H_4CH_3)_2Cl \qquad (5.72)$$

and its hydrolysis with water introduced into the reaction zone with cresol:

$$POCl_3 + 3H_2O \xrightarrow[3HCl]{} H_3PO_4 \qquad (5.73)$$

Raw stock: phosphorus chlorooxide (at least 99% of $POCl_3$); technical cresol, the mixture of three isomers, the most chemically active of which is *m*-cresol, and the least active is *p*-cresol. *o*-Cresol causes persistent lesions if it contacts skin. Techical cresol contains up to 30% of *o*-isomer; due to this technical cresol is a toxic substance and dealing with it requires the use of all safety measures.

The chemical activity of cresol largely depends on the content of *м*-isomer and impurities in it. E.g., if etherification is carried out with recirculating cresol where *м*-isomer is virtually absent, the reaction will either not happen or will proceed very slowly, with an insignificant yield of the target product. The presence of organic sulfur compounds in cresol is also very harmful, because they contribute to the disintegration of tricresyl-phosphate during vacuum distillation. On the other hand, the presence of phenol is desirable, because it increases the stability of tricresylphosphate during distillation and storage. The best cresol is the one that contains virtually no sulfure compounds, up to 20% of phenol and not more than 3% of o-cresol.

The production process of tricresylphosphate (Fig. 98) comprises two main stages: the etherification of phosphorus chlorooxide with cresol; the extraction of tricresylphosphate.

First, technical cresol is dried in steel cylindrical apparatus *15* lined with diabase tile. Cresol is heated to 90 °C in the apparatus and held at the residual pressure of 130 GPa until the water content is less than 0.2%. Water vapours and cresol vapours carried away with them condense in cooler *14* and collect in vacuum receptacle *16* (from there the solution is sent to biochemical purification). The dried cresol is sent into batch box *1* and then to etherification.

The etherification of phosphorus chlorooxide with cresol is carried out in etherificator 5, a cast iron cylindrical apparatus lined with two layers of diabase tiles. The reactor is filled with a necessary amount of phosphorus chlorooxide from batch box *2*; magnesium chloride is loaded through a hatch and dried cresol is loaded out of batch box *1*. After the reactants have been loaded, the apparatus is heated to 80 °C. The reaction is started at this temperature, but then it is gradually raised to 170 °C. At 170 °C the mixture is held for about 5 hours; after that, the acidity number is determined. When the acidity number is 30 mg KOH/g, the reaction is stopped.

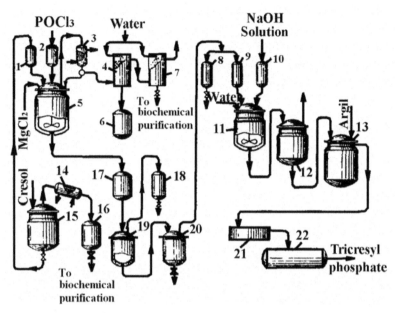

POCl₃   Water   NaOH
                      Solution

Fig. 98. Production diagram of tricresylphosphate: *1, 2, 10-* batch boxes; *3, 14-* coolers; *4, 7-* towers; *5-* etherificator; *6, 17-* collectors; *8, 9, 16 –* vacuum receptacles; *11 -* flusher; *12, 15 -* dehydrator; *13 –* "clarification" apparatus; *18 -* receptacle; *19, 20-* tanks; *21 –* pressure filter; *22-* container

The released hydrogen chloride, which carries vapours of phosphorus chlorooxide and cresol, enters through a porcelain pipeline into cooler *3*. There vapours of phosphorus chlorooxide and cresol condense and flow back into the reactor, and the uncondensed hydrogen chloride passes through porcelain piping into absorption tower *4* filled with Raschig rings and flushed with water. The 27-28% hydrochloric acid formed here is poured into collector *6*, and discharge gases enter tower *7*, which is also flushed with water. The reaction in this tower also forms 2.5-3% hydrochloric acid, which is then sent to biochemical purification.

The mixture from reactor *5* enters collector *17*, from where it sent to the distillation of tricresylphosphate as soon as it accumulates. The distillation is carried out in tanks *19* and *20*, which are thick-walled cylindrical apparatuses with fire-resistant lining of chamotte bricks. The first fraction is distilled into receptacle *18* below 200 °C. It contains cresol with an impurity of tricresylphosphate. After the first fraction has been completely distilled, the mixture from tank *19* enters tank *20* to distil the second and third fractions at the residual pressure of 130 GPa. The second fraction (tricresylphosphate with significant impurities of cresol and ethers of phosphoric acid) is distilled at 260-280 °C into vacuum receptacle *8*. This fraction can be sent for repeated distillation. After the second fraction has been distilled, the third fraction is separated into vacuum receptacle *9*. This fraction, which is tricresylphosphate, is distilled in the 280-310 °C range; then it is sent for purification. After several distillation operations tank *20* is heated to coke tank residue. After that, the apparatus is cooled and the obtained coke is removed.

Tricresylphosphate from receptacle *9* enters flusher *11*, which is a cylindrical apparatus with an agitator. First of all, tricresylphosphate is washed with 3% sodium hydroxide to neutralise residual hydrochloric acid and separate cresol (the allowable cresol content in tricresylphosphate is 0.6 g/l). It is also recommended to treat the product with diluted solution of potassium permanganate to improve the colour and oxidation stability of tricresylphosphate.

After that, apparatus *11* is filled with water to flush the alkali (there should be not more than 0.02 of NaOH in tricresylphosphate). The flushed product is dried in apparatus *12* at 90-100 °C and 130 GPa; by the end of the drying the moisture content in tricresylphosphate should not exceed 0.008%. Dry tricresylphosphate is sent into apparatus *13* for "clarification" where it is mixed for 1 hour at 80-90 °C with kil or active coal. This process eliminates acidic impurities and tarring products. After "clarification" the pulp is sent by a plunger pump to pressure filter *21*. Filtered from clay, the tricresylphosphate enters container *22*.

Tricresylphosphate can also be obtained by the continuous technique. In this case etherification is carried out in a cascade of consequtive reactors operating at increasing temperatures; the maximum temperature can reach 200 °C if the reaction is catalysed with metal halogenides. To reduce the losses of phosphorus chlorooxide with gaseous hydrogen chloride, the process should be carried out at reduced pressure.

Technical *tricresylphosphate* is a colourless, transparent, oily liquid without visible mechanical impurities (the density is 1.17 g/cm^3). It boils at 410 °C (decomposing). It hardly dissolves in water but can be easily dissolved in alcohols, ethers, chloroform, benzene, fats and oils. It combines well with cellulose acetate and various organic polymers. Industry manufactures two types of the product:

	Type I	Type II
Flash point, °C, at least	222	217
Acidity number, mg KOH/g, not more than	0.17	0.40
Content, %, not more than		
of volatile substances	0.10	0.12
of free cresol	0.08	Not standardised

Similarly to tricresylphosphate, one can also produce tributyl-, triphenyl- and other alkyl and aryl ethers of phosphoric acid.

Triorganophosphates are widely used as plastifiers used in the production of motion-picture film, linoleum, imitation leather, quick-drying varnishes based on modified cellulose nitrate, as well as plastics, PVC, etc. These plastifiers are much less volatile than organic plastifiers. E.g., if the weight loss at 85 °C within 96 hours is 7% for dibutylphthalate, 20% for dibutylladipinate and 40% for dipentiladipinate, for tricresylphosphate this characteristic amounts only to 0.5%. Tricresylphosphate and other triorganophosphates are also used as lubricants; adding approximately 1% of tricresylphosphate to carbon tetrachloride in fire extinguishers prevents the formation of phosgene.

5.6.3. Preparation of diethers of dithiophosphoric acid

Diethers of dithiophosphoric acid are bis(alkylphenoxy)dithiophosphates:

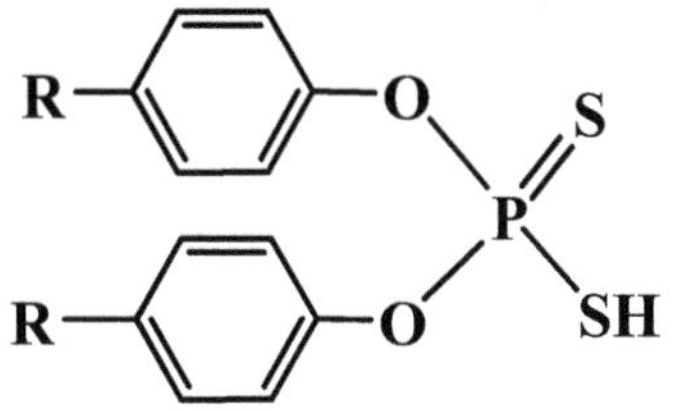

R - alkyl C_6 - C_{10} etc.
Recently they have acquired practical importance as so-called multifunctional additives for improving performance characteristics of lubricant oils used in intensive operation engines.
The synthesis of bis(alkylphenoxy)derivatives of dithiophosphoric acid is based on the interaction of alkylphenols with phosphorus pentasulfide:

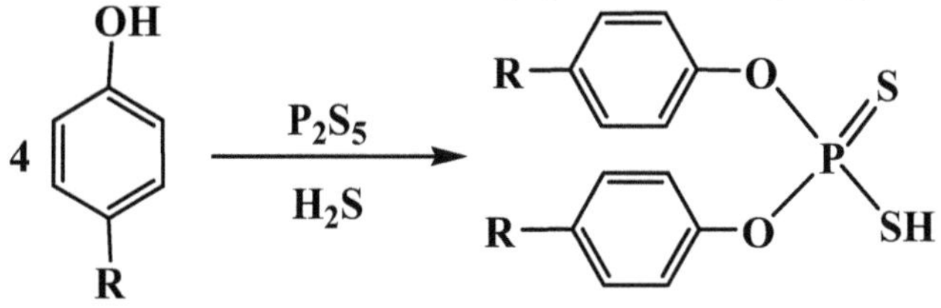

$$(5.74)$$

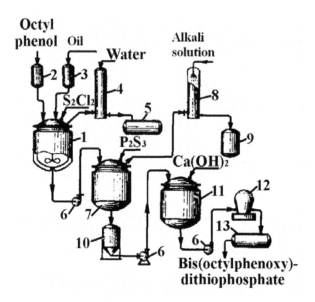

Fig. 99. Production diagram of bis(octylphenoxy)dithiophosphate: *1* – sulfiding reactor; *2, 3* - batch boxes; *4, 8*- absorbers; *5, 13* - containers; *6* - pumps; *7*- phosphorisation reactor; *9*- collector; *10*- settling box; *11* - neutraliser; *12* - ultracentrifuge

The industrial production of bis(octylphenoxy)dithiophosphate (Fig. 99) begins with the activation of octylphenol: it is sulfided, or treated with univalent sulfur chloride. To do this, reactor *7* is loaded with octylphenol (1 weight part), bobbin oil AU (2 weight parts) and then at 25-30 °C and agitation sulfur chloride is gradually added. The released hydrogen chloride is absorbed with water in absorber *4* and collected in containers *5* in the form of 8% hydrochloric acid. After sulfiding the mixture is pumped into reactor *7* where phosphorus pentasulfide is introduced at agitation. The hydrogen sulfide released in the process of phosphorisation, is neutralised with alkali in absorber *8* and collected in collector *9* in the form of the solution of sodium hydrosulfide.

After phosphorisation the mixture is sent into settling box *10* to be separated from tarlike products and unreacted phosphorus pentasulfide; then it is pumped into apparatus *11* to be neutralised with calcium hydroxide and dehydrated (at heating). The dehydrated mixture is sent to centrifuge *12*, where   bis(octylphenoxy)dithiophosphate   is   separated   from   mechanical impurities.

Bis(octylphenoxy)dithiophosphate is a viscous liquid ranging in colour from light yellow to brown; it dissolves well in mineral oils.

Similarly to bis(octylphenoxy)dithiophosphate, one can obtain other bis(alkylphenoxy)derivatives of dithiophosphoric acid. As mentioned above, these ethers are very efficient additives for lubricant oils used in intensive operation engines. An addition of 1-5% of these substances greatly increases thermal oxidative stability and anticorrosion properties of lubricant oils and reduce varnish formation in engines.

5.6.4. Preparation of aryl and alkylaryl ethers of phosphorous acid

Aryl and alkylaryl ethers of phosphorous and pyrocatechinphosphorous acids are efficient nonstain stabilisers for many polymers.

These ethers are obtained by the etherification of phosphorus trichloride or pyrocatechinphosphorus monochloride with phenols or naphthols. They have the formula R_2POAr, where $Ar = C_6H_5$, $C_6H_2(C_4H_9\text{-iso})_2CH_3$, $C_{10}H_7$, etc.

In particular, the synthesis of α-naphthyl ether of pyrocatechinphosphorous acid is conducted in two stages. At the first stage pyrocatechinphosphoromonochloride is synthesised from pyrocatechin and phosphorus trichloride:

$$(5.75)$$

The process is complicated and is accompanied by secondary reactions which lead to the formation of diphosphites. At the second stage pyrocatechinphosphoromonochloride is etherified with α-naphthol:

$$(5.76)$$

Raw stock: phosphorus trichloride (at least 97.5% of PCl_3); pyrocatechin (the melting point is 104 °C, the boiling point is 245-246 °C); α-нафтол (the melting point is 96.1 °C, the boiling point is 278-280 °C); carbon tetrachloride (the boiling point is 76-77 °C, $d_4^{20} = 1.593$).

The production of α-naphthyl ether of pyrocatechinphosphorous acid comprises three main stages: (Fig. 100): the synthesis of pyrocatechinphosphoromonochloride; the etherification of pyrocatechinphosphoromonochloride with α-naphthol; the crystallisation, filtering and drying of the α-naphthyl ether of pyrocatechinphosphorous acid.

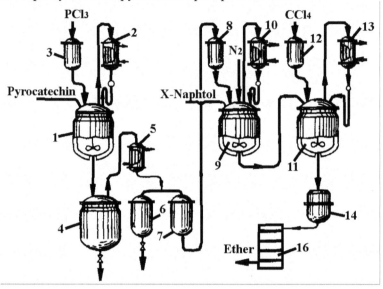

Fig. 100. Production diagram of α-naphthyl ether of pyrocatechinphosphorous acid: *1* - reactor; *2, 5, 10, 13* - cooler; *3, 8, 12* - batch boxes; *4* – vacuum distillation tank; *6, 7* - receptacles; *9* - etherificator; *11* - crystalliser; *14* - nutsch filter; *15* - vacuum draft

Pyrocatechin is loaded into steel enameled apparatus *1* with a water vapour jacket, agitator and inverse cooler *2*. It is also filled at agitation with phosphorus trichloride from batch box *3*. The mole ratio of trichloride and pyrocatechin is 1.5:1. After that the reactive mixture is held for 6-7 hours; the temperature is gradually raised to 90 °C. Then the temperature is increased to 140 °C within 2-3 hours and the reactive mixture is held for 10-11 more hours. The released hydrogen chloride is sent into the absorption system; the liquid products are sent into tank *4* for distillation.

The first distillate is the excess of phosphorus trichloride; its vapours condense in cooler *5* and flow into receptacle *6*. The distillation is begun at atmospheric pressure and 80 °C; then the vacuum pump is switched on and the remaining phosphorus trichloride is distilled at a residual pressure of 240-295 GPa below 80 °C. After the whole of trichloride has been distilled, the temperature is increased to 150-153 °C and pyrocatechinphosphoromonochloride is distilled at the same residual pressure. It is collected in receptacle *7* and from is sent to etherification through batch box *8*.

For etherification, steel enameled reactor *9* is loaded through a hatch with α-naphthol and heated to the melting point of α-naphthol (96-98 °C);

after that it is gradually filled at agitation with pyrocatechinphosphoro-monochloride from batch box *8*. After that, dehydrated nitrogen is passed through the reactive mixture and the temperature is increased to 120 °C. After 8 hours of holding at 120-130 °C and continuous supply of nitrogen the mixture is sent out of apparatus *9* into crystalliser *11*.

Special attention should be given to maintaining the given temperature during etherification, since an increase above 160 °C will form a noncrys-tallisable liquid product instead of crystalline α-naphthyl ether of pyro-catechinphosphorous acid.

The mixture in crystalliser 11 is cooled down to 70 °C and is gradually supplied with carbon tetrachloride; the temperature is reduced to 50 °C. At further gradual cooling and agitation α-naphthyl ether of pyrocatechin-phosphorous acid crystallises. After that the mixture is completely cooled to 20 °C. The formed sediment is filtered in nutsch filter *14* and (to elimi-nate carbon tetrachloride completely) dried for some time in vacuum draft *15*.

α-Naphthyl ether of pyrocatechinphosphorous acid can be extracted not only by crystallisation, but also by vacuum distillation at a residual pres-sure of 1.3-6.66 GPa. In this case at a temperature below 200 °C distilla-tion produces unreacted α-naphthol (2-3% of the loaded quantity), and be-low 230 °C, the target ether.

α-Naphthyl ether of pyrocatechinphosphorous acid is a white crystalline substance (the melting point is 86-87 °C); it dissolves well in chloroform, dichloroethane, benzene and diethyl ether and is easily hydrolysed with water. It is used as a nonstain stabiliser for many polymers, e.g. polyolefi-nes.

Similarly to α-naphthyl ether, one can also obtain β-naphthyl ether, phenyl ether and other aromatic ethers of pyrocatechinphosphorous acid.

In the production of some alkylaryl ethers, e.g. 2,6-di-*tret*-butyl-4-methylphenyl ether of pyrocatechinphosphorous acid,

$$(5.77)$$

etherification can be carried out in an inert solvent (benzene) in the presence of pyridine, which helps to greatly reduce the temperature of the process.

For the industrial etherification of pyrocatechinphosphorusmonochlo-ride with 2,6-di-*tret*-butyl-4-methylphenol in a solvent (Fig. 101), steel enameled reactor 6 with a propeller agitator and inverse cooler *1* is loaded at 30-40 °C with 2,6-di-*tret*-butyl-4-methylphenol, pyridine and benzene.

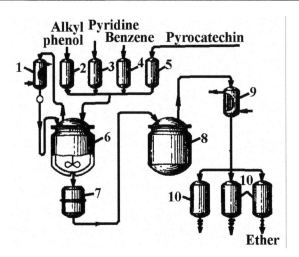

Fig. 101. Production diagram of 2,6-di-*tret*-butyl-4-methylphenyl ether of pyro-catechinphosphorous acid: *1,9* - coolers; *2-5*- batch boxes; *6*- reactor; *7*- nutsch fil-ter; *8* – vacuum distillation tank; *10* – receptacles

It is also gradually filled at agitation with pyrocatechinphosphoro-monochloride from batch box *5*. The residue of muriatic pyridine is fil-tered in nutsch filter *7*, and the solution is subjected to vacuum distillation in tank *8*. First, benzene is distilled; it is followed by unreacted products and finally by the target product, ether.

2,6-Di-*tret*-butyl-4-methylphenyl ether of pyrocatechinphosphorous acid is a white crystalline substance (the melting point is 86-89 °C); it dis-solves well in organic solvents and is easily hydrolysed with water.

5.6.5. Preparation of organic derivatives of phosphonitrilechloride

Recently, organic derivatives of phosphonitrilechloride:

R = OAlk, OAr, NHAlk, NHAr, etc.

have been widely used as plastifiers for various polymers, varnishes, elastomers and glues, additives for hydraulic liquids and drying oils; modifiers of organic and silicone polymers; as lubricant oils and impregnating compositions for fireproof cellulose films and fibres (fire retardants)

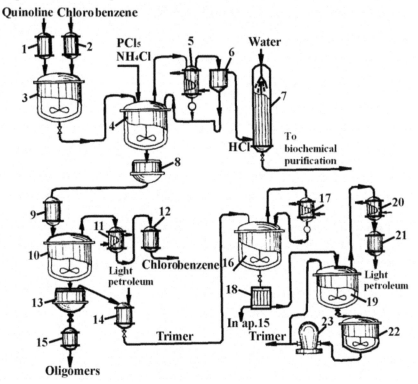

Fig. 102. Production diagram of phosphonitrilechloride trimer: *1,2-* batch boxes; *3* - agitator; *4-* reactor; 5, *11, 17, 20-* coolers; *6-* separator; *7* - tower; *8, 13* - nutsch filters; *9, 12, 21* - receptacles; *10, 19* - distillation tanks; *14, 15* -collectors; *16-* extractor; *18-* pressure filter; *22-* crystalliser; *23* – ultracentrifuge

The main raw stock for the production of organic phosphonitrilechloride s is the cyclic trimer of phosphonitrilechloride .

Preparation of phosphonitrilechloride trimer. Phosphonitrilechloride trimer is synthesised by partial ammonolysis of phosphorus pentachloride with ammonia chloride in the presence of quinoline as a catalyst:

$$3PC_5 + 3NH_4CI \xrightarrow[12HCI]{} [PNCI_2]_3 \qquad (5.78)$$

In fact, the process is more complicated and alongside with the polymer forms an oillike product consisting of higher oligomer homologues $[PNCI_2]_n$, where $n>4$. Tetramer is formed only in small quantities (less than 4%), mostly as an impurity in trimer.

The production process of phosphonitrilechloride   trimer (Fig. 102) comprises two main stages: the partial ammonolysis of phosphorus penta-chloride; the separation, purification and crystallisation of trimer.

Enameled reactor *4* with inverse cooler *5*, an agitator and a jacket is filled with a solution of quinoline in chlorobenzene from agitator *3* and loaded with ammonia chloride and phosphorus pentachloride through a hatch. The synthesis is conducted at 128-130 °C until the quantity of re-leased hydrogen chloride is noticeably reduced. Hydrogen chloride is ab-sorbed with water in packed tower *7*. After the process is finished, the re-active mixture is cooled and filtered in nutsch filter *8* to separate muriatic quinoline and the excess of ammonia chloride. Phosphonitrilechloride can also be conducted in tetrachloroethane medium; in this case the process is carried out at 135-140 °C.

The filtrate from the nutsch filter enters receptacle *9*, and from there into vapour-heted tank *10* for distilling chlorobenzene in vacuum (chloroben-zene is sent back to synthesis). The cooled oily residue from the tank is sent into nutsch filter *13* to be separated into crystals (raw phosphoni-trilechloride trimer) and oily liquid (phosphonitrilechloride oligomers). Trimer crystals are sent into collector *14* (it is also filled with light petro-leum, which does not dissolve oily oligomers). The emulsion formed in the collector is sent into extractor *16*. A temperature equal to the boiling point of light petroleum is maintained in this apparatus.

The obtained saturated solution is filtered in pressure filter 18, which heated with vapour to avoid crystallisation. The oily liquid remaining on filter cloth is collected and added to the product in collector *15*; the hot fil-tered crystal solution enters tank *19* for partial distillation of light petro-leum and preparation of oversaturated solution. The hot solution is sent from the tank into crystalliser *22*.

Apparatus *22* is cooled with salt solution (-15 °C) to crystallise phos-phonitrilechloride trimer; the mixture is held for some time to let crystals grow and precipitate. Then the crystals are separated in centrifuge *23*, and the mother solution is sent back to tank *19*. The obtained crystals of phos-phonitrilechloride trimer can be given additional purification, i.e. recrys-tallisation from fresh light petroleum or vacuum distillation at 127 °C and a residual pressure of 17 GPa.

Phosphonitrilechloride trimer with an impurity of tetramer is a crystal-line product (the melting point is . 108-114 °C) with a light odour. It has a slight irritating effect on mucuous membranes. Trimer and tetramer can be separated by fractional vacuum distillation, fractional crystallisation or vacuum sublimation.

The described technique forms approximately even quantities of trimer and oily product. Since the oily product is a mixture of higher oligomer homologues of phosphonitrilechloride, after purification it can be used to produce various substituted phosphonitrilechlorides.

The oily product can be purified by the reprecipitation technique. Oil-like oligomers are dissolved in benzene, treated with light petroleum, settled and distilled at 80-100 °C to avoid phosphonitrilechloride polymerisation.

The mixture of phosphonitrilechloride oligomers is a thick oily liquid ranging in colour from dark yellow to brown. It dissolves well in benzene, toluene, xylene, acetone and some other organic solvents; it does not dissolve in water, acetic acid and light petroleum.

The etherification of phosphonitrilechlorides with alcohols, phenols and naphthols can produce various alkoxy, phenoxy- and naphthyloxyderivatives of phosphonitrilechlorides; the interaction of these derivatives with amines or aniline yields various phosphonitrilamines and phosphonitrielanylide.

5.6.6. Preparation of organophosphorus pesticides

Organophosphorus pesticides are specific efficient contact and systemic (acting inside a plant) insecticides and acaricides. Among them are O,O-dimethyl-(1-hydroxy-2,2,2-trichloroethyl)phosphonate   (trichlorfon, or chlorophos)*, O,O-diethyl-O-(4-nitrophenyl)thiophosphate (thiophos)**, O,O-dimethyl-O-(ethylmercaptoethyl)thiophosphate   (methylmercaptophos)***. Besides, some organophosphorus compounds are efficient defoliants, i.e. substances which cause defoliation of cotton and other plants. These substances include S,S,S-tributyltrithiophosphate(butyphos), etc.

*dipterex (Germany), dilox (USA).

**parathion, E-605, folidol (Germany), niran, genithion (USA), fosferno (UK), SNP (France).

***metasystox, methyldemeton (Germany).

Preparation of trichlorfon

Trichlorfon can be synthesised in three stages: the preparation of trichloroacetic aldehyde (chloral); the production of O,O-dimethyl ether of phosphorous acid (dimethylphosphite); the synthesis of O,O-dimethyl(1-hydroxy-2,2,2-trichloroethyl)phosphonate (trichlorfon).

Chloral is obtained by direct chlorination of 96% ethyl alcohol with gaseous chlorine:

$$CH_3\text{-}CH_2OH + 4CI_2 \xrightarrow[5HCI]{} CCI_3\text{-}CHO \qquad (5.79)$$

In fact, the process is much more complicated and is accompanied by secondary reactions. Thus, raw chloral usually contains various impurities. To obtain the pure product, raw chloral should be distilled over concentrated (approximately 92%) sulfuric acid.

The distillation allows to separate two fractions. The first fraction, which is distilled below 92 °C in vapours, is a mixture of chloral, products of incomplete chlorination and hydrogen chloride (the latter in large quantities); this fraction is returned to repeated distillation. The second fraction, which is distilled in the 92-110 °C temperature range, contains mostly chloral (the boiling point is 97 °C); it is sent to the production of trichlorfon.

Dimethylphosphite is obtained by the interaction of methyl alcohol with phosphorus trichloride:

$$3CH_3\text{-}OH + PCI_3 \longrightarrow (CH_3O)_2POH + CH_3CI + 2HCI \qquad (5.80)$$

Trichlorfon is synthesised by the reaction of dimethylphosphite with chloral:

$$(CH_3O)_2POH + CCI_3\text{-}CHO \longrightarrow \underset{\underset{O}{\|}}{(CH_3O)_2P}\text{---}\underset{\underset{OH}{|}}{CH\text{-}CCI_3} \qquad (5.81)$$

At room temperature trichlorfon is stable; however, in the process of production and purification it is exposed to high temperature for a length of time, which reduces the yield of trichlorfon due to the secondary formation of O,O-dimethyl-O-(2,2-dichlorovinyl)phosphate:

$$\underset{\underset{O}{\|}}{(CH_3O)_2P}\text{---}\underset{\underset{OH}{|}}{CH\text{-}CCI_3} \xrightarrow{HCI} \underset{\underset{O}{\|}}{(CH_3O)_2P}\text{---}OCH{=}CCI_2 \qquad (5.82)$$

The yield of trichlorfon is also reduced if the reactive mixture contains methylchloride, since methylchloride interacts with trichlorfon:

$$\underset{\underset{O}{\|}}{(CH_3O)_2P}\text{---}\underset{\underset{OH}{|}}{CH\text{-}CCI_3} \xrightarrow[HCI]{CH_3CI} \underset{\underset{O}{\|}}{(CH_3O)_2P}\text{---}\underset{\underset{OCH_3}{|}}{CH\text{-}CCI_3} \qquad (5.83)$$

Thus, to produce purer trichlorfon, the synthesis should be conducted at low temperature and methylchloride should be thoroughly removed from dimethylphosphite.

Raw stock: ethyl alcohol (96%); chlorine (not more than 0.6% of hydrogen); sulfuric acid (at least 92.5% of monohydrate); phosphorus trichloride ($d_4^{20} = 1.575 \div 1.585$); ethyl alcohol (the boiling point is 64-65 °C, $d_4^{20} = 0.791 \div 0.793$).

The production of trichlorfon comprises three main stages: the synthesis and distillation of chloral; the synthesis of dimethylphosphite and distillation of raw dimethylphosphite; the synthesis of trichlorfon and the stripping of by-products and unreacted substances.

Preparation of chloral. Chloral is prepared in continuous apparatuses consisting of bubble tower *1* (Fig.103) and prechlorinator *2* (for prechlorination of ethyl alcohol). The prechlorinator is a steel apparatus lined with acid-resistant tiles. The bubble tower, more than twice as large as the prechlorinator, is lined with two layers of diabase tiles, filled with Raschig rings and fashioned with a cooling jacket. Ethyl alcohol is sent into the lower part of the prechlorinator; gaseous chlorine is heated to 50-80 °C and sent into the lower part of the bubble tower.

Chlorine should not contain impurities, because they impair the chlorination of ethyl alcohol; if the impurity content is about 15%, the reaction virtually does not happen.

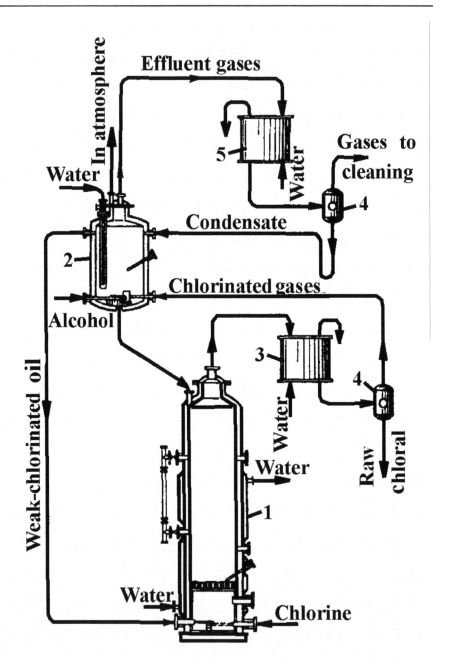

Fig. 103. Production diagram of chloral: *1* – bubble tower; *2* - prechlorinator; *3,5*-graphite coolers; *4* – separators

A temperature of 55-70 °C is maintained in the prechlorinator (owing to released heat); the chlorination degree of alcohol is monitored by the density of the reaction liquid. Weak-chlorinated oil is prepared there (the density is 1.15-1.25 g/cm³) and sent into tower *1*. Gases from the prechlorinator [hydrogen chloride with up to 3% (vol.) of chlorine impurity, vapours of unreacted alcohol and products of its incomplete chlorination] are sent into water-cooled graphite cooler *5*, where vapours of alcohol and products of its incomplete chlorination condense and flow back into the prechlorinator through separator *4*. The uncondensed gases (mostly hydrogen chloride) are sent into a water-flushed tower (not shown in the diagram), where 28% hydrochloric acid is prepared.

Weak-chlorinated oil from the prechlorinator continuously moves under the distribution graphite grate into the lower part of bubble tower *1*. There, at 88-92 °C ethyl alcohol is further chlorinated to reach the density of the chlorinated product of 1.48-1.50 g/cm³. Chlorination extracts a great deal of heat, the excess of which is withdrawn in the tower due to the evaporation of chloral hydrate formed (its boiling point is much lower than that of other components of the reactive mixture).

The hot mixture of gases (approximately 80% of hydrogen chloride and 20% of unreacted chlorine, ethyl alcohol, ethylchloride, chloral hydrate and products of the incomplete chlorination of alcohol) from the top part of tower *7* enters graphite cooler *3* to be cooled down to 30-40 °C. To avoid the supercooling of the cooler and freezing of the products, the cooler is filled with water heated to 30-40 °C. Part of the products condenses in the cooler and is sent back to tower *1* (not shown in the diagram) through separator *4* (or a phase separator); cooled effluent gases (chlorine, hydrogen chloride) are sent back into the prechlorinator. The obtained raw chloral (the density is 1.60-1.63 g/cm³) from the bubble tower is sent through the cooler and separator to distil over concentrated sulfuric acid.

Pure *chloral* is a colourless liquid (the boiling point is 97 °C; $d_4^{20}=$ 1.512) with a sharp odour and sweetish taste.

Preparation of dimethylphosphite Its synthesis is carried out in reactor *3* (Fig. 104), which is an enameled apparatus with an agitator and water vapour jacket, at 20 °C and a residual pressure of 480-520 GPa, continuously sending methanol and phosphorus trichloride into the reactor. The given temperature is maintained by sending liquefied methylchloride, which withdraws the released heat when it evaporates. The supply of methylchloride is regulated by the temperature in the reactor. Effluent gases (CH_3CI, HCl) are sent through trap *4* to be washed from hydrogen chloride first into tower *5* (flushed with water), then into tower *6* (flushed with 10% alkali solution); methylchloride moves further into the drying and compression system.

In the production of trichlorfon, the purity of dimethylphosphite is very important: the purer dimethylphosphite, the better is the yield and quality of trichlorfon. Therefore, much attention is given to the stripping and distillation of raw dimethylphosphite. The raw substance from reactor 3 is continuously sent to distil volatile substances (HC1, CH₃Cl) into the top-part of vertical cylindrical tower 7 with a tank.

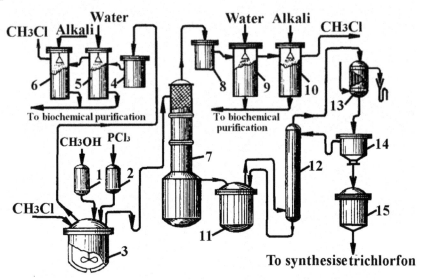

Fig. 104. Production diagram of dimethylphosphite: *1, 2* - batch boxes; *3* - reactor; *4, 8* - trap; *5-7,9, 10, 12* - towers; *11* - tank; *13* - cooler; *14* – gas separator; *15* – collector

Distillation is carried out at a residual pressure of 50-90 GPa and constant heating of the raw stock in the tank. The temperature is distributed through the tower thus: 22-30 °C on top, 60-70 °C in the tank. The raw dimethylphosphite continuously flows from reactor *3* through the tower packing and comes in contact with products which evaporate out of the tank; thus, dimethylphosphite vapours condense because they give heat to cold raw dimethylphosphite. The heating of raw dimethylphosphite evaporates its volatile components. As raw dimethylphosphite flows down the tower, its temperature grows, completely eliminating HC1 и CH₃Cl. To capture the evaporated dimethylphosphite, the top of tower *7* has a bumper filled with Raschig rings.

Effluent gases from tower *7* subsequently enter through trap *8* into towers *9* and *10*, which are flushed accordingly with water and 10% alkali solution. Due to the difference in pressure, evaporated dimethylphosphite continuously flows out of distillation tower *7* into tank *11* of rectification tower *12*. There is it subjected to vacuum distillation (the temperature in the tank is 130-135 °C, the temperature on top is 80 °C; the residual pressure is 25-55 GPa). Evaporating dimethylphosphite from tower *12* continuously enters water cooler *13* and condenses. It is collected in collector *15* through gas separator *14* and sent to the synthesis of trichlorfon.

Pure dimethylphosphite can also be extracted by rotory membranous vertical apparatus with the heat exchange surface of 0.8 m² and the rotor speed of 307 rotation per minute. In this case the optimal parameters are the following: the temperature is 105-110 °C, the residual pressure is 40 GPa. The purity of dimethylphosphite can reach 98%.

Preparation of trichlorfon Enameled reactor *3* with an agitator and water vapour jacket (Fig. 105) receives chloral and dimethylphosphite from batch boxes. The synthesis is carried out at 60-70 °C. The products from the reactor continuously enter ager *4* to make the reaction more complete.

Due to the difference of pressures in the appratuses raw trichlorfon continuously flows out of the ager into the system where secondary unreacted products are distilled. First, raw trichlorfon is heated to 80 °C in heater *5* by sending hot water into the space between the pipes. Then, the mixture leaves the heater and enters distillation tower *6*, filled with Raschig rings and fashioned with a coil with hot water. Distillation is conducted at 80-90°C and a residual pressure of 26-27 GPa. The distilled unreacted products (mostly chloral) flow through cooler *7* into collector *8* and are loaded into the distillation tank as they accumulate. There chloral is distilled over sulfuric acid, and the ready product, trichlorfon, continuously flows from tower *6* into collector *9*. It is heated with 40 °C water (through the jacket).

Trichlorfon can also be obtained in one stage:

$$CCl_3\text{-}CHO + PCl_3 + 3CH_3OH \xrightarrow[2HCI]{} (CH_3O)_2\underset{\underset{O}{\|}}{P}\text{---}\underset{\underset{OH}{|}}{CH}\text{-}CCl_3 + CH_3CI \quad (5.84)$$

This combined process is carried out in a solvent medium (carbon tetrachloride, chlorobenzene), withdrawing the released heat.

Technical trichlorfon should meet the following requirements:

Content, %:
of phosphorus	≥ 11
of dimethylphosphite	≤ 1
Solubility of 10 g in 100 ml of water at 25 °C	Complete

Trichlorfon can be purified by recrystallisation from water or organic solvent. Pure *trichlorfon* is a white crystalline substance (the melting point is 83-84 °C) with a pleasant odour. It is soluble in alcohol, benzene and most chlorinated hydrocarbons.

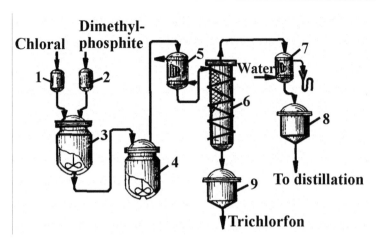

Fig. 105. Production diagram of trichlorfon: *1,2*- batch boxes; *3* - reactor; *4* - ager; *5* - heater; *6*- distillation tower; *7* - cooler; *8, 9* – collectors

It dissolves worse in diethyl ether and carbon tetrachloride. If stored for a long time, its aqueous solutions become acidic. The stability of trichlorfon largely depends on the pH of the medium: at pH > 5.5 it slowly transforms into O,O-dimethyl-O-(2,2-dichlorovinyl)phosphate, which finds an increasing application in agriculture and household.

The industrial production of O,O-dimethyl-O-(2,2-dichlorovinyl)phosphate is usually based on the action of caustic alkali on trichlorfon in an aqueous solution at 40-50 °C:

$$(CH_3O)_2\underset{\underset{O}{\|}}{P}\underset{\underset{OH}{|}}{-}CH\text{-}CCl_3 \xrightarrow[\text{KCl, H}_2\text{O}]{\text{KOH}} (CH_3O)_2\underset{\underset{O}{\|}}{P}-OCH=CCl_2 \qquad (5.85)$$

The synthesis can use mother solutions after the recrystallisation of trichlorfon.

Trichlorfon is widely used in agriculture to protect plants, in sanitation and veterinary medicine. More information about trichlorfon applications can be found in Part 6. Trichlorfon is toxic; however, it is less toxic than DDT, hexachlorane, thiophos and other pesticides.

Preparation of thiophos

The production of thiophos comprises four main stages: the preparation of O-ethyldichlorothiophosphate (ethyl dichloride); the preparation of O,O-diethylchlorothiophosphate (ethyl monochloride); the production of sodium *p*-nitrophenolate (*p*-NphNa) and the synthesis of O,O-diethyl-O-(4-nitrophenyl)thiophosphate (thiophos).

Ethyl dichloride is obtained by the interaction of phosphorus thiotrichloride and ethyl alcohol at 20 °C and atmospheric pressure:

$$PSCI_3 + C_2H_5OH \longrightarrow C_2H_5OP(S)CI_2 + HCI \qquad (5.86)$$

The process is continuous; supplementary operations at this stage are the flushing of hydrogen chloride. the separation of ethyl dichloride from flush waters and the neutralisation of flush waters (aqueous alcohol solution).

Ethyl monochloride is obtained by the reaction of dichloride with ethyl alcohol and sodium hydroxide:

$$C_2H_5OP(S)CI_2 + C_2H_5OH + NaOH \longrightarrow (C_2H_5O)_2P(S)CI + NaCI + H_2O \quad (5.87)$$

The process is conducted periodically; the supplementary operations at this stage are the flushing of unreacted alcohol and sodium chloride and the separation of monochloride from the flush waters.

Sodium p--nitrophenolate is obtained by the reaction of p-nitrophenol with sodium hydroxide:

$$O_2N\text{—}⟨⟩\text{—OH} + NaOH \longrightarrow O_2N\text{—}⟨⟩\text{—ONa} + H_2O \quad (5.88)$$

The process is periodic; the supplementary stage is the preparation of water suspension of p-nitrophenol.

Thiophos is obtained from ethyl monochloride and sodium p-nitrophenolate:

$$O_2N\text{—}⟨⟩\text{—ONa} + (C_2H_5O)_2P(S)CI \longrightarrow O_2N\text{—}⟨⟩\text{—OP(S)(OC}_2H_5)_2 \quad (5.89)$$

The process is carried out periodically, in the presence of OP-7 catalyst. The supplementary reactions are the flushing of the reactive mixture, the separation of the target product from the flush waters, the drying of the moist product and the rectification and absolution of ethyl alcohol.

Raw stock: phosphorus thiotrichloride (the boiling point is 124-125 °C, PCl_3 content is not more than 3%); absoluted ethyl alcohol (the alcohol content is at least 99.7%, aldehyde content is not more than 20 mg/l); sodium hydroxide is flaky solid NaOH, type A (NaOH content is at least 95%); liquid caustic (the density is 1.44 g/cm³, NaOH concentration is at least 42%); benzene (the boiling point is 80.1 °C, the density is 0.875-0.88 g/cm³); p-nitrophenol (the melting point is 114 °C, the density is 1.479 g/cm³; it dissolves badly in water but well in organic solvents; in the dry form explodes by shock); hydrochloric acid (the density is 1.19 g/cm³, HC1 content is 27.5%); soda Na_2CO_3 (the density is 2.53 g/cm³, it dissolves well in water).

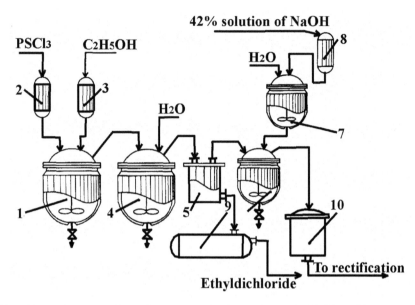

Fig. 106. Diagram of continuous production of O-ethyldichlorothiophosphate (ethyl dichloride) *1* – synthesis reactor; *2, 3, 8* - batch boxes; *4* - flusher; *5* – Florentine flask; *6* -neutraliser; *7*- agitator; *9*- dichloride collector; *10*- collector

Because the process of thiophos synthesis is so complex, the production diagram is presented in stages.

Preparation of ethyl dichloride. Synthesis reactor *7* (Fig. 106) is continuously fed through dosing pumps with phosphorus thiotrichloride and absoluted ethyl alcohol from batch boxes *2* and *3* accordingly.

Reactor *1* is a steel enameled cylindrical apparatus with a welded spherical bottom and removable lid, fashioned with a jacket and an agitator.

The reaction occurs at 20 °C and atmospheric pressure at constant agitation. The temperature is regulated automatically by feeding -15 °C salt solution into the jacket of apparatus *1*. The reactive mixture is continuously fed into flusher *4* out of reactor *1*. The flushing occurs continuously at agitation at 5-8 °C. The temperature is regulated automatically by feeding -15 °C salt solution into the jacket of apparatus *4*. The calculated amount of water for flushing is sent into flusher *4* and is regulated automatically with a gate and rotameter.

After flushing the reactive mixture is continuously sent to separate in Florentine flask *5*. The top acid aqueous alcohol layer from Florentine flask (the alcohol content ≈23%) is sent through a siphon into neutraliser *6* at agitation. The acid aqueous alcohol layer is neutralised with 10% alkali solution, which is prepared in agitator *7* by mixing 42% alkali solution with industrial water. Neutralised in this way, the aqueous alcohol layer self-flow enters collector 10 and then is sent to rectification to extract absoluted ethyl alcohol.

The lower layer, flushed ethyl dichloride, continuously moves out of Florentine flask *5* through a siphon into dichloride collector *9*.

Preparation of ethyl monochloride. Ethyl monochloride is synthesised (Fig. 107) in synthesis reactor 1, which is a steel cylindrical enameled apparatus with a welded spherical bottom and removable lid, fashioned with cooling coil *4* and an agitator. There are two synthesis reactors *1*. They operate ahead of each other by the time equal to half of the duration of the whole operation.

Ethyl dichloride is loaded from collector *9* into flow batch box *2* (see Fig. 106) and then pressed with compressed nitrogen into reactor *1*. After that, apparatus *1* is loaded at agitation and 0-5 °C is loaded with absoluted ethyl alcohol from batch box *3* at such speed that the temperature in the reactor does not exceed 5 °C. The temperature is regulated by feeding salt solution (-15 °C) into the jacket of the apparatus and into coil 4 through a regulating gate. The alcohol dichloride mixture obtained in the reactor is held for 10-15 minutes; at agitation and 0-5 °C it receives crystalline caustic through feeder *5* from bin *4*.

Caustic is fed for 3.5-4 hours.

In case the temperature in reactor 1 increases during the supply of caustic, its flow should be stopped, the temperature should be reduced to normal and only then caustic can be supplied again.

After all the components have been added, the mixture is held for 1 hour; after that, the reactive mixture is moved by vacuum into flushing settling box *7*.

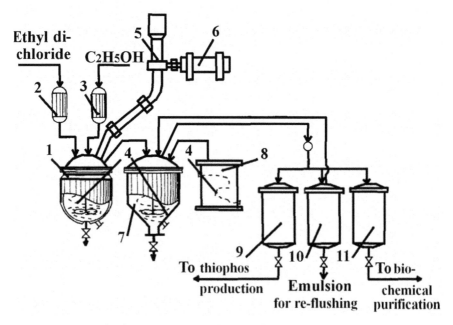

Fig. 107. Production diagram of O,O-diethylchlorothiophosphate (ethyl monochloride): *1* – synthesis reactor; *2, 3* - batch boxes; *4* - coils; *5* – feeder; *6* - bin; *7*- flushing settling box; *8*- collector for cooled condensate; *9-11* - collectors

The reactive mixture in the flusher is cooled with salt solution (-15 °C) down to 5 °C. At agitation flusher *7* is filled (pumped with a centrifugal pump) with condensate cooled to 5-8 °C from collector *8*. The collector is fashioned with coil *4*. The flushing process is conducted periodically at 5 °C.

NaCl, unreacted alcohol, salts of thiophosphoric acid, caustic and other impurities are washed from ethyl monochloride in flushing settling box *7*.

After loading the condensate, the mixture is agitated for 15-20 minutes. The agitator is stopped and the reactive mixture is held for 0.5-1 hours. Within that time the mixture splits into three layers: the lower layer is ethyl monochloride; the middle layer is emulsion; the top layer is aqueous alcohol solution. Ethyl monochloride is slowly poured through a run-down box into collector *9*; the emulsion is poured into collector *10* until there is a light layer in the run-down box; the aqueous alcohol solution is poured into collector *11*.

As soon as collector *10* is filled with emulsion, its contents are loaded into flusher *7* for repeated flushing and separation of the emulsion; after that, cooled condensate is sent out of collector *8* into flusher 7 in 1:1.5 ratio (1 volume of emulsion per 1.5 volumes of condensate).

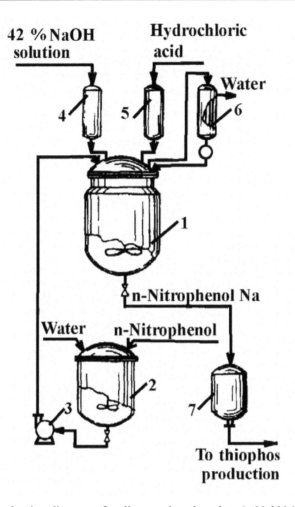

Fig. 108. Production diagram of sodium *p*-nitrophenolate (*p*-NphNa) : *1* -synthesis reactor; *2* - agitator; *3* - pump; *4, 5* - batch boxes; *6* - cooler; *7* – collector

After 15-20 minutes of agitation the mixture is settled for 1 hour without agitation. The separation is carried out in the normal way.

The lower layer (ethyl monochloride) from collector 9 is sent to thiophos production; the top layer (the aqueous alcohol solution) is sent to biochemical purification.

Preparation of sodium *p*-nitrophenolate. The production of sodium *p*-nitrophenolate (*p*-NphNa) is conducted periodically in reactor 1 (Fig. 108). First of all, *p*-nitrophenol is loaded out of wooden cylinders into agitator *2* at agitation. Agitator *2* has been filled with a necessary amount of hot water (75-80 °C). In the agitator pieces of *p*-nitrophenol are ground, forming a suspension of p-nitrophenol in water. The obtained suspension is pumped with centrifugal pump *3* out of agitator *2* into reactor *1*. At agitation the reactor is loaded with a necessary amount of 42% alkali solution out of batch box *4*. Sodium p-nitrophenolate (p-NphNa) is prepared at atmospheric pressure and 80 °C. The synthesis takes 4 hours. In the beginning of the reaction the reactor is heated by sending 90-95 °C water into the jacket of the reactor; after that, the temperature of the reaction is maintained due to the heat released by the reaction. Excess heat is withdrawn by sending cold water into the jacket of the reactor. The end of the synthesis is determined by the absence of p-nitrophenol in the reactive mixture (the allowable content does not exceed 0.5%). Excess of alkali in the reactive mixture is neutralised by hydrochloric acid sent from batch box *5*.

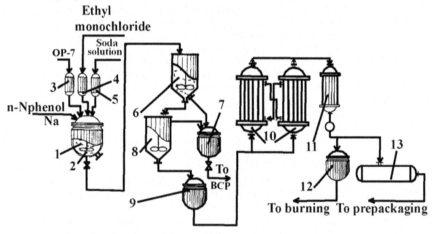

Fig. 109. Production diagram of O,O-diethyl-O-(4-nitrophenyl)thiophosphate (thiophos): *1* – synthesis reactor; *2* - coil; *3-5* - batch boxes; *6-8* - flushing settling boxes; *7, 9, 12* - collectors; *10* - shell-and-tube heat exchangers ; *11* – condenser; *13* – container for dry thiophos

The obtained reactive mixture consisting of sodium p-nitrophenolate (20%) and water (80%) is sent with compressed nitrogen into collector 7 and on to the production of thiophos.

Preparation of thiophos. The synthesis of thiophos is carried out in reactor *1* (Fig. 109) at atmospheric pressure and 85-90 °C.

The reactor is a steel cylindrical apparatus with a welded spherical bottom and removable lid; it is lined with two layers of diabase tiles and fashioned with coil *2* and an agitator.

A 20% solution of sodium *p*-nitrophenolate (*p*-NphNa) from collector *7* (see the previous diagram) is pressed by compressed nitrogen into reactor *1*. Reactor *1* is also loaded with OP-7 catalyst out of batch box *3*. The contents of the reactor are heated with water sent into coil *2*. After that the hot water is switched off and the coil is filled with cold water to withdraw the heat of the reaction. Reactor *1* receives ethyl monochloride from batch box *4*. 1 hour after ethyl monochloride begins to flow out of batch box *5* into reactor *1*, it is supplemented with an 18% soda solution to support the neutral medium in the reactor. The synthesis takes ~5 hours. The end of the reaction is determined by the analysis of thiophene by the refraction index ($n_D{}^{20} = 1.47 \div 1.50$).

After the reaction is finished, the contents of reactor *1* are sent by vacuum into flushing settling box *6* where the mixture splits.

The flushing settling box is a steel cylindrical apparatus with a welded spherical bottom and removable flat lid; it is lined with two layers of diabase tiles and fashioned with coil *2* and an agitator.

The top aqueous layer is poured out of flusher *6* into colllector of flush waters *7*; the lower layer, thiophos, is flushed with water in apparatus *6*; for that purpose the agitator is switched on and the apparatus is filled with a calculated amount of heated water (60-70 °C). The mixture in the flusher is agitated for 15-20 minutes; the agitator is stopped and the mixture is held for 30 minutes. During this time the mixture splits. The top aqueous layer is poured into collector *7*, and the lower layer, thiophos, is sent into flushing settling box *8*, where thiophos is re-flushed. Water for flushing enters flusher *8* through apparatus *6*. Flusher *8* has an agitator. It also requires 30 minutes of holding. After splitting, the bottom layer, thiophene, is poured into raw thiophene collector *9*, and the top layer, aqueous, is sent into collector *7*. Moist thiophos is sent out of collector *9* with vacuum to drying. Thiophos is dried in shell-and-tube heat exchangers *10* at 55-60 °C and in vacuum (the residual pressure is 65 GPa). Apparatuses *10* are heated with hot water (90-95 °C), which is sent with a centrifugal pump out of the heater (not shown in the diagram) into the space between the pipes of apparatuses *10* and circulates in the system. Water vapour from drying heat exchangers *10* enter condenser *11*; from there the condensate is poured into container *12*. The thiophene-contaminated waters are sent to burning. Dried thiophos self-flows through a run-down box into dry thiophos container *13*.

Thiophos

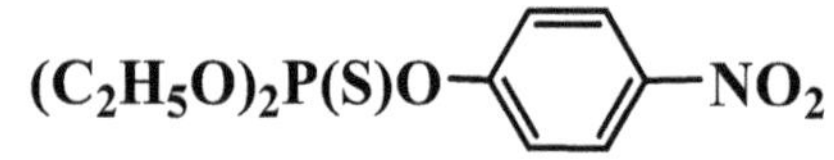

$$(C_2H_5O)_2P(S)O-\langle\!\!\langle\ \rangle\!\!\rangle-NO_2$$

is an oily dark brown or black liquid with an unpleasant specific odour.
Technical thiophos should meet the following requirements:

Density at 20 °C, g/cm^3, not less than   1.20
n_D^{20} 1.5200-1.5400
Thiophene content, %, at least   90

If heated above 105 °C, thiophos decomposes into coallike substance.
Thiophos is used in agriculture as a pesticide.

Preparation of methylmercaptophos

O,O-dimethyl-O-(ethylmercaptoethyl)thiophosphate   (methylmercaptophos) can be synthesised in three stages: the preparation of O,O-
dimethylchlorothiophosphate (methyl monochloride); the preparation of
ethyl mercaptan and from it 2-oxydiethylsulfide (2-ODES); the production
of O,O-dimethyl-O-(ethylmercaptoethyl)thiophosphate (methylmercaptophos).

Methyl monochloride is obtained by the interaction of phosphorus
thiotrichloride   with   methyl   alcohol.   This   forms   O-
methyldichlorothiophosphate (methyl dichloride) according to the reaction:

$$PSCl_3 + CH_3OH \xrightarrow{\text{HCl}} CH_3OP(S)Cl_2 \quad (5.90)$$

which in its turn enters a reaction with the nearest molecule of methyl
alcohol in the presence of a 40-42% solution of sodium hydroxide and
forms monochloride:

$$CH_3OP(S)Cl_2 + CH_3OH + NaOH \xrightarrow{\text{NAOH, H}_2O} (CH_3O)_2P(S)Cl \quad (5.91)$$

Ethyl mercaptan is raw stock for the synthesis of 2-oxydiethylsulfide. It
is formed by the interaction of ethylchloride with sodium hydrosulfide according to the reaction:

$$C_2H_5Cl + NaHS \longrightarrow C_2H_5SH + NaCl + Q \quad (5.92)$$

This reaction is complex. The main reaction is accompanied by some
by-processes, such as:

$$2C_2H_5SH + 1/_2O_2 \longrightarrow (C_2H_5)_2S + S + H_2O \quad (5.93)$$

$$NaHS + H_2O \longrightarrow NaOH + H_2S \quad (5.94)$$

$$C_2H_5SH + NaOH \longrightarrow C_2H_5SNa + H_2O \quad (5.95)$$

$$C_2H_5SNa + C_2H_5Cl \longrightarrow (C_2H_5)_2S + NaCl \quad (5.96)$$

Besides, impurities of sodium sulfide in sodium hydrosulfide react with ethylchloride:

$$Na_2S + 2C_2H_5CI \longrightarrow (C_2H_5)_2S + 2NaCI \quad (5.97)$$

These secondary reactions account for the presence of hydrogen sulfide and organic sulfides in technical ethyl mercaptan.

Obtained in this way, ethyl mercaptan is used in the reaction with ethylene oxide to produce 2-oxydiethylsulfide (2-ODES):

$$C_2H_5SH + C_2H_4O \longrightarrow C_2H_5SC_2H_4OH + Q \quad (5.98)$$

Methylmercaptophos is prepared from methyl monochloride of O,O-dimethylchlorothiophosphate and 2-oxydiethylsulfide in the presence of a 40-42% sodium hydroxide solution according to the reaction:

$$(CH_3O)_2P(S)CI + C_2H_5SC_2H_4OH + NaOH \longrightarrow$$
$$\longrightarrow (CH_3O)_2P(S)OC_2H_4OSC_2H_5 + NaCI + H_2O + Q \quad (5.99)$$

Conducting this reaction at 0-5 °C mostly forms the thion isomer of methylmercaptophos.

Methylmercaptophos can consist of two isomeric forms:

$$(CH_3O)_2P\overset{S}{\underset{OC_2H_4SC_2H_5}{<}} \qquad (CH_3O)_2P\overset{O}{\underset{SC_2H_4SC_2H_5}{<}}$$

$$\text{thion} \qquad\qquad \text{thiol}$$

The thion isomer can transform into the thiol one easily and to some extent happens even at room temperature. The speed of isomerisation grows along with the temperature increase to 80 °C; however, above 80 °C the product undergoes thermal decomposition and forms tarlike products.

Raw stock: phosphorus thiotrichloride $PSCI_3$ (the melting point is 124 °C, $d_4^{20} = 1.63 \div 1.64$, the PCl_3 content is not more than 3%); methyl alcohol [the content of the fraction boiling at 64-65 °C is not lower than 99% (vol.), $d_4^{20} = 0.791 \div 0.793$]; caustic in the form of syrupy liquid (the concentration of NaOH is 40-42%, $d_4^{20} = 1.44$); sodium hydrosulfide NaSH (the allowable Na_2S content does not exceed 2.5%); ethylchloride C_2H_5Cl (the 12,5-13,5 °C fraction content is not less than 98%, $d_4^{20} = 0.916 \div 0.925$); ethylene oxide C_2H_4O (the content of ethylene oxide is at least 98.5%, of chlorine derivatives is not more than 0.2%, of acetaldehyde is not more than 0.8% and of moisture is not more than 0.4%).

The production of methylmercaptophos comprises the following main stages: the production of O-methyldichlorothiophosphate (methyl dichloride); the production of O,O-dimethylchlorothiophosphate (methyl monochloride); the synthesis and rectification of ethyl mercaptan; the synthesis and vacuum rectification of 2-oxydiethylsulfide; the production of methylmercaptophos and its flushing.

Preparation of methyl dichloride. The preparation of methyl dichloride is carried out in reactor *1* (Fig. 110).

The reactor is a steel apparatus with a spherical lid and bottom, lined with two layers of graphite tile. Heat is withdrawn with the help of a jacket and a coil; better contact of reactants is ensured by a propeller agitator, which is covered with textolite against corrosion.

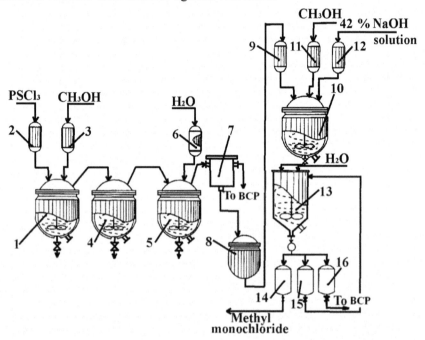

Fig. 110. Production diagram of O-methyldichlorothiophosphate (methyl dichloride) and O,O-dimethylchlorothiophosphate (methyl monochloride): *1, 10* – synthesis reactors; *2, 3, 9, 11, 12* - batch boxes; *4* - ager; *5* - flusher; *6* - cooler; *7*- Florentine flask; *8, 14-16* - collectors; *13* - flushing settling box

Methyl dichloride is obtained by the interaction of phosphorus thiotrichloride with methyl alcohol. The process is carried out in an excess of alcohol (2 moles of CH_3OH per 1 mole of $PSCl_3$), which serves to absorb released hydrogen chloride. Phosphorus thiotrichloride from batch box *2* and methyl alcohol from batch box *3* are pumped with batching pumps through siphons into reactor *1*. The synthesis occurs when the components are fed simultaneously at 0-5 °C. To withdraw the heat and maintain the temperature in reactor 1, the coil and jacket of the reactor are filled with -15 °C salt solution. The products of the reaction, which include dichloride, unreacted phosphorus thiotrichloride, a small amount of monochloride and excess methyl alcohol with hydrogen chloride dissolved in it, are sent into ager 4 through a constant flow choke and a siphon. The ager, like reactor 1, is a steel apparatus with a spherical lid and bottom, lined with two layers of graphite tile. It has a propeller agitator, a jacket and a coil.

Out of ager 4 raw dichloride enters through the constant flow choke and a siphon into flushing apparatus 5 with a blade agitator. To wash dichloride from hydrogen chloride, flusher 5 is filled with water cooled down to 4-6 °C in cooler 6. Excess heat is withdrawn with salt solution sent into the jacket and the coil of flusher 5. From the flusher, dichloride enters Florentine flask 7 through a constant flow choke; there dichloride and aqueous methanol solution split due to their different densities and mutual insolubility. The bottom layer, dichloride from apparatus 7, flows into collector 8; the top layer, acid aqueous methanol solution, is sent to biochemical purification. The temperature in Florentine flask 7 is maintained in the 4-6 °C range. Dichloride, which is a light yellow liquid with the boiling point of 44 °C (at 12-13 GPa) and density of 1.511-1.521 g/cm^3, leaves collector 8 and enters batch box 9 to be used for the production of methyl monochloride.

Preparation of methyl monochloride. Methyl monochloride is synthesised in reactor 10 (see Fig. 110), which is made of stainless steel and has a propeller agitator, a jacket and a coil. Methyl dichloride is sent from batch box 9 by compressed nitrogen into reactor 10 and is cooled there to 0 °C by sending -15 °C salt solution into the coil and the jacket of the apparatus. Then, reactor 10 is filled at agitation with a necessary amount of methyl alcohol from batch box 11. After mixing dichloride and methyl alcohol, the temperature in reactor 10 is brought to 0 °C; at this temperature and agitation the apparatus receives metered amounts of 40-42% solution of sodium hydroxide from batch box 12.

Methyl monochloride is synthesised at 0-5 °C. The temperature in the reactor is maintained by regulating the supply of salt solution into the coil and the jacket of the apparatus, as well as by regulating the speed at which sodium hydroxide enters the reactor. After the necessary amount of alkali solution has been supplied, the reactive mixture is agitated for 30 more minutes at 0 °C. The end of the reaction is determined by the refraction index of the product. For this purpose, an average sample is taken and flushed with water; the organic layer is used to determine the refraction index, which should not exceed 1.4820 or fall below 1.4810.

High refraction index signifies a lack of alkali in the reaction, i.e. high dichloride content. Low refraction index signifies an excess of alkali and high content of trimethyl ether of thiophosphoric acid, triether,

The reactive mixture from reactor *10* is sent to flush in flushing settling box *13*. Apparatus *13* is filled with a necessary amount of water, which is cooled to +5 °C before adding the reactive mixture at agitation from reactor *10*. During the flushing the temperature of 8-10 °C is maintained with the help of salt solution circulating through the coil.

The flushing is carried out approximately for 20 minutes. During the flushing the reactive mixture loses sodium chloride, unreacted alcohol and salts of thiophosphoric acid. After agitation the reactive mixture is settled for 1 hour. The bottom layer, methyl monochloride with the density of 1.311-1.321 g/cm^3 is poured into collector *14*; the middle layer, the water emulsion of monochloride, is sent into collector *15* and then on into apparatus *13* for additional settling; the top layer, the aqueous methyl solution with the density of 1.05-1.08 g/cm^3 is sent to biochemical purification from collector *16*.

The O,O-dimethylchlorothiophosphate (methyl monochloride) obtained in this way is sent to the production of methylmercaptophos. O,O-dimethylchlorothiophosphate is a yellowish liquid with the boiling point of 60.5-61.5 °C at 15 GPa.

Synthesis and rectification of ethyl mercaptan. The basic diagram of the synthesis and rectification of ethyl mercaptan is given in Fig. 111.

Ethyl mercaptan is synthesised in reactor 1, which is a carbon steel apparatus with a water vapour jacket and an agitator. Reactor *1* also has a protective membrane for approximately 20 atmospheres.

Before the synthesis reactor 1 should be tested for hermeticity by pressurising it with nitrogen at 0.3 MPa. The system is considered hermetic if for 30 minutes the pressure in the reactor does not fall more than by 50 GPa.

Reactor *1* is loaded with a necessary amount of sodium hydrosulfide from batch box 2. Then, batch box *3* with a cooling jacket filled with salt solution (-15 °C) receives ethylchloride. After raw stock has been loaded, the agitator is switched on and the contents of the reactor are heated by sending vapour into the jacket of the apparatus to initiate the reaction. The heating is carried out until the pressure in the reactor reaches 0.8-0.9 MPa. When this pressure has been achieved, the heating is stopped; however, due to the exothermicity of the reaction, the pressure and temperature continue to rise spontaneously.

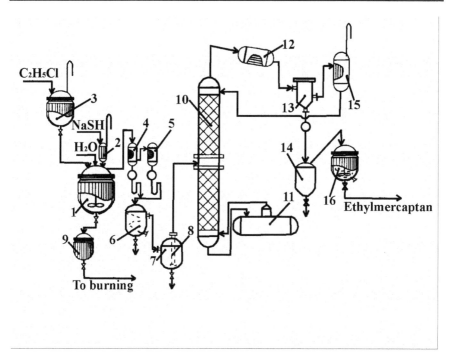

Fig. 111. Production diagram of ethyl mercaptan: *1* – synthesis reactor; *2, 3* - batch boxes; *4, 5, 15* - coolers; *6, 7, 9, 14* - collectors; *8* – immersed batching pump; *10* - rectification tower; *11* - boiler; *12* - refluxer; *13* – phase exchanger; *16* – dehydrator

When the pressure in the reactor is 1.9-2 MPa, the jacket of the reactor is filled with cold water. Cooling is regulated so that the pressure in the reactor does not rise above 2 MPa. By regulating the supply of vapour and cold water into the jacket of the reactor, the temperature in the reactor is maintained at 110-130 °C and the pressure, at 1.9-2 MPa. The synthesis at these parameters should be continued for at least 30-50 minutes.

Before the start of the synthesis of ethyl mercaptan, coolers 4 and 5 should be filled with -15 °C salt solution.

When the synthesis ends, as shown by pressure fall and constant temperature in the reactor, the valve between reactor *1* and cooler *4* opens, and ethyl mercaptan is distilled from the reactor. It self-flows into collector *6* through coolers *4* and *5*. Collector *6* serves as a kind of settling box for tank residue. From there, ethyl mercaptan enters apparatus *7* and is pumped with immersed batching pump *8* into rectification tower *10*. After ethyl mercaptan has been distilled, the tank residue in reactor *1* is cooled down; after that, it is pressurised with nitrogen under 3 atmospheres into collector 9 with the water added into the reactor for flushing, and is sent to burning.

Raw ethyl mercaptan is pumped with immersed pump 8 out of apparatus 7 into the middle part of rectification tower 10. The rectification tower is a vertical cylindrical apparatus consisting of three ring sections filled with Raschig rings. The tower has a remote boiler, 11. Ethyl mercaptan vapours move through the filling of the tower and enter refluxer 12, which is cooled with salt solution (-15 °C). They condense there are enter phase separator 13; from there part of the distillate is sent back to tower 19 as reflux; the other part is separated into collector 14. The uncondensed part of ethyl mercaptan from phase exchanger 13 is sent into tail cooler 15 to condense; then it is sent into tower 10 with the reflux. The uncondensed part is sent out of tail cooler 15 though a fire-resistant apparatus and hydraulic gate to burning. Ethyl mercaptan is rectified at the temperature of 32-34 °C at the top of the tower and 45-65 °C at the bottom, when the pressure in the tower is not higher than 0.02 MPa.

To dehydrate ethyl mercaptan, it is pumped out of collector 14 into dehydrator 16. Apparatus 16 has a coil for cooling with salt solution, an agitator and an insert cartridge with burnt calcium chloride. The dehydration is carried out at 7-10 °C and agitation.

The finished ethyl mercaptan, which is a colourless transparent liquid with the boiling point of 34.7 °C, is sent to the production of 2-oxydiethylsulfide.

Ethyl mercaptan can be used as a so-called "odorant", an additive to natural fuel gas to give it an odour.

Technical ethyl mercaptan should meet the following requirements:

Density at 4 °C, g/cm^3	0.865-0.846
Content, %:	
of ethyl mercaptan	≥ 90
of ethylchloride	≤ 3
of diethylsulfide	≤ 5
of hydrogen sulfide	≤ 2
Moisture	Absent

Synthesis and vacuum rectification of 2-oxydiethylsulfide. As stated above, 2-oxydiethylsulfide is obtained by the interaction of ethylene oxide with ethyl mercaptan. The synthesis is carried out in vertical bubble tower 1 consisting of three ring sections. Each ring section has a jacket, the lower two also have coils.

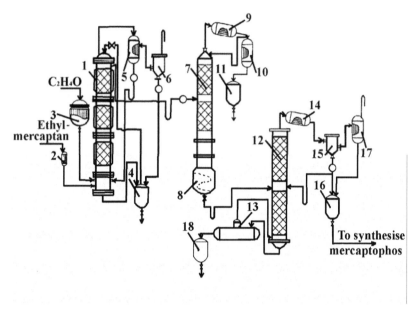

Fig. 112. Production diagram of 2-oxydiethylsulfide (2-ODES): *1* – synthesis tower, *2, 3* - batch boxes; *4, 11, 16, 18*- collectors; *5, 10, 17* - coolers; *6, 15* – phase separators; *7, 12* – distillation towers; *8* - distillation tank; *9, 14* - refluxers; *13* – boiler

Figure 112 presents a basic production diagram of 2-oxydiethylsulfide (2-ODES).

The lower part of tower *1* is simultaneously filled with ethyl mercaptan and ethylene oxide from batch boxes *2* and *3* accordingly in the (1.55÷1.85):1 ratio. The ratio of ethyl mercaptan and ethylene oxide is established depending on the percentage of ethyl mercaptan in raw stock.

Before synthesis, tower *1* is loaded with 2-oxydiethylsulfide up to the flow line from collector *4* with compressed nitrogen. Ethylene oxide and ethyl mercaptan vapours bubble through a layer of 2-oxydiethylsulfide, dissolve in it and interact with each other. The temperature of the synthesis is maintained within 70-80 °C and is regulated by sending cold and hot water into coils and jackets of the ring sections. The formed 2-oxydiethylsulfide is sent by vacuum through the flow line and hydraulic gate into the middle part of tower *7*. Abgases from tower *1* enter cooler *5* where they are cooled. Condensed ethyl mercaptan and ethylene oxide are sent back into the lower part of tower *1*, and the uncondensed part moves through phase separator *6* and is released through an air vent into the atmosphere; the liquid part flows out of the phase separator into collector *4*. Tower *7*, which is filled with Raschig rings, the light-boiling fraction is distilled. It consists of unreacted raw stock and reaction products, such as ethyl mercaptan, ethylene oxide, water, diethylsulfide. The rectification is continuous. Tank *8* of the tower is heated with vapour sent into the coil. During the distillation of the light-boiling fraction the temperature in tank *8* should be 90-102 °C, on top of tower *7* , 40-50 °C, and the vacuum should be 80-90 GPa. Vapour from tower *7* enters refluxer *9*, which is cooled with salt solution, and partially condenses. The condensed part is sent back into tower *7* as reflux; the uncondensed vapours of the light-boiling fraction (ethylene oxide and ethyl mercaptan) enter cooler *10*, which is cooled with salt solution, to condense and flow into receptacle *11*. Tank liquid, consisting of 2-diethylsulfide and polysulfides, continuously flows out of tank 8 into the middle part of tower *12*, where 2-diethylsulfide is subjected to vacuum rectification. Tower *12* is filled with Raschig rings and has vapour-heated remote boiler *13*. During the rectification the temperature in boiler *13* should be 120-125 °C, on top of tower *12* , 102-103 °C, and the vacuum should be 40-52 GPa. Vapours of 2-oxydiethylsulfide enter water-cooled refluxer *14* and condense. The condensate is sent into phase separator *15*; part of it is sent back into tower *12* as reflux; part is sent into collector *16* and on to the synthesis of mercaptophos. The uncondensed part from phase separator *15* enters tail cooler *17* to condense and flow into collector *16*; the abgases are sent through a fire-resistant apparatus and hydraulic gate to chimney fire. Tank residue. which consists mostly of polysulfides, is continuously withdrawn out of tank *13* into collector *18*.

2-Oxydiethylsulfide used in the production of methylmercaptophos should meet the following technical requirements:

Density, g/cm^3	1.0155-1.0175
Content, %:	
of 2-oxydiethylsulfide at 61-63 °C and 6.65 GPa	≥95
of ethyl mercaptan	≥0.2
Refraction index	1.484-1.486

Preparation and flushing of methylmercaptophos. Methylmercaptophos is obtained by the reaction of O,O-dimethylchlorothiophosphate (methyl monochloride) with 2-oxydiethylsulfide (2-ODES) in the presence of alkali solution.

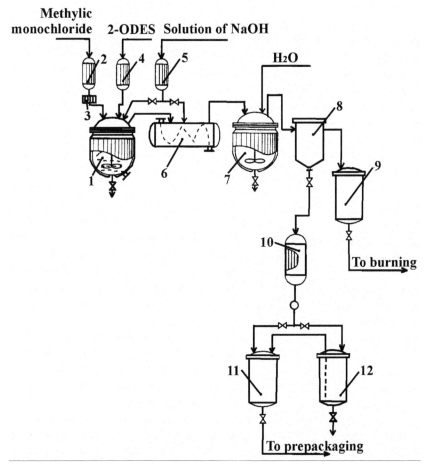

Fig. 113. Production diagram of methylmercaptophos: *1* – synthesis reactor; *2, 4, 5* – batch boxes; *3* - filter; *6* - ager; *7* - flusher; *8* – phase separator; *9, 11* - collectors; *10* - cooler; *12* – dehydrator

Methylmercaptophos is produced by the continuous technique in a reactor, which is a steel apparatus lined with two layers of graphite tile. The apparatus has a propeller agitator and a coil.

Figure 113 shows a basic technological diagram for the production of methylmercaptophos.

Reactor *1* receives methyl monochloride and 2-oxydiethylsulfide from batch boxes *2* and *4* accordingly. The substances are sent with batching pumps. There is filter *3* at the outlet of batch box *2* to purify methyl monochloride from mechanical impurities. Approximately 40-42% NaOH solution is sent with a batching pump from batch box *5* into reactor *7* and ager *6*. Before the production of methylmercaptophos starts, apparatus *1* receives necessary amounts of methyl monochloride and 2-oxydiethylsulfide. The parent substances should be mixed at 0-15 °C, because after an hour of standing the process of spontaneous methylmercaptophos production begins in the mixture. It releases a great deal of heat, which may cause a burst of the reactive mixture. The mixture continuously receives at 0-10 °C NaOH solution at the speed of about 130 liters per hour. The reactive mixture continuously flows through a cross-flow from reactor *1* into the ager, which is continuously filled with alkali solution at the speed of about 30 liters per hour. The temperature in reactor 1 and ager 6 is 0-10 °C and is maintained by sending -15 °C salt solution into the coils of the apparatuses. The reactive mixture consisting of methylmercaptophos, sodium chloride, water and a small amount of unreacted components flows through a cross-flow from ager *6* into flusher *7*, which is a steel apparatus lined with graphite tile and fashioned with a jacket for cooling with salt solution and an agitator. Here methylmercaptophos is flushed with filtered water sent into flusher *7* with a batching pump. The reactive mixture is flushed at 15-20 °C. From the flusher the reactive mixture continuously flows through a cross-flow siphon into Florentine flask *8*, which is a vertical steel apparatus lined with two layers of graphite tile. In the Florentine flask the reactive mixture splits. The top layer, the aqueous salt solution with the density of $1.06 - 1.07$ g/cm^3, continuously flows into collector *9* and is sent to burning; the lower layer, methylmercaptophos, continuously flows into water-cooled cooler *10* and is sent into methylmercaptophos collector *11*.

In case the product entering collector *11* is cloudy, it is sent into collector *12* for additional settling, and then sent back into collector *11*. Out of collector *11* the product is sent to packaging.

Methylmercaptophos $(CH_3O)_2P(S)OC_2H_4SC_2H_5$ is a colourless liquid with the boiling point of 99-102 °C at 0.3 GPa. Methylmercaptophos mostly consists of thion isomer $(CH_3O)_2P(S)OC_2H_4SC_2H_5$ with an impurity of thiol isomer $(CH_3O)_2P(O)SC_2H_4SC_2H_5$. The thion isomer can transform into the thiol one easily even at room temperature. Technical methylmercaptophos should meet the following requirements:

Density, g/cm^3	1.204-1.207
n_D^{20}	1.4063-1.5065
Solubility in water at 20 °C, %	0.033-0.038

Methylmercaptophos is used as a pesticide in the form of water emulsion.

5.6.7. Safety measures in the production and application of organophosphorus compounds

Organophosphorus compounds belong to the strongest known poisons. Their exceptional biological activity is caused by the suppression of a specific ferment, *choline esterase*, which is found in the body in very small quantities. Because of their high toxicity, the production of organophosphorus compounds requires special attention to safety measures.

First and foremost, direct contact of workers with the substances should be excluded. This is effectively achieved by total automation. Besides that, the content of organophosphorus compounds in air should be continuously monitored. Periodic examination of workers dealing with organophosphorus compounds allows them to be suspended from working in harmful workshops at the first symptoms of poisoning (heightened excitability, muscle twitching, sickness, perspiration) before there is any danger to their health.

Correct application of organophosphorus insecticides also removes the possibility of accidents. These may occur when plants are treated upwind or when the substance is mixed with water for better dissolution.

Plants that produce organophosphorus insecticides should strictly observe all the necessary safety measures.

6. Use of elementorganic substances

Research into the elementorganic chemistry started only 135-145 years ago, the elementorganic industry appeared as late as 45-50 years ago; however, the spheres of their application have proved very wide. Nowadays it would be hard to find an industry which does not use elementorganic compounds; their application continuously grows.

6.1. Use of silicone substances

Silicone oligomers and polymers, as well as materials based on them, are increasingly used almost in all industries (Fig. 114) owing to their unique properties; they are often implemented where other materials cannot. This wide range of applications is accounted for by the fact that silicone compounds greatly improve the quality of materials, give them long life and produce noticeable techical and economic effects.

Depending on their chemical composition, molecular structure and molecular weight, silicone compounds are used as liquids, oils and lubricants of various consistency, as elastomers (for sealants, compounds and rubbers), as well as polymers for varnishes, plastic laminates and films.

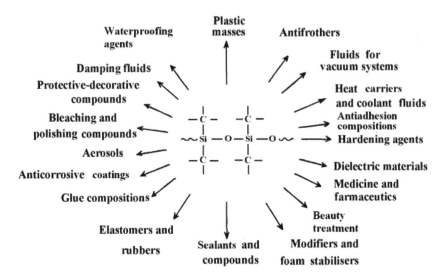

Fig. 114. Important applications of polyorganosiloxanes

6.1.1. Silicone liquids

Oligoorganosiloxanes are widely used in various fields of industry. As a rule, they are transparent, colourless, stable liquids with a rather low molecular weight (from 160 to 25 000); according to their molecular structure they are divided into cyclic and linear oligomers. Because of their valuable characteristics, oligoorganosiloxanes are often more practical than organic liquids.

The most important characteristics of oligoorganosiloxanes are the small change of viscosity with temperature, the low solidification point (from -60 to -135 °C), improved heat resistance, chemical inertness (towards various metals and alloys, many organic polymers, plastic masses and elastomers even at 150 ^{0}C for several weeks), corrosion stability, good dielectric and damping properties. These liquids resist prolonged heating to 200-250 °C in the presence of oxygen in air and to 250 °C in the absence of air; an addition of oxidation inhibitors helps to achieve the same stability in air.

The most essential and specific property of silicone liquids is their small change of viscosity in a wide range of temperatures. E.g., the increase of viscosity with the reduction of temperature for oligomethylsiloxanes is approximately 50 times less than for mineral oils:

Temperature, °C   105   50   0   -18   -88
Viscosity, mm^2/s:

| of oligomethylsiloxanes | 40 | 100 | 350 | 600 | 1570 |
| of mineral oil | 11 | 105 | 11500 | 22100 | - |

Owing to such a small change in viscosity, oligoorganosiloxanes are widely used in various hydraulic systems, in brake and shock gear, etc. The viscosity of silicone liquids also slightly changes with time. E.g., the viscosity of oligoorganosiloxane oils stored for a year in glass or tin containers changes less than by 5%.

As a rule, silicone liquids do not dissolve in water and in low-molecular aliphatic alcohols; however, they dissolve well in many aromatic and chlorinated hydrocarbons. These liquids are not affected by diluted acids and alkali and interact only with concentrated acids and alkali. They burn much less energetically than hydrocarbon oils and most organic liquids; the products of their complete combustion are carbon dioxide, water vapour and silicon dioxide (in the form of very thin powder).

Organic radicals at the silicon atoms largely determine such essential characteristics of oligoorganosiloxanes as the solidification point and heat resistance. E.g., oligomethylsiloxanes start to oxidise quickly only at 200 °C, oligoethylsiloxanes at 138 °C, and oligobutylsiloxanes as soon as at 120 °C. At the same time, it should be noted that replacing some methyl radicals with phenyl ones increases the heat resistance of oligomers. Oligomethylphenylsiloxanes do not gelate even after 1500 hours of standing in air at 250 °C.

Thanks to their exceptionally wide range of operating temperatures and other valuable characteristics mentioned above, silicone liquids are widely used as waterproofing agents, anti-foaming agents, hydraulic, shock and damping liquids. They are also used as dielectrics, heat carriers, coolants, anticorrosion coatings, polishers and bases for oils and lubricants. Silicone liquids are used in the production of paints, in medicine, pharmaceutics and beauty treatment, for precision molding, etc. The most important applications of liquid oligoorganosiloxanes are given below in more detail.

Waterproofing liquids Protecting various materials and articles from the destructive influence of water seems one of the most important uses for silicone liquids; however, waterproofing should be achieved with liquids which do not impair other characteristics of the material.

E.g., inorganic materials (ceramics, glass, porcelain) can be waterproofed with easily hydrolysed alkylchlorosilanes (methyltrichlorosilane, dimethyldichlorosilane, ethyltrichlorosilane, diethyldichlorosilane). Metals and porous materials (paper, leather, textile, plaster, cement, gypsum, etc.) should not be waterproofed with alkylchlorosilanes, because they release hydrogen chloride, which destroys these materials. Alkylchlorosilanes can be successfully replaced with silicone oligomers with aminogroups or hydrogen atoms in the molecule.

Materials can be treated with vapours of silicone liquids or immersed in their compositions diluted with organic solvents. However, it has been found that the most convenient waterproofing agents are various brands of GKZh-liquids. An important advantage of GKZh-liquids is their ability to form aqueous emulsions easily. The emulsions are prepared according to the following approximate formula: 50 weight parts of GKZh-liquid, 49 weight parts of water and 1 weight part of technical gelatin. The mixture is agitated for about 2 hours. Wool, artificial silk, nylon and cotton textiles can be waterproofed with a 2-5% aqueous emulsion of silicone liquid and held at 120 °C for a short time to "set" the surface layer.

Waterproofing is especially important in construction engineering. Water penetrates pores in construction materials, wedges them and thus reduces their durability. In winter the capillary moisture freezes; since ice takes more room than water, the pressure inside construction materials grows to 200 MPa. It does not happen at once, but slowly and surely ends in destruction. The use of waterproofing silicone liquids for the treatment of construction materials (natural and calcined gypsum, marble, limestone, sandstone, tufa) and construction parts increases the durability of the material, makes it more decorative and prevents from the detrimental effect of water. Treatment with silicone liquids makes brick or stone masonry waterproof. Treated asbestos cement slabs warp under the influence of water 50-60 times less than untreated slabs. After 24 hours in rain white marble increases its weight due to absorbed moisture by 1.2%; if it is waterproofed, the weight grows only by 0.04%; therefore, this kind of marble absorbs 30 times as little moisture.

The addition of 0.02% of an aqueous emulsion of silicone liquid into sand fibre slate reduces its water absorption by two and noticeably increases its cold resistance. Silicone liquids can be also put on finished constructions in the form of a 5% aqueous solution, which penetrates to the depth of 3-6 mm and after drying forms a durable waterproof surface for 5-10 years. This treatment can be used on various works of art, famous architectural buildings, etc. Plastered facades can also be waterproofed with good results. Treated plaster does not absorb rain drops; on the other hand, common plaster completely absorbs all rain drops after 30 seconds.

A runway treated with GKZh will never be covered with ice. Water slides like mercury off its surface. Covering windshields of airplanes and automobiles with a thin layer of silicone liquid or varnish prevents ice and mist formation. Silicone compositions are used to cover lenses and optical glass to improve light permeability and resistance to atmospheric effects. Oligomethylsiloxanes are used for the outer covering of glass fluorescent lamps to keep dust off the surface.

Waterproofing treatment can be given not only to glass but also to other inorganic materials, such as ceramics, porcelain, etc. Ceramic parts are waterproofed mostly for electric insulation, which is used in high humidity or at low temperatures. The electric resistance of ceramic parts, which are widely used as panel material in various radio equipment, sharply falls on contact with moisture. Condensed moisture deposits on the surface and gathers in large drops which form a solid conductive film. If these panels have been moisturised and are held for 15-20 minutes in vapours of dimethyldichlorosilane or other alkylchlorosilanes, then for some minutes in air, and heated at 120 °C (to remove hydrogen chloride that is formed), the electric resistance of treated material on contact with moisture will be 1000 and more times as big as that of the untreated material. A drop of water on the surface of ceramic tile treated with silicone liquid is round in shape and does not spread due to low wettability; a drop on untreated tile spreads.

Silicone liquids protect parts of various sensitive equipment and devices (electrical and radiotelephone equipment), field equipment, etc. from water.

Silicone liquids can also be used to protect metal parts from corrosion. It should be noted, however, that a hydrophobic film chemically bound with a metal surface (steel, copper, etc.) requires a substrate before silicone treatment. The substrate should fix the hydrophobic film chemically and at the same time ensure strong binding with metal. A steel surface, in particular, can be prepared for waterproofing by phosphating, i,e, creating a phosphate film with very high adhesion to metal on it. The phosphated surface is treated with vapours or solutions of alkylchlorosilanes (or alkylaminosilanes), and then the article is heated to set the film and completely remove the released hydrogen chloride. After waterproofing, the corrosive stability of phosphated metal parts is improved approximately 25-fold.

Water sharply decreases the durability of paper and can destroy it altogether. This is how valuable manuscripts, books, papers, drawings are sometimes lost. If manuscript pages or a drawing are impregnated with silicone liquids, the image will not be erasable either with an eraser or with water. After waterproofing, even filter paper does not absorb water or aqueous solutions.

Textiles treated with silicone compounds are not wettable with water; therefore, waterproof clothing can be manufactured from them. It is especially important that waterproofing does not decrease the air permabilility of textiles, which is necessary to let the body breathe. It is also important that waterproof textiles do not lose their properties even after 10 dry cleanings, to say nothing of normal washing with long boiling. Treated cloth does not differ in appearance from untreated.

The use of methylchlorosilanes for the treatment of cellulose materials (glazed cotton) gives a high waterproof effect (water absorption is 6%) and at the same time decreases their mechanical characteristics and air permeability. An addition of 0.5% of γ-aminopropyltriethoxysilane (AHM-9) to the finishing composition based on a monomer of the CH_3SiX_3 type (where X $=OC_2H_5$, $OCOCH_3$, $OCOCF_3$) makes cellulose materials highly waterproof without impairing any other properties.

The treatment of textiles with organohydrosiloxanes with phenyl, ethyl and butoxyl groups in the backbone helps to give them both waterproof and stain-repellent properties. An oligomer of the common formula:

$$HO-\underset{\underset{C_6H_5}{|}}{\overset{\overset{C_6H_5}{|}}{Si}}-O-\left[\underset{\underset{H}{|}}{\overset{\overset{C_2H_5}{|}}{Si}}-O-\underset{\underset{OC_4H_9}{|}}{\overset{\overset{C_2H_5}{|}}{Si}}-O\right]_n-\underset{\underset{C_6H_5}{|}}{\overset{\overset{C_6H_5}{|}}{Si}}-OH$$

(n=3÷4)
is applied in the form of a 1-7% solution in an organic solvent.

Leather can also be waterproofed. Treated with silicone compounds, leather is very resistant to water and is not subject to rot, mildew, etc. Waterproof leather is used to manufacture waterproof shoes, clothes, sports goods.

It is very important to waterproof various equipment and laboratory glassware. E.g., waterproof glass measuring devices (pipets, burettes, graduated flasks) are much easier to use; mistakes in analysis because of wettability or leakage can be eliminated. It is especially advisable to waterproof microburettes and micropipets used for ultramicroanalysis. This glassware does not have to be washed; sediment can be agitated with a pipet tip without the risk of losses because the sediment sticks to it; the salts do not "creep" onto the outer walls of the flask; the evaporation speed from small quantities of substances is noticeably reduced because the meniscus is strengthened; the precision of measuring solutions in capillaries grows; the loss of the substance adsorbed on the walls is reduced.

It has also been suggested using silicone liquids to determine the molecular weight of substances by the cryoscopic technique. They allow the device to be cooled with tap water instead of the commonly used mixture of salt with snow or ice. Covering glass articles with protective silicone film not only makes them waterproof, but also greatly increases their heat resistance and mechanical durability: in automatic filling machines the quantity of broken glassware is reduced (thanks to waterproofing) from 0.3-1 to 0.014%; when glass bottles are transported, the breakage is reduced from 1 to 0.00017%. The waterproofing of glassware is very important in medicine, since it prevents blood from clotting. Waterproofing of microscope slides helps to put very small drops on them which do not spread and can be easily moved along the glass.

Adding 1-2% of inert powder (e.g., ceyssatite) treated with methyltrichlorosilane vapours prevents clump formation in fertilisers (especially containing ammonia nitrate), which is essential for agriculture.

Waterproofing is also used for sports equipment and fishing gear. Glass fibre waterproofed with silicone liquids is used to manufacture floats, buoys, etc.

All this demonstrates that the use of silicone liquids for waterproofing is very wide and is growing wider still. It should be noted that efficient waterproofing of materials requires relatively small amounts of silicone substances: e.g., 1 m^2 of the façade requires only 5-10 g of the substance; fibre materials need about 1% of the quantity of the material. At the same time, they are very efficient: the waterproofing of various materials reduces their water absorption 5-10 times and increases their serviceability by several times.

Lubricants. The recent development of new fields of science and technology and the implementation of high and ultra low temperatures set a serious task before the researchers who were dealing with the synthesis of lubricants. They had to design synthetic oils with little change of viscosity at large temperature variations. These oils are oligoorganosiloxanes, stable transparent liquids with little change of viscosity in a wide range of temperatures (from -90-100 to +250 °C). In other words, silicone oils, which have approximately the same viscosity at room temperature as petroleum oils, solidify at a temperature 50-55 °C lower than petroleum oils; moreover, the viscosity of silicone oils does not change with temperature decrease as much as the viscosity of petroleum oils. At the same time, silicone oils and lubricants are effective at 40-60 °C higher than petroleum oils.

Due to these properties, oligoorganosiloxane oils are especially usable for lubricating mechanisms which operate in very low or high (sometimes sharply changing) temperatures, e.g. for ball bearings, automatic distribution valves and stuffing boxes, for lubricating pressure molds in the production of plastic masses, rubber technical articles, metalware, etc.

At car factories it is very important to attend to the conveyor system that transports varnished parts through the furnace at 230 °C. Even the best quality petroleum oils used to lubricate the ball bearings quickly solidify at such a high temperature. That is why the conveyor has to be stopped every two days for repeated lubrication. It is very unpractical. To launch the conveyor again, the motor should work at full capacity and additional workforce is required. Besides, at such heavy loads the steel connecting links of the conveyor often break. On the contrary, if a silicone lubricant is used, the ball bearings do not jam, steel links do not break, power consumption is greatly reduced, and, most importantly, another lubrication is necessary only in 3 months.

The main advantages of silicone lubricants are their prolonged serviceability at high temperature, prevention of the contamination of the parts (the oil does not leak), small attendance costs, low power consumption, as well as the total elimination of breaks in the production caused by a failure of ball bearings. Comparison tests with petroleum and oligoorganosiloxane oils showed a considerably larger wear of gear pumps in the former: after 100 hours of the test the wear of one gear in petroleum oil was 0.015 g against 0.0009 g in silicone oil. The lesser wear of parts in oligoorganosiloxane seems to be caused by the formation of the protective film on their surface.

Plastic lubricants can be obtained by solidifying liquid silicone oils with an addition of graphite, lithium stearate or black. These lubricants can operate in the 160 - -50 °C temperature range; if certain ethers of carboxylic acids are added, even to -70 °C. Plastic silicone lubricants are widely used in ball bearings of shafts operating at temperatures above 175 °C; as a packing material in systems operating in vacuum, at high temperatures and in oxidising media, as well as in pipings for strong mineral acids, in taps, valves and gates of vacuum systems. These lubricants have proved very efficient in valves for hot water, steam and many corrosive chemical reactants.

Silicone lubricants are widely used in a wide range of devices, including photographic, optical, sighting, X-ray equipment, etc.

Antiadhesion liquids. Owing to their nonvolatility, resistance to high temperatures and oxidants, as well as immiscibility with most organic polymers, silicone liquids are superior agents which can be used to prevent the sticking of various materials to molds during molding operations. This role can be played by silicone liquids in the form of water-soluble emulsions or dispersions. The low surface tension of these substances accounts for their efficiency even at very small concentrations of oligoorganosiloxanes in the emulsion, e.g. 0.25%.

The use of silicone emulsions for detaching rubber articles from pressure molds greatly increases the productivity of pressing, reduces spoilage (by 90%) and costs of cleaning the pressure molds (by 80%). If plastic masses are pressed, the introduction of 1-2% of silicone liquid into the composition helps to detach the article from the mold easily, improves its surface and, besides, keeps the pressure mold in use without cleaning for 4 months. When gypsum molds for ceramic articles are manufactured, lubricating the models with solutions of methyl- or ethylsiliconates of sodium ensures good detachment of the molded goods from the molds.

Because of their biological inertness and intoxicity, silicone liquids (as well as some silicone varnishes) are widely used in the food industry as antiadhesion surfaces. A great problem in breadmaking and confectionary is that baked goods and sweets stick to pans and conveyor belts. This leads to inefficient losses of large amounts of flour, fats and oils. If steel and aluminum pans for baking bread or making confectionary, as well as conveyor belts, are treated with silicone varnish (e.g., KO-919) or the solution of GKZh-94 in carbon tetrachloride, durable antiadhesion coating is formed, which prevents the food from sticking for a long time. Apart from the considerable economy of valuable foodstuffs, it improves the working conditions at food factories, to say nothing of the quality and appearance of baked goods, confectionary, sausages, etc.

Silicone liquids and emulsions are successfully used for prolonged storage and mould protection of fruit and vegetables . They form an airproof colourless film on the surface, which hampers the growth of mould and puts off the expiry date. It does not spoil the taste of the produce.

Shock and damping fluids. Little change of viscosity of silicone liquids as temperature changes is convenient when the liquids are used in vibration damping devices. The damping ability of silicone liquids with the temperature varying from -40 to +70 °C changes only by three times, whereas the figure for highly viscous petroleum oil changes by 2.5 thousand times.

Little change of viscosity makes silicone liquids indispensable damping devices and parts of various measuring devices. A drop of silicone liquid put on a pointer index in the cockpit prevents vibration and keeps the index at rest; this kind of damping is more efficient than complicated mechanical devices. Silicone damping liquids are also used in speedometers, amperemeters, voltmeters and gasometers, in relays and breakers, in acoustic pickups.

Good compressibility of silicone liquids (up to 14%) is essential for their use as shock absorbers. In particular, they are used for filling the unions of cylinders and pipelines, for shock absorption for airplace chassis, for sheet metal dyes, for artillery weapons, etc. Apart from absorbing shocks, these liquids damp any ensuing oscillations, i.e. act as damping agents.

Heat carriers and coolants. As technological processes intensify, the role of high temperature heat carriers is becoming even more important. The main requirements to contemporary heat carriers are: thermal and corrosive stability; high coefficient of heat transfer; low melting point and high boiling point; fire and exposion safety; intoxicity. All these requirements are met by silicone liquids PMS and PFMS, as well as by certain aromatic ethers of orthosilicon acid. Thus, oligomethylphenylsiloxanes PFMS-4, PFMS-5 and PFMS-6 used as heat carriers operate for a long time (more than 1000 hours) at 350-360 °C and 0.4-0.6 MPa; at 300 °C heat carrier PFMS-6 functions steadily for 10 thousand hours. Such aromatic ethers of orthosilicon acid as tetracresoxy- and tetraxyleneoxysilanes do not decompose at 350 °C for 900 hours. On the whole, thermal stability of aromatic ethers of orthosilicon acid with various substituents at the silicon atom dwindles according to this sequence:

$C_6H_4C_6H_5 > C_6H_5 > OC_6H_5 > C_6H_4OC_6H_5 > OC_6H_4C_6H_5 > C_6H_4C_6H_{11} > C_6H_4C_6H_4C_6H_5 > C_6H_4C_nH_{2n-1}$

Thermoregulating systems and other devices often require coolants (low-temperature heat carriers) which function at low temperatures. In these cases branched oligomethylsiloxanes are used, such as PMS-1b, PMS-1,5b, PMS-2,5b and PMS-10b, which remain highly motile at -100 °C and below.

Dielectrics and sealants. High dielectric characteristics of silicone liquids allow them to be widely used in condensers and other electric and radio equipment, in airplanes and radars. High arc resistance of organosiloxanes is due to the formation of carbon dioxide, a good dielectric, when they disintegrate.

Owing to their small volatility and high thermal stability, some silicone liquids are used as binding agents for sealing compositions in a wide temperature range (from -70 to +250 °C). E.g., arc resistant plastic silicone dielectrics are widely used to seal airplane spark plugs and to prevent corona discharge.

Polishing liquids. Immiscibility, inertness, nonvolatility and hydrophobicity of silicone liquids allow them to be used in the preparation of furniture and automobile polishing liquids or pastes. E.g., Russia produces polishing compositions with an addition of 2-5% of some brands of oligoorganosiloxanes (PMS-500, PMS-200A, PES-5, GKZh-94, etc.) used to preserve the polished look of varnish coating on the automobile body and to prevent corrosion. They make the treated surface shiny and waterproof; some also have a detergent effect.

Anti-foaming agents. Silicone liquids have an exceptionally low surface tension, which only slightly depends on viscosity. E.g., liquids with the viscosity of 0.65 mm^2/s have the surface tension of 1.6 Pa, at 20 mm^2/s 2 Pa; at any further increase of viscosity to 1000 mm^2/s the surface tension grows only to 2.1 Pa. The highest-molecular silicone liquids (and therefore the most viscous) have a lower surface tension than many organic solvents (e.g., 2.9 Pa for benzene, 4.8 Pa for ethylene glycol). It is this characteristic of silicone liquids in combination with high surface activity and immiscibility on the surface of the liquid and gas separation (in the foam) that makes them useful as antifoaming agents.

It is especially important to use these antifoaming agents for lubricating oils; a very small addition of the liquid is enough (0.0005-0.001%). In aqueous media the necessary amount is smaller than in organic media. The best results for organic liquids are achieved when oligomethylsiloxanes are used, and for aqueous media, when silicone oligomers with longer radicals are added. To prevent the foaming of petroleum, it is enough to add just 0.5 l of oligomethylsiloxane per tank.

Some technological processes in the textile industry create much foam; this greatly reduces the efficiency of the equipment and often the quality of the products. E.g., foam formation in the processes of treatment and dyeing of textiles and crocheted goods impairs their quality and increases fouling. The role of antifoaming agent can be played by various substances, such as aliphatic acids, alcohols and ethers, sulfonated oils, turpentine, organophosphorus compounds, etc. However, they are efficient only when introduced in significant quantities (from 0.3 to 3%). It increases the costs, impairs the quality of the products and complicates their purification. On the other hand, silicone antifoaming agents are economical, indifferent, nonvolatile, do not lose their properties at high temperatures and do not impair the quantity of the products.

When some foodstuffs like sugar or medicines (penicillin, streptomycin and other antibiotics), it is necessary to evaporate large quantities of liquids to extract the diluted substance in the crystallic form. The evaporation forms a lot of foam, which interferes with the process, causes losses of the products and makes it necessary to increase the volume of the apparatuses. Experiment shows that a neglible amount of silicone liquid added to these solutions (1 weight part per 100 thousand or even 1 mln of weight parts of the solution) completely prevents the formation of foam during evaporation.

Silicone liquids are valuable antifoaming agents in the production of various soft drinks; the additions of these liquids have absolutely no effect on taste and other characteristics of the drinks, considerably facilitate the production technology and help to use the equipment more economically. Organosiloxanes are used as antifoaming agents for cooking fat, condensing milk, producing juice and other foodstuffs.

Antifoamers are vital in surgery. During operations on heart and large vessels even the smallest bubble of air in the vascular system can block a thin vessel in the brain or other vital organ. Treating medical instruments with silicone liquids greatly facilitates the work of surgeons.

To extinguish foam in aqueous media, it is advisable to use water emulsion silicone liquids PMS-154A, PMS-200A, etc. to ease the distribution of the antifoaming agent on the surface of the foaming mixture.

Silicone liquids in the production of paints. The use of silicone liquids as additives for paints gives them specific properties. E.g., an introduction of a small amount of silicone liquid (3 weight parts per 100 weight parts of the paint) prevents the foaming of paints when they are applied, improves the flow properties of the paint and gives gloss to the painted surface. This beneficial effect of oligoorganosiloxanes is caused by their low surface tension.

Silicone liquids in medicine, pharmaceutics and beauty treatment. In these fields silicone liquids are widely used thanks to their inertness to organic and inorganic reactants, stability at high and low temperatures, resistance to weather effects, small surface tension, absence of smell and colour and very low changeability of properties even after prolonged heating (e.g. after a year at 150 °C the weight loss of oligomethylsiloxanes at evaporation did not exceed 2%).

Silicone liquids are used to protect skin. Nutrient facecreams, theatre makeup, lipstick containing these liquids are resistant to the effects of moisture and warmth. Protective creams used to protect hands from salts, acids, alkali and other irritants contain up to 25% of liquid oligoorganosiloxanes. Oligoorganosiloxanes are also used as waterproofing preparations for keeping hair in shape. Liquid oligoorganosiloxanes are used to treat skin (e.g. for professional dermatitis); there are instances of positive effect silicone preparations have on eczema and other conditions.

Because of their many characteristics, especially harmlessness for human body, they have been widely used in pharmaceutics (bases and components for medicines, antifoaming agents in some processes, waterproofers for equipment and containers).

High resistance of silicone liquids to oxidation helps to use them to lubricate valves in oxygen apparatuses for artificial breathing. Another medical application of organosiloxanes is lubricating surgical instruments; this lubrication is preserved even when instruments are boiled. The low solidification point of silicone liquids allows them to be used in the Arctic for bandaging wounds, because treated bandages remain elastic in the extreme cold and do not stick to the wound.

Silicone liquids are used by dentists in the preparation of dentures. Dentures become waterproof, their material is not leached by saliva. Food particles do not stick to such dentures; mucous membranes in the mouth are not irritated, which often happens when dentures are produced from organic plastic masses. If silicone toothpaste is used, teeth are covered with a colourless film that prevents them from tartar.

Waterproofing glass with silicone compositions gives some advantages when glassware is used in pharmaceutics. E.g., transfused blood coagulates more slowly, medical glassware can be fully emptied, the tips of ampules do not form carbon when melted.

Disinfectant and bactericidal liquids. Silicone compounds which are have a simular structure to diphenyldichloro(trichloromethyl)methane (DDT) have much stronger disinfecting properties than DDT.

Some silicone compounds are also bactericidal, i.e. they can withstand the destructive effect of microorganisms. E.g., an addition of 0.3-1% of al-kylisothiocyanatesilanes $R_nSi(NCS)_{4-n}$ (R=CH$_3$, C$_2$H$_5$, etc., n =1÷3) to the nutrient medium delays the development of mildew and bacteria approximately for a year. Vapours of these substances are an insecticide for the corn-bug.

This chapter dwells only on the main applications for silicone liquids. However, the possibilities they offer are virtually limitless and will surely be further implemented as the chemistry and production of these substances continue to develop.

6.1.2. Silicone elastomers, silicone elastomer rubbers and sealants

Silicone elastomers were developed not so long ago; however, they help to solve many complicated issues of modern technology.

These elastomers are resistant to high and low temperatures: they can function in the -90 - +270 °C range and at 350-400 °C for a short time. The resistance to low temperatures is explained by the fact that mobile molecules of elastomer in the normal state are wound spirally and the Si—O—Si chains are shielded by organic radicals on the outside (Fig. 115). This explains the hydrophobicity of the elastomer, weakens intermolecular interactions and therefore reduces the intermolecular mobility of the chains. If the temperature is increased, the spirals unwind, some parts of the Si—O—Si chain lose the protection of organic radicals, which increases adhesion forces between the molecules and makes the molecules more mobile. As for the high (compared to organic elastomers) resistance of polyorganosiloxane elastomers to high temperatures, it is caused by the considerable strength of the Si—C bond. There are silicone elastomers which sustain heating up to 400 ^{0}C for several dozens or even hundreds of hours without a considerable change of properties.

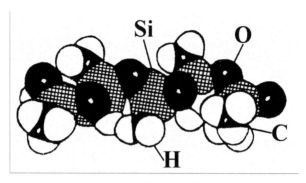

Fig. 115. Model of the structure of the dimethylsiloxane chain

Of all the known elastomers, polyorganosiloxane elastomers are the most resistant to weather effect; they are insensitive to oxidation with oxygen in air and ozone, as well as to UV rays. That is why they do not age even in very harsh conditions. E.g., if natural rubber decomposes under the influence of ozone within 5 minutes at 20 °C and within 6 seconds at 100 °C, polydimethylsiloxane elastomer does not decompose even after 60 minutes in ozone at 100 °C. If heated in air to 320 °C, elastomers based on polydimethylsiloxanes, polydimethyl(metylphenyl)siloxanes, etc. only slowly oxidise; on the other hand, natural rubber and synthetic organic elastomers decompose at once.

An elastomer filled with Aerosil, technical carbon (lamp or acetylene black), iron and titanium oxides and other ingredients including a vulcaniser is raw rubber used to manufacture various products. The elasticity and resilience of silicone rubbers depend on the number of siloxane links in the chain and on the number of cross links. The higher the molecular weight of the elastomer and its elasticity; the more the quantity of cross links (to a certain extent), the greater its mechanical strength.

Silicone elastomers are divided into low-molecular (the molecular weight is 25-75 thousand) and high-molecular (the molecular weight is from 400 thousand to 1 million and more). Rubbers based on high-molecular silicone elastomers have small residual deformations, i.e. they can go back to their original size after the load has been taken off in the - 60 - +250 °C range; all organic rubbers at these temperatures become rigid and brittle. E.g., an article from silicone rubber, which has been subjected to compression to 2/3 of the original thickness and has been in this state for several hours at 150 °C, after taking off the compressing load restores its original size by 90%. The tensile strength of silicone rubbers (5-5.5 MPa) is smaller than that of organic rubbers (approximately 130 MPa). However, there have already been obtained polyorganosiloxane rubbers with tensile strength up to 13 MPa.

Rubbers based on silicone elastomers have high resistance to many solvents and oils. It is especially true for rubbers based on elastomers with up to 8% of methyl-γ-trifluoropropylsiloxy-, methyl-γ-cyanpropylsiloxy- or methyl-β-cyanethylsiloxy links.

By their resistance to swelling under the influence of solvents, these rubbers compare well even with the most resistant chloroprene rubbers. If the rubbers swell under the influence of hydrocarbons (petrol, kerosene), carbon tetrachloride and other solvents, they usually restore their properties when the solvent is removed from the sphere of influence. In some cases silicone rubbers endure hot water and steam (at <0.7 MPa). At temperatures above 100 °C modified polydimethylsiloxane rubber is even more resistant to petroleum oil than rubbers based on butadien nitryl and chloroprene elastomers. E.g., after 24 hours of petroleum oil influence at 180 °C the tensile strength of chloroprene rubber is reduced by 50%, whereas for modified polydimethylsiloxane rubber the figure is only 15%. At the same time its relative elongation at rupture slightly grows (300% before swelling, 330% after swelling), and the figure for chloroprene rubber abruptly falls (400 in the original state against 140% after swelling).

Rubbers based on elastomers with methylphenylsiloxy- or diphenylsiloxy links in the dimethylsiloxane chain have improved resistance to cold and radiation.

Silicone rubbers have one more very significant advantage in comparison with rubbers based on organic elastomers, and that is high dielectric characteristics. E.g., rubbers based on silicone elastomers do not conduct electric current even at 250-300 °C, whereas rubbers based on organic elastomers become conductive already at 120-150 °C. Insulating properties of silicone rubbers are preserved even at contact with water.

Silicone rubbers burn if the temperature of the flames exceeds 600-700 °C. However, their combustion does not release toxic products, and there is an isolating layer of carbon dioxide on the product. If this rubber is sealed in a glass or asbestos shell, the cable can endure operating voltage and ensure normal functioning of the electric circuit even in a fire. These properties help to reduce the requirement for wires and cables in most cases by 20% and noticeably increase the safety of operation in case of overloads and fires.

Finally, silicone rubbers are superior to organic rubbers because they have no harmful effect on living organisms; this gives them a whole range of applications in the production of medical materials and devices.

All these valuable characteristics of organosiloxane elastomers and organosiloxane elastomer rubbers determine their wide application in vairous spheres of modern technology and economy. They help to increase the temperature range of elastic materials: from -90 - -60 to 250-350 °C (subjected to the temperature for a long time). Besides, such positive qualities as high resistance to aging, nontoxic decomposition, antiadhesion ability, etc. make silicone rubbers more desirable than organic rubbers even when the product is used in the normal operating temperature range (from -50 to +150 °C).

One of the most essential spheres of their application is aircraft engineering. Silicone rubbers are used as seals, membranes, profile parts, flexjoints, etc. and operate at very low temperatures at high altitudes, in large concentrations of ozone and under various weather effects.

Silicone rubbers have proven to be successful inserts for sea projectors. These inserts are meant to keep coal projector electrodes dry. In harsh conditions of the Arctic Sea no insert from organic rubber can stand the test, because the ambient temperature is well below zero, whereas the insert is heated by the voltage arc to 300 °C. Only an insert from silicone rubber successfully operates at such temperature changes.

Like organic elastomers, organosiloxane elastomers can be given a porous structure. Porous elastic rubber is used to produce light inserts, shock absorbers in various devices, etc. Because of its resistance to negative temperatures, it is used in airplanes.

Glass fibre impregnated with silicone elastomer and shaped in the form of corrugated tubes is used to join pipes in the blasts of power units. Diaphragms based on silicone rubbers can be used in gasometers and pressure regulators where a sufficient amount of heat is released. This material can also be used to produce sacks for high-temperature molding of large-sized articles from plastic laminates. Lines from silicone rubber and glass fibre are used as transporters in drying ovens. Pipes from rubberised glass fibre in some cases can replace aluminum pipes.

Silicone rubber in combination with asbestos fibre can be used in pressure reducers in receivers. In comparison tests all diaphragms from organic rubbers and asbestos cease operation due to aging after 200-360 thousand cycles; on the other hand, diaphragms from silicone rubber are still in perfect state after 1 million cycles. We should also mention the use of silicone rubbers in industrial furnaces and various apparatuses operating at high temperatures (oil cracking towers, gas pipelines, recuperation installations).

Their high resistance to aging, moisture and ozone allows silicone rubbers to be used in lighting and signalling facilities, as well as in special-purpose electrical installations, as seals in meteorological devices and lights for airports, as shock absorbers, etc.

The resistance to heat and steam in combination with resistance to aging allows silicone rubbers to be used as seals in steam-heated irons, in safety valves in boilers, etc.

Silicone elastomers do not dissolve in oils, petrol and other hydrocarbons; that is why they can be successfully used in printing and as protective coatings for glass, enameled, ceramic, steel and aluminum articles. The success of silicone rubber is also due to its complete absence of corrosive effect.

The absence of adhesion to plastic masses allows silicone rubbers to be used in the production of transporter lines and belts, various presses or for lining metal cylinders. At present pressure cylinders used for corrugating, glueing or changing the thickness of articles from plastic masses are coated with rubbers from synthetic organic elastomers (mostly from neopren), and cease to operate after a short time, sometimes after a week. If coatings are silicone, cylinders do not have to be cleaned, since plastic masses do not stick to them; besides, these coatings are very long-lived.

Their chemical inertness makes silicone rubbers a good choice for connective seals in various chemical equipment.

Of considerable interest is the use of silicone rubbers for insulation in electrotechnical equipment. This is accounted for by superior heat resistance of elastomers and their good dielectric properties. E.g., the dielectric permeability of polyorganosiloxane elastomers at 500 V and 60 Hz is 3.5-5.5, their electric strength at 60 Hz is 15-20 KV/mm, and   the dielectric loss tangent, which characterises the losses of electric energy in insulation, at 500 V and 60 Hz amounts only to 0.001. It is very important that these characteristics are preserved in a much wider temperature range than in the case of natural and synthetic organic elastomers.

These high dielectric characteristics in combination with high heat resistance account for the wide application of silicone elastomers in the production of rubberised wires and cables. At present there are several types of cables manufactured from these materials:

1. feeder cables which can be used in various climatic conditions;
2. aircraft cables which are similar to Teflon in their properties but are less expensive; besides, silicone elastomers do not release toxic substances when they burn, and articles made from them are covered with an isolating layer of carbon dioxide;
3. ignition cables which are not only resistant to high temperatures around engines but are also insensitive to the electric charge ("corona") and do not split on contact with moisture;
4. contact wires for highly heat-resistant electric machines;

5. feeder cables for street lights, luminescent lamps, medical equipment, powerful projectors;
6. defrosters for refrigerators;
7. electroheating cables which can function at a temperature below 200 °C and are used to heat pipelines and tanks (for storing black oil, oils and other liquids), for melting snow, heating soil, etc.
8. marine cables of a smaller diameter, which considerably save space on board and ensure operation even in case of fire or submersion (even after 8 hours of being kept in flames at 950 °C the cable can still transmit power or signals);
9. wires and cables meant to be used for operations near furnaces and other industrial high-temperature installations.

Silicone rubbers are often used in the production of medical equipment. This happens because they are biologically inert, can help to produce transparent products, have no smell, are not rigid, can be sterilised and easily processed to manufacture various extruded and molded articles.

Silicone rubbers are used as a basis for various pipes used for the transfusion of blood and other physiological solutions, as well as in the production of catheters, probes, vessel prostheses, etc. E.g., multichannel catheters based on silicone rubbers can be easily inserted into vascular beds; they can be used for simultaneous registration of blood pressure, injection of medicine and making tests.

Silicone elastomers are used to produce numerous molded products which become parts of artificial organs (artificial hearts and assist circulation systems, joint prostheses, etc.). The use of silicone vulcanisates for coating the walls of the artificial ventricle allows it to be used for prolonged assist circulation and prevents the formation of clots on its surface.

Silicone SKTV and SKTV-1 elastomers are used to produce gas-permeable membrane oxygenators for the apparatuses for blood oxygenation and dioxide carbon flushing.

The chemical, biological and physiological inertness of silicone rubbers account for their use in contact with medicines (plugs, seals); their combinability with living tissues, absence of inflammation, allergic and cancerogenic effects, resistance to aging and microorganisms, similarity to living tissues by their density and elasticity have made them indispensable for long-term or constant implantation in the body as a carrier for medications.

Of great practical interest are sealants and compounds from low-molecular silicone elastomers, which in the presence of such solidifying agents as alkoxyaminosilanes (ADE-3, AHM-9), acetoxyorganosilanes, etc. solidify at room temperature in contact with humidity in air. They have good moisture and weather resistance, endure high and low temperatures, are highly adhesive to metal, glass, ceramics and various plastics. All this has made them very important in construction, electrical engineering. microelectronics and other fields of technology, as well as in medicine and the food industry.

Low-molecular silicone rubbers are used to insulate electrodes for electric heart stimulation and measuring brain biocurrents.

Low-molecular silicone elastomers are a basis for rubber membranes which are used in the production of containers for prolonged storage of fruit, vegetables and other kinds of food. Sealants from low-molecular silicone elastomers are more and more often used to copy articles with complex configurations. The sealant is poured to make a mold which is then filled with molding mixture. This technique is used to copy works of art, in criminalistics (to fix prints), in dentistry (to make models for dentures), etc.

30-35 years ago silicone elastomers were just a laboratory curiosity; nowadays silicone elastomers., rubbers, sealants and compounds are indispensable in many fields of modern technology, and the demand for them grows even more every day.

6.1.3. Varnish silicone polymers

Varnish polyorganosiloxanes are high-molecular polymers with long molecular chains; however, unlike elastomers, they have a branched, ladder or spirocyclic structure. As a rule, these polyorganosiloxanes dissolve well in organic solvents. This is one of the reasons why they are used as solutions in organic solvents, although they can exist in the form of compounds, i.e. solid brittle polymers or highly viscous liquids without solvents. Polyorganosiloxane solutions sprayed on a surface leave a film on it after the solvent evaporates. After drying and baking at increased temperatures the film becomes solid and acquires all the properties characteristic of silicone polymers, such as resistance to water, moisture, cold, heat and aggressive media. The properties of polyorganosiloxanes can be regulated to a wide extent by changing their structure as well as the number and nature of the organic radicals around the siloxane chain.

Polyorganosiloxanes are very stable chemically: the siloxane chain is preserved in many chemical reactions, whereas the thermal oxidative destruction of the molecule is generally connected only with the detachment of lateral radicals. It is essential that the decomposition product is polymer $(SiO_2)_x$, which keeps all its dielectric properties and some strength, unlike the decomposion products of organic polymers. E.g., at 200 °C the dielectric characteristics of silicone polymers keep 100 times longer than those of organic polymers.

The thermal resistance of silicone polymers is also considerably greater than that of organic polymers. E.g., the weight loss of polyorganosiloxanes within 24 hours at 250 °C is (depending on the type of the polymer) 2-8%; in the same conditions the weight loss of capron is 55.5%, of polystyrene is 65.6%, of glyptal polymer is 93.4%. Within the same time at 350 °C organic polymers burn by 70-90%, silicone polymers lose not more than 20% of their weight, while polymethylsiloxanes lose only 3-7%.

The high resistance of polyorganosiloxanes allows them to be used as heat- and weather-resistant coatings to protect steel, aluminum and other metals and materials from corrosion. Polyorganosiloxane coatings can be used at 200 °C and higher.

Silicone polymers have one more remarkable characteristic. They combine well with various organic and inorganic substances, which makes it possible to increase their strength, elasticity and especially glueing and adhesive ability to glass, metals and other materials.

Silicone enamels (i.e. polyorganosiloxanes in combination with fillers and pigments) can endure very high temperatures. The pigments can be aluminum, zinc, zink chromate, titanium dioxide, oxides and salts of other metals. The metal pigments added to polyorganosiloxanes (especially powder aluminum) give maximum thermal capacity to enamel coating. It can function at temperatures up to 550-600 °C. These coatings are used to paint electric furnaces and other heaters, chimney and exhaust tubes, aircraft and automobile equipment, electric engines at chemical plants, etc. The serviceability of these coatings is much larger than that of organic coatings: the coatings based on pigmented polyorganosiloxanes, which are used for chimneys in waste disposal plants, preserve for more than 18 months, whereas organic paints decompose after 1-2 weeks.

Protective polyorganosiloxane coatings pigmented with aluminum powder increase the longevity of steel parts operating at high temperatures. If low-carbon steel is covered with these enamels, it can be used in the temperature range at which unprotected steel usually oxidises to the point of destruction. Tests show that after 380 hours of being held at 465 °C the weight of samples from unprotected steel increased (due to oxidation) by 14%, and the weight of samples coated with polyorganosiloxane enamel, only by 2%; even after 1000 hours of heating no damage of enamel coating was found. The high thermal resistance of these coatings is due to the presence of hydroxyl groups in polyorganosiloxane chains, which react with aluminum and form polyalumorganosiloxanes, more heat-resistant polymers. This releases hydrogen, but in small amounts, which does not lead to the destruction of the film.

Like organic polymers, polyorganosiloxanes can be foamed with gas-forming agents in the process of polymer solidification. After 200 hours of keeping at 270 °C the polymers almost completely preserve their initial resistance to compression (they start to deform only above 350 °C), do not decompose after a short time at red heat and are fire-resistant. These polymers are used in aircraft engineering to make light sandwich constructions, fire-resistant isolating layers, etc.

Like silicone liquids, polyorganosiloxane varnishes are used as coatings which prevent the sticking of various materials to metals. They are used in the food industry (baking pans, trays for roasting meat, troughs for freezing fruit) and in various technological processes (the molding of rubbers, plastic masses, etc.). If paper is impregnated with a weak solution of silicone varnish, it does not stick to adhesive materials. This paper is used as packaging material for foodstuffs, as the nonadhesive strip for the adhesive insulating tape. A film of silicone varnish on the inner surface of glass ampules helps to avoid losses of valuable preparations.

Polyorganosiloxane varnishes are often used in electrical engineering to produce insulating materials. There are special requirements to insulating materials in electrical engineering: they should have good dielectric characteristics and high heat resistance (often up to 250 °C and higher), be resistant to sparks, arcs, coronas and oils.

The insulation of contemporary electric machines and apparatuses is subject to considerable overheating; that is why the thermal stability of insulation is of particular importance. The heat resistance of dielectrics limits the allowable temperature to which machines and apparatuses can be heated. If synthetic organic polymers are used for insulation, their heat resistance is often not enough; besides, it is very difficult to raise, since organic polymers can oxidise. The higher the temperature, the more intensive is the oxidation.

Of all organic polymers used to produce insulation materials, glyptal and phenol-formaldehyde polymers are the most thermal resistant. They can function for a long time in electrotechnical devices at temperatures up to 130 °C. At higher temperatures insulation from organic polymers burns. Its dielectric properties considerably decrease, because the carbon formed is a good conductor.

The use of silicone polymers for insulation has helped to build electric machines and apparatuses which can operate for a long time at 180-200 °C and for a limited period of time at 450-500 °C and higher. Owing to the high moisture resistance of silicone polymers, electric engines with silicone insulation can operate for a long time even underwater. The experience shows that the application of silicone insulating materials helps to considerably reduce the size of machines and increase the time of their service, as well as to solve an array of specific issues connected with operations in very difficult conditions (in coal mines, steelworks, in all kinds of vehicles, in tropical climates, at increased humidity, under vibration loads, etc.). The use of insulating materials based on silicone polymers gives a considerable technical and economic effect: the power of electric engines of the same dimensions grows by 1.5 times, the power of railway motors, by 35%. The longevity of engines for mining machines and coal-plough machines is increased 6-fold; the time of operation without repair grows 2-fold, and the overhaul costs are reduced by several times.

In the wide range of silicone polymer materials for the manufacture of electric insulation of particular interest are silicone varnishes, elastomers and liquids, as well as composition materials such as varnish glass fibre (glass fibre which has been repeatedly impregnated with a solution of silicone elastomer), glass micanates (a laminated material from plucked mica or mica paper and glass fibre glued with silicone varnish), glass textolites (glass fibre which has been pressed at heating and impregnated with silicone varnish) and silicone plastic masses.

This chapter gives a brief description only of the industries where silicone products have been most widely used. However, the applications of silicone polymers and materials are growing; there is no doubt that the possibilities they offer will continue to increase with further research into this exciting field of chemistry.

6.2. Use of other elementorganic substances

Recent years have seen the implementation of many various elementorganic compounds in economy. Their applications have proved to be extraordinarily wide: for stereospecific polymerisation of olefines, for stabilisation of polymer and lubricant materials, as antiknock additives and additives for engine fuels, as antiseptics, pesticides and so on.

6.2.1. Organoboron compounds

Among organoboron compounds, the greatest practical interest is presented nowadays by trialkylborates. They are used as additives to hydrocarbon oils, to fuels and lubricants, as well as a protective medium which prevents hot metal in the process of molding from oxidation (trialkylborate vapours are sent over the surface of the metal).

Trialkylborates can also be used as raw stock to obtain higher aliphatic alcohols; e.g.. there is a continuous process for producing these alcohols by the hydrolysis of corresponding trialkylborates.

Of great practical interest is also trimethylborate; it is used as raw stock for the synthesis of various boron compounds (sodium and potassium boranes, trimethoxyboroxol, etc.) and as a flux for acetylene welding of metals. We should also note the possibility of using lower trialkylborates in the distillation of alcohols, and of triethanolamineborate as a catalyst for the solidification of epoxy polymers.

Apart from trimethylborate, other trialkyl- and triarylborates, as well as alkylboron acids, are used. E.g., trialkylborates are good catalysts for the polymerisation of acrylonitrile, methylmethacrylate, styrene, vinylacetate, vinylchloride, vinylidenchloride, etc. The polymerisation of these monomers requires traces of oxygen. Oxygen seems to transform part of trialkylborate into peroxide, which reacts with unoxidised trialkylborate and forms free radicals initiating polymerisation.

Alkylboron acids and their ethers are used as additives to engine fuels. Some alkylboron acids (nonyl- and dodecylboron) are bactericides, fungicides, surface active agents. They can be used as additives to petrol, as reactants in the synthesis of polymers and as stabilisers of aromatic amides.

Recently a lot of attention has been given to the possible use of organoboron compounds as biologically active substances. E.g., researchers deal with their derivatives of the following type:

Boronsilicone compounds of the composition:

$x = 1 \div 3$; $n = 1 \div 3$.

are efficient hardeners for epoxy and silicone polymers.

6.2.2. Organoaluminum compounds

During the last 30-35 years the interest to organoaluminum compounds has particularly grown. This is due to the fact that aluminumtrialkyls are used as components of the catalytic system in polymerisation (Ziegler-Natta catalysts). However, the practical application of aluminumtrialkyls is not limited to catalytic systems. Recently aluminumtrialkyls have been widely used for the industrial synthesis of higher fatty alcohols. In this case the mixture of aluminumtrialkyls and olefines is oxidised with air; as a result, aluminum alcoholates are formed, which interact with water to decompose and form aluminum oxide and primary fatty alcohols. If the process is thoroughly monitored, one can provide the conditions for preferable formation of one or other product; thus, the process becomes especially important for the industrial manufacture of detergents.

Aluminumtrialkyls are also used as raw stock in the production of alkylderivatives of silicon, tin, lead, zinc, boron, arsenic, antimony and bismuth. These derivatives are obtained from aluminumtrialkyls and halogenides of the corresponding elements (by the direct replacement of alkyl groups with halogens) or by the electrolytic technique. Another sphere of application of aluminumtrialkyls is the production of ultrapure aluminum. The easy release of metal aluminum from alkyl derivatives of aluminum helps to use them for galvanostegy and metal spraying.

Trimethyl- and triethylaluminum proved to be active additives to gaseous and liquid fuels. E.g., for propane+air and cerosene+air mixtures these additives give a very slight ignition delay at a very low ignition temperature. Aluminumtrialkyls are also used as fuels in their own right, which helps to decrease fuel consumption considerably. This fuel ensures more engine power at lower fuel:air ratios than hydrocarbon fuels, and is more motile at low temperatures.

Aluminum alkylderivatives can also be used as sublayers, e.g. for fluoro- and fluorochlorine-containing polymers. The adhesion of these polymers to various materials (including metals) is considerably stronger, if the material has been immersed for 10-60 monutes into an aluminumtrialkyl solution in an inert solvent.

Like aluminumtrialkyls, alkylaluminumhalogenides are quite widely used as components of catalytic systems during polymerisation. To polymerise unsaturated compounds, it is preferable to use alkylaluminumchlorides with titanium tetrachloride. Alkylaluminumhalogenides, and ethylaluminumbromides in particular, are also efficient catalysts for the alkylation of ethylbenzene and cyclohexene. Besides, alkylaluminumhalogenides, like aluminumtrialkyls, are used to spray metal aluminum on various surfaces and to make electroplate aluminum coatings.

Alkylaluminumhydrides are mostly reducers. They are nearly as active as lithiumaluminumhydride, but the simplicity of their production and rather low costs give them some advantages in comparison with lithiumaluminumhydride.

6.2.3. Organotitanium compounds

Among organotitanium compounds of greatest practical interest nowadays are ethers of orthotitanium acid H_4TiO_4 and the products of hydrolytic polycondensation of these ethers, polyalkoxytitaniumoxanes.

E.g., tetrabutoxytitanium and polybutoxytitaniumoxane are components of heat-resistant paints. The use of polybutoxytitaniumoxane helps to increase the corrosive resistance of painted steel in high humidity. The most practical component for heat-resistant paints is the product of hydrolysis of tetrabutoxytitanium with water. In order to protect steel from corrosion in high humidity and at the same time preserve a tough film of paint at temperatures up to 650 °C, one should have two layers of coating, both with polybutoxytitaniumoxane (the lower layer is with zinc dust, the upper layer is with aluminum powder). Before painting the surface should be cleaned with a sand blaster; after painting the film should be solidified at 300-350 °C. Paint based on polybutoxytitaniumoxane pigmented with zinc dust can be used in rocket launchers, because this coating is stable under short-term flames which appear when rockets are launched.

An introduction of small amounts of ethers of orthotitanium acid into solutions or emulsions of waterproofing silicone liquids increases water-repellent properties of these liquids.

Ethers of orthotitanium acid are good catalysts for the solidification of epoxy polymers; they improve the quantity and strength of polymers. Thus, solidified polymers can be used as protective coatings or isolation materials. Organic orthotitanates are good solidifiers for polyorganosiloxanes. E.g., polymethylphenylsiloxane solution solidifies after 1 hour at 200 °C in the presence of catalytic quantities of a mixture of tetrabutoxytitanium and lead naphthenate. This forms a solid, heat-resistant and solvent-resistant coating. The solidification of polyorganosiloxanes modified with tetrabutoxytitanium produces insulation material for wires. Linear polydiorganosiloxanes can be solidified at room temperature with tetrabutoxytitanium (1-2.5%). The process is conducted by means of hydrolysis; it is efficient if polydiorganosiloxane contains at least one HO-group per 1000 Si atoms.

Ethers of orthotitanium acid are also used as catalysts for the polymerisation of unsaturated compounds (as components of the Ziegler-Natta catalytic system) and directly, for the polymerisation of α-olefines, butadiene, isoprene, styrene, etc. Ethers of orthotitanium acid are good catalysts during re-etherification. Tetraisopropoxytitanium can be used in the re-etherification of esters with alcohols or other ethers as a catalyst. Various tetraalkoxytitaniums can also be used in the re-etherification of ether of orthosilicon acid with alcohols, phenols or esters.

Olefines can be polymerised not only with ethers of orthotitanum acid, but also with halogenised cyclopentadienyl titanium compounds, bis(cyclopentadienyl)titaniumdichloride, as well as diphenyl-bis(cyclopentadienyl) titanium compounds combined with aluminumtrialkyls, or tetrakys(trimethylsiloxy)titanium combined with aluminumtrialkyl and aluminum alkylhalogenides.

Cyclopentadienyl titanium compounds, particularly bis(cyclopentadienyl)titaniumdichloride and cyclopentadienyltitanium-dibutoxychloride, are advisable as antiknock substances in oil fuels. These additives stimulate fuel combustion and reduce soot deposit. Some siloxy- and stannoxyderivatives of titanim with triethanolamine are fungicidal.

Triorganosiloxyderivatives of titanium have been used to produce thermally stable polymers.

6.2.4. Organotin compounds

Organotin compounds are widely used in various fields of technology and agriculture. Their world production is 45-50 thousand tonnes a year, coming the fourth in the group of other elementorganic compounds (after silicone, organoaluminum and organolead). Contemporary research gives opportunities for using organotin compounds as protection for various materials and articles from biodamage, for highly efficient nonfouling coatings for sea and river vessels, as well as underwater constructions. They are also used, though to a lesser extent, as catalysts and cocatalysts in the industrial production of various polymers (polyurethanes, polyepoxides) in etherification and other processes. They are also used in the production of low-molecular silicone elastomers (SKTN) of cold vulcanisation as cocatalysts in the amount of 0.1-1% (mixed with a cross-linking agent, tetraethoxysilane).

Organotin compounds are widely used as stabilisers for PVC. They are usually stabilised with R_2SnX_2 compounds, where R is the alkyl, and X is oxygen, ester, mercaptide, acetyl and other groups.

To make PVC stable under the influence of heat and light, the stabiliser should meet the following requirements: 1) prevent the detachment of hydrogen chloride or be its acceptor; 2) possess antioxidant properties; 3) react with double-link compounds; 4) protect against UV rays. Organotin compounds of the given type meet these requirements to the fullest. These compounds, as well as nonsymmetric R_3SnX compounds, are good anti-oxidants for elastomers, which prevent them from cracking.

Fundamentally new types of PVC stabilisers are dyalkyltinmaleates of the following structure:

$[$—SnR_2—OOCCH=CHCOO—$]_n$, and β-carboxyethyltinchlorides obtained according to the following reaction:

Sn + 2HCI + 2CH$_2$=CHCOOH ⟶ CI$_2$Sn(CH$_2$CH$_2$COOH)$_2$

These compounds are excellent PVC thermostabilisers. They have less smell, leach less out of the polymer and have higher light-stabilising properties than alkyltinmercaptides.

Research by Ziegler and Natta showed that organoaluminum compounds in combination with titanium halogenides are active catalysts of the polymerisation of olefines at low pressure. This inspired the study of other organometal compounds (e.g., organotin compounds like tetraalkyltin) in similar catalytic systems. It has been proved experimentally.

Tin organohalogenides can be used as catalysts to produce ethers of phosphoric acid and to polymerise lactons forming colourless polyesters. Various halogen derivatives of dibutyltin have been put forward as catalysts for the solidification of silicone elastomers, as agents preventing the cracking of polystyrene, as inhibitors of metal corrosion in silicone polymers.

Dibutyltin and hexabutylditin oxides are used in the vulcanisation of compositions with silicone elastomers with high filler content. These organotin compounds help to form rubbers with low residual compressive strain and high relative elongation at rupture).Organotin compounds like R_3SnX (X is halogen, hydroxyl or carboxyl) are active fungicides. Halogen derivatives of organotin compounds are also used to treat glass and obtain conductive films.

At present many studies deal with the possibility of obtaining polyorganotinoxanes from various organotin monomers. E.g., there have been reports about preparing a polymer from tributyltin metacrylate. The production of polyorganotinsiloxanes has also been described. It is quite possible that these substances will be presently given industrial application.

6.2.5. Organolead compounds

Lead tetraalkyl derivatives are still used as antiknock substances. They are mainly tetraethyl- and tetramethyllead. The efficiency of these additives as antiknock substances has contributed to the development and improvement of piston engines, increasing their heat-and-power characteristics. The use of organolead additives helps to obtain high-octane petrols from rather low-quality raw stock, which is equal to saving 20% of crude oil. For petrols with high content of aromatic substances, tetramethyllead proved to be more efficient, since it is distributed through fuel more evenly and noticeably increases its octane number.

However, due to the high toxicity of lead-containing antiknock substances and exhaust fumes from doped fuel, as well as the increasing demands to protect the environment (although antiknock substances cannot be totally eliminated due to economic reasons), many countries are looking for comprehensive solutions of this vital and complicated problem. Thus, in our country in 1979 the ethylation rate was reduced by 40%, although it had been the lowest around the world. The countries that consume large amounts of car petrol have also considerably reduced the specific consumption of lead additives. Thus, the general production of organolead antiknock substances is expected to decrease, although it will remain rather high. E.g., only in the USA it will soon amount to approximately 150 thousand tonnes per year.

Lead tetraalkyl derivatives are used in catalytic systems to polymerise olefines, as catalysts of re-etherification and polycondensation, to speed up the alkylation of lateral chains of alkylbenzenes with ethylene and its derivatives. An addition of lead tetraalkyl derivatives (0.05-2% of alkylbenzene quantity) to catalysts of the liquid-phase oxidation of alkylbenzenes speeds up the oxidation. Tetraethyllead proved to be a good initiator for Diels-Alder reactions to join polymers with alkenylsiloxy chains and can be used as an additive to reduce the attrition and wear of rubbing metal parts. Tetrabutyllead is an active cross-linking agent for polyethylene and modifying agent for plastics.

Some organolead compounds have high biological activity. E.g., alkyl- and arylacetates, alkyl- and arylhalogenides of lead, which have the largest biological activity, are used in the production of nonfouling paints, bactericides, fungicides, etc. There are about 600 different kinds of plants and 1300 living species which foul the bottom of ships in seawater. This reduces the speed characteristics of the ship and damages the anticorrosion coating of the hull. To prevent the fouling, nonfouling paints are used with toxic substances in them. The most efficient paints are based on vinyl polymers with copper oxide or lead triphenyl- or tributylacetates.

Before applying nonfouling paints, the cleaned surface of the hull is covered with an adhesive sublayer or a layer of adhesion promoter, and anticorrosion paint. In seawater tests panels covered with these paints do not foul even after 27 months of exposition; panels covered with normal paints fouled by 100% already after 2 months. The serviceability of paints with an addition of lead triphenylacetate is approximately 50% longer than that of paints with organotin additives. Very active in nonfouling paints are lead triphenyl- and tributyllaurates.

Considerable bactericidal and fungicidal activity is exhibited by R_3PbX compounds; their efficiency depends on radical R. Highly active fungicides are organolead compounds containing sulfur and nitrogen. Lead triethylphthaloilhydrazide is efficient against potato and peach diseases. Additions of lead thiomethyl-, thioethyl- and thiophenylacetates (4.5-5%) make cotton articles resistant to rot and give them high physicochemical characteristics.

Thiobenzyl- and thiophenylacetyl derivatives of lead have a noticeable antiinflammation effect, which makes them useful in medicine.

6.2.6. Organophosphorus compounds

The study of organophosphorus chemistry began more than 130 years ago. However, these substances found practical application only after World War II. At present the most widespread application for organophosphorus compounds is pest control in agriculture. Biological activity is characteristic of the following derivatives of phosphorus organic acids:

$$\begin{matrix} \mathbf{RR'PX} \\ \| \\ \mathbf{O(S)} \end{matrix}$$

besides, the activity is the greatest when the following requirements are met:

1. the oxygen (or sulfur) atom is connected directly with the atom of quinquivalent phosphorus;
2. R and R' are alkoxygroups, alkyl groups or aminogroups;
3. X is the residue of inorganic or organic acid (the fluorine atom, cyanogroup, thiocyan) or acid residue (phenol, mercaptan, etc.). At present hundreds of organophosphorus insecticides are known. Some of them have a very wide action spectrum and kill many species of insects. Organophosphorus insecticides differ by their toxicity for warm-blooded animals, stability, period of action, etc.

Some substances affect insects when they come in contact with them. They are so-called *contact* insecticides. They include, e.g., thiophos and methaphos, a substance similar to thiophos, but with O-methyl groups instead of O-ethyl groups.

$$(H_3CO)_2\overset{\displaystyle\|}{\underset{\displaystyle S}{P}}-O-\!\!\!\left\langle\!\!\!\bigcirc\!\!\!\right\rangle\!\!\!-NO_2$$

Methaphos is slightly less efficient than thiophos, but considerably less toxic for warm-blooded animals. That is why methaphos is often given preference. Trichlorfon, which is practically not toxic for people and animalsl, is a very promising contact insecticide. It seems to be one of the safest organophosphorus insecticides; that is why it is used for insect control not only in fields, but also indoors (e.g. to kill flies in homes, stables and barns). By killing flies and other insects, the preparation limits the spread of dangerous diseases. Trichlorfon proved to be especially efficient in controlling such insects as caterpillars, may-bugs, chafers, nun moths.

Trichlorfon is mostly used to kill mangold fly larvae; it is also very efficient against the dangerous rice weevil, which can destroy whole harvests. Trichlorfon is successfully used in viticulture (to kill grapevine moths) and pomiculture (to kill apple, pear and plum sawflies, apple ermine moths and gooseberry sawflies). Trichlorfon is very efficient in cotton-growing (to kill cotton worms). The low toxicity of trichlorfon for warm-blooded animals accounts for its use in veterinary medicine to combat parasites on large animals.

Of great interest is contact insecticide malathion (carbophos):

$$(H_3CO)_2\overset{\displaystyle\|}{\underset{\displaystyle S}{P}}-S-\overset{\displaystyle\quad}{\underset{\displaystyle H_2C-COOC_2H_5}{CH}}-COOC_2H_5$$

This substance is less toxic for warm-blooded animals than trichlorfon and is meant for pest control in gardens and on farmlands.

Another group of organophosphorus insecticides consists of preparations with the so-called *systemic* (inside the plant) effect. They are absorbed by plant tissue, making the plant poisonous for insects feeding on it. These insecticides are mercaptophos (systox) and M-74 (pisyston), derivatives of mono- and dithiophosphoric acid. However, both substances are very toxic for people and cattle and are even more dangerous than thiophos; that is why they are not used in Russia. Other systemic preparations, such as methylmercaptophos (metasystox) synthesised by G.Schrader, and M-81 (intrathion) obtained in the laboratory of Academician M.I.Kabachnik, are less toxic than methaphos but more toxic than trichlorfon.

An important property of ethers of phosphoric acids is that they can mix with many insecticides, fungicides and acaricides which do not contain phosphorus, while their activity is preserved. These mixtures are used to destroy various agricultural pests. It is often necessary to combine insecticides with fungicides. These combined preparations are essential, because none of the substances used at present has both insecticidal and fungicidal properties.

Organophosphorus compounds are used not only as pesticides, but also as medicines. Organophosphorus medicines are most often used to treat glaucoma. Organophosphorus glaucoma medicines include paraoxon (mintakol) synthesised by G.Schrader, and armin, which was developed in the laboratory of Academician A.E.Arbuzov:

$$(C_2H_5O)_2\overset{\displaystyle O}{\underset{\displaystyle \|}{P}}-O-\underset{}{\bigcirc}-NO_2 \qquad \overset{\displaystyle C_2H_5O}{\underset{\displaystyle C_2H_5}{\diagdown}}\overset{}{\underset{\displaystyle \|}{P}}-O-\underset{}{\bigcirc}-NO_2$$

Later armin was introduced into gynecological treatment (as birth stimulator).

A characteristic property of organophosphorus compounds with P=O or P=S bonds is their ability to form complexes. This was the basis for the industrial use of these substances as flotation agents in nonferrous metallurgy. Dibutyl- and dicresyldithiophosphates proved efficient collectors of sulfide or sulfidised nonferrous metals during the flotation of oxide ores.

$$(RO)_2\overset{\displaystyle S}{\underset{\displaystyle \|}{P}}-SH$$

Organophosphorus compounds are more and more often used in hydrometallurgy (as *extractants*). Among organophosphorus extractants the most widespread is tributylphosphate. It is especially often used to extract uranium from highly acid media. Tributylphosphate forms a complex with uranyl salt, which easily transforms into the organic phase. Organophosphorus extractants have become so important due to their high selectivity, resistance to aggressive media and convenient re-extraction (i.e. reverse preparation of salt from the solution of the complex in an organic solvent). Recently researchers have found new organophosphorus extractants, which are even more efficient. It can be supposed than the role of organophosphorus extractants in hydrometallurgy will continue to grow.

Organophosphorus compounds are also becoming essential as antioxidants of petroleum lubricant oils, as synthetic lubricants and plastifiers for various polymers. Organophosphorus compounds are widely used to obtain substances which do not sustain combustion, because they are highly efficient fire retardants.

Organophosphorus polymers have also attracted great interest, especially organophosphorus polyesters, polydichlorophosphazenes and polyorganophosphazenes. This is due to the fact that polymers and copolymers of many unsaturated ethers of phosphoric acid have good transparency, adhesion, chemical and fire resistance, and polyorganophosphazenes have superior thermal stability and fire resistance. That is why polymers and copolymers based on mono- and diallyl ethers of alkyl- and arylphosphine acids, as well as vinyl- and isopropenylphosphin acids are used to manufacture transparent plastic laminates, organic glass (conventional or reinforced with glass fibre) for aviation, optical devices, etc., as well as for the preparation of glues, varnishes and coatings. Polyorganophosphasenes are often used to produce glues, impregnating compositions, elastic insulating coatings and polyceramic materials which are stable up to 500-600 °C. They are more and more often used as modifying additives for various organic polymers to improve their fire resistance and thermal plasticity.

This part of the book considers only the most essential applications of the most widespread elementorganic compounds. We can say that chemistry, and the elementorganic industry in particular, are at the peak of their development, and further research in these fields will undoubtedly lead to many exciting and unexpected results and possibilities for using elementorganic compounds in science and technology.

References

1.

Andrianov KA (1968) Methods of elementorganic chemistry. Silicon. Nauka Moscow, p 28 [in Russian]

Andrianov KA ed. (1976) Synthesis and modification of polymers. Nauka Moscow, pp 5-15, 48-72 c

Berlin AA, Lariny VP (1949) The success of chemistry vol 18, edn 3, p 546 [in Russian]

Ignatyeva GM, Rebrov EA, Myakushev VD, Chechenskaya TB, Muzafarov AM (1997) The universal scheme for the synthesis of silicone dendrimers.. High-Molecular Compounds, 1997, vol 39A, edn 8, pp 1271-1280 [in Russian]

Khananashvili LM (1998)Chemistry and technology of elementorganic monomers and polymers. Khimiya Moscow [in Russian]

Korneev NN, Scherbakova GI, Kolesov VS (1988) Application of aluminum- and silicon-based elementorganic compounds to obtain ceramic and glass materials.. In series: Elementorganic compounds and their application. NIITEKhim Moscow [in Russian]

Korshak VV ed. (1969) Progress of polymer chemistry. Nauka, Leningrad, pp 32-37 [in Russian]

Kreshkov AP (1950) Silicone compounds in engineering. Promstroyizdat Moscow [in Russian]

Mukbaniani OV, Zaikov GE (2003) New Concepts in Polymer Science. Cyclolinear Organosilicon Copolymers: Synthesis, Properties, Application. The Netherlands, VSP, Utrecht, Boston, pp. 1-499

Topchiev AV, Andrianov KA (1953) Izv. USSR Ac.Sci. OKhN, 1953, vol 3, p491 [in Russian]

Voronkov MG (1952) Silicone chemistry in Russian and Soviet research. LGU Leningrad [in Russian]

2.4.

Chernyshev EA ed. (1994) Plant equipment and the basics of design. Lecture notes. Lomonosov MGATKhT Moscow [in Russian]

Gelperin YaI, Ainshtein VG, Kvasha VB (1967) The basics of the fluidising technique. Khimiya Moscow, pp 446-452 [in Russian]

Gorbunov AI, Bely AP, Golubtsov SA (1969) The direct synthesis of alkyl(aryl)chlorosilanes and hydridechlorosilanes. Kinetics and mechanism. In series: Elementorganic compounds and their application. NIITEKhim Moscow [in Russian]

Molokanov YuK et al (1974) Separation of mixtures of silicone compounds. Khimiya Moscow [in Russian]

Series: Elementorganic compounds and their application. NIITEKhim Moscow 1972 edn 1, pp 5-63, 176-202 [in Russian]

2.5

Motsaryov GV, Sobolevsky MV, Rosenberg VR (1990) Carbofunctional organosilanes and organosiloxanes. Khimiya Moscow [in Russian]

3.1

Andrianov KA (1968) Methods of elementorganic chemistry. Silicon. Nauka Moscow, p 175-201[in Russian]

Skorokhodov II (1978) Physicochemical and performance characteristics of polydimethylsiloxane liquids. GNIIKhTEOS Moscow, edn2-3 [in Russian]

Sobolevsky MV (1985) Oligoorganosiloxanes: characteristics, preparation, application. Khimiya Moscow [in Russian]

Voronkov MG, Dyakov VM (1971) Silatranes. Nauka Novosibirsk [in Russian]

4.1.

Kirpichnikov PA, Averko-Antonovich LA, Averko-Antonovich YuO (1975) Chemistry and technology of synthetic resin. Khimiya Leningrad [in Russian]

Reikhsfeld VO ed (1973) Chemistry and technology of silicone elastomers. Khimiya Leningrad [in Russian]

Shets M (1975) Silicone resin. Khimiya Leningrad [in Russian, translated from Czech]

Sobolev VM, Borodina IV (1977) Industrial synthetic resins. Khimiya Moscow, pp 203-219 [in Russian]

4.2

Andrianov KA (1962) Polymers with inorganic backbones. Ac.Sci. Moscow. pp 106-184, 202-236 [in Russian]

Continuous processes of the synthesis of organomagnesium compounds and organomagnesium synthesis of silicone monomers (1984)   In series: Elementorganic compounds and their application. NIITEKhim Moscow [in Russian]

4.3

Andrianov KA (1962) Polymers with inorganic backbones. Ac.Sci. Moscow. pp
106-184, 245-287 [in Russian]

4.4

Ryabov IV ed (1970) Fire hazard of substances and materials used in the chemical
industry. Khimiya Moscow [in Russian]
Maximum allowable concentrations of harmul substances in air and water (1975)
Khimiya Leningrad [in Russian]

5.1

Gerard V (1965) Chemistry of boron compounds Khimiya Moscow [in
Russian, translated from English]

5.2

Korneev NN (1979) Chemistry and technology of organoaluminum compounds.
Khimiya Moscow [in Russian]

5.3

Field R, Kove P (1969) Organic chemistry of titanium. Mir Moscow [in Russian,
translated from English]

5.4

Ingam R, Rosenberg S, Gilman G, Rikens F (1962) Organotin and
organogermanium compounds. Izdatinlit Moscow [in Russian, translated from
German]
Situation and prospects for organotin applications (1988) In series:
Elementorganic compounds and their application. NIITEKhim Moscow [in
Russian]

5.5

Samarin KM, Gorina FA, Zhitaryova LV (1972) Surveys of separate industrial
productions. In series: Elementorganic compounds and their application.
NIITEKhim Moscow, edn 18, pp 23-41 [in Russian]

5.6

Melnikov NN (1974) Chemistry and technology of pesticides. Khimiya Moscow
[in Russian]

Purdela D, Vylchanu R (1972) Chemistry of organic phosphorus compounds. Khimiya Moscow [in Russian, translated from Romanian]

6.1.

Andrianov KA (1964) Heat-resistant silicone dieletrics. Energia Leningrad [in Russian]

Dolgov ON, Voronkov MG, Grinblat MP (1976) Silicone resins and materials based on them. Khimiya Leningrad [in Russian]

Kardashov DA (1976) Synthetic glues. Khimiya Moscow [in Russian]

Kharitonov NP, Shentenkova IA (1977) Heat-resistant organosilicate sealant materials. Nauka Leningrad [in Russian]

Kharitonov NP, Veseloye PA, Kuzinets AS (1976) Vacuum-sealed compositions based on polyorganosilanes. Nauka Leningrad [in Russian]

Orlov NF, Androsova MV, Vvedensky NV (1966) Silicone compounds in textile and light industry. Lyogkaya Industria Moscow [in Russian]

Pashenko AA et al (1973) Waterproofing. Naukova Dumka Kiev in Russian]

Sobolevsky MV ed (1985) Oligoorganosiloxanes: properties, production, application. Khimiya Moscow [in Russian]

Sobolevsky MV, Muzovskaya OA, Popeleva GS (1975) Properties and applications of silicone products. Khimiya Moscow [in Russian]

Voronkov MG, Makarskaya VM (1978) Dressing textiles with silicone monomers and polymers. Nauka Novosibirsk [in Russian]

Application of silicone compounds in light industry (1986) In series: Elementorganic compounds and their application. NIITEKhim Moscow, [in Russian]

Application of silicone compounds in medicines (1984) In series: Elementorganic compounds and their application. NIITEKhim Moscow, [in Russian]

Regulators for plant growth and development (1997). Papers of the 4th internation conference. GNIIKhTEOS Moscow [in Russian]

Silicone application in dermatology and farmacology. A survey (1981). In series: Elementorganic compounds and their application. NIITEKhim Moscow, [in Russian]

Silicone products manufactured in USSR (1975).Khimiya Moscow [in Russian]

6.2.

Harwood G. (1970) Industrial applications of metalorganic compounds. Khimiya Leningrad [in Russian, translated from English]

Shiryaev VI, Stepina EI (1988) Situation and prospects of organotin compounds, A survey. In series: Elementorganic compounds and their application. NIITEKhim Moscow, [in Russian]

Production of organolead antiknock substances. A survey (1980). In series: Elementorganic compounds and their application. NIITEKhim Moscow, [in Russian]

For Product Safety Concerns and Information please contact our EU
representative GPSR@taylorandfrancis.com
Taylor & Francis Verlag GmbH, Kaufingerstraße 24, 80331 München, Germany